# Environmental and Microbial Biotechnology

**Series Editor**
Ram Prasad
Department of Botany
Mahatma Gandhi Central University
Motihari, Bihar, India

Innovative and novel advances in microbial biotechnology are providing great understandings in to the machineries of nature, presenting fascinating prospects to apply principles of biology to different arenas of science. Sustainable elucidations are emerging to address the concerns on improving crop productivity through microbes, depleting natural resources, environmental pollution, microbial degradation of pollutants, nanomaterials, nanotoxicity & safety issues, safety of food & agricultural products etc. Simultaneously, there is an increasing demand for natural bio-products of therapeutic and industrial significance (in the areas of healthcare, environmental remediation, microbial biotechnology). Growing awareness and an increased attention on environmental issues such as climate change, energy use, and loss of non-renewable resources have carried out a superior quality for research that provides potential solutions to these problems. Emerging microbiome approaches potentially can significantly increase agriculture productivity & human healthcare and henceforth can contribute to meet several sustainable development goals.

The main objectives have provided an impetus for research on plants and microorganisms that produce novel bio-products with variable properties and understanding their mechanisms of action at cellular and molecular level. Hence, research activities of the environmental and microbial Biotechnology are comprehensively focused up on major sectors viz., bioresources, biorefining, bioremediation of organic and inorganic pollutants, environmental risk analysis of microorganisms, environmental assessment using microbiological indicators, enzymes for environment, food & industrial applications, nanomaterials & nanotoxicity, sustainable ecobiotechnology, biofertilizer, biocontrol agents for agriculture improvement and natural products for healthcare applications.

This book series is a state-of-the-art for a wide range of scientists, researchers, students, policy makers and academician involve in understanding and implementing the knowledge on environmental and microbial biotechnology to develop biologics for proper health care to continue life in smooth and sustainable strategy without any adverse effect.

More information about this series at http://www.springer.com/series/16324

Joginder Singh • Ashish Vyas •
Shanquan Wang • Ram Prasad
Editors

# Microbial Biotechnology: Basic Research and Applications

Springer

*Editors*
Joginder Singh
Department of Microbiology
Lovely Professional University
Jalandhar, Punjab, India

Ashish Vyas
Dept of Microbiology and Biochemistry
Lovely Professional University
Jalandhar, Punjab, India

Shanquan Wang
School of Civil and Environmental Engineering
Sun Yat-Sen University
Guangzhou, Guangdong, China

Ram Prasad
Department of Botany
Mahatma Gandhi Central University
Motihari, Bihar, India

ISSN 2662-1681 ISSN 2662-169X (electronic)
Environmental and Microbial Biotechnology
ISBN 978-981-15-2816-3 ISBN 978-981-15-2817-0 (eBook)
https://doi.org/10.1007/978-981-15-2817-0



This Springer imprint is published by the registered company Springer Nature Singapore Pte Ltd.
The registered company address is: 152 Beach Road, #21-01/04 Gateway East, Singapore 189721, Singapore

# Preface

In the quest to technological advancement in the field of microbial technology in the last several decades to counteract health-related issues, microbial infections, plant–microbe interactions, and environmental sustainability, several important issues were explored. Microbial biotechnology is an important array that promotes advanced research into using microbes for value-added products, human nutrition, food-grade components, and the sustainable development of agriculture and environment. The endeavor of book entitled Microbial Biotechnology: Basic Research and Applications is to present state-of-the-art techniques used to harness microbial biotechnological traits on development of new industrial microorganisms, improved microbial agents for biological control of plants and animals, development of new microbial agents for bioremediation of contaminants and wastewater treatment, and biosensors for monitoring and diagnosis. Gathering contributions from authoritative researchers in the field, it addresses recent advances in microbial biotechnological approaches that offer sustainable options for future generations. Exploring an extensive collection of microbial products and their uses, this book specifically emphasizes the application of microorganisms in health care, the environment, and industry. It also discusses human nourishment and functional foods, plant and animal safety, and furthering fundamental research in the agricultural sciences. Following a general approach to recent advances in the utilization of various microbes as biotechnological tools, the book also covers traditional uses and explores emerging strategies to promise their full potential.

This volume would serves as an excellent reference book for microbial science scholars, especially microbiologists, biotechnologists, researchers, technocrats, and agriculture scientists of microbial biotechnology. We have been honored the leading scientists who have extensive, in-depth experience and expertise in microbial technology and took time and effort to develop outstanding chapters.

We wish to thank Dr. Naren Aggarwal, Editorial Director; Ms. Aakanksha Tyagi, Senior Editor, Springer; Mr. Ashok Kumar, Project Coordinator, Springer; Ms. Immaculate Jayanthi, Production Editor; and Ms. Metilda Nancy, SPi Global, for their generous assistance, constant support, and patience in initializing the volume. Dr. Ram Prasad is particularly very thankful to Honourable Vice Chancellor

Professor Dr. Sanjeev Kumar Sharma, Mahatma Gandhi Central University, Bihar for constant encouragement. The editors are also very grateful to our esteemed friends and well-wishers and all faculty colleagues of the Lovely Professional University, India; Sun Yat-sen University, China, and Mahatma Gandhi Central University, Motihari, Bihar, India.

| | |
|---|---|
| Jalandhar, India | Joginder Singh |
| Jalandhar, India | Ashish Vyas |
| Guangzhou, China | Shanquan Wang |
| Motihari, India | Ram Prasad |

# Contents

# Editors and Contributors

## About the Editors

**Joginder Singh** is presently working as Professor at the Department of Microbiology, Lovely Professional University, Punjab, India. Previously, he worked as Young Scientist at Microbial Biotechnology and Biofertilizer Laboratory, Department of Botany, Jai Narain Vyas University in Department of Science and Technology, Govt. of India. He is an active member of various scientific societies and organizations including Association of Microbiologists of India, Indian Society of Salinity Research Scientists, Indian Society for Radiation Biology, and European Federation of Biotechnology. He has more than 65 research and review articles in the peer-reviewed journals, edited four books entitled "Arbuscular Mycorrhizal Fungi" and "Microbes: In Action" published by AgroBios, India and "Microbial Bioprospecting for Sustainable Development" and "Fungi and their Role in Sustainable Development: Current Perspectives" published by Springer International Publishing, and authored/co-authored 35 chapters in edited books. He serves as reviewer for many prestigious journals, including Science of the Total Environment, Environmental Monitoring and Assessment, Pedosphere, Soil and Sediment Contamination, International Journal of Phytoremediation, Ecotoxicology and Environmental Safety, Annals of Agricultural Sciences, and Annals of the Brazilian Academy of Sciences. He attended several international and national seminars, symposia, and conferences and chaired technical sessions and presented papers in them.

**Ashish Vyas** is PhD from Dr. HS Gour University (a central university) Sagar, Madhya Pradesh, India. His areas of research include the multidisciplinary field of fermentation technology with special reference to microbial enzymes, pilot-scale production, and optimization on low-value lignocelluloses. Additional research area includes biocontrol strategies to mitigate plant pathogen and microbial pigments, green nano synthesis for therapeutic applications. He has more than 40 peer-reviewed national and international publications and has delivered numerous oral and poster presentations in numerous platforms. Dr. Ashish is the Member Secretary of Association of Microbiologists of India-LPU unit. He is presently serving as Professor and Head, Department of Microbiology, Lovely Professional University, Phagwara, Punjab.

**Shanquan (Alan) Wang** is currently an associate professor at School of Environmental Science and Engineering, Sun Yat-sen University, Guangzhou, China. His research focuses on environmental microbiology, especially on organohalide-respiring bacteria and their conversion of halogenated persistent organic pollutants (POPs). He integrates microbial cultivation, metagenomics, molecular techniques, and bioreactor operation to gain fundamental insights into complex biosystems (e.g., bioremediation sites and anaerobic digesters), specifically from molecular-, cellular-, community- to system-levels. The generated knowledge on these reductive processes will be further employed to devise novel methods, techniques, and products for environmental engineering purposes. He was awarded as SYSU Bairen Scholar and Zhujiang Youth Scholar in 2016 and 2018, respectively.

**Ram Prasad** is associated with Department of Botany, Mahatma Gandhi Central University, Motihari, Bihar, India. His research interest includes applied microbiology, plant–microbe interactions, sustainable agriculture, and nanobiotechnology. Dr. Prasad has more than one hundred and fifty publications to his credit, including research papers, review articles and book chapters, and five patents issued or pending, and edited or authored several books. Dr. Prasad has twelve years of teaching experience and has been awarded the Young Scientist Award (2007) & Prof. J.S. Datta Munshi Gold Medal (2009) by the International Society for Ecological Communications; FSAB fellowship (2010) by the Society for Applied Biotechnology; the American Cancer Society UICC International Fellowship for Beginning Investigators, USA (2014); Outstanding Scientist Award (2015) in the field of Microbiology by Venus International Foundation; BRICPL Science Investigator Award (ICAABT-2017); and Research Excellence Award (2018). He has been serving as editorial board members: Frontiers in Microbiology, Frontiers in Nutrition, Academia Journal of Biotechnology including Series editor of Nanotechnology in the Life Sciences, Springer Nature, USA. Previously, Dr. Prasad served as Assistant Professor, Amity University, Uttar Pradesh, India; Visiting Assistant Professor, Whiting School of Engineering, Department of Mechanical Engineering at Johns Hopkins University, USA; and Research Associate Professor at School of Environmental Science and Engineering, Sun Yat-sen University, Guangzhou, China.

## Contributors

**Monika Barkodia** Department of Microbiology, CCS Haryana Agricultural University, Hisar, Haryana, India

**Nisha Choudhary** School of Nanosciences, Central University of Gujarat, Gandhinagar, Gujarat, India

**Shivika Datta** Department of Zoology, Doaba College Jalandhar, Jalandhar, India

**Daljeet Singh Dhanjal** Department of Biotechnology, Lovely Professional University, Phagwara, India

**Anshul Dubey** Department of Botany and Microbiology, St. Aloysius College, Jabalpur, India

**G. Gnanamoorthy** Department of Inorganic Chemistry, University of Madras, Chennai, Tamil Nadu, India

**Haresh Kalasariya** Department of Microbiology, S.S. Nootan Degree College of Science and Commerce, Sankalchand Patel University, Mehsana, Gujarat, India

**Divya Kapoor** Department of Microbiology, CCS Haryana Agricultural University, Hisar, Haryana, India

**Arun Karnwal** Department of Microbiology, School of Bioengineering and Biosciences, Lovely Professional University, Phagwara, Punjab, India

**Ram S. Kaulgud** Multi-Disciplinary Research Unit, Karnataka Institute of Medical Sciences, Hubballi, Karnataka, India

Department of General Medicine, Karnataka Institute of Medical Sciences, Hubballi, Karnataka, India

**Gurleen Kaur** Department of Biotechnology, Lovely Professional University, Phagwara, Punjab, India

**Intelli Kaur** Department of Nutrition & Dietetics, National Health Services, Edinburgh, UK

**Pardeep Kaur** Department of Biotechnology, Sri Guru Granth Sahib World University, Fatehgarh Sahib, India

**Robinka Khajuria** Department of Biotechnology, Harlal Institute of Management and Technology, Greater Noida, India

**Samreen Heena Khan** School of Nanosciences, Central University of Gujarat, Gandhinagar, Gujarat, India

**Bhupendra Koul** Department of Biotechnology, School of Bioengineering and Biosciences, Lovely Professional University, Phagwara, Punjab, India

**Rajeev Kumar** Division of Forensic Science, SBAS, Galgotias University, Greater Noida, India

**Vijay Kumar** Regional Ayurveda Research Institute for Drug Development, Gwalior, India

**Vijay Kumar** Department of Basic and Applied Sciences, National Institute of Food Technology Entrepreneurship and Management, Sonipat, Haryana, India

**Mahantesh M. Kurjogi** Multi-Disciplinary Research Unit, Karnataka Institute of Medical Sciences, Hubballi, Karnataka, India

**Parth Mallik** School of Nanosciences, Central University of Gujarat, Gandhinagar, Gujarat, India

**Mukesh Meena** Department of Botany, Mohanlal Sukhadia University, Udaipur, India

**Himesh Namdeo** Department of Botany and Microbiology, St. Aloysius College, Jabalpur, India

**Poondla Naresh** Graduate School of Biomedical Science and Engineering, Hanyang University, Seoul, South Korea

**Ram Prasad** Department of Botany, Mahatma Gandhi Central University, Motihari, Bihar, India

**Chayanika Putatunda** Om Sterling Global University, Hisar, Haryana, India

**Kavita Rani** Department of Microbiology, CCS Haryana Agricultural University, Hisar, Haryana, India

**Raman Kumar Ravi** School of Environment and Sustainable Development, Central University of Gujarat, Gandhinagar, Gujarat, India

**Namita Anant Raytekar** Department of Microbiology, Sandip University, Nasik, India

**Priyanka Roy** Department of Basic and Applied Sciences, National Institute of Food Technology Entrepreneurship and Management, Sonipat, Haryana, India

**Mahipal Singh Sankhla** Division of Forensic Science, SBAS, Galgotias University, Greater Noida, India

**Saurabh Satija** Department of Pharmaceutical Sciences, Lovely Professional University, Phagwara, Punjab, India

**Anjana Sharma** Bacteriology Laboratory, Rani Durgavati University, Jabalpur, India

**Divakar Sharma** CRF, Mass Spectrometry Laboratory, Kusuma School of Biological Sciences (KSBS), Indian Institute of Technology-Delhi (IIT-D), New Delhi, India

**Ganga Sharma** Central Research Institute, Kasauli, Himachal Pradesh, India

**Juhi Sharma** Department of Botany and Microbiology, St. Aloysius College, Jabalpur, India

**Mayur Mukut Murlidhar Sharma** Department of Agriculture and Life Industry, Kangwon National University, Chuncheon, South Korea

**Pankaj Sharma** Department of Microbiology, CCS Haryana Agricultural University, Hisar, Haryana, India

**Rashmi Sharma** Department of Microbiology, DAV University, Jalandhar, Punjab, India

**Shefali** Department of Zoology, DPG Degree College, Gurugram, Haryana, India

**Anjuvan Singh** Department of Biotechnology, School of Bioengineering and Biosciences, Lovely Professional University, Phagwara, Punjab, India

**Digvijay Singh** Department of Biotechnology, Lovely Professional University, Phagwara, Punjab, India

**Dilbag Singh** Department of Microbiology, CCS Haryana Agricultural University, Hisar, Haryana, India

**Joginder Singh** Department of Biotechnology, School of Bioengineering and Biosciences, Lovely Professional University, Phagwara, Punjab, India

Department of Microbiology, Lovely Professional University, Phagwara, Punjab, India

**Shalini Singh** School of Bioengineering and Biosciences, Lovely Professional University, Phagwara, Punjab, India

**Simranjeet Singh** Department of Biotechnology, Lovely Professional University, Phagwara, Punjab, India

Punjab Biotechnology Incubator (PBTI), Mohali, Punjab, India

Regional Advanced Water Testing Laboratory, Mohali, Punjab, India

**Preeti Solanki** Multidisciplinary Research Unit (MRU), Pt. BD Sharma PGIMS, Rohtak, Haryana, India

**R. Suriyaprabha** School of Nanosciences, Central University of Gujarat, Gandhinagar, Gujarat, India

**Prashant Swapnil** International Centre for Genetic Engineering and Biotechnology, New Delhi, India

**Anamika** International Centre for Genetic Engineering and Biotechnology, New Delhi, India

**Anju Thappa** School of Biotechnology, Shoolini University, Solan, Himachal Pradesh, India

**Siddharth Tiwari** National Agri-Food Biotechnology Institute (NABI), Department of Biotechnology, Ministry of Science and Technology (Government of India), Mohali, Punjab, India

**R. S. Upadhyay** Department of Botany, Banaras Hindu University, Varanasi, India

**Vaishali Vishwakarma** Department of Botany and Microbiology, St Aloysius College, Jabalpur, India

**Ashish Vyas** Department of Microbiology, School of Bioscience and Bioengineering, Lovely Professional University, Phagwara, Punjab, India

**Abhishek Walia** Department of Microbiology, CSK HPKV, Palampur, Himachal Pradesh, India

**V. K. Yadav** Department of Microbiology, S.S. Nootan Degree College of Science and Commerce, Sankalchand Patel University, Mehsana, Gujarat, India

**Abdulhadi Yakubu** Department of Microbiology, School of Bioscience and Bioengineering, Lovely Professional University, Phagwara, Punjab, India

Department of Science Laboratory Technology, College of Science and Technology, Jigawa State Polytechnic, Dutse, Nigeria

**Andleeb Zehra** Department of Botany, Banaras Hindu University, Varanasi, India

# Chapter 1
# The Contribution of Microbial Biotechnology for Achieving Sustainable Development

**Juhi Sharma, Divakar Sharma, Anjana Sharma, Vaishali Vishwakarma, Anshul Dubey, and Himesh Namdeo**

**Abstract** Microbes are requisite constituent of biotic diversity that maintain sustainable ecosystem. They are chief customs of life which have progressed into environmentally, metabolically and genetically diverse species. In ecosystem, microbial diversity strives to comprehend innumerable metabolic courses to maintain resolute integrity for sustainable ecology. Utility of microbial communities has better indulgent of the bio-network. Until now, only 0.1–10% of microbial species are recognized and, the rest being uncultured, inhabit noteworthy niches in biomes and are accountable for several loom based on molecular genetics, systems and synthetic biology, genomics, proteomics and metagenomics. Exploring biotechnological applications and understanding their mechanism of alteration permit the progress on the circumstances necessary for various microbial applications with stare to sustainable development, community structure and environmental processes. Most appreciated tools for investigating the microbial resistance to antibiotics and search for new antimicrobials can be done using molecular techniques. Therefore, currently, metagenomics and meta-proteomics studies have been utilized effectively to get novel microbes as well as their by-products from uncultured microorganisms. Microbes can be used for a variety of biotechnological appliance such as food products, therapeutic protein, recombinant microbes, vaccine and diagnostic tool. Even though microbe's inventorying and cataloguing are discouraging tasks, requiring skills and creativity but imparting considerable pecuniary import.

J. Sharma (✉) · V. Vishwakarma · A. Dubey · H. Namdeo
Department of Botany and Microbiology, St. Aloysius College, Sadar Cantt, Jabalpur, India

D. Sharma
CRF, Mass Spectrometry Laboratory, Kusuma School of Biological Sciences (KSBS), Indian Institute of Technology-Delhi (IIT-D), New Delhi, India

A. Sharma
Bacteriology Laboratory, Rani Durgavati University, Jabalpur, India



J. Singh et al. (eds.), *Microbial Biotechnology: Basic Research and Applications*, Environmental and Microbial Biotechnology,
https://doi.org/10.1007/978-981-15-2817-0_1

**Keywords** Microbial engineering · Sustainable · Niche · Meta-proteomics

## 1.1 Introduction

The richness, variability and complexity among the living organisms are designated under the term biodiversity or biological diversity. It is determined by different plants, animals and microbes in natural ecosystems. Biodiversity is classified in terms of three fundamental levels, viz. genetic, species and ecosystem diversity.

Biodiversity of the microbial world has focused on the evolution of all forms of life on earth. Microbial diversity covers a wide range of inconsistency between prokaryotes, eukaryotes and viruses in each possible habitat on the planet. They are found in almost every nook and cranny yet in that environment where all forms of life cannot exists (Vitorino and Bessa 2018). They thrive in extreme conditions, viz. sites of hydrothermal vent, hot springs, ocean, sea ice, hypersaline environment and extreme pH and temperature that are unfavourable for survival. These organisms are called as extremophiles that flourish in environments which are lethal for survival of other living beings (Rampelotto 2013).

Microbes are incredibly minute and constitute maximum proportion among living beings on the globe. However, particularly small portion of this huge variety has been searched for the development of microbial diversity. It has been reported that greater part of microbes cannot be cultured in laboratories (Sharma et al. 2018). It is a crucial part of microbial rallies to craft these organisms and those isolated microbes accessible to the research community. Microorganisms can be cultured for conservation and utilization after they are isolated from their native environment. Microbial activities make earth liveable and have limitless commercial applications principally in the field of life science (Arrigo 2005).

Gases such as oxygen and nitrogen which are the result of microbial activities make habitable climate. They occupy significant part in remediation of harmful chemical compounds (Kostka et al. 2011). Microorganisms also offer primary and secondary metabolites which have potent antimicrobials, immunosuppressants and anti-inflammatory and antitumour properties (Challis and Hopwood 2003). More than 104 metabolites produced from microbes have been explored for these compounds in the last decades. Bioplastics of microbial origin are promising substitute to chemical-based plastics, and these possess medical importance (Verlindin et al. 2007). Microbial diversity studies can be classified on the basis of culture-dependent (culturable) and culture-independent (unculturable) methods.

## 1.2 Assortment of Nonculturable and Culturable Microbes

### *1.2.1 Culture-Noncontingent and Culture-Contingent Method*

The word 'unculturable microbes' designates microbes that have hitherto been cultured in vitro on non-natural media (Hugenholtz et al. 1998). This method involves mining of DNA from the environmental illustration and afterwards, examined through molecular biology-based methods, or it involved another method for unculturable microbes in vitro by mimicking natural environment for the cultivation of unculturable microbes which show resistance to grow on cultivation media (Vartoukian et al. 2010).

### *1.2.2 Intents for 'Unculturability'*

In the history of science, the microbial life endurance was recognized from more than 300 years ago (Gírio et al. 2010). It was found that some microbes have yet not been known by cultural analysis. This might be owing to the piece of evidence that low occurrence and sluggish growers of microbes have been ignored. Moreover, in traditional biochemical identification methods, many characteristics of microbes are overlapped which makes their identification difficult (Schmeisser et al. 2007).

In contrast, certain microbes require particular nutrients for their fastidious growth. They had reported various substances and growth requirements of microbes of marshy sediments and found that microbes are specific to particular cultivation method. Therefore, only certain groups of microbes are identified on the basis of traditional method; however, the rest remain unidentified. In a mixed population, growth of a specific group of microorganisms is suppressed by the microbial product of other organism in the medium (Tamaki et al. 2005). Culture-independent approach has been extensively used to study microbes in different habitat (Yashiro et al. 2016).

### *1.2.3 Perception of Nonculturability*

Metagenomics study revealed the genetics of uncultured microbes which aims to know microbial environment as well as enhance biotechnological aspects. It is well known that uncultured microbes are known for their novel compounds which are yet to be discovered (Schmeisser et al. 2007). Some microorganisms are described on the basis of special habitat in spite of the microbial ubiquity theory, due to which their distribution is more restricted, as they tend to be giving rarer results in intricacy to culture these species in the samples.

Alain and Querellou (2009) found 30 cultured groups among 100 phyla all the way through phylogenetic analysis. Till date, only a small proportion of microbes (0.1–10%) have been cultivated of this vast diversity (Leadbetter 2003). Molecular studies make uncultivated microbes—oligotrophs and fastidious organisms to grow. The difficulty faced in cultivation of organisms includes slow or late growth of microbes on nutrient-rich media, lack of knowledge about novel media formulation and inefficiency of individuals, trained in the field of microbiology (Leadbetter 2003). This can be overcome by exploring microbes with well-equipped approaches and knowledge about microbes (Gest 2001).

The effect and response of microbial diversity to long-term environmental change are not properly understood, and it is also not clear that how much local microbial communities have impact on the environment (Tripp et al. 2008). *Pelagibacter ubique* SAR11 is widely distributed among heterotrophs cultured using sea water which was ameliorated with traces of phosphorus and ammonium ions. Many bacterial strains have to be cultured by maintaining solidifying agents in natural media, for example—Acidobacteria incubates for longer duration (Kuske et al. 2002). *Nitrosopumilus maritimus* is the first mesophile that belongs to Crenarchaea which is abundant in nature (Stevenson et al. 2004).

Cultivation of microbes is a very tedious task incorporated with many complexities. For the inhibition of microbial growth on Petri dish, many growth factors are responsible such as nutritional shock (Overmann 2006). Therefore, it is very imperative to discern the difficulties faced during the cultivation of new microorganisms. Although there is significant progress in the development of cultivation techniques, meticulous strategies are required for new media formulation. It is impossible to culture samples of all habitats; therefore, microbiologists are encouraged to explore novel microbes especially in extreme conditions. Different investigation showed that different microbes are able to colonize in cold environment of the planet from north to south poles.

Microorganisms are divided into two groups—fast growers (r-strategist) and slow growers (k-strategist) according to the microbial growth pattern and their potential of survivability (Overmann 2006). Nutritional shock is one of the parameter responsible for the growth inhibition or lethal to microbial cell. Apart from nutritional shock, in short duration, excessive growth of fast growers inhibits the growth of slow grower's types of microbes. Sometimes, undesired microbial growth also inhibits the growth of desired microbes in absence of inhibitory compounds. As a result of which, only few microbes are cultivated in the Petri dishes (Zengler 2009).

Even though all present acquaintance on the variety of microorganisms, it is believed that investigations in unexplored sites may result via additional evidence. Broad and unidentified speciations, mainly in the bacteria and archaea domains, are still unexplored. Usually, for the cultivation of bacteria and fungi, antifungal and antibacterial antibiotics are used, respectively. Environmental samples are serially diluted, and, by plating different dilutions, broad range of microbes are obtained (Tripp et al. 2008). Using traditional cultivation strategies, relevant but slow-growing microorganisms were not yet cultured and are then known as unculturable microbes. Due to several complications in cultivation of microbes, many endeavours have been made for new microorganism cultivation (Gest 2008).

The biotic interactions provide nutrients to the plants, increase soil fertility and showed adverse effects on pathogens but essential for the sustainability of natural ecosystems. Till now, various types of nutrient-rich media have been used for fastidious organism cultivation over slow growers (Koch 1997) and may be repressed by substrate-rich conventional media. For the cultivation of oligotrophs, the incubation period has increased; as a result of which, fastidious organisms progressively die off in mixed cultures. Davis et al. (2005) isolated most rare species after 12 weeks. Similarly, results have been reported for the segregation of strains from SAR11 clade after 24 weeks (Song et al. 2009). Many microbes require chemicals for their growth. For instance, *Abiotrophia* and *Granulicatella* required pyridoxal or L-cysteine for their augmentation; on the other hand, *Tannerella* is nutritionally dependent on *N*-acetyl muramic acid for their growth.

Another method is mimicking the natural environment in laboratory conditions for the culture of as-yet-uncultivated organisms. Kaeberlein et al. (2002) have intended a diffusion chamber for the marine bacteria that were precedently uncultivated. These organisms are completely dependent on other bacteria for their existence in a media. Since the mid-1980s, molecular biology practices have been focussed on microbial diversity and their ecology in their natural habitat. Molecular approaches revealed the molecular sequences of many uncultivated microbes which have many potential applications. An additional pioneering method resembling natural environment involves microcolony development of uncultured soil bacteria on soil substrate membrane system (Ferrari et al. 2008). This method involved viability staining and micromanipulation techniques for the detection and isolation of live microcolonies (Ferrari and Gillings 2009). However, colony hybridization method involves the isolation of colony containing a plasmid from a mixed microbial population (Salama et al. 1993). Culturing of the uncultured member of phylum utilizes this approach for research. Synergistetes had been isolated from dental plaque samples (Vartoukian et al. 2010). Flow cytometry and cell sorting (FACS) is a method that has also been used for the cultivation of cultured as-yet-uncultivated organisms (Zengler et al. 2002).

Genomic analysis of cultured as-yet-uncultivated organisms helps in identifying these organisms as well as gives some more information about the organism which will help in cultivation of previously uncultivated microbes in the vicinity of prospect (Tripp et al. 2008). Ghosh et al. (2010) used cultivation-independent molecular approach to study microbial diversity in the mangrove sediment of the Sundarbans, India. Proteobacteria (alpha, beta, gamma and delta), Flexibacteria (CFB group), Actinobacteria, Acidobacteria, Chloroflexi, Firmicutes, Planctomycetes and Gemmatimonadetes were the major divisions of detected bacterial phyla. Several reports suggested that *Bacillus* is an efficient tissue colonizer in different plants including *Coffea arabica* L., sunflower, cotton, potato, strawberry, *Panax notoginseng* and citrus plants (Vega et al. 2005). *Microbacterium* sp. was indigenous to plants such as maize, rice and wheat (Rijavec et al. 2007). Genus *Pseudomonas* is an extensively disseminated plant-associated bacterium reported activity of growth promotion in plants such as alfalfa (Gagné et al. 1987), clover (Sturz et al. 1997), potato (Reiter et al. 2002) and pea. Some minor groups such as Enterobacteriaceae,

Moraxellaceae, Xanthomonadaceae *and* Burkholderiaceae were also observed from proteobacterial phylum. Shivaji et al. (2004) recognized bacterial community including culturable bacteria from soil in the vicinity of Lake Zub, Schirmacher Oasis, Antarctica, has its place in the genera such as *Pseudomonas, Sphingobacterium, Arthrobacter, Micrococcus, Brevundimonas, Rhodococcus and Microbacterium.* Jiang et al. (2006) used culture-dependent and culture-independent techniques to examine microbial assortment. It was reported that gram-positive bacteria were predominant among the bacterial strains isolated from Lake Chaka. Analyzed bacterial diversity includes *Aeromonas hydrophila, Escherichia coli, Chryseomonas luteola, Chromobacterium violaceum, Pseudomonas aeruginosa, Serratia rubidaea, Klebsiella pneumoniae* and *Enterobacter cloacae* in mangrove soil at Tanjung Lumpur. They have also investigated their resistance against several antibiotics.

## 1.3 Microbes Role in Habitat/Environment

The environment is an essential perception because microbes are greatly affected by the atmosphere. Microorganisms are involved in many biogeochemical processes in different habitat. They are richest repertoire and considered as pillars of existence in nature. On earth, more than four billion years ago, microbes have been evolved and play numerous and important roles for maintaining sustainable biosphere that includes nutrient (elemental) cycling and detoxification of hazardous compounds present in the atmosphere. The microbial world is a treasure in itself and covers broad range of discrepancy of microbes among all types of microorganisms (bacteria, archaea, eukaryotes and viruses) in each possible habitat and is linked with plants and human on the planet. They are proficient in exploiting broad spectrum of energy sources and inhabitant of different environments like normal as well as extreme hot mainsprings, hydrothermal vent sites, drought, ocean and sea, polar ice, hypersaline and extremes pH that is lethal, or in other environments that are unfavourable for survival. Microorganisms have become an important part of the natural elemental cycle and played significant roles in biogeochemical cycles and converted the oxidized forms of molecules into reduced forms. Unicellular and filamentous cyanobacteria are mainly accountable for the fixation of nitrogen. Microbial study in different surroundings has confirmed that assessment of metabolically effective group is the key to explain microbial activities (Baldrian et al. 2012).

### *1.3.1 Role in Terrestrial Ecosystem*

This type of ecosystem is surrounded by forests, cropping systems and grazing lands. Soil acts as source of micro- as well as macronutrients. These nutrients are necessary for the plants, insects, protozoa, nematodes, worms and microbial growth (Staben et al. 1997). This biological diversity is responsible for the formation, maintenance and degradation of soil. Among this vast community, microbes constitute the major proportion and are versatile in their action. Bacterial community counts approximately $10^8$–$10^9$ cell $g^{-1}$ dry weight of soil in surface of the soil microscopically, while fungi can be contemporary up to numerous metres of hyphae in $g^{-1}$ of soil. Plant actions also augmented microbes in the soil. Microbes are associated with plants via roots (rhizosphere and rhizoplane) and leaves (phyllosphere and phylloplane). Rhizosphere acts as a reservoir for microbial diversity (Singh et al. 2019). Some may induce resistance or suppress the development of plant pathogens (Lanteigne et al. 2012) and exhibit positive as well as negative impact on plant growth. Microorganisms are versatile in nature and play significant role in increasing soil fertility. Microbial action contributes to nutrients cycling in soil such as carbon, nitrogen, sulphur, iron and manganese cycles. They act as biofertilizers and fix atmospheric nitrogen, phosphorus, and sulphur and other elements which are unavailable for plants and finally contribute to plant's nutrition (Yadav and Saxena 2018). The degradation of hydrocarbons and dead and decayed plants and animal matter along with its involvement in the formation of humus is an important role played by bacterial community in the soil. Actinomycetes imparts soils their characteristic earthy odour by producing a compound called eosin by *Streptomyces* species.

### *1.3.2 Role in Mangrove Ecosystems*

High load of biological diversity belonging to plants, animals and microorganisms occurs at mangrove forests occurring at the border of terrestrial and marine environment. Mangroves cover nearly 70% of the world's tropical and subtropical coastal regions, which are identified to be highly fecund ecosystems of huge ecological value. These ecosystems are highly productive all over the world despite they are fragile and sparsely distributed. In this habitat, microbes transform nutrients and detoxify pollution-causing agents, and as biocontrol of pests, a unique environment harbouring diverse groups of microbes such as bacteria, fungi, cyanobacteria, microalgae, macroalgae and protists is provided by them. They are abundantly nitrogen and phosphorus deficient (Holguin et al. 1999).

Amid the microbial distribution, bacteria and fungi represent the foremost proportion after algae and protozoa. The most common bacteria are sulphate reducers belonging to the genera *Desulfovibrio, Desulfotomaculum, Desulfosarcina* and *Desulfococcus,* nitrogen fixers and methane producers (genera *Azospirillum, Azotobacter, Rhizobium, Clostridium, Klebsiella, Methanococcoides methylutens*), phosphate solubilizers (genera *Bacillus, Paenibacillus, Xanthobacter, Vibrio proteolyticus, Enterobacter, Kluyvera, Chryseomonas and Pseudomonas*) and photosynthetic anoxygenic bacteria (genera *Chloronema, Chromatium, Beggiatoa, Thiopedia, Leucothoe* bacteria) (Das et al. 2009). Moreover, fungi, such as ligninolytic, cellulolytic, pectinolytic, amylolytic and proteolytic fungi, as well as actinomycetes are present in mangrove ecosystems. Among the algae, Chlorophyta,

Chrysophyta, Phaeophyta, Rhodophyta and Cyanophyta are the dominant groups of the mangrove ecosystem. They harbour unique microbial composition which contains major source of therapeutic enzymes, antimicrobial and antitumour agents, insecticides, etc.

### *1.3.3 Role in Aquatic Environment*

Aquatic habitat classified into fresh water, marine and both. Fresh water consists of lakes and rivers. Open ocean, coral reefs, and intertidal zones constitute the marine environment. Ecosystems that are considered both marine and freshwater system is composed of estuaries and salt marshes. Microbes are widespread, well-adapted in fresh water and involved in diverse biogeochemical processes, such as petrification of organic compounds; nutrients can be remineralized in maintenance of water ecosystem (Newton and McLellan 2015). They have its place to the group of photosynthetic oxygenic and anoxygenic organisms that include bacteria, algae and cyanobacteria. Further 70% of the earth is roofed by ocean, and microbes are accounted for more than 98% of ocean biomass. Marine microbes are called as 'the canary in the coal mine'. The marine microbial diversity constitutes microalgae, bacteria and archaea, fungi and viruses (Fuhrman and Noble 1995). The Antarctic bacterium *Pseudoalteromonas haloplanktis* TAC125 (PhTAC125) is studied for their enormously fast growth and its wide temperature array from −2.5 to 25 °C (Wilkins et al. 2013). They offer a tremendous biodiversity and potential for drug sighting and delivery of novel marine-derived products in therapeutic claims. They play diverse role in the marine environment such as in food chain, transformation of nutrients and sustaining marine ecosystem for the survival of marine organisms.

### *1.3.4 Role in Extreme Environments*

On the basis of stability, environment can be classified into two types—normal and extreme environment. Stable condition can be considered as 'normal', while in extreme environment, organisms experience dramatic changes. The inhabitant organisms of extreme ecology are known as extremophiles. High and low temperature classify the nature of microbe's adaptability for extremophiles; these factors are the basis of their classification (thermophiles and psychrophiles), high salt concentration, high and low pH (acidophiles and alkaliphiles) and low water activity ($a_w$). The microbial product produces from extremophiles are of immense importance. Many information on microbial diversity from thrilling environments, for instance, low temperature (Yadav 2015), high temperature, saline soil, drought, acidic soil and alkaline soil, have been reported. In an unstable environment, the costs to survive in stress condition may increase for some organisms, while most will probably die off. Extreme environments are well known for novel microbial diversity.

Microbes at high temperature make a hydrophobic environment for their survival (Acharya and Chaudhary 2012). Complicated zig-zag structure of proteins provides microbial cells to withstand denaturation and proteolysis.

### *1.3.5 Role in Saline Environment*

Microorganisms are widely distributed in hypersaline environment from solar salterns to deep salt mines (Selvarajan et al. 2017). Most of the Indian saline ecosystems, such as Sambhar Lake in Rajasthan, Chilika Lake in Odisha, the Great Rann of Kutch in Gujarat and Lonar Lake in Maharashtra, are known for novel and potential applications. In this environment, microbes play vital character in the remineralization of organic matter (Joshi et al. 2008). The microorganisms in saline environments that have been isolated and identified mainly belong to the family Halobacteriaceae. Halophilic microbes have been described from different phylum including Actinobacteria, Bacteroides, Euryarchaeota, Firmicutes, Proteobacteria and Spirochaetes (Yadav and Saxena 2018).

### *1.3.6 Role in Cold Environment*

Cold environments cover the largest region on the earth. The term 'psychrophiles' is used for the microbe that are living/inhabitant in cold condition. In India, microbes are widely distributed and explored in the Himalayan region. Psychrophiles are potential sources for production of extracellular proteins. Polyhydroxyalkanoates (PHAs) are chiefly produced by psychrophiles which increase survivability in stress conditions (Tribelli and López 2018). It includes diverse groups of microorganisms, i.e. archaea, bacteria and fungi. had reported many species from high-altitude and low-temperature environments of Indian Himalayan region belonging to genera *Aurantimonas, Bacillus, Disemia and Paenibacillus.*

### *1.3.7 Role in Drought Environment*

The desert microbiota potentially is known for their efficiency in maintaining the harmony of recycling of different nutrients and ecological balance as well as for the development of soil structure. In rain-fed conditions, microorganisms that are tolerated in drought environment have been secluded and characterized for plant growth promoters (PGP) and have its place to the family Halobacteriaceae and genera such as *Haloarcula argentinensis, Halobacterium* sp., *Halococcus hamelinensis, Haloferax alexandrinus, Haloferax larsenii, Haloferax volcanii, Halolamina pelagic, Halostagnicola kamekurae, Haloterrigena thermotolerans, Natrinema* sp. and *Nanoarchaeum mannanilyticum.*

### *1.3.8 Role of Microbes in Human Health*

The human body is immensely colonized with microbes in different tissues and body parts. Approximately, thousands of different bacterial species exists side-by-side together in the intestinal tract of human. Among various organs of human, the digestive tract heavily occupied with enormous bacteria represents the dominant genera *Lactobacillus* sp., *Escherichia coli*, *Klebsiella* sp. and *Proteus* sp. that perform various roles in metabolic processes of substrates, build up defence mechanism against various infections, synthesize vitamins and various cofactors for their development, support in degradation of fats and polysaccharides and also have antioxidant properties of foodstuffs which in turn enhance the nutritional value (Odonkor and Ampofo 2013). Probiotics (live microorganisms) predominately belonging to the genera *Lactobacillus* and *Bifidobacterium* are dietary supplements added to the foodstuff impacting their nutritional and therapeutic value (Kumar et al. 2012). Important roles played by microbes in the gut are energy generation, production of cellular constituents and processing of nutrients (metabolism). In certain circumstances, microbiota may result in diverse health issues such as diarrhoea, human gastritis, typhoid, gastroenteritis, bacterial vaginosis, chronic peptic ulcers, urinary tract infections and gastric adenocarcinoma (Peris-Bondia et al. 2011).

## 1.4 Potential Applications

Divergence among microbes is imperative for the endurance of all life forms and offers enormous reservoirs that exploit for human welfare. They have become reservoirs of many substances. Microbes have been used in beer, wine, acetic acid, cheese and yoghurt production and involved in many industries, viz. baking, leather, paper pulp and textile industries (Acharya and Chaudhary 2012). Methanogens plays a significant role in the biogas production; however, psychrophiles are being subjugated in biodiesel production (Bernard et al. 2012). We have summarized the application of microbial biotechnology to maintain the sustainable development of the ecosystem in Fig. 1.1.

### *1.4.1 Environmental Applications (Bioremediation)*

Grouping of diverse technologies, such as designed biosensors for assessing the level of contamination, mining of the large number of polluted spots and designing of geohydrobiological engineering models, via polishing the spots with microbe-assisted flora (Pilon-Smits 2005), is the most competent and cost-effective way of bioremediation. Bioremediation has capability to fix polluted environments. For the retrieval of degraded lands, the integrative attempt might provide an evidence to be

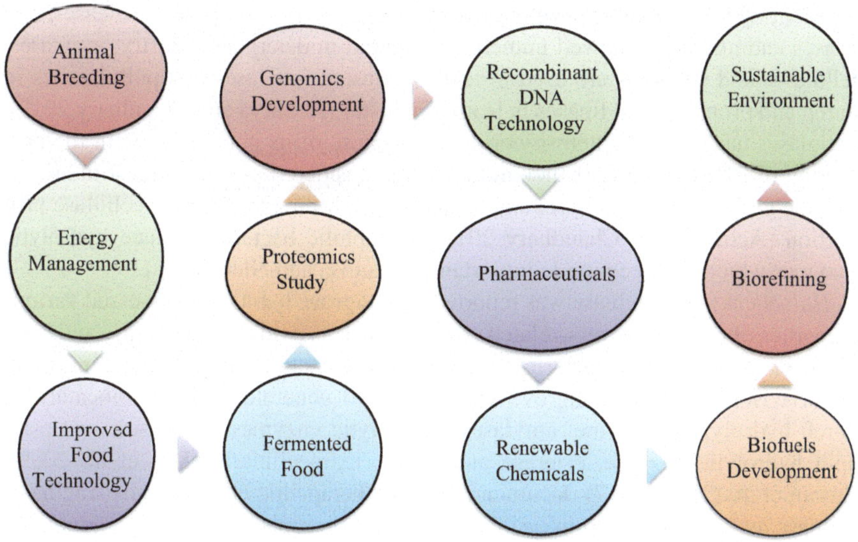


**Fig. 1.1** Application of microbial biotechnology to maintain the sustainable development of the ecosystem

one of the pre-eminent ecological practices. Microbes actively participate in the removal of toxic compounds and oil biodegradation. Microbial-based biosensors provide application in monitoring of toxic compounds.

### *1.4.2 In Industry (Novel Biotechnological and Pharmaceutical Products)*

Soil has proven to be the major resource of microbes from where they are extracted and used in industries, food processing and production, biocontrol agents, advancement of biocides, drugs and other natural products. Microorganisms can occur naturally or even through human ingenuity, i.e. it can be genetically engineered. Mangrove ecosystem has comprehended many biotechnological importances.

### *1.4.3 Enzymes*

The enzymes have many advantages over chemical-based industries due to its high efficiency and negligible substrate loss (Acharya and Chaudhary 2012). Microbes have a varied array of enzymatic activities and are proficient in catalyzing numerous biochemical reactions with novel enzymes. Microbes in marine environment offer microbial proteins which have therapeutic importance for human welfare. Enzymes contributing to sustainable development of industries such as lipases, proteases, cellulases and amylases showed numerous potential in detergent industry; amylases, cellulases and catalases are used in textile industry, amylases and pullulanases in starch and proteases and lipases in leather industry (Acharya and Chaudhary 2012). Cellulases have gained interest worldwide because of its potential role in the production of transportation of fuel and also are considered as third largest industrial enzyme globally. Mostly, fungi and bacteria have been exploited for cellulase production (Acharya and Chaudhary 2012). Halophilic bacteria produce hydrolytic enzymes which have economic importance (Ventosa and Nieto 1995).

*Halococcus* asparaginase was reported in mangrove habitat and assessed various properties of oxidative stress-related enzymes such as oxidase, peroxidase and catalase from gram-negative bacteria in mangrove ecosystem of Bhitarkanika. In Brazil, bacteria isolated from mangrove were found to generate different biocatalysts, starch hydrolyzing enzyme, amylases; proteolytic enzymes, proteases; and ester lipid hydrolytic enzyme, esterase and lipase, extracellularly (Das et al. 2009). Husain et al. (2016, 2017) documented chemotherapeutic enzymes (asparaginase, arginase and arginine deiminase) isolated from rhizospheric soil, and endophytic bacteria revealed that *Aspergillus niger* isolated from mangrove ecosystem can generate an enzyme xylanase which can withstand high temperature and pH and carry out biobleaching of paper pulp. Carbonic anhydrase (CA) is a biocatalyst exploited for the sequestration of carbon dioxide. In fresh water, Sharma et al. (2018) optimized various factors for increasing CA production from the genera *Enterobacter* sp. and *Aeromonas* sp. and purified and compared *P. fragi* CA, *M. lylae* CA and *M. luteus* 2 CA against commercial bovine carbonic anhydrase (BCA). In food industry, enzymes play a vital role to process food.

### 1.4.4 Biosurfactants

Biosurfactants exhibit several therapeutic significances such as antibacterial, antifungal, antiviral and anticoagulation properties. High surface and emulsifying activity of microbial molecules are categorized as biosurfactants or bio-emulsifiers. Due to lower toxicity, mild production conditions, environmental compatibility and higher biodegradability, biosurfactants have gained interest to a large extent as compared to chemical surfactants (Mulligan et al. 2011). All these biosurfactant properties have driven their importance in protecting the environment and have been utilized in many industries, viz. food, cosmetics, biopesticides and pharmaceuticals. On the basis of microbial origin of biosurfactants and their chemical configuration characteristics, biosurfactants can be classified as glycolipids, phospholipids, lipopeptides and polymeric surfactants. The most common biosurfactants among these four groups assessed are glycolipids and lipopeptides produced by *P. aeruginosa* and *B. subtilis*, respectively (Pornsunthorntawee et al. 2008). Properties of biosurfactants, i.e. degradation of substances, less toxic and efficient at low/high pH or temperature, make them more valuable than chemical surfactants and display prominent eco-friendly compatibility by enhancing bioremediation efficiency. Mulligan et al. (2011) reported and assessed the promising biosurfactant production from *Leucobacter komagotae* 183 strain isolated from mangrove ecosystem of Thailand.

## 1.5 Medical Importance (Antimicrobial Substances)

The necessity for variety and expansion of novel classes of antimicrobial agents is growing due to resistance shown against several antibiotics by diverse groups of bacteria, fungi and other microorganisms that causes severe complications in repression of contagious diseases. Many documents have reported antifungal substances from mangrove ecosystem. Two despidones (auranticins A and B) that exhibit antimicrobial activity were generated from fungus *Preussia aurantiaca* (Poch and Gloer 1991). In mangrove environment, isolated *Aigialus parvus* BCC-5311 synthesized aigialomycins A–E, resorcylic macrolides and hypothemycin. Few studies reported a novel compound—enniatin G—extracted from *Fusarium* sp. which showed antibiotic, antitumour, phytotoxic and insecticidal activity. Lin et al. (2008) described an actinomycete *Streptomyces* sp. that strongly constrains the growth of gram-negative as well as gram-positive bacteria.

### *1.5.1 Bio-mediated Compounds*

Study revealed that plants, fungi, bacteria and actinomycetes were found to produce bioactive compounds. Actinomycetes act as promising candidate for the treatment of diabetes and neurodegenerative diseases and are likely to be the rich cause for the detection of antitumour and anti-inflammatory compounds after few genetic modifications. Microorganisms living in the mangrove ecosystems are considered as a natural 'hotspot' for producing novel and superior drugs. It was reported that 2000 microbes, viz. fungi, bacteria and actinomycetes, that have potential to synthesize secondary metabolites were also having anticancer, antitumour and anti-inflammatory properties. *Streptomyces albidoflavus* isolated from the Pichavaram mangrove that exhibited antitumour properties was reported by Marine algae are the key source of phycocolloids such as agar, carrageenan and alginate (Shanmugam and Mody 1999). They were also reported to have anticomplementary, antimutagenic, blood anticoagulant, antiviral, hypolipidemic, hypoglycaemic, immunomodulating, anti-inflammatory and antitumour activities.

Microorganisms are in diverse form and a vital constituent of biotic diversity that has played a momentous role in origin of life on the earth and in maintaining ecology of various habitats. They are the abundant janitors across the globe occurring in all climatic regions, including Arctic, Antarctic, deep within rocks and oceanic hot vents. The extensive genetic variation encompasses the spectrum of variability among various species. Microbial diversity comprises broad group of microbes that are useful for food production and global environmental protection as well as have many applications: for example, as immunosuppressants; as antimicrobial and antiproliferative drugs; as immunomodulators; as anthelmintics in pharma industries; as a fermentation product in food industries; as food processing agents, antiparasitic agents and biopesticides in agricultural region; as a microbial product intended for manufacturing organic compounds, vitamins, amino acids, biocatalyst and bioconversion agents; as detergents by chemical industries; as bioenergy and bioremediation agents in environmental industries using biotechnological approaches to clean up the contaminated sites, in recycling process of nutrients and in maintenance of ecosystem health in biosphere.

The traditional techniques for cultivating microbes and advanced culture-independent methods might be deliberated as a principal approach to understand how microbes live as well as their role in extreme habitats. The studies of microbial diversity pave a healthier thought of the role and purpose of microbial communities in terrestrial, aquatic and marine environments and a better understanding of the consequences of extinction of plant and animal species and of trepidations on ecosystem. Therefore, microbial communities are excellent models for studying and examining fundamental biological interactions for maintainable ecology of plants and animals and improved dimensions to uphold water quality and soil fertility.

## 1.6 Conclusion

The usage of microbial enzymes is dispersed in several fields such as in preparing enantiomer, pure drugs from racemic mixture; for the production of robust drugs, as a therapeutic agent; etc., while pathogenic microbes causing disease to humans, plants and animals pose a threat to health, food safety and security. Recent advances in molecular genetics are currently gaining attention which is being supplemented by culture-dependent analysis. Diversity among microbes can be used to monitor and predict the changes in the environment as microbes are the major sources that have been involved in sustainable development. The ecosystem may function as a key parameter that controls various global cycles (nitrogen, carbon, sulphur, phosphorus and heavy metal cycles) by maintaining the dynamic equilibrium and integrity of our planet. The extensive industrial development has led to the exploitation of diverse forms of microbial communities by gradual changes in existing biotic and abiotic factors. Therefore, metagenomics is useful in exploiting unidentified microorganisms in various environments to reveal genomic content of new species and biomarkers for detecting several metabolic activities; it also delivers innovative techniques to obtain products from microorganisms without culturing them in laboratories and can be helpful in understanding complexity within microbial communities. The inability of traditional culturing techniques has shown the diversity of microorganisms and also that the species assortment in terrestrial and aquatic habitats is far superior than expected. The mainstream of microbial diversity (>90%) remains to be revealed. The extensive genetic variation encompasses the spectrum of alteration among various species. Nevertheless, the information regarding microbial physiology and genetics is crucial for transcriptomic and gene-level studies.

## 1.7 Future Perspectives

Most assorted group of organisms found in any form of environment are microbes. Till date, investigation has focused on those microbes that are culturable; however, an affluence of evidence is now being collected from unculturable microbes. A grander consideration of the microbial world can benefit ecological organization. These organisms are the root for revolutions that empower life to endure; thus, acquaintance of their interfaces, characters and tasks is vivacious to our understanding. We are in the middle-of-the-road to explore, identify, conserve and use microbes to support mankind specifically and ecology. Marine organisms thrive not just in the surface of ocean but also in addition in the lower and deep profundities from coastal to the offshore regions of a particular habitat. This community of microbes is still unknown and might have evolved with many mysteries which are still to be answered. Plant microbiome-based solutions could achieve a change in perspective of their role in health and illness and have significant results for biocontrol and medical problem. Targeted microbiome engineering for crops is an upcoming inclination. Biodiversity should be a biomarker for these microbiome cadences. Higher plant-associated miscellany can be achieved not only through the enactment of biological control agents which shifts the microbiome but also by the solicitation of microbial consortia. Half-life of gut microbes could be managed through probiotics, and estrogen replacement therapy can be benefited by long-term users through altering properties of estrogen without increasing the risk of reproductive cancers. Microbiota populations have been found in the human skin, mucosal membranes, and gut where they can influence several disorders, such as diabetes, obesity, cancer and colitis. Thus, a multiomics approach to identify and sequence different microbial populations in the body can provide valuable information. Gut bacteria markers can be called as 'smoking gun' for liver disease as they can spot the early stages of liver disease by releasing chemical compounds produced by the bacteria in our gut. Microbiologists are finding new-fangled ways to sightsee the new places and new biotechnological intervention in the hunt of new medicine and new techniques to help mankind. Some longstanding mysteries about microbial diversity are their diversity and stability in various ecosystems which can solve many queries about evolution and could also help to understand the future and make it easy.

## References

Acharya S, Chaudhary A (2012) Bioprospecting thermophiles for cellulase production. Braz J Microbiol 43:844–856

Alain K, Querellou J (2009) Cultivating the uncultured: limits, advances and future challenges. Extremophiles 4:583–594

Arrigo K (2005) Marine microorganisms and global nutrient cycles. Nature 437:349–355

Baldrian P, Kolařík M, Stursová M, Kopecký J, Valášková V, Větrovský T, Zifčáková L, Snajdr J, Rídl J, Vlček C, Voříšková J (2012) Active and total microbial communities in forest soil are largely different and highly stratified during decomposition. ISME J 6(2):248–258

Bernard L, Chapuis-Lardy L, Razafimbelo T, Razafindrakoto M, Pablo AL, Legname E, Poulain J, Brüls T, O'Donohue M, Brauman A, Chotte JL, Blanchart E (2012) Endogeic earthworms shape bacterial functional communities and affect organic matter mineralization in a tropical soil. ISME J 6(1):213–222

Challis G, Hopwood DA (2003) Synergy and contingency as driving forces for the evolution of multiple secondary metabolite production by Streptomyces species. Proc Natl Acad Sci U S A 2:14555–14561

Das P, Mukherjee S, Sen R (2009) Substrate dependent production of extracellular biosurfactant by a marine bacterium. Bioresour Technol 100:1015–1019

Davis KE, Joseph SJ, Janssen PH (2005) Effects of growth medium, inoculum size, and incubation time on culturability and isolation of soil bacteria. Appl Environ Microbiol 71:826–834

Ferrari B, Gillings MR (2009) Cultivation of fastidious bacteria by viability staining and micromanipulation in a soil substrate membrane system. Appl Environ Microbiol 75:3352–3354

Ferrari B, Winsley T, Gillings M, Binnerup S (2008) Cultivating previously uncultured soil bacteria using a soil substrate membrane system. Nat Protoc 3:1261–1269

Fuhrman JA, Noble RT (1995) Viruses and protists cause similar bacterial mortality in coastal seawater. Limnol Oceanogr 40(7):1236–1242

Gagné S, Richard C, Rousseau H, Antoun H (1987) Xylem-residing bacteria in alfalfa roots. Can J Microbiol 33(11):996–1000

Gest H (2001) Evolution of knowledge encapsulated in scientific definitions. Persp Biol Med 44:556–564

Gest H (2008; posting date) The modern myth of "unculturable" bacteria/scotoma of contemporary microbiology. http://hdl.handle.net/2022/3149

Ghosh K, Roy M, Kar N, Ringø E (2010) Gastrointestinal bacteria in rohu, Labeo rohita (Actinopterygii: Cypriniformes: Cyprinidae): scanning electron microscopy and bacteriological study. Acta Ichthyol Piscatoria 40:129–135

Gírio FM, Fonseca C, Carvalheiro F, Duarte LC, Marques S, Bogel-Łukasik R (2010) Hemicelluloses for fuel ethanol: a review. Bioresour Technol 13:4775–4800

Holguin G, Bashan Y, Mendoza-Salgado RA, Amador E, Toledo G, Vazquez P, Amador A (1999) La Microbiologia de los manglares. Bosques en la frontera entre el mar y la tierrra. Ciencia Desarrollo 144:26–35

Hugenholtz P, Goebel BM, Pace NR (1998) Impact of culture independent studies on the emerging phylogenetic view of bacterial diversity. J Bacteriol 180:4765–4774

Husain I, Sharma A, Kumar S, Malik F (2016) Purification and characterization of glutaminase free asparaginase from *Pseudomonas otitidis*: induce apoptosis in human leukemia MOLT-4 cells. Biochimie 121:38

Husain I, Bala K, Wani A, Makhdoomi U, Malik F, Sharma A (2017) Arginase purified from endophytic *Pseudomonas aeruginosa* IH2: induce apoptosis through both cell cycle arrest and MMP loss in human leukemic HL-60 cells. Chem Biol Interact 274:35–49

Jiang HL, Tay STL, Maszenan AM, Tay JH (2006) Physiological traits of bacterial strains isolated from phenol-degrading aerobic granules. FEMS Microbiol Ecol

Joshi AA, Kanekar PP, Kelkar AS, Shouche YS, Vani AA, Borgave SB, Sarnaik SS (2008) Cultivable bacterial diversity of alkaline Lonar Lake, India. Microb Ecol 55:163–172

Kaeberlein T, Lewis K, Epstein SS (2002) Isolating 'uncultivable' microorganisms in pure culture in a simulated natural environment. Science 296:1127–1129

Koch AL (1997) Microbial physiology and ecology of slow growth. Microbiol Mol Biol Rev 61:305–318

Kostka JE, Prakash O, Overholt W, Green S, Freyer G, Canion A, Delgardio J, Norton N, Huettel M (2011) Hydrocarbon-degrading bacteria and the bacterial community response in Gulf of Mexico beach sands impacted by the Deepwater Horizon oil spill. Appl Environ Microbiol 77:7962–7797

Kumar S, Narayani S, Shankar T, Pandian V (2012) Optimization of cultural conditions for exopolysaccharides production by Frateuria aurantia. Int J Appl Biol Pharmaceut Technol 3:133–143

Kuske C, Ticknor LO, Miller ME, Dunbar JM, Davis JA, Barns SM, Belnap J (2002) Comparison of soil bacterial communities in rhizospheres of three plant species and the interspaces in an arid grassland. Appl Environ Microbiol 68:1854–1863

Lanteigne C, Gadkar VJ, Wallon T, Novinscak A, Filion M (2012) Production of DAPG and HCN by Pseudomonas sp. LBUM300 contributes to the biological control of bacterial canker of tomato. Phytopathology 102:967–973

Leadbetter JR (2003) Cultivation of recalcitrant microbes: cells are alive, well and revealing their secrets in the 21st century laboratory. Curr Opin Microbiol 6(3):274–281

Lin DX, Wang ET, Tang H, Han TX, He YR, Guan SH, Chen WX (2008) Shinella kummerowiae sp. nov., a symbiotic bacterium isolated from root nodules of the herbal legume Kummerowia stipulacea. Int J Syst Evol Microbiol 58(Pt 6):1409–1413

Mulligan C, Fischer M, Thomas GH (2011) Tripartite ATP-independent periplasmic (TRAP) transporters in bacteria and archaea. FEMS Microbiol Rev

Newton RJ, McLellan SL (2015) A unique assemblage of cosmopolitan freshwater bacteria and higher community diversity differentiate an urbanized estuary from oligotrophic Lake Michigan. Front Microbiol 6:1028

Odonkor S, Ampofo JK (2013) Escherichia coli as an indicator of bacteriological quality of water. Microbiol Res 4:5–11

Overmann J (2006) Principal of enrichment, isolation, cultivation and preservation of prokaryotes. Prokaryotes 1:80–136

Peris-Bondia F, Latorre A, Artacho A, Moya A, D'Auria G (2011) The active human gut microbiota differs from the total microbiota. PLoS One:e22448

Pilon-Smits E (2005) Phytoremediation. Annu Rev Plant Biol 56:15–39

Poch GK, Gloer JB (1991) Auranticins A and B: two new depsidones from a mangrove isolate of the fungus Preussia aurantiaca. J Nat Prod 54(1):213–217

Pornsunthorntawee O, Wongpanit P, Chavadej S, Abe M, Rujiravanit R (2008) Structural and physicochemical characterization of crude biosurfactant produced by Pseudomonas aeruginosa SP4 isolated from petroleum-contaminated soil. Bioresour Technol 99:1589–1595

Rampelotto PH (2013) Extremophiles and extreme environments. Life (Basel) 3(3):482–485

Reiter B, Pfeifer U, Schwab H, Sessitsch A (2002) Response of endophytic bacterial communities in potato plants to infection with Erwinia carotovora subsp. atroseptica. Appl Environ Microbiol 68(5):2261–2268

Rijavec T, Lapanje A, Dermastia M, Rupnik M (2007) Isolation of bacterial endophytes from germinated maize kernels. Can J Microbiol 53(6):802–808

Salama MS, Sandine WE, Giovannoni SJ (1993) Isolation of Lactococcus lactis subsp. cremoris from nature by colony hybridization with rRNA probes. Appl Environ Microbiol 59:3941–3945

Schmeisser C, Steele H, Streit WR (2007) Metagenomics, biotechnology with non-culturable microbes. Appl Microbiol Biotechnol 75(5):955–962

Selvarajan R, Sibanda T, Tekere M, Hand N, Taylor SM (2017) Diversity analysis and bioresource characterization of halophilic bacteria isolated from a South African Saltpan. Molecules 22(4)

Shanmugam MH, Mody K (1999) Heparinoid-active sulphated polysaccharides from marine algae as potential blood anticoagulant agents. Curr Sci 79

Sharma A, Pal PB, Rehman MB, Sharma J (2018) Monitoring of phenolic compounds and surfactants in waste water mixing with Narmada river at Jabalpur, (India). Int J Adv Res 6(4):97–102

Shivaji S, Reddy G, Aduri P, Kutty R, Ravenschlag K (2004) Bacterial diversity of a soil sample from Schirmacher Oasis, Antarctica. Cell Mol Biol (Noisy-le-Grand, France) 50:525–536

Singh D, Raina TK, Kumar A, Singh J, Prasad R (2019) Plant microbiome: a reservoir of novel genes and metabolites. Plant Gene. https://doi.org/10.1016/j.plgene.2019.100177

Song J, Oh HM, Cho JC (2009) Improved culturability of SAR11 strains in dilution-to-extinction culturing from the East Sea, West Pacific Ocean. FEMS Microbiol Lett 295:141–147

Staben ML, Bezdicek DF, Fauci MF, Smith JL (1997) Assessment of soil quality in conservation reserve program and wheat-fallow soils. American Society of Agronomy, Crop Science Society of America, Soil Science Society of America

Stevenson BS, Eichorst SA, Wertz JT, Schmidt TM, Breznak JA (2004) New strategies for cultivation and detection of previously uncultured microbe. Appl Environ Microbiol 70:4748–4755

Sturz AV, Christie BR, Matheson G, Nowak J (1997) Biodiversity of endophytic bacteria which colonizer clover nodules, roots, stem sand foliage and their influence on host growth. Biol Fertil Soils 25:13–19

Tamaki H, Sekiguchi Y, Hanada S, Nakamura K, Nomura N, Matsumura M, Kamagata Y (2005) Comparative analysis of bacterial diversity in freshwater sediment of a shallow eutrophic lake by molecular and improved cultivation-based techniques. Appl Environ Microbiol 71(4):2162–2169

Tribelli PM, López NI (2018) Reporting key features in cold-adapted bacteria. Life (Basel) 8(1):8

Tripp HJ, Kitner JB, Schwalbach MS, Dacey JW, Wilhelm LJ, Giovannoni SJ (2008) SAR11 marine bacteria require exogenous reduced sulphur for growth. Nature 452:741–744

Vartoukian SR, Palmer RM, Wade WG (2010) Cultivation of a Synergistetes strain representing a previously uncultivated lineage. Environ Microbiol 12:916–928

Vega FE, Pava-Ripoll M, Posada F, Buyer JS (2005) Endophytic bacteria in Coffea arabica L. J Basic Microbiol

Ventosa A, Nieto J (1995) Biotechnological applications and potentialities of halophilic microorganisms. World J Microbiol Biotechnol 11:85–94

Verlindin RA, Hill DJ, Kenward MA, Williams CD, Radecka I (2007) Bacterial synthesis of biodegradable polyhydroxyalkanoates. J Appl Microbiol 102:1437–1449

Vitorino LC, Bessa LA (2018) Microbial diversity: the gap between the estimated and the known. Laboratory of Agricultural Microbiology, Goiano Federal Institute, Rio Verde Campus, Rio Verde GO 75901-970, Brazil

Wilkins D, Sebille EV, Stephen R, Rintoul SR, Lauro FM, Cavicchioli R (2013) Advection shapes Southern Ocean microbial assemblages independent of distance and environment effects. Nat Commun 4:2457

Yadav AN (2015) Bacterial diversity of cold deserts and mining of genes for low temperature tolerance. https://doi.org/10.13140/RG.2.1.2948.1283/2

Yadav AN, Saxena AK (2018) Biodiversity and biotechnological applications of halophilic microbes for sustainable agriculture. J Appl Biol Biotechnol

Yashiro E, Figueroa EP, Buri A, Spangenberg JE, Adatte T, Hirzel HN et al (2016) Local environmental factors drive divergent grassland soil bacterial communities in the western Swiss Alps. Appl Environ Microbiol 82:6303–6316

Zengler K (2009) Central role of the cell in microbial ecology. Microbiol Mol Biol Rev 73:712–729

Zengler K, Toledo C, Rappe M, Elkins J, Mathur EJ, Short JM, Keller M (2002) Cultivating the uncultured. Proc Natl Acad Sci USA 99:15681–15686

# Chapter 2
# Microbe-Mediated Genetic Engineering for Enhancement of Nutritional Value in Food Crops

**Bhupendra Koul and Siddharth Tiwari**

**Abstract** Malnutrition is a severe public health challenge in several underdeveloped countries worldwide. Thus, sustainable plant productivity and its easy availability in the coming years shall be a major constraint for food and nutritional security for the teeming millions. Moreover, various abiotic and biotic stresses in plants contribute to yield penalty. The conventional breeding techniques for improvement/enhancement of growth, yield, and quality traits are tedious, time-consuming, and impermanent. On the contrary, microbe-assisted genetic manipulation of crop plants has revolutionized the crop improvement through incorporation of value-added traits of agronomic and nutritional importance. It is now possible to transfer genes(s) of interest, irrespective of its origin to crop plants through direct or indirect (vector or vectorless) approaches. Indeed, *Agrobacterium* has become the most effective vector for gene transfer in the arena of transgenic technology. The success of a transgene of nutritional importance depends upon its high expression level and stability in plant system. To give effect to this hypothesis, various strategies have been deployed including elimination of destabilizing elements of the transgene; removal of putative polyadenylation sequences, cryptic splicing sites, and codon biasness; and incorporation of elements for high-level expression (strong promoter(s), 5′ untranslated leader sequence, translation initiation context). In this chapter, we have discussed the importance of *Agrobacterium* and various genetic engineering approaches for enhancing the expression of foreign genes including the current scenario and advancements in the biofortification of crops. This chapter also summarizes state of the art of nutrition enhancement in crops, major challenges, and future prospects.

---

B. Koul
Department of Biotechnology, School of Bioengineering and Biosciences, Lovely Professional University, Phagwara, Punjab, India

S. Tiwari (✉)
National Agri-Food Biotechnology Institute (NABI), Department of Biotechnology, Ministry of Science and Technology (Government of India), Mohali, Punjab, India
e-mail: siddharth@nabi.res.in



J. Singh et al. (eds.), *Microbial Biotechnology: Basic Research and Applications*, Environmental and Microbial Biotechnology,
https://doi.org/10.1007/978-981-15-2817-0_2

**Keywords** Transgene · Codon optimization · Biofortification · Agrobacterium · Crop improvement

## 2.1 Introduction

Scientists are of the opinion that "presence of water makes life possible on the planet." In continuation to it, we must remember that "plants sustain life on the planet." Almost half of the world population does not have regular access to fresh fruits, vegetable, meat, and fish. Therefore, the only option that remains for them is to rely on the food crops like rice, wheat, and maize (Christou and Twyman 2004). Unfortunately, these major food crops do not possess ample vitamins (vitamin A, folic acid), minerals (zinc, iron, and selenium), and amino acids (lysine) required for normal growth and metabolism. Therefore, in the near future, malnutrition and the related nutrition deficiency shall pose serious public health issues, worldwide (United Nations 2016, 2017). As a consequence of malnutrition and inadequate nutrition, one-third of the infant deaths occur throughout the world. Keeping in view of the forthcoming worldwide demand for quality foods to feed the teeming millions and to manage the nutritional-deficiency-induced diseases, the United Nations (UN) has proposed its Agenda 2030 for sustainable agriculture which shall strive to "eradicate hunger, attain food security and enhanced nutrition, and encourage sustainable agriculture."

Biofortification is a research-based procedure involving the amalgamation of genetic engineering, plant breeding, and agronomic practices to augment the nutritional and economic worth of crops in a profitable and feasible way (Chattha et al. 2017; Camak and Kutman 2018; Lockyer et al. 2018; Connorton and Balk 2019). The targeted crop improvement for abiotic and biotic stress management, increasing and improving the oil content, increasing the nutritional quality, and removing the anti-nutritive and allergenic components, is possible through the advancements in plant biotechnology (transgenic techniques) (Pérez-Massot et al. 2012). Several strategies are being followed to raise such genetically modified (GM) crops: (1) enriching the functional ingredients in plants (high xanthophyll content in fruits, provitamin A in crops, and plant sterols in cereals), (2) synthesizing valuable ingredients in plants (polyunsaturated fatty acids [PUFAs], vitamin E in oil crops, increasing the starch content in plants), (3) increasing the crop yield through enhancement of photosynthesis, (4) delaying the ripening process in fruits and vegetables (postharvest stability), (5) modifying the fruit color, and (6) improving the plant protein (amino acids) composition (Zhang et al. 2016). The examples of biofortification projects include zinc (Zn) biofortification of maize, wheat, rice, sweet potato, and beans; iron (Fe) biofortification of rice, legumes, cassava, sweet potato, and beans; provitamin A carotenoid biofortification of maize, sweet potato, and cassava; and amino acid and protein biofortification of cassava and sorghum. However, the biofortification of crop plants is dependent on certain criteria, for instance, (1) enhancement of crop productivity to ensure farmer acceptance (*high yielding*), (2) significant impact of micronutrient enrichment levels on human health (*effective*), (3) stability of enriched levels (*stability*), (4) testing of bioavailability of nutrients from enriched lines in humans consuming them (*efficacious*), and (5) testing of consumer acceptance (*cooking quality and taste*).

Several strategies have evolved to transfer foreign DNA to a heterologous host. The DNA transfer does occur in nature by conjugation, bacterial transformation, *Agrobacterium*-mediated transfer, transposition, phage transduction, and retroviral transduction. There are two strategies for DNA transfer (Table 2.1): (1) vector-mediated or indirect DNA transfer (*Agrobacterium*-mediated transformation and *Agrobacterium*-mediated virus infection) and (2) vectorless or direct DNA transfer (physical, chemical gene transfer method and DNA imbibition by cells, tissues, embryos, and seeds). The physical gene transfer methods involve (1) electroporation/ultrasound/sonication, (2) particle bombardment (microprojectile/biolistics), (3) macroinjection, (4) microinjection, (5) lipofection, (6) silicon carbide fiber (SCF)-mediated transformation, and (7) DNA transfer via pollen. On the other hand, the chemical methods of gene transfer include (1) PEG-mediated gene transfer, (2) Ca phosphate coprecipitation, (3) polycation-DMSO procedure, and (4) DEAE-dextran procedure. Among all these techniques, the *Agrobacterium*-mediated transformation ensures stable genetic transformation (gene remains functional and expresses in subsequent generations) due to less number of integrations of the transgene and without any rearrangement or alteration.

## 2.2 *Agrobacterium* as a Natural Genetic Engineer

*Agrobacterium tumefaciens* is a ubiquitous, Gram-negative, non-sporing, motile, rod-shaped soil bacterium that forms nitrogen-fixing nodules like the rhizobium, on leguminous plants. In nature, *Agrobacterium* causes tumorous growths in 331 genera and more than 645 plant species including several dicots and some monocot angiosperms and gymnosperms (De Cleene and De Ley 1976). It can also transform other organisms like bacterium *Streptomyces lividans*, fungi, algae, human cells, and sea urchin embryos (de Groot et al. 1998; Kunik et al. 2001; Hooykaas 2004; Pelczar et al. 2004; Bulgakov et al. 2006). The *Bacterium tumefaciens* (now *A. tumefaciens*) as the cause of "crown gall disease" of plants was first reported by Smith and Townsend (1907), which laid the foundation of plant transgenics. Apart from gall formation, it has the capability to transfer a segment of its plasmid DNA into the host (dicots) genome (Gelvin 2003; Tzfira and Citovsky 2006). Therefore, this soil bacterium has become the most effective/reliable vector to transfer genes of diverse origin (trans-kingdom DNA) in plants.

The genes responsible for crown gall disease are present in tumor-inducing (Ti) plasmid (~200 kb). Only the transfer DNA (T-DNA) enters the host plant. The T-DNA is involved in (a) Ti plasmid conjugation, (b) production of cytokinins and auxins, (c) production of unique plant metabolites called opines and agrocinopines, and (d) initiation, transfer, and processing of the T-DNA. The cytokinins and auxins

**Table 2.1** Techniques of plant transformation

| | Technique | Physical/chemical technique | Advantages | Limitation(s) |
|---|---|---|---|---|
| Indirect DNA transfer | Electroporation | Physical | Offers transient expression in plants<br>Convenient, simple, and fast | Limited to protoplasts or few plant cell types; difficult to regenerate viable plants |
| | Microprojectile bombardment | | Applicable to several plants and animal tissues. An excellent and efficient system for transient expression, simultaneous delivery of large numbers of different genetic elements, effective system for organelle transformation | Multiple copies of DNA insertions, fragmentation and loss of DNA integrity, possible emergence of chimeric plants |
| | Macroinjection | | Technically simple. Instrument used is simple and inexpensive; it is useful for cereal transformation which does not regenerate from cultured cells easily | It is less specific and less efficient with a low transformation frequency |
| | Microinjection | | Not applicable to all plants. Only few successful reports in rye, barley, and other plants | Costly technique, requires highly skilled personnel, more advantageous for animal cells, requires protoplasts and not useful for the walled cells, and hard to regenerate viable plants |
| | Liposome-mediated transformation | | Applicable to several eukaryotic cells and offers high transfection efficiencies. It is simple to perform and ensures consistently reproducible results. It works well with those cell lines which are normally resistant to transfection by other methods. Transfect effectively with cationic lipid reagents | Very low transfection efficiency in suspension cells, not applicable to all cell types, depends on cell division and on high rate of endocytosis |
| | Silicon carbide fiber (SCF)-mediated transformation | | Easy technique. Does not require protoplast isolation; works well with walled cells, low equipment cost | Results in low transformation efficiency; SCF has some carcinogenic properties |
| | DNA transfer via pollen | | Elimination of tissue culture systems and is not limited by the ability to regenerate from transformed plant tissues, cells, or protoplasts. Only few successful reports in rice, maize, cotton, soybean, wheat, and other plants | Difficulty in establishing standard protocols and low success rates, only works with plants that flower and readily produce seeds, can only be performed during the flowering period |

| | | | | |
|---|---|---|---|---|
| | PEG-mediated gene transfer | Chemical | Direct gene transfer to cells circumvents the host range limitation. Easy to perform with relatively low cost | Confined to protoplasts or few plant cell types, difficult and slow process in regeneration of viable plants |
| | Calcium phosphate coprecipitation | | Increased transformation efficiency due to the use of DMSO | Very low frequency of transformation Leads to gene alteration; DNA-calcium phosphate forms complex which deposit on the surface of the cell |
| | Polycation-DMSO technique | | Use of polycation polybrene (less toxic than other polycations) increases the transformation efficiency. Requires very small quantities of plasmid DNA | DMSO is toxic to the cells |
| | DEAE-dextran procedure | | DEAE-dextran exhibits higher transfection efficiency than calcium phosphate | Cause cellular toxicity due to DMSO, DEAE-dextran itself is toxic to cells at high concentration, does not result in stable transformants |
| Direct DNA transfer | *Agrobacterium*-mediated transformation | | Minimal equipment and facility required and less expensive. Easy to set up and use, generally low transgene copy number, minimal DNA rearrangements, fewer undesired mutations and somaclonal variation, relatively higher stability of transferred gene | Not successful with several elite monocot crops, legumes, and woody plants |

initiate rapid cell growth and produce galls. Wild-type *A. tumefaciens*, harboring the Ti plasmid, parasitizes the plant for its nitrogen and carbon source in the form of specific opines (octopine, nopaline, and succinamopines) which are unique amino acid derivatives. *Agrobacterium* uses these compounds as the source of carbon and energy. The tumor-inducing (Ti) plasmid with the genes for synthesis of three main types of opines as well as genes for virulence (*vir*) governs the specificity for pathogenicity. Wild-type Ti plasmid containing a T-DNA region limited by a 25 bp RT and LT border inverted repeats has genes for host-directed opine synthesis and bacterial-type plant hormones which ultimately direct tumorigenicity (Van Larebeke et al. 1974). This T-DNA when engineered with desired genes is directed by the bacterium to integrate illegitimately into plant genome at random sites with the help of a cascade of proteins of vir operons (VirA–J). Wounded plant cells secrete polysaccharides and various phenolic signals that induce the expression of various bacterial chromosomal genes like *chvA*, *chvB*, and *pscA*, followed by induction in turn of various genes of vir operons (six essential (VirA, B, C, D, E, G) and four additional operons (VirF, I, J, H)), present on Ti plasmid. T-DNA molecule associated with $VirD_2$ and $VirE_2$, the two essential products of vir operons, is responsible for directing the single-stranded (ss) transfer DNA into the nucleus (Tzfira et al. 2000). Ti plasmid is engineered so as to disarm *A. tumefaciens* strain by the removal of genes for opine synthesis and plant hormone (cytokinins and auxin) synthesis. These disarmed strains are supplied with extra copies of vir operons to increase its virulence and, hence, the extent of infection. It is only the engineered T-DNA (containing the desired construct) that is ultimately exported by the bacterial cell into the host cell nucleus (Chilton et al. 1978; Depicker et al. 1978). Illegitimate recombination with the aid of known and unknown bacterial as well as cellular host factors is reported in the case of plants, but homologous recombination (HR) is reported in the case of unicellular yeast.

## 2.3 *Agrobacterium* as a Vehicle for Transformation

The use of *A. tumefaciens* strains (Table 2.2) for plant genetic modification began extensively in 1983 when Robert Fraley introduced a gene for kanamycin resistance in tobacco genome. *Agrobacterium* is quite unique in this aspect as it facilitates interkingdom transfer of genes (Stachel and Zambryski 1989) in dicots (Horsch et al. 1985) and monocots species as well as yeasts (Piers et al. 1996), filamentous fungi, and mushrooms (de Groot et al. 1998) and also in human cells (Kunik et al. 2001). It has several advantages over direct transformation techniques because it is feasible and cost-effective and facilitates few and well-defined transgene integrations (Shou et al. 2004; Zhang et al. 2005; Gao et al. 2008). On the contrary, the traditional methods of raising transgenic plants (electroporation, microinjection, or PEG fusion) involving protoplast transformation are not reliable because of slow regeneration of protoplasts and low transformation efficiency (Newell 2000; Banta and Montenegro 2008). Biolistics technique, although an alternative to

**Table 2.2** Common *Agrobacterium* strains used in plant transformation

| Strain name | Ti plasmid type | Antibiotic resistance | Reference |
|---|---|---|---|
| LBA4404 | A disarmed octopine-type Ti plasmid pAL4404 | Rifampicin | Hoekema et al. (1983) |
| EHA101 | Disarmed nopaline-type Ti plasmid pEHA101 (pTiBo542 DT-DNA) | Rifampicin, kanamycin | Hood et al. (1986) |
| EHA105 | Disarmed agropine-type Ti plasmid pEHA105 (pTiBo542 DT-DNA) | Rifampicin | Hood et al. (1993) |
| C58C1 | Cured of Ti plasmid | Rifampicin | Deblaere et al. (1985) |
| C58C1(pTiB6S3ΔT, pCH32) | Disarmed octopine-type Ti plasmid pTiB6S3 DT-DNA and a helper plasmid pCH32 | Rifampicin, carbenicillin, tetracycline | McBride and Summerfelt (1990) |
| GV3101 | Cured of Ti plasmid | Rifampicin | Holsters et al. (1980) |
| GV3101(pMP90) | A disarmed nopaline-type pTiC58 DT-DNA | Rifampicin, gentamycin | Koncz and Schell (1986) |
| A136 | Cured of Ti plasmid | Rifampicin | Watson et al. (1975) |
| AGL-1 | Disarmed pTiBo542 DT-DNA | Rifampicin, carbenicillin | Lazo et al. (1991) |
| C58-Z707 | Disarmed nopaline-type pTiC58-Z707 | Kanamycin | Hepburn et al. (1985) |
| NTL4(pKPSF2) | Disarmed chrysopine-type pTiChry5 DT-DNA | Erythromycin | Palanichelvam et al. (2000) |

*Agrobacterium*-mediated plant transformation, has disadvantages of multiple-copy gene integration and epigenetic silencing of the transgene and has a limitation of size of the DNA to be transferred (Lorence and Verpoorte 2004; Altpeter et al. 2005).

The T-DNA binary vector system has made the gene transfer process feasible and flexible. It consists of two plasmids within *A. tumefaciens*, a shuttle vector that contains gene of interest between the T-DNA borders and a helper Ti plasmid which offers *vir* gene products which are required for transfer of T-DNA. The presence of multiple cloning sites (mcs) within the transfer DNA region and a broad host range origin of replication functional in both *E. coli* and *A. tumefaciens* make the binary vector system an ideal DNA delivery system.

## 2.4 Factors Influencing *Agrobacterium*-Mediated Genetic Modification of Plants

The factors which greatly influence *Agrobacterium*-mediated genetic modification of plants include the (1) *Agrobacterium* strain, (2) type of plant material (explant genotype, age, size, etc.), (3) agro-inoculation conditions (bacterial density, time, temperature, pH, agitation, sonication, presence of acetosyringone, etc.), (4) co-cultivation conditions (time, temperature, light, etc.), and (5) plant growth conditions (preincubation, light intensity, temperature, subculturing, type of selection marker (Table 2.3), presence of endogenous phytohormones, exogenous addition of hormones, media composition, etc.) (Frary and Earle 1996; Gelvin 2003; Gao et al. 2008). However, the following rate-limiting steps are considered as the major bottlenecks controlling the overall rate of transformation during the T-DNA transfer

**Table 2.3** Commonly used selection marker genes in plants

| Genes | Substrate(s) |
|---|---|
| Neomycin phosphotransferase (*npt*II) | G 418, kanamycin, neomycin, paromomycin |
| Hygromycin phosphotransferase (*hpt*) | Hygromycin B |
| Chloramphenicol acetyltransferase (*cat*) | Chloramphenicol |
| Gentamycin acetyltransferase (*gat*) | Gentamycin |
| Streptomycin phosphotransferase(*sph*) | Streptomycin |
| Bleomycin hydrolase (*bmh*) | Bleomycin |
| Blasticidin deaminase gene (*bsd*) | Blasticidin |
| Dihydrofolate reductase (*dhfr*) | Methotrexate |
| Phosphinothricin acetyltransferase (*pat*) | L-Phosphinothricin (PPT) |
| 5-Enolpyruvylshikimate-3-phosphate (EPSP) synthase (*aro*A) | Glyphosate |
| Bromoxynil nitrilase (*bxn*) | Bromoxynil |
| Acetolactate synthase mutant form (*als*) | Sulphonyl urea, imidazolinones |
| Threonine dehydratase (*tdh*) | L-Threonine |
| Phosphomannose isomerase (*pmi*) | Mannose-6-phosphate |

into the host plant genome: (1) *Agrobacterium* plant compatibility and host cell competence; (2) T-strand generation; (3) T-complex transfer into the plant cell; (4) T-strand stability and its entry into plant nucleus; (5) T-DNA integration; (6) copies of T-DNA inserts into genomic DNA, position effect, and posttranscriptional transgene silencing; and (7) regeneration of transformed cells.

## 2.5 Molecular Mechanism of T-DNA Transfer

*Agrobacterium tumefaciens*-mediated plant transformation has been explained through a number of reviews which have focused on giving the very recent and updated information with respect to the molecules involved in the entire process, factors influencing the process, and also the various possibilities in the light of present research in this area. It is extremely exciting to see how this primitive form of life (bacteria) manages and also can be manipulated to infect and colonize wide range of higher life forms (plants, fungi, yeast, and human cells). The whole scenario can be hypothetically visualized in steps which is summarized in Fig. 2.1.

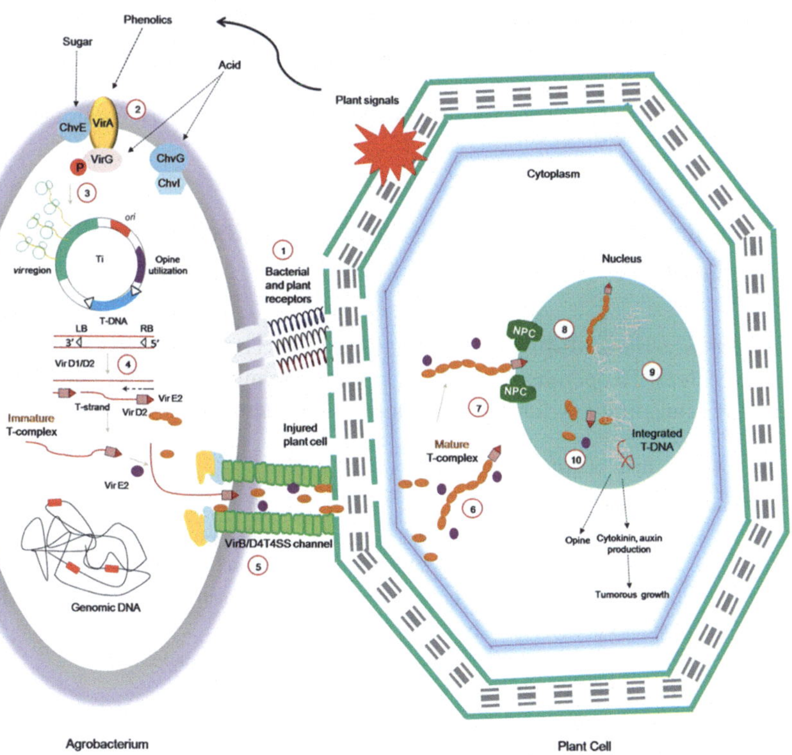


**Fig. 2.1** Model for *Agrobacterium*-mediated genetic transformation of plants (Tzfira and Citovsky 2006; Hwang et al. 2017)

The transformation process involves ten steps. It starts with recognition and attachment of the *Agrobacterium* to the plant cells, [1, 2] sensing of plant signals by *Agrobacterium* and activation of *vir* genes followed by transmission of the sensed signals, [3, 4] generation and transfer of T-DNA and virulence proteins from the bacterial cell into plant cell, [5, 6] nuclear import of T-DNA and effector proteins in the plant cell, and [7–10] T-DNA integration and expression in the plant genome.

## 2.6 Colonization and Bacterial Infection

Many capsular polysaccharides (like K-antigen) and lipopolysaccharides (LPS) (like O-antigen) form the outer membrane of the bacterium which is supposed to link the bacterium and the plant cell during infection. *Agrobacterium tumefaciens* gets attached to the plant cell surface with the help of an acidic polysaccharide (Bradley et al. 1997). The role of LPS was demonstrated with the help of transposon insertion mutants and other mutagenic studies at the chromosomal 20 kb *att* locus, which contains the gene for bacterial attachment (Whatley and Spiess 1977). The region 10 kb upstream and 10 kb downstream of *att* locus was mutated which resulted in the total loss of bacterial attachment, and the region is supposed to code for the synthesis of fundamental components. The ABC transport system (periplasmic binding protein-dependent transport system) appears to be involved in secretion of some compounds that are essential for bacterial attachment (Higgins et al. 1990).

## 2.7 Formation of T-Complex and Its "Interkingdom" Transfer

An *Agrobacterium* 30–40 kb *vir* region located on Ti plasmid outside T-DNA consists of at least 10 operons, each having different number of genes (Table 2.4). VirA and VirG constitutively form a two-component system, which receives the "signal"

**Table 2.4** Functions of different *vir* genes

| *Vir* region | Function | No. of genes |
|---|---|---|
| *vir*A | Encodes a sensor protein receptor for acetosyringone and functions as an autokinase; also phosphorylates *vir*G protein; constitutive expression | 1 |
| *vir*B | Membrane proteins; has a role in conjugal tube formation | 11 |
| *vir*C | Exhibits helicase activity | 2 |
| *vir*D | *Vir*D1 has topoisomerase activity and *vir*D2 is an endonuclease | 4 |
| *vir*E | Single-stranded binding protein (SSBP) | 2 |
| *vir*F | Not well understood | 1 |
| *vir*G | DNA binding protein, induces the expression of all *vir* operons; constitutive expression | 2 |
| *vir*H | Not well understood | 2 |

(wounding) and transients the signals to activate the consequential expression of other vir operons. Plant wounding creates an environment of acidic pH, secreted phenolics such as acetosyringone (Winam et al. 1986), and many monosaccharides, which turn on the expression of *Agrobacterium* chromosomal genes, which in turn activates VirA. VirA protein is formed of three domains, which help in receiving the signals. A transmembrane, dimeric VirA protein senses the signals from a cascade of chromosomal gene products like chvA, chvB, and pscA (Ankenbauer and Nester 1990; Cangelosi et al. 1990). TM-2 domain of VirA is a kinase domain which gets phosphorylated on a conserved histidine residue at position 474. Activated VirA transfers its phosphate group to a conserved aspartate residue on VirG, which acts as a transcription factor regulating the downstream regulation of *vir* genes.

The generation of single-stranded (ss) T-DNA from the excision of bottom strand of double-stranded (ds) Ti plasmid is accomplished by the help of $VirD_1$-$VirD_2$ complex. $VirD_2$ recognizes the 5′ right border (RB) repeats, where it remains attached after nicking the bottom strand. $VirD_1$ helps in the endonuclease activity in uncoiling of ds Ti plasmid at its 5′ end (Zupan and Zambryski 1995; Christie 1997). Ensuring active nuclear import in a polar fashion, staring with the 5′ end of ss T-DNA where $VirD_2$ is attached through a bipartite but only one functional NLS, ss T-DNA-$VirD_2$ complex is generated. The $VirD_2$ molecules are the key regulators for directing the T-complex to the plant nucleus for its integration. With the help of yeast two-hybrid protein assay, some highly conserved proteins belonging to cyclophilin family of peptidyl-prolyl cis-trans isomerase were found. DIP1 of *Arabidopsis* (Fields and Song 1989; Hollenberg et al. 1995; Deng et al. 1998) that interacts with $VirD_2$ is proposed to maintain proper confirmation of $VirD_2$ in *Agrobacterium* as well as host cells. The $VirD_2$ pilot protein of *Agrobacterium* T-DNA interacts with the TATA box binding protein (TBP) and a nuclear protein kinase (NPK). It is reported in alfalfa cells that $VirD_2$ interacts with a conserved plant ortholog of plant cyclin-dependent protein kinases (Bako et al. 2003). In transformed *Arabidopsis* cells, TBP is found to be in tight association with $VirD_2$. It is phosphorylated by conserved cyclin-dependent protein kinase (CAK2Ms) (Bako et al. 2003). $VirD_2$ interacts with multiple plant cyclophilins and α-importin nuclear receptors. It is also reported that AtKAP-α (nuclear localization signal-binding protein) specifically mediates the transport of ss T-DNA-VirD2 complex into the nucleus for integration of T-DNA in host genomic DNA (Ballas and Citovsky 1997).

The ss T-DNA attached at 5′ end to $VirD_2$ is transferred to the plant cells after crossing a triple barrier of membranes and cellular spaces, which contains our desired DNA. This $VirD_2$ prevents the exonucleolytic cleavage of 5′ end (Durrenberger et al. 1989). To accomplish this strand transfer the needed apparatus is constructed with the products of VirB operons (9.5 kb region), which has 11 different components (Christie et al. 1989), with 10 essential proteins, which are integral proteins of the inner and outer membrane. VirB4 and VirB11 span the membrane and have ATPase activity for active transport of T-complex. VirB constitutes the channel-connecting link between a bacterium and a plant and hence can be thought to be a reminiscent of conjugation pili of bacteria (Lal and Kado 1998). VirB7 and

VirB9 associated with VirD4 proteins help to give anchorage and hence maintain this VirB channel. Other VirB proteins are reported to play a role in peptidoglycan digestion of bacterial cell wall. Thus, VirB forms the "delivery system" between the cells of two different kingdoms.

It is not very clear that ss T-DNA enters into the plant cell coated with many molecules of $VirE_2$, or both T-DNA and $VirE_2$ enter independently and are associated afterward in the plant cell cytoplasm. But many facts favor the latter hypothesis. $VirE_2$ with the essential help of $VirE_1$ protects the single-stranded T-DNA molecule from endonucleolytic cleavage by various enzymes. $VirE_2$ when mutated was still able to transfer T-DNA strand which confirms that T-DNA strand can also be independently transferred (Sundberg et al. 1960; Binns et al. 1995). $VirE_2$ and ss T-DNA both independently travel, and when it reaches plant cell cytoplasm, $VirE_2$ coats the T-DNA strand.

## 2.8 Integration of T-DNA in Plant Genome

The T-DNA coated with $VirE_2$ and bound to $VirD_2$ at its 5′ end integrates illegitimately at random but preferential sites in plant genome (Gheysen et al. 1991; Puchta 1998). These are mostly transcriptionally active sites with a few base pair (bp) similarity with the border regions, specially the right border, which is mostly found conserved due to bound $VirD_2$ at 5′ end. The microhomologies provide minimum specificity between host DNA and T-DNA in which many host proteins help. These host, nonhomologous end joining (NHEJ) proteins Yku70, Rad52, Rad50, Mre11, Xrs2, and lig4 are required for integration at telemetric regions as shown in *S. cerevisiae* (Van Attikum et al. 2001; Van Attikum and Hooykaas 2003). T-DNA integration is mainly governed by host factors, which can drive toward homologous recombination (HR) in yeast (Offringa et al. 1990) as compared to nonhomologous recombination (NHR) in plants (Gallego et al. 2003).

T-DNA integration creates rearrangement of the target site in plant genome in which its right border sequence is more conserved than its left border. This T-DNA is supposed to displace the strand from plant genomic DNA through the $VirD_2$ at 5′ end RB. It is very frequent to have multiple copies of T-DNA integrated at the same position in either inverted or direct repeats (De buck et al. 1999). It is demonstrated that it does not involve replication, but recombination between two or more distinct T-DNA that forms T-DNA repeats. *Arabidopsis thaliana* was frequently studied to analyze the junctions between linked T-DNA along RB-RB-, RB-LB-, and LB-LB-type orientations. RB-RB end-to-end ligations of ds T-DNAs occur commonly, but LB T-DNA junctions are formed through illegitimate recombination on the basis of microhomologies, deletions, repair mechanism, and insertions of filler DNA.

Inverted repeats cannot be amplified by PCR due to intrastrand annealing (De Buck et al. 1999). RB-RB junctions are formed with minor deletions, but RB-LB and LB-LB junctions, when formed, were found to be present with a stretch of DNA called "filler DNA" that varied in length between 7 and 235 bp and originated from either the host genome or T-DNA near the break pairs (imprecise insertions). The precise insertions or target placements involve replacement of target deletion with T-DNA of same length with accuracy. Ultimately to visualize the possible mechanism of T-DNA integrations (single or multiple copies), the following two ways are possibly accepted (Gheysen et al. 1991; Mayerhofer et al. 1991): (1) Two distinct T-DNAs integrate via single-stranded annealing with the double-stranded genomic DNA followed by interaction and annealing of $VirD_2$-coated 5′ RB end (Tinland 1996). However, $VirD_2$, on the contrary, does not possess general ligase activity (Ziemienowie et al. 2000), and (2) right T-DNA borders recognize the similar sequence in plant genome (preinsertion target site) and are found to be partly truncated. Double-stranded break (DSB) repair mechanisms are involved in such integration events (Salomon and Puchta 1998).

When analyzed in *Arabidopsis*, tobacco (Gheysen et al. 1987; Mayerhofer et al. 1991; Gheysen et al. 1991), and in tree species—aspen (Kumar and Fladung 2002)—the class (1) type of T-DNA integration and $VirD_2$ was found to preserve the conserved right T-DNA border as compared to much variable left border repeat. But, in some transgenic lines in the same experimental setup, when analyzed, sequence analysis of preinsertion sites showed that these recombination hot spots were mostly AT-rich regions ranging from 54% in dicots, like tobacco (Gheysen et al. 1987; Salinas et al. 1988) and *Arabidopsis* (Mayerhofer et al. 1991), to 60–70% in aspen (Kumar and Fladung 2001). T-DNA integration occurs through DSB repair mechanism as in genomic DNA. The origin of filter DNAs can be explained by abortive gap repair mechanism through synthesis-dependent strand annealing (SDSA) pathway.

It has been shown through transient expression kinetics that T-DNA strand is converted to ds T-DNA (Janssen and Gardner 1989) and, then in case of T-DNA repeat formation, recombines before integration in genome (De Buck et al. 1999). Inter-chromatid and inter-homologue recombination mechanism has been explained in *A. thaliana* by Molinier et al. (2004). To summarize, the integration is phenomenon driven and governed by multiple sets of host factors, which are still not well defined.

Here we have summarized the available knowledge about the happenings till the transfer of T-complex from *Agrobacterium* cell to the plant cell, but there is still a scope of basic study to understand about the possible host cell factors and phenomenon. To understand the complex phenomenon of T-DNA integration illegitimately into plant genome, some preferential sites need to be analyzed with respect to its interaction with nuclear receptor proteins, kinases, histone, and particular DNA stretches of definite sequence. Complex DNA–DNA, DNA–protein, and protein–protein interactions are needed to be studied to understand the crests and thoughts of T-DNA during its journey and ultimate union into the most like destination—the plant cell genome.

## 2.9 Direct Editing of Genomic DNA by CRISPR Tool

The bacteria-derived clustered regularly interspaced short palindromic repeats (CRISPR)-Cas genome editing system has empowered our ability in gene manipulation, detection, imaging, and annotation in living cells of diverse species. This revolutionary genome engineering technology is highly advanced because of its simplicity, stability/robustness, efficiency, and specificity of the target in both animal and plant systems (Ran et al. 2013; Pickar-Oliver and Gersbach 2019). In CRISPR/Cas9, Cas9 is a bacterial DNA endonuclease which protects it from virus attack, etc. The "guide RNA" guides the endonuclease to the invading/targeting DNA, and then the Cas9 cleaves the double-stranded DNA. The DNA double-stranded breaks (DSBs) get repaired by two different mechanisms: (1) homologous recombination (HR) mechanism—it facilitates the addition of a donor DNA into the endogenous gene at the break site, and (2) nonhomologous end joining (NHEJ)—it can cause a small deletion or random DNA insertion that generates a truncated gene or a knockout (Fig. 2.2). CRISPR/Cas9 based genome-editing technique is being used by the biotechnologists to devlop useful traits in crops. Recently, Kaur et al. (2020) successfully deployed the same technique for metabolic engineering of the Cavendish banana cultivar Grand Naine to edit *lycopene epsilon-cyclase* (*LCYε*) gene and showed enhanced accumulation of β-carotene content up to 6-fold (~24 μg/g) in fruits compared to unedited plants.

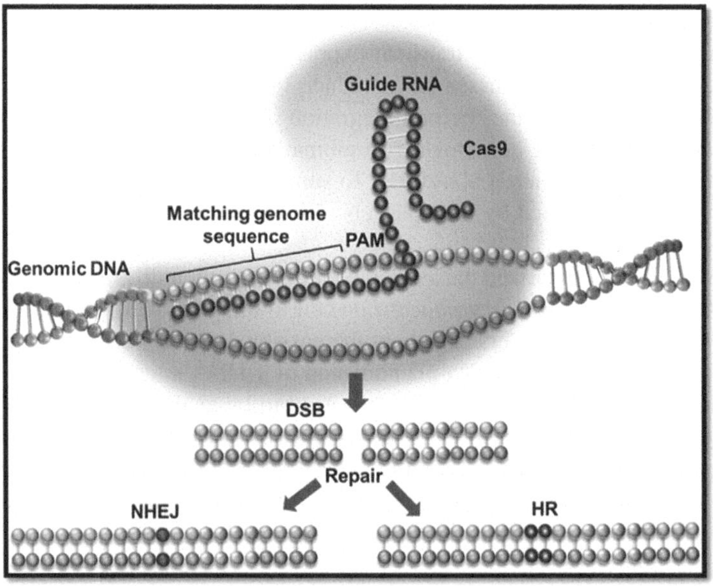


**Fig. 2.2** Representation of CRISPR-Cas9 gene-editing technique

## 2.10 Strategies to Enhance the Expression and Stability of the Transgene

It is challenging to express a foreign gene from an AT-rich organism to a GC-rich organism. Several factors control the transgene expression at different stages including transcription, translation, posttranslational modifications, or accumulation of recombinant protein, in heterologous environment of the plant cell. Several structural genomics approaches have been developed and applied to boost the expression of the heterologous genes in plants. These include (1) avoiding sequence motifs and codons that lead to mRNA degradation and low-level expression ((a) putative polyadenylation and mRNA instability sequences, (b) RNA polymerase II termination signals, (c) cryptic splicing sites, (d) secondary structures, etc.), (2) targeting protein to cellular compartments suitable for accumulation and stability (ER, vacuole, apoplast, etc.), and (3) incorporating elements directing high-level expression ((a) strong promoters, either modified natural or synthetic; (b) 5′ untranslated leader sequence; (c) translation initiation context; etc.).

## 2.11 Malnutrition and Its Impact

The global food problem is expected to simultaneously increase along with the increasing global population by 2030 (8.6 billion) or 2050 (9.8 billion) (United Nations 2016, 2017). There is a crucial requirement for sustainable food production strategies to eradicate the global inequalities and to feed the teaming billions (Lockyer 2018).

Nutrition is defined as a food intake as per body's dietary requirements. It includes major nutrients (carbohydrates, proteins, fibers) and micronutrients (minerals and vitamins) as well. On the one hand, an adequate and balanced diet intake leads to good health, while an imbalanced diet or poor nutrition causes serious health problems, like increased susceptibility to diseases and impaired physical and mental development. Malnutrition is caused by a deficiency or excesses of nutrient intake (WHO (World Health Organization) 2017). It includes two major forms: (1) undernutrition and (2) micronutrient-related malnutrition. The third form of malnutrition is overweight, obesity, and diet-related noncommunicable diseases (cancers, heart diseases, and strokes). Undernutrition children are more prone to infectious diseases (diarrhea) and death. These children do not reach their physical and cognitive potential. Inadequacy of micronutrients causes insufficient production of enzymes, hormones, and other constituents that are required for proper growth and development. Iron, vitamin A, iodine, and zinc are the important micronutrient deficiencies that cause a threat to the health of children, pregnant women, and adults worldwide. Furthermore, micronutrient deficiencies affect society for a long duration (Anderson et al. 2008). The world is facing a severe threat for nutrition. Malnutrition is responsible for the death of nearly half of children under the age of

**Table 2.5** Selected reports on transgenic plants with improved nutritional value

| Crop name | Gene | Gene source | Useful trait introduced | Reference(s) |
|---|---|---|---|---|
| Carrot (*Daucus carota*) | $H^+/Ca^{2+}$ transporter cation exchanger 1 | *Arabidopsis thaliana* | Calcium (Ca) content ↑ | Park et al. (2004) |
| | *Beta-carotene ketolase gene* | *Haematococcus pluvialis* | Expression of β-carotene hydroxylase ↑ | Jayaraj et al. (2008) |
| | *Bacterial phytoene synthase* | *Bacteria* | Carotenoid biosynthesis ↑ | Maass et al. (2009) |
| Cassava (*Manihot esculenta*) | *Pantoea agglomerans* | *Pantoea agglomerans* | Vitamin A content ↑ | Welsch et al. (2010) |
| | *Iron-specific assimilatory protein* | *Chlamydomonas reinhardtii* (codon optimized) | Iron (Fe) content ↑ | Sayre et al. (2011) |
| | ZAT *transporter* | *A. thaliana* | Zinc (Zn) content ↑ | Sayre et al. (2011) |
| | ZIP plasma membrane zinc transporter | *A. thaliana* | Zinc (Zn) content ↑ | Sayre et al. (2011) |
| Corn (*Zea mays*) | *Rice dehydroascorbate reductase* | *Oryza sativa* | Vitamin C content ↑ | Naqvi et al. (2009) |
| | *Gm ferritin* and *Af phytase* | *Glycine max*, *Aspergillus* | Iron (Fe) content ↑ | Drakakaki et al. (2005) |
| | *Phytoene synthase* | *Zea mays* | Vitamin A content ↑ | Zhu et al. (2008) |
| | *β-Carotene hydroxylase* | *Gentiana lutea* | Vitamin A content ↑ | Zhu et al. (2008) |
| | *Lycopene ε-cyclase* | *G. lutea* | Vitamin A content ↑ | Zhu et al. (2008) |
| | *β-Carotene ketolase* | *Paracoccus* | β-Carotene and other carotenoids ↑ | Zhu et al. (2008) |
| | *Phytoene desaturase* | *P. ananatis* | Vitamin A content ↑ | Zhu et al. (2008) |
| | *Phytoene synthase* | *P. ananatis* | Vitamin A content ↑ | Aluru et al. (2008) |
| | *Phytoene desaturase* | *P. ananatis* | Vitamin A content ↑ | Aluru et al. (2008) |
| | *ζ-Carotene desaturase* | *P. ananatis* | β-Carotene content ↑ | Aluru et al. (2008) |
| | *Phytoene synthase* | *Z. mays* | Vitamin A content ↑ | Naqvi et al. (2009) |
| | *Carotene desaturase* | *P. ananatis* | β-Carotene content ↑ | Naqvi et al. (2009) |
| Indian mustard (*Brassica juncea*) | Plastidic ATP sulfurylase | *A. thaliana* | Selenium (Se) content ↑ | Pilon-Smits et al. (1999) |
| | Selenocysteine methyltransferase | *Astragalus bisulcatus* | Selenium (Se) content ↑ | LeDuc et al. (2004) |

| | | | | |
|---|---|---|---|---|
| Lettuce (*Lactuca sativa*) | $H^+/Ca^{2+}$ transporter cation exchanger 1 | *A. thaliana* | Calcium (Ca) content ↑ | Park et al. (2009) |
| Potato (*Solanum tuberosum*) | Phytoene desaturase, phytoene synthase, and lycopene beta-cyclase | *Erwinia uredovora* | Vitamin A content ↑ | Diretto et al. (2007) |
| | *Or* gene | *Brassica oleracea* | Vitamin A content ↑ | Lopez et al. (2008) |
| | Zeaxanthin epoxidase | *A. thaliana* | Vitamin A content ↑ | Romer et al. (2002) |
| | *Phytoene synthase* (*psy*) *crtB* | *E. uredovora* | Vitamin A content ↑ | Ducreux et al. (2005) |
| | GDP-l-galactose phosphorylase | *A. thaliana* | Vitamin C content ↑ | Bulley et al. (2011) |
| | H+/Ca2+ transporter cation exchanger 1 | *A. thaliana* | Calcium (Ca) content ↑ | Park et al. (2005) |
| | *Lycopene* epsilon-cyclase (*lcy-e*) | *P. ananatis* | β-Carotene content ↑ | Diretto et al. (2006) |
| | Beta-carotene hydroxylase | *Solanum tuberosum* | β-Carotene and lutein content ↑ | Van Eck et al. (2007) |
| | *Orange cauliflower* | *Brassica oleracea* | Carotenoid accumulation ↑ | Lopez et al. (2008) |
| Rice (*Oryza sativa*) | Mammalian GTP cyclohydrolase I gene, aminodeoxychorismate synthase | *Mammal*, *A. thaliana* | Folic acid content ↑ | Storozhenko et al. (2007) |
| | Nicotianamine synthase | *O. sativa* | Iron (Fe) content ↑ | Johnson et al. (2011) |
| | Nicotianamine synthase | *O. sativa* | Iron (Fe) content ↑ | Lee et al. (2009) |
| | Nicotianamine synthase 2 | *O. sativa* | Zinc (Zn) content ↑ | Lee et al. (2011) |
| | Ferritin genes, nicotianamine synthase | *O. sativa* | Iron (Fe) and zinc (Zn) content ↑ | Johnson et al. (2011) |
| | *nas*1, *ferritin*, *phytase* | *A. thaliana*: *nas*, *Phaseolus vulgaris*: *ferritin*, *Aspergillus fumigates*: *phytase* | Iron (Fe) content ↑<br>Zinc (Zn) content ↑ | Wirth et al. (2009) |
| | *Phytoene synthase* | *Zea mays* | Total carotenoids ↑ | Paine et al. (2005) |
| | *Phytoene desaturase* | *E. uredovora* | Total carotenoids ↑ | Paine et al. (2005) |
| | *Phytoene synthase* | *Narcissus pseudonarcissus* | Total carotenoids ↑ | Ye et al. (2000) |
| | *Phytoene desaturase* | *Erwinia uredovora* | Total carotenoids ↑ | Ye et al. (2000) |

(continued)

**Table 2.5** (continued)

| Crop name | Gene | Gene source | Useful trait introduced | Reference(s) |
|---|---|---|---|---|
| Tomato (*Solanum lycopersicum*) | GDP-l-galactose phosphorylase | *Actinidia chinensis* | Vitamin C content ↑ | Bulley et al. (2011) |
| | GTP-cyclohydrolase I, aminodeoxychorismate synthase | *A. thaliana* | Folic acid content ↑ | Storozhenko et al. (2007) |
| | NADP-dependent glutamate dehydrogenase | *Aspergillus nidulans* | Fruit taste and glutamate levels ↑ | Kisaka and Kida (2003) |
| | Phytoene synthase | *E. uredovora* | Carotenoid content ↑ | Fraser et al. (2002) |
| | Yeast S-adenosylmethionine decarboxylase | *Saccharomyces cerevisiae* | Fruit phytonutrient, carotene ↑ | Mehta et al. (2002) |
| | *Phytoene desaturase* | *E. uredovora* | β-Carotene content ↑ | Romer (2000) |
| | *Lycopene β-cyclase* | *A. thaliana* | Total carotenoid content ↑ | Rosati et al. (2000) |
| | Beta-carotene hydroxylase | *Capsicum annuum* | β-Carotene, β-cryptoxanthin, and zeaxanthin content ↑ | Dharmapuri et al. (2002) |
| | *Phytoene synthase* | *E. uredovora* | Total carotenoid content ↑ | Fraser et al. (2002) |
| | *Lycopene β-cyclase* | *Solanum lycopersicum* | β-Carotene content ↑ | D'Ambrosio et al. (2004) |
| | *3-Hydroxymethylglutaryl CoA* | *A. thaliana* | Phytosterol content ↑ | Enfissi et al. (2005) |
| | *1-Deoxy-D-xylulose-5-phosphate synthase* | *Escherichia coli* | Carotenoid content ↑ | Enfissi et al. (2005) |
| | *Cryptochrome* | *Solanum lycopersicum* | Fruit lycopene content ↑ | Giliberto et al. (2005) |
| | *Fibrillin* | *Capsicum annuum* | Carotenoid content ↑ | Simkin et al. (2007) |
| | *Lycopene β-cyclase* | *E. herbicola, Narcissus pseudonarcissus* | Provitamin A and total carotenoid accumulation ↑ | Apel and Bock (2009) |
| | *SlNCED1* | *Solanum lycopersicum* | Carotenoid content and flavor volatiles ↑ | Sun et al. (2012) |
| Wheat (*Triticum aestivum*) | *Phytoene desaturase* | *E. uredovora* | Vitamin A content ↑ | Cong et al. (2009) |
| | *Phytoene synthase* | *Z. mays* | Vitamin A content ↑ | Cong et al. (2009) |

| | | | | |
|---|---|---|---|---|
| Canola (*Brassica napus*) | *Phytoene synthase* | *E. uredovora* | Carotenoids ↑ | Shewmaker et al. (1999) |
| | Geranylgeranyl diphosphate synthase + phytoene synthase | *E. uredovora* | Total carotenoids ↑ | Ravanello et al. (2003) |
| | Phytoene synthase + *lycopene β-cyclase* | *E. uredovora* | Total carotenoids ↑ | Ravanello et al. (2003) |
| | *Lycopene ε-cyclase* | *A. thaliana* | β-Carotene, zeaxanthin, violaxanthin, and lutein ↑ | Yu et al. (2008) |
| | *Isopentenyl pyrophosphate isomerase* | *Paracoccus* spp. | Ketocarotenoids ↑ | Fujisawa et al. (2009) |
| | *β-Carotene ketolase* | *Brevundimonas* spp. | Ketocarotenoids ↑ | Fujisawa et al. (2009) |
| | *β-Carotene hydroxylase* | *P. ananatis* | Ketocarotenoids ↑ | Fujisawa et al. (2009) |
| | *Phytoene desaturase* | *P. ananatis* | Ketocarotenoids ↑ | Fujisawa et al. (2009) |
| | *Lycopene β-cyclase* | *P. ananatis* | Ketocarotenoids ↑ | Fujisawa et al. (2009) |
| | *Lycopene ε-cyclase* | *P. ananatis* | Ketocarotenoids ↑ | Fujisawa et al. (2009) |
| | *Phytoene synthase* | *P. ananatis* | Ketocarotenoids ↑ | Fujisawa et al. (2009) |
| | *microRNA miR156b* | *A. thaliana* | Carotenoid content ↑ | Wei et al. (2010) |
| Citrus (*Citrus sinensis* L. Osbeck) | β-Carotene hydroxylase | *Citrus sinensis* | β-Carotene content ↑ | Pons et al. (2014) |

5 years. Nearly 462 million people are underweight and 1.9 million are obese throughout the world (WHO (World Health Organization) 2017). According to WHO, 88% of 140 countries are suffering from one or the other forms of malnutrition. This problem occurs primarily in low/middle-income countries like in South Asia and sub-Saharan Africa. These countries lose 11% of its gross domestic income every year due to malnutrition. Nearly two million people are reported to be deficient for key micronutrients, i.e., iron and vitamin A, worldwide (WHO (World Health Organization) 2017).

## 2.12 Enhancement of Nutritional Value

An offshoot of the crop improvement strategies is the concept of "biofortification" which is an amalgam of plant breeding, agronomic practices, and genetic engineering that increases the nutritional value of crop plants in a cost-effective and sustainable way (Chattha et al. 2017; Camak and Kutman 2018; Lockyer et al. 2018; Connorton and Balk 2019). With the aim of crop improvement, several strategies are being followed (Table 2.5):

1. Enriching the functional ingredients in food:
    (a) *High xanthophyll content in fruits*
    (b) *Plant sterols in cereals*
    (c) *Provitamin A in crops*
2. Synthesizing valuable ingredients in plants:
    (a) *Polyunsaturated fatty acids (PUFAs) and vitamin E in oil crops*
    (b) *Increasing the starch content in plants*
3. Increasing the crop yield through enhancement of photosynthesis
4. Delaying the ripening process in fruits
5. Modification of the fruit color
6. Improvement in the plant protein (amino acids) composition

## 2.13 Vitamin A Deficiency (VAD)

Vitamin A is an indispensable micronutrient and is crucial for numerous physiological functions in the human body. It is a lipid-soluble vitamin available in the liver or fatty tissues. It is needed for the synthesis of rhodopsin which is required for night vision. It is also crucial for immunity; aids in bodily growth and skin maintenance, reproduction, and genetic coding; and promotes growth and normal development of teeth, soft, and skeletal tissues. It acts as an antioxidant which protects against free radicals and decreased the risk of chronic diseases. Watson (2014) demonstrated that vitamin A prevents type II diabetes and protects the heart. In the human body, β-carotene is transformed to retinal/retinaldehyde by β-carotene 15, 15′-monooxygenase (BCMO)/dioxygenase in intestinal lumen. β-Carotene dioxygenase cleaves double bond of β-carotene symmetrically into two molecules of all-trans-retinal. Retinal is further converted to retinol by retinal reductase and to retinoic acid by retinal oxidase in the intestine. The conversion and absorption potential of retinol depend on many factors including the need for vitamin A, bile production, intestinal health, and the amount of dietary fat in the intestines (Harrison et al. 2015). The β-carotene is stored within subcutaneous fat without cleavage if retinol is not needed by the body. Retinoic acid is essential for cell differentiation, proliferation, and immunity (Pino-Lagos et al. 2010). Retinol is converted by retinol acyltransferase to retinyl ester that is stored in the liver. Free retinol which is chemically unstable is available in the form of esters mainly retinyl palmitate in food (Bender 2003). Vitamin A forms micelles in the intestinal lumen, and the mucosal cells of duodenum absorb it. Vitamin A precursors are present in the form of PVACs in plants and preformed vitamin A (retinol and retinyl ester) in animals. The sources of PVA include broccoli and spinach (leafy vegetables), pumpkin, carrots and sweet potatoes, and red- or orange-colored fruits like orange, apricot, papaya, and mango. Vitamin A sources from animals include beef liver, butter, cheddar cheese, egg yolk, full cream milk, and fish. The main plant-based source of PVA is crude red palm oil (*Elaeis guineensis*) (Rukmini 1994; Rao 2000).

VAD is a result of prolonged insufficient intake of vitamin A. It is usually associated with poor socioeconomic conditions, which affect body functions negatively (WHO (World Health Organization) 2017). VAD is considered if vitamin A level goes below 20 μg/deciliter in the liver. Serum retinol (biochemical indicator) reflects vitamin A status. In developing countries, people depend upon plant sources for vitamin A requirements (Akhtar et al. 2013). In children, insufficient breast-feeding, poor diet, and poor maternal health lead to VAD (Ahmed and Darnton-Hill 2004). VAD causes exophthalmia (blindness—earliest symptom of VAD and caused due to progressive degeneration of mucous membrane) and reduced immunity (increases the possibility of infections like diarrhea, measles, and malaria) (WHO (World Health Organization) 2017). Acute VAD may lead to irreversible blindness due to eye ulcers, inflammation, and interior infections. VAD will lead to increased mortality rates. In developing countries, 25,000–50,000 children get blinded every year. It has also been observed that half of them die within 12 months of eyesight loss (WHO (World Health Organization) 2018). The maternal mortality due to VAD in South Africa has gone up to 11% (Dorrington and Bradshaw 2011). The mortality rate of children having 6–59-month age group has touched 48% in the Saharan belt in Africa and 44% in South Asia (UNICEF (United Nations Children's Fund) 2018). VAD affected 190 million children who had not started schooling yet and 19 million pregnant women. It is estimated that worldwide 2.8 million preschool children are at the risk of blindness (WHO (World Health Organization) 2017).

Vitamin A supplements are provided to infants and pregnant women to combat VAD (WHO (World Health Organization) 2011). Vitamin A supplements coverage increased rapidly from 2000 to 2015 but declined again in 2016. The mortality rate of the children aged 6–59 months increased from 19 million to 62 million as the coverage of VAD children reduced more than half in 2015–2016. Vitamin A supplementation reduced the mortality rate by 12–24%, but it has been noticed that 1/3 of children still are not getting the benefits of the aforementioned (UNICEF (United Nations Children's Fund) 2018).

## 2.14 Golden Rice to Combat Vitamin A Deficiency

As it is obvious that rice grains cannot accumulate provitamin A through breeding, therefore, "Humanitarian Golden Rice project" was initiated. Golden rice is an appropriate example of biofortification crop to combat VAD. It was obtained by engineering the rice plant to produce and accumulate provitamin A (β-carotene) in the grain. PVA biofortification in golden rice was the first reported success story of genetic engineering in food grains (Ye et al. 2000; Beyer 2010). The Gates Foundation also financed four biofortification projects (rice: golden rice, sorghum: ABS, banana: Banana21, and cassava: BioCassava Plus) to compete for the global health issues in 2005.The first generation golden rice (GR1) expressed the carotene desaturase gene (*crt*I) from the bacterium *Pantoea ananatis* and phytoene synthase gene (*psy*) from daffodil. The average levels of carotenoids accumulated were 6 μg/g. GR2 was developed by incorporating *psy*-maize (replacing the *psy*-daffodil gene) along with *crt*I gene from *P. ananatis* that used to generate the original GR1. There was a 23-fold increase (37 μg/g) in the accumulation of provitamin A in GR2 (Paine et al. 2005). However, the time taken for regulatory clearance (GE regulation) and trials was the main stumbling block in the release of golden rice (Beyer 2010). The extreme precautionary attitude and nonscientific and unanswerable political delays in the regulatory clearance are like an offense against humanity as it affects only the poor and not the rich (Potrykus 2005).

## 2.15 Banana Biofortification to Combat Vitamin A Deficiency

Banana is cultivated across the tropical and subtropical areas. It is a monocotyledonous perennial herbaceous plant that grows in damp and humid places. These regions contained more than 75 wild species. It belongs to the Musaceae family of order Zingiberales. Banana is a climacteric fruit in which ethylene production continues even after harvesting. Banana is considered as a regular fruit in several developing countries which make it an appropriate target for PVA enhancement. Banana eaten in these regions are deficient in PVA and ultimately led to malnutrition (Fungo and Pillay 2011). Previously, food fortification, dietary diversification, and supplementation were tried to reduce VAD (Tanumihardjo and Furr 2013). Despite these initiatives, the problem of VAD remains the same in South Asia and sub-Saharan belt of Africa due to budget constraints and other reasons (Hamer and Keusch 2015). These limitations lead to the emergence of new strategy like biofortification for nutrient improvement of the staple crops.

Availability of banana crop throughout the year, high nutritional value, and reduction in gene flow due to its ploidy level make banana as an excellent target for biofortification. Genetic modification of banana through conventional breeding is not feasible as all commercial cultivars are triploid, hence sterile in nature, and develop parthenocarpic fruit along with long life cycle (Shivani et al. 2017). The cross-pollination of banana is not possible due to its sterile nature which eliminates the chances of transgene flow. Therefore, genetic transformation is an appropriate substitute for the incorporation of desirable attributes into banana genome. It also overcomes intergeneric or interspecific and interkingdom gene-transfer-related issues (Tiwari and Tuli 2012).

Three approaches, namely, particle bombardment, electroporation, and *A. tumefaciens*-mediated transformation, had been used for genetic transformation (Tripathi et al. 2015). *Agrobacterium*-mediated transformation is the most preferable method because it leads to a limited number of transgene integrations and it can carry an ample segment of DNA with high efficiency (Gelvin 2003). The embryogenic cell suspensions have been used as explant for *Agrobacterium*-mediated genetic modification in several studies (Cote et al. 1996; Khanna et al. 2004; Tripathi et al. 2015). Although genetic modification and regeneration are extremely genotype dependent, the other factors such as age and type of tissue, postinfection time effect, and additives affect the transformation efficiency (Khanna et al. 2004). Therefore, transformation protocol optimization is a prerequisite for improving or introducing any agronomic trait in a particular species.

There are several other reports regarding improvement of provitamin A content in plants. In the carrot, expression of *ketolase* gene with *Ubi* and *CaMV35S* promoters increased β-carotene content to 39 μg/g FW, in roots (Jayaraj et al. 2008). Concurrent performance of the seven genes, namely, *Brevundimonas β-carotene ketolase* (*CrtW*), *Paracoccus isopentenyl pyrophosphate isomerase*, *β-carotene hydroxylase*, *crtB*, *crtI*, *lycopene ε-cyclase*, and *lycopene β-cyclase* derived from *Pantoea ananatis* with promoters *CaMV35S* and *napin* enhanced β-carotene level to 214 μg/g FW (30-fold) in canola (Fujisawa et al. 2009). In Korean soybean (*Glycine max* L. cv. Kwangan), bicistronic systems involving *PSY from capsicum, 2A from foot-and-mouth disease virus*, and *crtl from Pantoea ananatis* generate *PAC (PSY-2A-crtl)* under *CaMV35S*, or soybean seed-specific β-conglycinin (β) promoter accumulated 112 μg/g β-carotene in seeds (Kim et al. 2012). *CrtW* gene derived from *Brevundimonas* regulated by soybean seed-specific lectin promoter enhanced carotenoids in soybean seeds (Pierce et al. 2015). The expression of capsicum *LCYβ* gene with the tomato-fruit-specific *PDS* promoter enhanced β-carotene content up to 57 μg/g FW and in combination with β-*CHX* raised β-carotene up to 63 μg/g FW in tomato (Dharmapuri et al. 2002). Other studies reported in tomato where the expression of tomato *LCYβ* and tomato *cryptochrome* (*CRY2*) genes under *CaMV35S* promoter boosted β-carotene up to 205 μg/g FW and 101 μg/g DW, respectively (D'Ambrosio et al. 2004; Giliberto et al. 2005). Overexpression of

*E. coli DXS* under the control of *CaMV35S* promoter raised carotenoid content in tomato (Enfissi et al. 2005).

Carotenoid content was also increased in crops using gene silencing technique through genetic engineering. In potato tuber, the silencing of the *LCYε* gene with *patatin* promoter increased the β-carotene level to 43.56 ng/g DW (Diretto et al. 2006). In another study on potato, silencing of the *β-CHX* gene under *CaMV35S* and *granule-bound starch synthase* (*GBSS*) promoters increased β-carotene up to 331 μg/100 g FW (Van Eck et al. 2007). Lopez et al. (2008) reported β-carotene (5.01 μg/g DW) accumulation in potato tubers after expressing *orange cauliflower* (*Or*) gene from *Brassica oleracea* with *GBSS* promoter. In canola seeds, β-carotene content increased up to 91 μg/g FW (42-fold) by silencing of the *LCYε* gene using RNA interference (RNAi) construct (Yu et al. 2008). The β-carotene was upgraded up to 40 μg/g FW by silencing of *Solanum lycopersicum NCED1* (*SlNCED1*) gene with fruit-specific ethylene inducible (*E8*) promoter (Sun et al. 2012). In citrus fruit, silencing of *β-CHX* placed under *CaMV35S* promoter also increased β-carotene (114 ng/g FW) (Pons et al. 2014). Targeted mutagenesis through CRISPR/Cas9 for *LCYε* gene was recently demonstrated in banana cultivars Grand Naine and Rasthali (Kaur et al. 2020). It showed 6-fold increased β-carotene (24μg/g) in Grand Naine banana fruits.

## 2.16 Biofortified Wheat/Rice with Micronutrient(s)

The wheat crop and the wheat products that are routinely consumed have inherently less amount of zinc than required by the human body (35–45 μg/g). This has led to zinc deficiency especially among the low-income populations (20% of the world population) (Cakmak 2008; Zou et al. 2012). Zinc is an indispensable micronutrient that is required for the metabolic activities (normal growth) of plant and humans (Tisdale et al. 1984; Hafeez et al. 2013). It is also required for the development and function of the components of the immune system (neutrophils and natural killer cells) (Shankar and Prasad 1998; Prasad 2008). Its role in stress tolerance, gene expression, and pollen tube growth is also reported (Cakmak 2008; Pandey et al. 2006). There are reports on increasing zinc deficiency in both food crops and humans (Alloway 2004; Cakmak 2008). With the commencement of "International HarvestPlus program" (sub-project: "HarvestZinc"), plant biotechnologists began to focus on the enhancement of the Zn concentrations in the economic part(s) of crop plants (Camak and Kutman 2018). To cope with the problem of less micronutrient uptake in plants and their bioavailability, biofortification can be done:

*Strategy 1*: It aims at planting new varieties of crops that are efficient accumulators of micronutrients such as Fe and Zn, coupled with the foliar spray(s) of micronutrients in the form of fertilizer(s) (Hotz and McClafferty 2007). This strategy can be effective, but the net efficiency of biofortification usually remains nonuniform. Moreover, these agronomic inputs shall add to the cost of raising a crop, which may not be acceptable and affordable to the world's rural population.

*Strategy 2*: Plant-produced phytic acids/phytate (inositol hexaphosphates and pentaphosphates) acts as anti-nutrients. These may bind to zinc forming insoluble complexes which culminates in reduced zinc absorption and its bioavailability in the humans. Phytate content varies among crop plants and is found in high levels in rice and beans. With the aid of RNAi technology, the level of phytates can be reduced in many crop species so as to augment the availability of zinc and elements (Lönnerdal 2003; Ali et al. 2010; Li et al. 2014; Sakai et al. 2015).

*Strategy 3:* Wheat seeds have high phytate content, which chelates the metal ions and reduces their bioavailability. Transgenic plants harboring phytase genes can degrade the seed-phytic acid content. Abid et al. (2017) have developed transgenic wheat expressing *phyA* gene from *Aspergillus japonicus* in wheat endosperm with 12–76% reduced phytic acid content, compared to the non-transgenic control. This strategy can be a boon to the malnourished human population.

*Strategy 4:* The metal transporter proteins found in plants are able to use multiple metal substrates, including iron and zinc for absorption from soil to the roots. Masuda et al. (2013) have increased the accumulation of the iron storage protein (ferritin) as well as enhanced iron translocation in rice endosperm, without any yield penalty, by overexpressing the iron (II)-nicotianamine transporter OsYSL2. The transgenic lines accumulated high levels of iron (4.4-fold) and zinc (1.6-fold) in rice. In the same year, Aung et al. (2013) generated a transgenic line of rice over-expressing three genes, (1) the nicotianamine synthase gene (*HvNAS1*), for enhancing iron transport; (2) Fe(II)-nicotianamine transporter gene (*OsYSL2*), for transporting iron to the endosperm; and (3) the Fe storage protein gene (*SoyferH2*), for enhancing iron storage in the endosperm. The transgenic rice plants accumulated 1.3- and 3.4-fold higher Zn and Fe contents, respectively, than the non-transformed controls. In another report, Tan et al. (2015) used the iron transporter gene (MxIRT1) from *Malus domestica* to develop transgenic rice which accumulated threefold higher Fe and Zn.

*Strategy 5:* Banakar et al. (2017) developed transgenic rice plants expressing nicotianamine and 20-deoxymugenic acid (DMA) for accumulating both Zn and Fe in rice endosperm. Transgenic plants accumulated fourfold higher Fe and twofold higher Zn content, compared to the non-transformed plants. These results are in consonance with earlier reports (Masuda et al. 2013; Paul et al. 2014; Trijatmiko et al. 2016). Trijatmiko et al. (2016) generated transgenic rice plants expressing nicotianamine synthase (*OsNAS2*) and soybean ferritin (*SferH-1*) genes that accumulated a higher Fe and Zn content in the endosperm. Masuda et al. (2013) reported the development of transgenic rice with high Fe and Zn accumulation capacity by expressing *SoyferH2*, *HvNAS1*, *HvNAAT-A*, *HvNAAT-B*, and *IDS3* (ferric iron chelator) genes. The transgenic plants accumulated 2.5-folds and fourfold iron when grown in iron-deficient and iron-sufficient soils, respectively. Paul et al. (2014) raised transgenic indica rice that expressed 2.6-fold higher ferritin content in the T4 generation, without affecting the yield. Even after milling, the rice seeds reflected a 2.54 and 1.54 times increase in Fe and Zn content, respectively. Thus, molecular plant breeding, genetic manipulation, and agronomic practices can be used for enhancing Zn and Fe content in food crops.

## 2.17 Conclusions

The discovery of *Agrobacterium* and the accumulation of tremendous knowledge in the field of plant-microbe interaction, cell biology, plant gene function, and biotechnologies have opened new vistas for generating genetically modified (GM) crops with desired traits. Actually, with an expanding range including many commercially vital flowers, crops, and tree species and due to its latest applications to the non-plant species such as yeast, mushroom, and human cells, *Agrobacterium* has got intransience and popularity almost in every laboratory of plant molecular biology. However, this new use of *Agrobacterium* requires designing and construction of host-specific binary plasmids and more suitable *Agrobacterium* strains.

CRISPR/Cas technique came into limelight in the year 2012 as a boon to the crop improvement strategies because the plants engineered through it are considered as non-GM. It is simpler, reliable, and functional to craft plant genomes compared to ZFNs/TALENs. In plants, it's observed that the editing mostly has been detected in the species such as rice, *Arabidopsis*, and tobacco (Jiang et al. 2013). Furthermore, this technology has been applied to more than twenty crops (Ricroch et al. 2017) including (1) maize, phytic acid synthesis (Liang et al. 2014); (2) tomato, fruit ripening (Ito et al. 2015); (3) soybean, carotenoid biosynthesis (Du et al. 2016); (4) wheat, increasing Fe content (Connorton et al. 2017); (5) cassava, carotenoid biosynthesis (Odipio et al. 2017); and (6) banana, carotenoid biosynthesis (Kaur et al. 2018, 2020)

In order to develop healthy crops across the globe, much efforts are required to ensure interaction between peoples from many disciplines including crop breeding, molecular biology, nutritional and social sciences. Before spending money, time, and labor on developing herbicide-tolerant or biofortified crops, dissemination of knowledge regarding their benefits to the younger populations is a prerequisite. Researches must be aimed at increasing the nutritional value of "orphan crops" like pearl millet, pigeon pea, and sorghum that are important for the world's rural communities but are underutilized by the developed countries. Thus, cooperative efforts between academia, industry, government, and nonprofit organizations can eradicate the problem of malnutrition and inadequate nutrition.

## References

Abid N, Khatoon A, Maqbool A, Irfan M, Bashir A, Asif I, Shahid M, Saeed A, Brinch-Pedersen H, Malik KA (2017) Transgenic expression of phytase in wheat endosperm increases bioavailability of iron and zinc in grains. Transgenic Res 26(1):109–122

Ahmed F, Darnton-Hill I (2004) Vitamin A. In: Public health nutrition. The nutrition, vol 355. Blackwell Sciences, Oxford, pp 192–215

Akhtar S, Ahmed A, Randhawa MA, Atukorala S, Arlappa N, Ismail T, Ali Z (2013) Prevalence of vitamin A deficiency in South Asia: causes, outcomes, and possible remedies. J Health Popul Nutr 31(4):413–423

Ali M, Shuja MN, Zahoor M, Qadri I (2010) Phytic acid: how far have we come? Afr J Biotechnol 9:1551–1554

Alloway BJ (2004) Zinc in soils and crop nutrition. IZA Publications, Brussels

Altpeter F, Baisakh N, Beachy R, Bock R, Capell T, Christou P, Daniell H, Datta K, Datta S, Dix PJ, Fauquet C, Huang N, Kohli A, Mooibroek H, Nicholson L, Nguyen TT, Nugent G, Raemakers K, Romano A, Somers DA, Stoger E, Taylor N, Visser R (2005) Particle bombardment and the genetic enhancement of crops: myths and realities. Mol Breed 15:305–327

Aluru M, Xu Y, Guo R (2008) Generation of transgenic maize with enhanced provitamin A content. Exp Bot 59:3551–3562

Anderson VP, Jack S, Monchy D, Hem N, Hok P, Bailey KB, Gibson RS (2008) Co-existing micronutrient deficiencies among stunted Cambodian infants and toddlers. Asia Pac J Clin Nutr 17(1):72–79

Ankenbauer RG, Nester EW (1990) Sugar-mediated induction of *Agrobacterium tumefaciens* virulence genes: structural specificity and activities of monosaccharides. J Bacteriol 172:6442–6446

Apel W, Bock R (2009) Enhancement of carotenoid bio-synthesis in transplastomic tomatoes by induced lycopene-to-provitamin A conversion. Plant Physiol 151:59–66

Aung MS, Masuda H, Kobayashi T, Nakanishi H, Yamakawa T, Nishizawa NK (2013) Iron biofortification of Myanmar rice. Front Plant Sci 4:158

Bako L, Umeda M, Tiburcio AF, Schell J, Koncz C (2003) The VirD2 pilot protein of Agrobacterium-transferred DNA interacts with the TATA box-binding protein and a nuclear protein kinase in plants. Proc Natl Acad Sci 100(17):10108–10113

Ballas N, Citovsky V (1997) Nuclear localization signal binding protein from *Arabidopsis* mediates nuclear import of *Agrobacterium* VirD2 protein. Proc Natl Acad Sci U S A 94(20):10723–10728

Banakar R, Alvarez Fernandez A, Díaz-Benito P, Abadia J, Capell T, Christou P (2017) Phytosiderophores determine thresholds for iron and zinc accumulation in biofortified rice endosperm while inhibiting the accumulation of cadmium. J Exp Bot 68:4983–4995

Banta LM, Montenegro M (2008) *Agrobacterium* and plant biotechnology. In: Tzfira T, Citovsky V (eds) *Agrobacterium*: from biology to biotechnology. Springer, Berlin, pp 73–147

Bender DA (2003) Nutritional biochemistry of the vitamins, 2nd edn. Cambridge University Press, Cambridge

Beyer P (2010) Golden rice and 'golden' crops for human nutrition. New Biotechnol 27(5):478–481

Binns AN, Beaupre CE, Dale M (1995) Inhibition of VirB-mediated transfer of diverse substrate from *Agrobacterium tumefaciens* by the IncQ plasmid RSF1010. J Bacteriol 177:4890–4899

Bradley LR, Kim JS, Matthysse AG (1997) Attachment of *Agrobacterium tumefaciens* to carrot cells and Arabidopsis wound sites is correlated with the presence of a cell-associated acidic polysaccharide. J Bacteriol 179:5372–5379

Bulgakov VP, Kiselev KV, Yakovlev KV, Zhuravlev YN, Gontcharov AA, Odintsova NA (2006) *Agrobacterium*-mediated transformation of sea urchin embryos. Biotechnol J 1:454–461

Bulley S, Wright M, Rommens C, Yan H, Rassam M, Lin-Wang K, Andre C, Brewster D, Karunairetnam S, Allan AC, Laing WA (2011) Enhancing ascorbate in fruits and tubers through over-expression of the L-galactose pathway gene GDP-L-galactose phosphorylase. Plant Biotechnol J 10:390–397

Cakmak I (2008) Enrichment of cereal grains with zinc: agronomic or genetic biofortification? Plant Soil 302:1–17

Camak I, Kutman UB (2018) Agronomic biofortification of cereals with zinc: a review. Eur J Soil Sci 69:172–180

Cangelosi GA, Ankenbauer RG, Nester EW (1990) Sugars induce the *Agrobacterium virulence* genes through a periplasmic binding protein and transmembrane signal protein. Proc Natl Acad Sci U S A 87:6708–6712

Chattha MU, Hassan MU, Khan I, Chattha MB, Mahmood A, Chattha MU, Nawaz M, Subhani MN, Kharal M, Khan S (2017) Biofortification of wheat cultivars to combat zinc deficiency. Front Plant Sci 8:1–8

Chilton MD, Drummond MH, Merlo DJ, Sciaky D (1978) Highly conserved DNA of Ti plasmid overlaps T-DNA maintained in plant tumors. Nature 275:147–149

Christie PJ (1997) *Agrobacterium tumefaciens T*-complex transport apparatus: a paradigm for a new family of multifunctional transporters in eubacteria. J Bact 179:3085–3094

Christie PJ, Ward JE Jr, Gordon MP, Nester EW (1989) A gene required for transfer of T-DNA to plants encodes an ATPase with autophosphorylating activity. Proc Natl Acad Sci U S A 86(24):9677–9681

Christou P, Twyman RM (2004) The potential of genetically enhanced plants to address food insecurity. Nutr Res Rev 17:23–42

Cong L, Wang C, Chen L, Liu H, Yang G, He G (2009) Expression of phytoene synthase1 and carotene desaturase crtI genes result in an increase in the total carotenoids content in transgenic elite wheat (*Triticum aestivum* L.). J Agric Food Chem 57:8652–8660

Connorton JM, Balk J (2019) Iron biofortification of staple crops: lessons and challenges in plant genetics. Plant Cell Physiol 60(7):1447–1456

Connorton JM, Jones ER, Rodriguez-Ramiro I, Fairweather-Tait S, Uauy C, Balk J (2017) Wheat vacuolar iron transporter TaVIT2 transports Fe and Mn and is effective for biofortification. Plant Physiol 174:2434–2444

Cote FX, Domergue R, Monmarson S, Schwendiman J, Teisson C, Escalant JV (1996) Embryonic cell suspensions from the male flower of musa AAA Cv. grand nain. Physiol Plant 97:285–290

D'Ambrosio C, Giorio G, Marino I, Merendino A, Petrozza A, Salfi L, Stigliani AL, Cellini F (2004) Virtually complete conversion of lycopene into β-carotene in fruits of tomato plants transformed with the tomato lycopene β-cyclase (tlcy-b) cDNA. Plant Sci 166:207–214

De Buck S, Jacobs A, Van Montagu M, Depicker A (1999) The DNA sequences of T-DNA junctions suggest that complex T-DNA loci are formed by a recombination process resembling T-DNA integration. Plant J 20:295–304

De Cleene M, De Ley J (1976) The host range of crown gall. Bot Rev 42(4):389–466

de Groot MJA, Bundock P, Hooykaas PJ, Beijersbergen AG (1998) *Agrobacterium tumefaciens*-mediated transformation of filamentous fungi. Nat Biotechnol 16:839–842

Deblaere R, Bytebier B, De Greve H, Deboeck F, Schell J, Van Montagu M, Leemans J (1985) Efficient octopine Ti plasmid-derived vectors for Agrobacterium-mediated gene transfer to plants. Nucleic Acids Res 13(13):4777–4788

Deng WY, Chan LH, Wood DW, Metcalfe T, Liang X, Gordon MP, Comai L, Nester EW (1998) *Agrobacterium* VirD2 protein interacts with plant host cyclophilins. Proc Natl Acad Sci U S A 95(12):7040–7045

Depicker A, Van Montagu M, Schell J (1978) Homologous sequences in different Ti plasmids are essential for oncogenicity. Nature 275:150–153

Dharmapuri S, Rosati C, Pallara P, Pallaraa P, Aquilani R, Bouvierb F, Camarab B, Giulianoa G (2002) Metabolic engineering of xanthophyll content in tomato fruits. FEBS Lett 519:30–34

Diretto G, Tavazza R, Welsch R, Pizzichini D, Mourgues F, Papacchioli V, Beyer P, Giuliano G (2006) Metabolic engineering of potato tuber carotenoids through tuber- specific silencing of lycopene ε-cyclase. BMC Plant Biol 6:13

Diretto G, Al-Babili S, Tavazza R, Papacchioli V, Beyer P, Giuliano G (2007) Metabolic engineering of potato carotenoid content through tuber-specific overexpression of a bacterial mini pathway. PLoS One 2:e350

Dorrington R, Bradshaw D (2011) Maternal mortality in South Africa: lessons from a case study in the use of deaths reported by households in censuses and surveys. J Popul Res 28(1):49–73

Drakakaki G, Marcell S, Glahn RP, Lund EK, Pariagh S, Fischer R, Christou P, Stoger E (2005) Endosperm-specific co-expression of recombinant soybean ferritin and Aspergillus phytase in maize results in significant increases in the levels of bioavailable iron. Plant Mol Biol 59:869–880

Du H, Zeng X, Zhao M, Cui X, Wang Q, Yang H et al (2016) Efficient targeted mutagenesis in soybean by TALENs and CRISPR/Cas9. J Biotechnol 217:90–97

Ducreux LJ, Morris WL, Hedley PE, Shepherd T, Davies HV, Millam S, Taylor MA (2005) Metabolic engineering of high carotenoid potato tubers containing enhanced levels of beta-carotene and lutein. J Exp Bot 56:81–89

Durrenberger F, Crameri A, Hohn B, Koukolikova-Nicola Z (1989) Covalently bound VirD2 protein of *Agrobacterium tumefaciens* protects the T-DNA from exonucleolytic degradation. Proc Natl Acad Sci U S A 86:9154–9158

Enfissi EMA, Fraser PD, Lois LM, Boronat A, Schuch W, Bramley PM (2005) Metabolic engineering of the mevalonate and nonmevalonate isopentenyl diphosphate-forming pathways for the production of health-promoting isoprenoids in tomato. Plant Biotechnol J 3:17–27

Fields S, Song O-K (1989) A novel genetic system to detect protein-protein interactions. Nature 340:245–246

Frary A, Earle ED (1996) An examination of factors affecting the efficiency of Agrobacterium-mediated transformation of tomato. Plant Cell Rep 16(3–4):235–240

Fraser PD, Susanne R, Cathie AS, Philippa BM, Kiano JW, Misawa N, Drake RG, Schuch BPM (2002) Evaluation of transgenic tomato plants expressing an additional phytoene synthase in a fruit-specific manner. Proc Natl Acad Sci U S A 99:1092–1097

Fujisawa M, Takita E, Harada H, Sakurai N, Suzuki H, Ohyama K, Shibata D, Misawa N (2009) Pathway engineering of *Brassica napus* seeds using multiple key enzyme genes involved in ketocarotenoid formation. J Exp Bot 60:1319–1332

Fungo R, Pillay M (2011) Carotene content of selected banana genotypes from Uganda. Afr J Biotechnol 10(28):5423–5430

Gallego ME, Bleuyard JY, Daoudal-Cotterell S, Jallut N, White CI (2003) Ku80 plays a role in non-homologous recombination but is not required for T-DNA integration in Arabidopsis. Plant J 35(5):557–565

Gao C, Long D, Lenk I, Nielsen KK (2008) Comparative analysis of transgenic tall fescue (*Festuca arundinacea* Schreb.) plants obtained by *Agrobacterium*-mediated transformation and particle bombardment. Plant Cell Rep 27:1601–1609

Gelvin SB (2003) *Agrobacterium*-mediated plant transformation: the biology behind the "gene-jockeying" tool. Microbiol Mol Biol Rev 67:16–37

Gheysen G, Van Montagu M, Zambryski P (1987) Integration of *Agrobacterium tumefaciens* transfer DNA (T-DNA) involves rearrangements of target plant DNA. Proc Natl Acad Sci U S A 84:6169–6173

Gheysen G, Villarroel R, Van Montagu M (1991) Illegitimate recombination in plants: a model for T-DNA integration. Genes Dev 5(2):287–297

Giliberto L, Perrotta G, Pallara P, Weller JL, Fraser PD, Bramley PM, Fiore A, Tavazza M, Giuliano G (2005) Manipulation of the blue light photoreceptor cryptochrome 2 in tomato affects vegetative development, flowering time, and fruit antioxidant content. Plant Physiol 137:199–208

Hafeez B, Khanif M, Saleem M (2013) Role of zinc in plant nutrition: a review. Am J Exp Agric 3:374–391

Hamer DH, Keusch GT (2015) Vitamin A deficiency: slow progress towards elimination. Lancet Glob Health 3:502–503

Harrison EH, dela Seña's C (2015) Assimilation and conversion of dietary vitamin A into bioactive retinoids. Retinoids: Biol Biochem Dis, 35–56

Hepburn AG, White J, Pearson L, Maunders MJ, Clarke LE, Prescott AG, Blundy KS (1985) The use of pNJ5000 as an intermediate vector for the genetic manipulation of Agrobacterium Ti-plasmids. Microbiology 131(11):2961–2969

Higgins CF, Hyde SC, Mimmack MM, Gileadi U, Gill DR, Gallagher MP (1990) Binding protein-dependent transport systems. J Bioenerg Biomembr 22:571–592

Hoekema A, Hirsch PR, Hooykaas PJ, Schilperoort RA (1983) A binary plant vector strategy based on separation of vir-and T-region of the Agrobacterium tumefaciens Ti-plasmid. Nature 303(5913):179–180

Hollenberg SM, Sternglanz R, Cheng PF, Weintraub H (1995) Identification of a new family of tissue-specific helix-loop helix proteins with a two-hybrid system. Mol Cell Biol 15:3813–3822

Holsters M, Silva B, Van Vliet F, Genetello C, De Block M, Dhaese P, Depicker A, Inzé D, Engler G, Villarroel R, Van Montagu M, Schell J (1980) The functional organization of the nopaline A. tumefaciens plasmid pTiC58. Plasmid 3(2):212–230

Hood EE, Helmer GL, Fraley RT, Chilton MD (1986) The hypervirulence of Agrobacterium tumefaciens A281 is encoded in a region of pTiBo542 outside of T-DNA. J Bacteriol 168(3):1291–1301

Hood EE, Gelvin SB, Melchers LS, Hoekema A (1993) NewAgrobacterium helper plasmids for gene transfer to plants. Transgenic Res 2(4):208–218

Hooykaas PJJ (2004) Transformation mediated by Agrobacterium tumefaciens. In: Tkacz JS, Lange L (eds) Advances in fungal biotechnology for industry, agriculture, and medicine. Springer, Boston, MA

Horsch RB, Fry JE, Hoffmann NL, Eichholiz D, Rogers SG, Fraley RT (1985) A simple and general method of transferring genes into plants. Science 227:1229–12231

Hotz C, McClafferty B (2007) From harvest to health: challenges for developing biofortified staple foods and determining their impact on micronutrient status. Food Nutr Bull 28:271–279

Hwang H-H, Yu M, Lai E-M (2017) Agrobacterium-mediated plant transformation: biology and applications. Arabidopsis Book 15:e0186

Ito Y, Nishizawa-Yokoi A, Endo M, Mikami M, Toki S (2015) CRISPR/Cas9-mediated mutagenesis of the RIN locus that regulates tomato fruit ripening. Biochem Biophys Res Commun 467(1):76–82

Janssen B-J, Gardner RC (1989) Localized transient expression of GUS in leaf discs following cocultivation with *Agrobacterium*. Plant Mol Biol 14:61–72

Jayaraj J, Devlin R, Punja Z (2008) Metabolic engineering of novel ketocarotenoid production in carrot plants. Transgenic Res 17:489–501

Jiang W, Zhou H, Bi H, Fromm M, Yang B, Weeks DP (2013) Demonstration of CRISPR/Cas9/sgRNA-mediated targeted gene modification in Arabidopsis, tobacco, sorghum and rice. Nucleic Acids Res 41(20):e188

Johnson AAT, Kyriacou B, Callahan DL, Carruthers L, Stangoulis J, Lombi E, Tester M (2011) Constitutive overexpression of the OsNAS gene family reveals single-gene strategies for effective iron- and zinc-biofortification of rice endosperm. PLoS One 6:e24476

Kaur N, Alok A, Shivani KN, Pandey P, Awasthi P, Tiwari S (2018) CRISPR/Cas9-mediated efficient editing in phytoene desaturase (PDS) demonstrates precise manipulation in banana cv. Rasthali genome. Funct Integr Genomics 18(1):89–99

Kaur N, Alok A, Shivani PK, Kaur N, Awasthi P, Chaturvedi S, Pandey P, Pandey A, Pandey AK, Tiwari S (2020) CRISPR/Cas9 directed editing of lycopene epsilon-cyclase modulates metabolic flux for β-carotene biosynthesis in banana fruit. Metab Eng 59:76–86

Khanna H, Becker D, Kleidon J, Dale J (2004) Centrifugation assisted agrobacterium tumefaciens-mediated transformation (CAAT) of embryogenic cell suspensions of banana (Musa Spp. Cavendish AAA and Lady Finger AAB). Mol Breed 14(3):239–252

Kim M-J, Kim JK, Kim HJ, Pak JH, Lee J-H, Kim D-H, Choi HK, Jung HW, Lee J-D, Chung Y-S, Ha S-H, El-Shemy HA (2012) Genetic modification of the soybean to enhance the β-carotene content through seed-specific expression. PLoS One 7(10):e48287

Kisaka H, Kida T (2003) Transgenic tomato plant carrying a gene for NADP-dependent glutamate dehydrogenase (*gdh A*) from *Aspergillus nidulans*. Plant Sci 164:35–42

Koncz C, Schell J (1986) The promoter of TL-DNA gene 5 controls the tissue-specific expression of chimaeric genes carried by a novel type of Agrobacterium binary vector. Mol Gen Genet MGG 204(3):383–396

Kumar S, Fladung M (2001) Gene stability in transgenic aspen (Populus). II. Molecular characterization of variable expression of transgene in wild and hybrid aspen. Planta 213(5):731–740

Kumar S, Fladung M (2002) Transgene integration in aspen: structures of integration sites and mechanism of T-DNA integration. Plant J 31(4):543–551

Kunik T, Tzfira T, Kapulnik Y, Gafni Y, Dingwall C, Citovsky V (2001) Genetic transformation of HeLa cells by *Agrobacterium*. Proc Natl Acad Sci U S A 98:1871–1876

Lal EM, Kado CI (1998) Proceed VirB2 is the major subunit of the promiscuous plus of *Agrobacterium tumefaciens*. J Bacteriol 180:2711–2717

Lazo GR, Stein PA, Ludwig RA (1991) A DNA transformation–competent Arabidopsis genomic library in Agrobacterium. Bio/technology 9(10):963–967

LeDuc DL, Tarun AS, Montes-Bayon M, Meija J, Malit MF, Wu CP, AbdelSamie M, Chiang CY, Tagmount A, deSouza M, Neuhierl B, Böck A, Caruso J, Terry N (2004) Overexpression of selenocysteine methyltransferase in *Arabidopsis* and Indian mustard increases selenium tolerance and accumulation. Plant Physiol 135:377–383

Lee S, Jeon US, Lee SJ, Kim YK, Persson DP, Husted S, Schjørring JK, Kakei Y, Masuda H, Nishizawa NK, An G (2009) Iron fortification of rice seeds through activation of the nicotianamine synthase gene. Proc Natl Acad Sci U S A 106:22014–22019

Lee S, Persson DP, Hansen TH, Husted S, Schjoerring JK, Kim YS, Jeon US, Kim YK, Kakei Y, Masuda H, Nishizhawa NK, An G (2011) Bio-available zinc in rice seeds is increased by activation tagging of nicotianamine synthase. Plant Biotechnol J 9:865–873

Li WX, Zhao HJ, Pang WQ, Cui HR, Poirier Y, Shu QY (2014) Seed-specific silencing of OsMRP5 reduces seed phytic acid and weight in rice. Transgenic Res 23:585–599

Liang Z, Zhang K, Chen K, Gao C (2014) Targeted mutagenesis in Zea mays using TALENs and the CRISPR/Cas system. J Genet Genomics 41(2):63–68

Lockyer S (2018) Who is shaping the food choices of the future? Nutr Bull 43:93–96

Lockyer S, White A, Buttriss JL (2018) Biofortified crops for tackling micronutrient deficiencies – what impact are these having in developing countries and could they be of relevance within Europe? Nutr Bull 43:319–357

Lönnerdal B (2003) Genetically modified plants for improved trace element nutrition. J Nutr 133:1490S–1493S

Lopez AB, Eck JV, Conlin BJ, Paolillo DJ, O'Neill J, Li L (2008) Effect of the cauliflower or transgene on carotenoid accumulation and chromoplast formation in transgenic potato tubers. Exp Bot 59:213–223

Lorence A, Verpoorte R (2004) Gene transfer and expression in plants. Methods Mol Biol 267:329–350

Maass D, Arango J, Wüst F, Beyer P, Welsch R (2009) Carotenoid crystal formation in Arabidopsis and carrot roots caused by increased phytoene synthase protein levels. PLoS One 4:e6373

Masuda H, Kobayashi T, Ishimaru Y, Takahashi M, Aung MS, Nakanishi H, Mori S, Nishizawa NK (2013) Iron-biofortification in rice by the introduction of three barley genes participated in mugineic acid biosynthesis with soybean ferritin gene. Front Plant Sci 4:132

Mayerhofer R, Koncz-Kalman Z, Nawrath C, Bakkeren G, Crameri A, Angelis K, Redei GP, Schell J, Hohn B, Koncz C (1991) T-DNA integration: a mode of illegitimate recombination in plants. EMBO J 10:697–704

McBride KE, Summerfelt KR (1990) Improved binary vectors for Agrobacterium-mediated plant transformation. Plant Mol Biol 14(2):269–276

Mehta RA, Cassol T, Li N, Ali N, Handa AK, Mattoo AK (2002) Engineered polyamine accumulation in tomato enhances phytonutrient content, juice quality, and vine life. Nat Biotechnol 20:613–618

Molinier J, Ramos C, Fritsch O, Hohn B (2004) CENTRIN2 modulates homologous recombination and nucleotide excision repair in Arabidopsis. Plant Cell 16(6):1633–1643

Naqvi S, Zhu C, Farre G, Ramessar K, Bassie L, Breitenbach J, Conesa D, Ros G, Sandmann G, Capell T, Christou P (2009) Transgenic multivitamin corn through biofortification of endosperm with three vitamins representing three distinct metabolic pathways. Proc Natl Acad Sci U S A 106:7762–7767

Newell CA (2000) Plant transformation technology: developments and applications. Mol Biotechnol 16:53–66

Odipio J, Alicai T, Ingelbrecht I, Nusinow DA, Bart R, Taylor NJ (2017) Efficient CRISPR/Cas9 genome editing of phytoene desaturase in cassava. Front Plant Sci 18:1780

Offringa R, De Groot MJ, Haagsman HJ, Does MP, Van Den Elzen PJ, Hooykaas PJ (1990) Extrachromosomal homologous recombination and gene targeting in plant cells after Agrobacterium mediated transformation. EMBO J 9(10):3077–3084

Paine J, Shipton C, Chaggar S, Howells R, Kennedy M, Vernon G, Wright S, Hinchliffe E, Adams J, Silverstone A, Drake R (2005) Improving the nutritional value of golden rice through increased pro-vitamin A content. Nat Biotechnol 23:482–487

Palanichelvam K, Oger P, Clough SJ, Chung C, Bent AF, Farrand SK (2000) A second T-region of the soybean-supervirulent chrysopine-type Ti plasmid pTiChry5, and construction of a fully disarmed helper plasmid. Mol Plant-Microbe Interact 13(10):1081–1091

Pandey N, Pathak GC, Sharma CP (2006) Zinc is critically required for pollen function and fertilisation in lentil. J Trace Elem Med Biol 20:89–96

Park SH, Kim C-K, Pike LM, Smith RH, Hirschi KD (2004) Increased calcium in carrots by expression of an Arabidopsis $H^+/Ca^{2+}$ transporter. Mol Breed 14:275–282

Park S, King TS, Kim CK, Han JS, Kim S, Smith RH, Pike LM, Hirschi KD (2005) Genetic manipulation for enhancing calcium content in potato tuber. J Agric Food Chem 53:5598–5603

Park S, Elless MP, Park J, Jenkins A, Lim W, Chambers E, Hirschi KD (2009) Sensory analysis of calcium biofortified lettuce. Plant Biotechnol J 7:106–117

Paul S, Ali N, Datta SK, Datta K (2014) Development of an iron-enriched high-yieldings indica rice cultivar by introgression of a high-iron trait from transgenic iron-biofortified rice. Plant Foods Hum Nutr 2014:203–208

Pelczar P, Kalck V, Gomez D, Hohn B (2004) *Agrobacterium* proteins VirD2 and VirE2 mediate precise integration of synthetic TDNA complexes in mammalian cells. EMBO Rep 5:632–637

Pérez-Massot E, Banakar R, Gómez-Galera S, Zorrilla-López U, Sanahuja G, Arjó G, Miralpeix B, Vamvaka E, Farré G, Rivera SM, Dashevskaya S, Berman J, Sabalza M, Yuan D, Bai C, Bassie L, Twyman RM, Capell T, Christou P, Zhu C (2012) The contribution of transgenic plants to better health through improved nutrition: opportunities and constraints. Genes Nutr 8(1):29–41

Pickar-Oliver A, Gersbach CA (2019) The next generation of CRISPR–Cas technologies and applications. Nat Rev Mol Cell Biol 20:490–507

Pierce EC, LaFayette PR, Ortega MA, Joyce BL, Kopsell DA, Parrott WA, Peña L (2015) Ketocarotenoid production in soybean seeds through metabolic engineering. PLoS One 10(9):e0138196

Piers KL, Health JD, Liang X, Stephens KM, Nester EW (1996) *Agrobacterium tumefaciens*-mediated transformation of yeast. Proc Natl Acad Sci U S A 93(4):1613–1618

Pilon-Smits EAH, Hwang S, Lytle CM, Zhu Y, Tai JC, Bravo RC, Chen Y, Leustek T, Terry N (1999) Over expression of ATP sulfurylase in Indian mustard leads to increased selenate uptake, reduction and tolerance. Plant Physiol 119:123–132

Pino-Lagos K, Guo Y, Noelle RJ (2010) Retinoic acid: a key player in immunity. Biofactors 36:430–436. https://doi.org/10.1002/biof.117

Pons E, Alquézar B, Rodríguez A, Martorel IP, Genoves S, Ramon D, Rodrigo MJ, Zacar L, Pena L (2014) Metabolic engineering of β-carotene in orange fruit increases its in vivo antioxidant properties. Plant Biotechnol J 12:17–27

Potrykus I (2005) GMO-technology and malnutrition – public responsibility and failure. Electron J Biotechnol 8:3

Prasad AS (2008) Clinical, anti-inflammatory and antioxidant role of zinc. Exp Gerintol 43:370–377

Puchta H (1998) Repair of genomic double-strand breaks in somatic plant cells by one-sided invasion homologous sequences. Plant J 13:331–339

Ran FA, Hsu PD, Wright J, Agarwala V, Scott DA, Zhang F (2013) Genome engineering using the CRISPR-Cas9 system. Nat Protoc 8(11):2281–2308

Rao BSN (2000) Potential use of red palm oil in combating vitamin A deficiency in India. Food Nutr Bull 21(2):202–211

Ravanello MP, Ke D, Alvarez J, Huang B, Shewmaker CK (2003) Coordinate expression of multiple bacterial carotenoid genes in canola leading to altered carotenoid production. Metab Eng 5:255–263

Ricroch A, Clairand P, Harwood W (2017) Use of CRISPR systems in plant genome editing: toward new opportunities in agriculture. Emerg Top Life Sci 1:169–182

Romer S (2000) Elevation of the provitamin A content of transgenic tomato plant. Nat Biotechnol 18:666–669

Romer S, Lubeck J, Kauder F, Steiger S, Adomat C, Sandmann G (2002) Genetic engineering of a zeaxanthin-rich potato by antisense inactivation and co suppression of carotenoid epoxidation. Metab Eng 4:263–272

Rosati C, Aquilani R, Dharmapuri S, Pallara P, Marusic C, Tavazza R, Bouvier F, Camara B, Giuliano G (2000) Metabolic engineering of beta-carotene and lycopene content in tomato fruit. Plant J 24:413–419

Rukmini C (1994) Red palm oil to combat vitamin A deficiency in developing countries. Food Nutr Bull 15(2):1–6

Sakai H, Iwai T, Matsubara C, Usui Y, Okamura M, Yatou O, Terada Y, Aoki N, Nishida S, Yoshida KT (2015) A decrease in phytic acid content substantially affects the distribution of mineral elements within rice seeds. Plant Sci 238:170–177

Salinas J, Matassi G, Montero LM, Bernardi G (1988) Compositional compartmentalization and compositional patterns in the nuclear genomes of plants. Nucleic Acids Res 16:4269–4285

Salomon S, Puchta H (1998) Capture of genomic and T-DNA sequences during double-strand break repair in somatic plant cells. EMBO J 17:6086–6095

Sayre R, Beeching JR, Cahoon EB, Egesi C, Fauquet C, Fellman J, Fregene M, Gruissem W, Mallowa S, Manary M, Maziya-Dixon B, Mbanaso A, Schachtman DP, Siritunga D, Taylor N, Vanderschuren H, Zhang P (2011) The BioCassava plus program: biofortification of cassava for sub-Saharan Africa. Annu Rev Plant Biol 62:251–272

Shankar AH, Prasad AS (1998) Zinc and immune infection: the biology of altered resistance to infection. Am J Clin Nutr 68:447–463

Shewmaker C, Sheehy J, Daley M, Colburn S, Ke DY (1999) Seed-specific overexpression of phytoene synthase: increase in carotenoids and other metabolic effects. Plant J 20:401–412

Shivani AP, Sharma V, Kaur N, Kaur N, Pandey P, Tiwari S (2017) Genome-wide analysis of transcription factors during somatic embryogenesis in banana (Musa spp.) cv. Grand Nain. PLoS One 12(8):e0182242

Shou H, Frame BR, Whitham SA, Wang K (2004) Assessment of transgenic maize events produced by particle bombardment or *Agrobacterium*-mediated transformation. Mol Breed 13:201–208

Simkin AJ, Gaffé J, Alcaraz JP, Carde JP, Bramley PM, Fraser PD, Kuntz M (2007) Fibrillin influence on plastid ultrastructure and pigment content in tomato fruit. Phytochemistry 68:1545–1556

Smith EF, Townsend CO (1907) A plant tumor of bacterial origin. Science 25:671–673

Stachel SE, Zambryski PC (1989) Genetic trans-kingdom sex. Nature 340:190–191

Storozhenko S, De Brouwer V, Volckaert M, Navarrete O, Blancquaert D, Zhang GF, Lambert W, Van Der Straeten D (2007) Folate fortification of rice by metabolic engineering. Nat Biotechnol 25:1277–1279

Sun L, Yuan B, Zhang M, Wang L, Cui M, Wang Q, Leng P (2012) Fruit-specific RNAi-mediated suppression of SlNCED1 increases both lycopene and β-carotene contents in tomato fruit. J Exp Bot 63:3097–3108

Sundberg C, Meek L, Carnell K, Das A, Ream W (1960) VirE1 protein mediated export the singles DNA binding protein VirE2 from *Agrobacterium tumefaciens* into plant cell. J Bacteriol 178:1207–1212

Tan S, Han R, Li P, Yang G, Li S, Zhang P, Wang WB, Zhao WZ, Yin LP (2015) Over-expression of the MxIRT1 gene increases iron and zinc content in rice seeds. Transgenic Res 24:109–122

Tanumihardjo SA, Furr HC (2013) International efforts to eradicate vitamin a deficiency. In: Tanumihardjo S (ed) Carotenoids and human health. Nutrition and health. Humana Press, Totowa

Tinland B (1996) The integration of T-DNA into plant genomes. Trends Plant Sci 1:178–184

Tisdale SL, Nelson WL, Beaten JD (1984) Zinc in soil fertility and fertilizers, 4th edn. Macmillan Publishing Company, New York, pp 382–391

Tiwari S, Tuli R (2012) Optimization of factors for efficient recovery of transgenic peanut (Arachis hypogaea L.). Plant Cell Tiss Organ Cult 109(1):111–121

Trijatmiko KR, Dueñas C, Tsakirpaloglou N, Torrizo L, Arines FM, Adeva C, Balindong J, Oliva N, Sapasap MV, Borrero J, Rey J, Francisco P, Nelson A, Nakanishi H, Lombi E, Tako E, Glahn RP, Stangoulis J, Mohanty PC, Johnson AAT, Tohme J, Barry G, Slamet-Loedina IH (2016) Biofortified indica rice attains iron and zinc nutrition dietary targets in the field. Sci Rep 6:19792

Tripathi JN, Ntui VO, Ron M, Muiruri SK, Britt A, Tripathi L (2015) CRISPR/Cas9 editing of endogenous banana streak virus in the B genome of Musa spp. Overcomes a major challenge in banana breeding. Commun Biol 2019:1–11

Tzfira T, Citovsky V (2006) *Agrobacterium*-mediated genetic transformation of plants: biology and biotechnology. Curr Opin Biotechnol 17:147–154

Tzfira T, Rhee Y, Chen M-H, Kunik T, Citovsky V (2000) Nucleic acid transport in plant-microbe interactions: the molecules that walk through the walls. Annu Rev Microbiol 54:187–219

UNICEF (United Nations Children's Fund) (2018) Coverage at a Crossroads: New Directions for Vitamin A Supplementation Programmes. Available at www.data.unicef.org/wp-content/uploads/2018/05/Vitamin-a-report-web.pdf

United Nations (2016) Sustainable development goal 2 – zero hunger. Available at www.undp.org/content/undp/en/home/sustainable-development-goals/goal-2-zero-hunger/targets/

United Nations (2017) World population prospects. Available at www.esa.un.org/unpd/wpp/Publications/Files/WPP2017_KeyFind ings.pdf

Van Attikum H, Hooykaas PJJ (2003) Genetic requirements for the targeted integration of *Agrobacterium* T-DNA in *Saccharomyces cerevisiae*. Nucleic Acids Res 31(3):826–832

Van Attikum H, Bundock P, Hooykaas PJJ (2001) Non-homologous end-joining proteins are required for *Agrobacterium* T-DNA integration. EMBO J 22:6550–6558

Van Eck J, Conlin B, Garvin DF (2007) Enhancing beta-carotene content in potato by RNAi-mediated silencing of the beta-carotene hydroxylase gene. Am J Potato Res 84:331–342

Van Larebeke N, Engler G, Holsters M, Van Den Elsacker S, Zaenen I, Schilperoort RA, Schell J (1974) Large plasmid in *Agrobacterium tumefaciens* essential for crown gall-inducing ability. Nature 252:169–170

Watson JD (2014) Type 2 diabetes as a redox disease. Lancet 383(9919):841–843

Watson B, Currier TC, Gordon MP, Chilton MD, Nester EW (1975) Plasmid required for virulence of Agrobacterium tumefaciens. J Bacteriol 123(1):255–264

Wei S, Yu B, Gruber MY, Khachatourians GG, Hegedus DD, Hannoufa A (2010) Enhanced seed carotenoid levels and branching in transgenic *Brassica napus* expressing the *Arabidopsis* miR156b gene. J Agric Food Chem 58:9572–9578

Welsch R, Arango J, Bar C, Salazar B, Al-Babili S, Beltran J, Chavarriaga P, Ceballos H, Tohme J, Beyer P (2010) Provitamin A accumulation in cassava (*Manihot esculenta*) roots driven by a single nucleotide polymorphism in a phytoene synthase gene. Plant Cell 22:3348–3356

Whatley MH, Spiess LD (1977) Role of bacterial lipopolysaccharide in attachment of *Agrobacterium* to moss. Plant Physiol 60:765–766

WHO (World Health Organization) (2011) Vitamin A supplementation in infants and children 6–59 months of age. Available at www.apps.who.int/iris/bitstream/handle/10665/44664/9789241501767_eng.pdf

WHO (World Health Organization) (2017) Malnutrition. Available at www.who.int/mediacentre/factsheets/malnutrition/en/

WHO (World Health Organization) (2018) Micronutrient deficiencies – vitamin A deficiency. Available at www.who.int/nutrition/topics/vad/en/

Winam SC, Ebert PR, Stachel SE, Lordon MP, Nester EW (1986) A gene essential for *Agrobacterium* is homologous to a family of positive regulator loci. Proc Natl Acad Sci U S A 83:8278–8282

Wirth J, Poletti S, Aeschlimann B, Yakandawala N, Drosse B, Osorio S, Tohge T, Fernie AR, Gunther D, Gruissem W, Sautter C (2009) Rice endosperm iron biofortification by targeted and synergistic action of nicotianamine synthase and ferritin. Plant Biotechnol J 7:631–644

Ye X, Al-Babili S, Kloti A, Zhang J, Lucca P, Beyer P, Potrykus I (2000) Engineering the provitamin A (beta-carotene) biosynthetic pathway into (carotenoid-free) rice endosperm. Science 287:303–305

Yu B, Lydiate DJ, Young LW, Schäfer UA, Hannoufa A (2008) Enhancing the carotenoid content of *Brassica napus* seeds by down-regulating lycopene epsilon cyclase. Transgenic Res 17:573–585

Zhang Y, Yin X, Yang A, Li G, Zhang J (2005) Stability of inheritance of transgenes in maize (*Zea mays* L.) lines produced using different transformation methods. Euphytica 144:11–22

Zhang C, Wohlhueter R, Zhang H (2016) Genetically modified foods: a critical review of their promise and problems. Food Sci Human Wellness 5:116–123

Zhu C, Naqvi S, Breitenbach J, Sandman G, Christou P, Capell T (2008) Combinatorial genetic transformation generated a library of metabolic phenotypes for the carotenoid pathway in maize. Proc Natl Acad Sci U S A 105:18232–18237

Ziemienowie A, Tinland B, Bryant J, Goeekler V, Hohn B (2000) Plant enzymes but not *Agrobacterium* VirD2 mediate T-DNA ligation *in vitro*. Mol Cell Biol 20:6317–6322

Zou CQ, Zhang YQ, Rashid A, Savasli E, Arisoy RZ, Ortiz-Monasterio I, Simunji S, Wang ZH, Sohu V, Hassan M, Kaya Y, Onder O, Lungu O, Yaqub Mujahid M, Joshi AK, Zelenskiy Y, Zhang FS, Cakmak I (2012) Biofortification of wheat with zinc through zinc fertilization in seven countries. Plant Soil 361:119–130

Zupan JR, Zambryski PC (1995) Transfer T-DNA from *Agrobacterium* to the plant cell. Physiology 107(4):1041–1047

# Chapter 3
# Role of Microbes for Attaining Enhanced Food Crop Production

**Pankaj Sharma, Mayur Mukut Murlidhar Sharma, Anamika, Divya Kapoor, Kavita Rani, Dilbag Singh, and Monika Barkodia**

**Abstract** The global production of food crops is on increase but the perpetual accrual of demographic strain has posed a challenge toward the attainment of global food security. In spite of covering several milestones in enhancing global food production, in a total of 821 million people, one out of nine still sleeps with an empty stomach each night, and one in three has to face the evil of malnutrition. Comprehensively, the world hunger is appraised to have augmented since 2014, as assessed in terms of both percentage and the absolute number of population. Unambiguously, Africa had the highest prevalence of undernourishment, representing 27.4% of its entire population, whereas Asia accounts for 64% of the total malnourished people across the world. Surprisingly, India also serves as a home to one-fourth of all malnourished people worldwide, which makes this country a key focus for confronting the hunger on an international scale. Therefore, in a quest to increase food production, microbes can play tremendous role due to their various incredible potentials. The microflora associated with plants have fabulous potential to improve plant resilience and produce in farming systems. The judicial employment of microbes or their metabolites can heighten nutrient uptake and yield, control pests, and mitigate plant stress responses. The unhidden potential of microbes makes them potent biocontrol agents and promotes their application as biofertilizers and their role as agents for improving soil health along with their plant growth promotion attributes that warrant their employment in agroecosystems for enhancing crop production. Therefore, the present review targets different stratagems which advocate the role of microbes for enhancing the quality and quantity of the produce.

P. Sharma · D. Kapoor · K. Rani · D. Singh · M. Barkodia (✉)
Department of Microbiology, CCS Haryana Agricultural University, Hisar, Haryana, India

M. M. M. Sharma
Department of Agriculture and Life Industry, Kangwon National University, Chuncheon, Republic of Korea

Anamika
International Centre for Genetic Engineering and Biotechnology, New Delhi, India



J. Singh et al. (eds.), *Microbial Biotechnology: Basic Research and Applications*, Environmental and Microbial Biotechnology,
https://doi.org/10.1007/978-981-15-2817-0_3

**Keywords** Food security · Farm productivity · Microbial inoculant · Nutrient bioavailability · Soil fertility

## 3.1 Introduction

The realm of agriculture has to confront an expansive gamut of challenges of climatic changes, stagnant crop yield, nutrient deficiency and deterioration of soil organic matter, availability of water and dwindling of cultivable land, increasing resistance toward GMOs, and paucity of labor. Moreover, the distress of lassitude instigated by the doings of green revolution is still being suffered by the agricultural systems. The reckless fertilizer applications in the last five decades has tremendously augmented from 0.5 tons to 23 million tons in a period of time from 1960 to 2008. Although the explicit usage of chemical fertilizers has resulted in a tremendous increase of around four times in the output of food grains, the continuous decline in the organic content of the soil due to this unchecked fertilization has also resulted in a stagnant yield of certain crops (Pandey 2018). In addition to it, the degradation of land is also promoted by a number of natural as well as anthropogenic activities including the declined fertility of soil, loss of soil organic matter, erosion, and the harmful consequence of toxic chemicals. Thus, all these deeds are posing a grave threat to the global environmental status (DeFries et al. 2012). Moreover, the food system of the present world is also facing severe challenges pertaining to environmental sustainability such as biodiversity loss, climate change, food insecurity, and water scarcity (Bilali and Allahyari 2018). The perpetual increase in the human population which further leads to a subsequent increase in the global consumption of food thus leads to an increase in the demand for affordable productive lands. The 2018 Global Hunger Index (GHI) report ranks India 103rd out of the 119 countries. The score of India is 31.1 which clearly denotes the serious hunger levels in the country. The country has the 17th highest hunger levels indicating alarming levels which has raised the food safety concerns. Therefore, there is a dire need to prevent any further degradation of such lands, and the restoration of the earlier degraded lands is also seeming to be extremely important. Moreover, it is quite unblemished that the development of orthodox agricultural habits in order to meet the future food loads is neither parsimoniously nor ecologically feasible. There should be involvement and implementation of the sustainable practices of land use along with the restoration as well as protection of the deteriorated or marginal soils for safeguarding the food security for this growing population demand (Ahmad et al. 2018). The global hunger has enlarged after nearly a decade of sustained waning. The population of malnourished people has augmented surprisingly from 777 million to 815 million from 2015 to 2016, respectively, on a worldwide basis. The global prevalence of stunt growth among children below five years of age is still 23% which accounts for 155 million children worldwide. The projection of current trends will lead to stunting of around 130 million children in 2025 which is 30 million above the World Health Assembly target (FAO 2017). The exertions required to combat the hunger and malnutrition in a sustainable way will be governed by restructuring the food systems. Thus, a number of such problems can be fixed by food, but such processes require the reshaping of food systems for health, inclusion, and nutrition along with the environmental sustainability (IFPRI 2018). In this regard, the development in agricultural sector is perceived as an obligation to fight food insecurity confronted by numerous agricultural families (Goshu et al. 2013). Therefore, the quest of this modern era is a sustainability transition. In agriculture, the perception of sustainability transition appertains to a swing from a system that is strictly focusing on enhancing productivity called as agri-food system to another system which is solely put together around the extensive ideologies of sustainable agriculture (Brunori et al. 2013). Thus, in order to upsurge the food security, the vital factor is the maintenance of soil fertility and sophisticated food productivity with admiration to the ecological challenges (Nkomoki et al. 2018). Thus, the present era accentuates the dire role of adoption of eco-friendly and sustainable agricultural practices in order to uphold the soil fertility as well as productivity.

An outsized fraction of various organisms in the terrestrial bionetworks reside underneath the ground in soils and are supposed to play an important part in the ecosystem services (Prashar et al. 2014). The engagement of such microorganisms seems to be a viable approach for uplifting the status of modern agriculture in terms of advocation of environmental sustainability as well as for enhancing the production efficacy. The plant-allied microbiota have an incredible capability of improving the plant resilience as well as produces in agricultural systems. There is accumulative indication that biological technologies employing microbes or their metabolites can augment the nutrient uptake and yield, check the pest dynamics, and also alleviate the plant stress responses along with the promotion of resistance toward disease in plants (Trivedi et al. 2017). Plants have evolved into a world of microbes. The most primitive plants stretched their roots into primeval soil where they come across a territory already crowded with bacterial as well as fungal life (Heckman et al. 2001). From the very first day, plants probably started prompting the rhizospheric microbiota. In fact, plants engineer their own rhizospheric environs by discharging explicit exudates so as to advance nutrient accessibility and to interact with definite advantageous microbes (Liu et al. 2016). Surprisingly, the furthermost primitive plant lines also display their sturdy ability to modify the comparative lavishness of diverse microbial groups in the soils neighboring the rhizospheric portion (Valverde et al. 2016). The microbial group supported by different plant species is found to be precise as well as unique (Gertsson and Alsanius 2001). These distinct microbiomes have been accredited as a result of variances in the chemistry of root exudates (Rasmann and Turlings 2016) and in plant nutrient uptake rates (Bell et al. 2015). There are also leading evidences that soil microbiomes have the ability of acclimatizing to their particular crops over time which results in improved plant–microbe dealings (Berendsen et al. 2012). This approach of engineering the microbiome which is greatly host-facilitated picks out the microbial groups ultimately through the host and controls host behaviors that developed to impact microbiomes. The approach of increasing the plant fitness coupled with the elevated yields by artificially selecting upon microbiomes and consequently engineering advanced microbes with specific effects on the plants has been anticipated (Mueller and Sachs 2015). Similarly, optimized microbiota that aid the plants to develop early or flower later may perhaps be employed as inoculants for conferring drought resistance since plants are acknowledged to implement transformed flowering time in reaction to numerous abiotic stresses (Kazan and Lyons 2015). The process of engendering the host-facilitated artificial selection of microbes seems to be an inexpensive method to curb plant ailments instead of the reckless application of different pesticides as well as antibiotics, or creating genetically modified organisms. Moreover, different discoveries on the overlying "functional core microbiome" in different plant species provide strong support for cross-compatibility of microbiome transfer with phylogenetically unrelated plant species (Trivedi et al. 2017). Therefore, the different lucrative beneficial attributes of the microbes make them superior agents for improving the crop yields. They benefit the plants in numerous ways. Microbes are largely involved in upsurging the nutrient mobilization, thereby augmenting the nutrient consumption efficiency of the plants. They also have several indirect beneficial attributes like they encourage the plant growth in an indirect way by averting the growth as well as activity of pathogens. Microbes are also directly involved in the growth promotion, for example, by fabrication of phytohormones. There has been a large body of literature describing potential uses of plant-associated bacteria as agents stimulating plant growth and managing soil and plant fitness (Welbaum et al. 2004). Such beneficial microbes can be judiciously employed for increasing the growth of crop plants in order to get increased yields for feeding this ever-increasing population. Therefore, the chapter largely focuses on the different beneficial attributes of microorganisms which make them superior agents for attaining heightened crop production.

## 3.2 Microbes as Agents for Soil Rejuvenation

The three-dimensional natural body on the Earth's surface is designated as soil which is considered to be of utmost significance to several ecosystem occupations comprising biomass production and net primary productivity, climate temperance, water purification, biodegradation of toxic pollutants, storage of water and plant nutrients, and recycling of elements (Lal 2009). Soil is the foundation of agricultural as well as of natural plant populations; thus, it is the core of all the terrestrial life forms. The tinny sheet of soil casing the Earth's surface characterizes the variance amid survival and annihilation for most land-based life. Nevertheless, records of productive aptitude of soil point toward human-induced deprivation on almost 40% of the global agronomic land as an outcome of soil erosion, atmospheric pollution, widespread soil cultivation, overgrazing, land clearing, salinization, and desertification. To be sure, the deprivation coupled with damage of fruitful farming land is among the demanding environmental trepidations, outdone only by the man-made environmental complications, for instance, global climate change, depletion of the protective ozone layer, and grave deteriorations in soil biodiversity (Doran and Zeiss 2000). The theatrical intensification in the usage of chemical fertilizers in the quest of attaining optimal harvests has greatly become a fundamental constituent of the current agronomic practices. Such a recurrent and unnecessary application not only is lavish but also deteriorates the environs at a very quick frequency and thereby turns the soils incongruous for agriculture. Additionally, the soil deterioration, instabilities in configuration coupled with the practical possessions of soil microbial populations, and, subsequently, forfeiture of the soil fertility following different soil management performs have further added to the agronomic difficulties (Khan et al. 2009). The dreadful conditions of soil also have a very strong effect on human nutrition as well as health due to its hostile influences on the quality as well as quantity food production. The perpetual reduction in the harvests and agronomic fabrication intensify the food timidity that at present distresses around 854 million people internationally, and the truncated level of protein as well as of micronutrients exacerbates undernourishment and concealed hunger that touches 3.7 billion individuals, specifically children (Lal 2009). Therefore, in order to mitigate such detrimental deeds, mankind is in a dire need of a viable alternate that could have the potential to address the prevailing glitches in a more effective way and in a sustained manner too. Thus, the microbiological approaches incorporating the employment of functionally varied microbial communities as the vivacious constituents of soil ecosystems offer an economically viable option (Khan et al. 2009) for improving soil health as well as soil quality.

Soil health may be defined as the capability of the soil to function as a vigorous living organization, within the ecosystem and land-use frontiers, to withstand plant as well as animal production, preserve or augment air and water quality, and encourage plant and animal health. Soil quality, on the other hand, denotes the aptitude of the soil to accomplish numerous of these ecosystem purposes. On the contrary, soil deterioration indicates reduction in the quality and capability of the soil as a result of natural or anthropogenic perturbations. It can also be stated that soil degradation is the reduction of soil's current or prospective aptitude of performing several ecosystem functions, conspicuously the food, feed, and fiber production as an outcome of one or more deprivation progressions. The chief degradation processes comprise physical, chemical, and biological changes in the soil (Lal 1993, 1997).

The services that play a key role in shaping the rhizospheric microbial populations cannot be entirely understood without discussing their impacts on the soil surroundings. The diversity of soil ecosystems can be assumed by the fact that a gram of soil is a home for around 10,000–50,000 microbial species (Schloss and Handelsman 2006). The unique bacterial as well as fungal groups are found to be allied with soils of variable texture (Frey et al. 2004) and variable nutrient composition (Chaparro et al. 2012). The different physical parameters of soil, for instance, soil pH, are found to be largely interconnected to the presence as well as composition of microbial communities (Rousk et al. 2010) where addition of beneficial microbes to those which are already inhabiting the soil can upsurge the nutrient uptake by plants, upturn the plant growth, deliberate resistance to several kinds of abiotic stresses, and subdue disease (Chaparro et al. 2012). This living portion of soil is highly dynamic and potentially self-sustaining, which reduces the requirement for recurrent applications that thereby can circumvent the problematic infestation of pests as well as pathogens developing resistance to the dealings (Lucas 2011). The advantageous microbiota flourishing in this atmosphere have the capability of taking up the space as well as nutrients made accessible for probable pathogenic intruders on a very quick basis as compared to the pathogens (Kaymak 2011). This act of microbial "sealing off" of undefended ecological positions is coupled with the increase in the soil's capability of resisting pathogenic conquest, amplified produces, nutrient procurement, stress tolerance, and disease resistance to the plant host (Lugtenberg and Kamilova 2009). The huge microbial diversity sheltered in soil ecosystems also works as strong peacekeepers for recycling, impounding, and supplying of different nutrients to plants. The other beneficial attributes of soil microbiota encompass mineral chelation, pathogen suppression, improvement of soil aggregation, and bioremediation of the soils (Sahu et al. 2018) which constitutively improves soil health. Therefore, in order to attain strong and productive plants, the maintenance of soil quality is of utmost prominence. Besides, the soil microbiota can also be employed as the indicator of soil quality attributable to its compassion to slight changes in the soil environment subsequent to different ecological strains or natural unrests (Sharma et al. 2010). The elevations in the levels of species richness as well as diversity yield extraordinary purposeful redundancy within the soil microbiome, which further promotes its quicker recovery throughout the stress. The extraordinary functional redundancy in the diversity of soil microbiota also deliberates shield against numerous soilborne diseases (Yin et al. 2000; Nannipieri et al. 2003). The richness of microbial range results in a stable microbiome that does not permit the flourishment of pathogenic microbes. There are numerous key factors convoluted in the soil health. Freshly, the community consistency has also been acknowledged as an imperative factor in community functioning, soil health, and plant productivity. The evenness of microbial community also warrants that no individual microbial taxum is able to take over and flourish, upsetting the ecological balance (Elliot and Lynch 1994).

## 3.3 Microbes for Degradation of Toxic Xenobiotic Compounds

The completion of the twentieth century marked a significant and perpetual hike in the worldwide grain production with an increase of 700 million tons to 500 million tons now (FAO 2018). Out of the major food crops, a greater portion equaling to 80% of human ingestion is represented by the cereal crops. The production of food crops has to encounter a number of challenges, and one among them is the invasion of crops by several pests throughout the usual growth or storage of crops. For instance, China, although principally being an agriculturally dominating nation, loses 8.8% equivalent to 40 million tons of the total grain production as a result of insect pests annually (Pimentel et al. 2001). India, being a major producer of cereals, produces around 250 million tons of grain annually, but it also has to sacrifice 11–15% of the total production equaling 27.5–37.5 million per year, due to different pests and other reasons (Walter et al. 2016). The developments in organic chemistry have largely contributed to the synthesis and development of plentiful novel organic composites, mostly xenobiotics. A greater portion of such xenobiotic compounds is represented by pesticides, which largely found application in agricultural zones (Duong et al. 1997). Therefore, for the purpose of circumventing such damages, pesticides are extensively applied for checking agronomic and domestic pests. The explicit usage of pesticides prevented the significant loss of food, but it led to the widespread distribution of pesticides in the different environments along with the agronomic harvests. Therefore, such an application of pesticides is responsible for a prodigious possible threat to the environment as well as human health (Chen et al. 2007). Although these have contributed much toward the improvement of quality of the human life, their widespread use which is coupled with the generation of unavoidable concomitant waste has largely intensified the complication of releasing toxic wastes in the environment. Though much xenobiotic composites can degrade swiftly in the soil systems, some persist in the environment for an extended period of time and thereby are potentially menacing (Conant 2005). The extent of pesticide contamination is not limited up to the soil and harvests, but they also further intrude the groundwater systems in addition to the marine environment, thereby unswervingly intimidating human as well as environmental health. The presence of such xenobiotic compounds is supposed to affect the ecosystems in a number of ways, such as dwindled soil fertility, soil acidification, nitrate leaching, increased resistance in flora and fauna, groundwater as well as surface contamination, and adulteration of agrarian soils (Kumar et al. 2018). A vast array of limitations caused by these xenobiotic compounds as well as the increasing environmental concerns coupled with the promotion of sustainability in agroecosystems have led to banning of several classes of pesticides in preceding years. Therefore, with the intention of solving the ambiguity amid elevated yields and their ill effects, the quest for alternate approaches is largely under consideration. There can be several ways like pesticides having lower harmfulness and extraordinary competence which also yields a lower residual pesticide ought to be established and industrialized (Kumar et al. 2018) which seems to be a hellacious job. On the flip side, the biological methods targeting degradation of residual pesticidal particles seem to be worthy of getting consideration. Studies on microbial degradation of pesticide residues originated in the 1940s, and as people pay more attention to the environment, the research on the degradation process and degradation mechanism of organic pollutants has been deeply studied (Rangasamy et al. 2018).

Bacteria in the natural world are capable of degrading the residual pesticidal particles, which seems to be an economically viable as well as ecologically responsive alternate which would not cause any kind of secondary pollution. However, the competence of microbial systems is found to be moderately slow as the natural environment is found to be multifaceted and unpredictable, which largely distress the practicability as well as efficacy of microbial degradation of pesticidal particles. Subsequently, a vast array of scientific community has devoted much time on the studies of bacterial approach of pesticide degradation and had a clear consideration of the different detoxification mechanisms of various organic pesticides. A great figure of microbes possessing the ability of degradation as well as conversion of pesticides have been isolated (Kumar et al. 2018). Therefore, the existing studies of biodegradable pesticides are principally focused on the role of microorganism in the soil, such as bacteria, fungi, and actinomycetes (Singh 2008), where the principal roles are played by bacteria and fungi. The bacterial ability of inducing mutant strains, endowed with a variety of biochemical capability to adaptive environs, and therefore allowing enhanced in-depth study makes them superior agents for bioremediation of such contaminated environments. Although a vast array of bacterial communities possess the ability of degrading pesticides, the research is largely focused on only some members of the bacterial community. A great diversity of *Pseudomonas* species encompassing *Pseudomonas stutzeri*, *P. putida*, *P. nitroreducens*, *P. aeruginosa*, and *P. fluorescence*, retrieved from different agronomic farms and polluted discharges through different sections, has established to be much effective in the biodegradation of pesticides. Some strains of lactic acid bacteria, for instance, *Leuconostoc mesenteroides*, *Lactobacillus brevis*, *L. plantarum*, and *L. sakei*, are also acknowledged for their ability of utilizing pesticides as a solitary source of carbon and phosphorus (Cho et al. 2009).

Several fungal species such as *Aspergillus niger*, *A. fumigatus*, *Cladosporium cladosporioides*, *Penicillium raistrickii*, and *A. sydowii* have also been retrieved from numerous polluted situates and are also confirmed to possess the aptitude of degrading diverse pesticides. Likewise, different genera of algae, for instance, *Stichococcus*, *Scenedesmus*, and *Chlorella*, and some cyanobacteria like *Nostoc*, *Anabaena*, and *Oscillatoria* have been recognized to have the efficiency of transforming different pesticides (Kumar et al. 2018). Therefore, the increasing evidences of the microbial capability of transforming pesticides has directed the focus of different researchers across the globe toward the exploration of microbial diversity, predominantly at the polluted sites. However, the mere occurrence of microbes is not sufficient, but an appropriate environment coupled with diverse degradation attitudes, for instance, hydrolysis and adsorption, is also required. Besides, the enzymatic solicitations targeting pesticidal degradation are also attracting much attention, and the genetically modified microorganisms have also been deliberated to upsurge the potentialities of these microbes and augment the proportions of biodegradation (Tang et al. 2009). Several academics have emphasized the occurrence of *oph* gene in the microbial systems that vitiate organophosphorus pesticides and hydrolase as the principal enzyme behind the process. The approach of using enzymes as well as the associated genes governing the degradation processes can prominently augment the consideration of the biodegradation process and therefore can largely benefit the bioremediation efforts.

## 3.4 Microbes as Biofertilizers

Plants necessitate nutrients in adequate and secure quantity in order to nurture optimally (Chen et al. 2006). The major constriction which limits crop productivity in emerging countries particularly in resource-deprived nations, that too, on a global basis, is the soil infertility. The truncated fertility does not allow the agrarian communities to get more benefit from the amended crop cultivars and more fruitful agricultural performs. The green revolution marked the unambiguous usage of fertilizers of chemical origin for augmenting plant development and production efficiency along with the replenishment of soil nutrient eminence (Mohammadi and Sohrabi 2012). Although they have contributed a lot toward the development of superior agronomic practices for obtaining an elevated level of production, their continuous usage has also awarded several, for instance, increased prices, inaccessibility of plants toward a major proportion of nutrients, and lethal and nonbiodegradable attitude, which further affects the environmental systems and turns the soil resources incongruous for farming practices. Thus, the employment of fertilizers of biological origin seems to be a competent substitute for enhancing the productivity as well as upgrading the nutrient status of agroecosystems. The approach of using biofertilizers is primarily centered upon the fertilizer's biological origin especially the microbes, counting bacteria as well as fungi. Since these are resources of biological origin, therefore, they also behave as eco-friendly elements and thus maintain the healthy status of the environment. The principal intention of using biofertilizers is to upsurge the organic contents of agricultural systems which further upgrades the structure of soil along with a reduction in the forfeiture of essential nutrients like zinc, phosphorus, iron, nitrogen, iron, and calcium (Lal and Greenland 1979). Biofertilizers are also known as "microbial inoculants" and can be usually defined as a formulation encompassing living or dormant cells of proficient microbial strains endowed with the capability of nitrogen fixation, phosphate solubilization, or cellulolytic microbes as well and are often employed for seed application, soil, or composting zones with the prime aim of augmenting the population of these microbes, along with the acceleration of definite microbial practices for improving the degree of accessibility of nutrients in a definite form which is effortlessly obtainable by the plant systems (Giri et al. 2019). Biofertilizers function as a basis of all the nutrients owing to their capability of solubilizing multifaceted form of nutrients into soluble and easily accessible form (Singh et al. 2018). There are numerous factors which should be strictly taken into consideration before the formulation of any kind of biofertilizer, such as the growth sketch of microorganism, optimal conditions required for growth, and preparation of inoculum. Furthermore, the survival in carrier method along with its efficacy in the field conditions is inevitable for preparing any kind of biofertilizer.

The marketable biofertilizers are established by coating of numerous bacterial members such as *Rhizobium, Azotobacter, Bacillus, Azospirillum,* and *Pseudomonas* on the seeds, and this process is said to be bacterization. These microbes secrete several compounds which assist in making their formulations. For example, azotobacterin is secreted by *Azotobacter chroococcum*, and phosphobacterin is secreted by *Bacillus megaterium* (Kumar and Bohra 2006). It is not a mandatory process that the bacteria will surely make a symbiotic relationship for benefiting the plants, but it also boosts the development of lateral root hairs which further aids in an enhanced level of mineral and water uptake, also upsurges nitrogen accessibility, and discharges numerous plant growth hormones and other growth motivating factors which collectively direct the plant to increase its photosynthetic capability, thereby eventually improving nutrient eminence of the plants.

The microbes that are mainly exploited as components of biofertilizer are nitrogen fixers, potassium solubilizer and phosphorus solubilizer. Most of the bacteria included in biofertilizer have close relationship with plant roots. Rhizobium has symbiotic interaction with legume roots, and rhizobacteria inhabit on root surface or in rhizosphere soil. Nitrogen is considered to be a very vital element for all living systems, and its unavailability strongly retards the plant growth. The rhizobial biofertilizers have got the unique ability of fixing around 50–150 N/ha/annum. Therefore, the process of biological nitrogen fixation (BNF) is deliberated as an essential process for the maintenance of nitrogen stability in soil environments. Thus, BNF is considered to be a unique way of transforming the elemental and unavailable nitrogen into a form that is easily accessible to plant (Gothwal et al. 2007). This approach is being utilized continuously by employing biofertilizers in the legumes and other crops to uplift their yield as well as quality (Kannaiyan 2002). The biofertilizers for nitrogen fixation are alive microbial inoculants capable of fixing atmospheric nitrogen either in a symbiotic way, for instance, *Rhizobia*, *Frankia*, and *Azolla*, or in an associative or free-living forms like *Azospirillum* and *Azotobacter* (Gupta 2004). There is a vast array of nitrogen-fixing microbes allied with nonleguminous plants which comprises species of *Achromobacter, Rhodopseudomonas, Alcaligenes, Rhodospirillum, Methylosinus, Arthrobacter, Xanthobacter, Klebsiella, Mycobacterium, Acetobacter, Corynebacterium, Azomonas, Herbaspirillum, Beijerinckia, Bacillus, Clostridium, Desulfovibrio, Enterobacter, Lignobacter, Erwinia, Derxia, Campylobacter*, and *Mycobacterium* (Wani 1990). Although a great number of microbes have been retrieved from the rhizospheric portion of cereal crops, the members of *Azotobacter* and *Azospirillum* are largely employed under field conditions.

The other group of microbes which are most often used are phosphate-solubilizing microorganisms. The application of such microbes especially bacteria and fungi makes available the insoluble forms of phosphorus to the plant systems (Gupta 2004). A number of bacteria existing in soil along with the population of some fungi solubilize the phosphate by secretion of organic acids (Gupta 2004). The secretion of such acids brings down the level of soil pH and thereby leads to the solubilization of the complex phosphatic forms. Therefore, these phosphate-solubilizing bacteria (PSB) are greatly acknowledged for transforming the inorganic and unavailable phosphorus to the soluble forms, i.e., $HPO_4^{2-}$ and $H_2PO_4^-$. This solubilization process is mediated through a myriad of mechanisms which include organic acid secretion, chelation, and ion exchange reactions as well. Consequently, the employment of PSB in agricultural systems can be used to combat the increasing manufacturing costs of phosphate fertilizers and would surely contribute in mobilization of the insoluble fertilizers to the plant bodies (Chang and Yang 2009; Banerjee et al. 2010). The most promising soil bacterial groups are represented by the ecto-rhizospheric species of *Bacillus* as well as *Pseudomonas* along with some endosymbiotic rhizobia for operative phosphate solubilization (Igual et al. 2001). Different microbes which display effective solubilization of phosphate are represented by *Pseudomonas*,

*Rhizobium*, *Bacillus*, and *Enterobacter* accompanied by some fungal members such as *Penicillium* and *Aspergillus* (Whitelaw 2000). *Bacillus megaterium, B. polymyxa, B. subtilis, B. circulans, B. sircalmous, Pseudomonas striata*, and *Enterobacter* are among the most potent strains of phosphate solubilizers (Subbarao 1988). Several species of *Bacillus*, for instance, *Bacillus mucilaginous*, have the capability of potassium solubilization also. Therefore, biofertilizers play an indispensable role in upsurging the productivity of food crops by the possession of numerous beneficial attributes and thus owe the immense capability of either replacing the synthetic fertilizers in part or a total replacement that can be carried out by the means of several targeted approaches.

## 3.5 Microbes as Biocontrol Agents

It is a tenacious matter that the global farming community has to experience a great economic loss due to the massive quantity of plant pathogens which varies from the minutest viroid comprising merely of a single-stranded RNA to further multifarious pathogens, for instance, viruses, bacteria, fungi, oomycetes, and nematodes, which are responsible for numerous significant plant diseases and thereby are also accountable for chief harvest damages. Even though there is a greater number of other factors also which are also held responsible for reduction in crop production, the damage caused by pests and pathogens plays an influential role in the harms on a global basis (Roberts et al. 2006). The plant ailments are responsible for an appraised loss of 40 billion dollars globally on an annual basis, either in a direct manner or in an indirect way. The pathogenic infection are liable for around 20–40% of fatalities in the crop yields (Savary et al. 2012). The magnitudes of the plant diseases vary from foremost destructions to the slight pains. Several plant diseases are found to be exceedingly damaging and disastrous on a great scale. For instance, the potato late blight caused by the fungi *Phytophthora infestans* resulted in food scarcities which led to a million deaths and relocation of around 1.5 million people from Ireland in the year 1840 (Donnelly 2002). The yearly harms of potato, the fourth largest food crop, as a result of late blight are conservatively assessed to be around US$6.7 billion per year. The other major notable example is of the disease brown leaf spot of rice which is instigated by *Helminthosporium oryzae* and is found to be prevailing largely in Asia, Africa, South America, and the USA. It was also a major rice fungal disease of historic attention (Padmanabhan 1973). It was responsible for causing austere destruction by plummeting rice harvests which resulted in the death of around two million people in Bengal as an outcome of the catastrophic scarcity in the year 1940s (Tatum 1971; Ullstrup 1972). The thing of utmost concern for mankind is the unparalleled current tendency of new fungus and fungus-resembling pathogenic signals that have amplified by over and above seven times since 2000 (Evers et al. 2007). The continued agricultural practices of precise monocultures improved global trade, and the practice of growing only some restricted cultivars is largely responsible for this increase in the population of pathogenic microbes.

These acts also endorse the development and progression of pathogenic strains with increased virulence, repeatedly with accumulative tolerance toward pesticides which not only distress the agricultural productivity but also affect several native wild species (Ab Rahman et al. 2018). Therefore, the mounting global population necessitates an effectual management as well as resistor for plant diseases in food crop production. Crop defense always plays an imperative part in shielding crop output contrary to competition from pathogens (Oerke and Dehne 2004). The modern agronomic practices now necessitate a robust impulse for mounting low-input and further sustainable agrarian practices that comprise substitutes to the chemicals employed for monitoring pests and diseases, which act as a chief and dynamic element responsible for substantial fatalities in agricultural production. The ever-increasing adverse effects of chemicals used for pest control on human health, the ecosystem, and the living creatures have directed the researchers' focus on potent biological control microorganisms as doable substitutes for managing pests and plant pathogens (Ab Rahman et al. 2018). There are ever-increasing confirmations that validate the capability of leaf as well as root-allied microbes to intensify the plant efficacy as well as yield in the cropping systems. It is much imperative to comprehend and appreciate the part of such microbes in stimulating growth and monitoring plant diseases as well; however, their presentation as biopesticides in the field conditions is still unpredictable (De Silva et al. 2019).

The employment of beneficial biocontrol microorganisms can be a viable option that could display potential of competing with the pathogenic microorganisms or can provide benefits by directly antagonizing pathogens by secreting antimicrobial composites (Mansfield 2000). The indigenous infection by pathogenic microbes can lead to the induction of systemic acquired resistance, but the capability of nonpathogenic rhizobacteria to prime plants for induced systemic resistance alongside different pathogens is much crucial and therefore is of substantial interest (Choudhary et al. 2007). This priming offers a better preparedness to plants that are able to respond faster and stronger to pathogen attack. The method of biologically controlling plant disease seems to be the preeminent choice for the manufacture of economically viable, environmental-friendly, and sustainability-promoting approach for shielding plants and other crops. Therefore, the biocontrol microorganisms are now widely being acknowledged as important tools for controlling plant diseases for an enhanced crop production in the sustainable agriculture (Azcón-Aguilar and Barea 1997). There is a great diversity of microbes which can be utilized as potent biocontrol agents. Nevertheless, a sound exploration of the multifaceted interactions amid plants, the environment, and the pathogens is quite indispensable for further consideration as the results may vary if the plants are already under an extensive load of disease (Mirzaee et al. 2015). The principal knowledge of piloting several research projects on the biocontrol was primarily meant for reduction in the dependence on the usage of agrochemicals due to several detrimental effects they lay on human health as well as on the environment. It was mainly accelerated around three decades due to increasing interest in the employment of useful microbes for suppressing of plant diseases, comprising the infections by plant parasitic nematodes as well (Cook and Baker 1983). The resultant biocontrol activity is often produced as a result of compound dealings, for instance, clampdown of the pest by use of other organisms or the use of antagonistic microbes to combat diseases and the introduction of host-specific pathogens. The employment of natural yields and chemical extracts of ordinary or modified microbes or gene products is another example of biological control. There is a vast array of interactions among the microbial populations such as mutualism, protocooperation, commensalism, neutralism, competition, amensalism, parasitism, and predation. All these biological control interactions amid plants and microorganisms happen unsurprisingly at a macroscopic as well as microscopic level (Gardener and Fravel 2002; Pal and Gardener 2006). The microbes inhabiting the rhizospheric portion of plants are deliberated to be superior biocontrol agents as the rhizospheric soil is considered to be microbiologically oppressive to numerous pathogens, attributable to the capability of this portion acting as a line of frontline defense against innumerable pathogenic occurrences. The occupation of roots by diverse beneficial microorganisms distributes their pathogen-alienating metabolic products into the root atmosphere where they result in direct or indirect suppression of pathogens (Shoda 2000). There is also the involvement of the phenomenon of antibiosis that happens as an outcome of the emission of diffusible volatile organic compounds, antibiotics, and toxins, along with the formation of extracellular cell wall-humiliating enzymes, for instance, pectin methylesterase, β-1,3-glucanase, chitinase, and β-xylosidase (Shoda 2000; Compant et al. 2005). The root systems of plants also deliberate a biological environment for progression of soil microbes that flourish on different root exudates as well as lysates which are utilized by the microbes as nutrients. The diversity of endophytic as well as free-living rhizobacteria harbored by the plants in their rhizospheric area utilize the nutrients secreted by the plant systems via their roots, and therefore grow and proliferate to secrete various metabolites in the soil systems which are found to control various plant diseases triggered by fungal or bacterial activities (Gray and Smith 2005; Kiely et al. 2006). In addition to it, the inhabitation of rhizospheric slot by different plant growth-promoting bacteria is further supported by the secretion of various kinds of allelochemicals, for instance, antibiotics, iron-chelating siderophores, biocidal volatiles, lytic enzymes, and detoxification enzymes (Glick 1995; Sturz and Christie 2003). Allelochemicals are also examples of secondary metabolites that are produced either in a direct way or in an indirect manner by plants and released into the root region as an outcome of various chemical and biochemical reactions (Tang et al. 1989; Shaw et al. 2006); however, they can also be released by the allied fungal and bacterial inhabitants. Several bacterial agents such as the nonpathogenic *Pseudomonas* and rhizobacteria are endowed with the potential of inducing systemic resistance in plants which honors the plant with the ability of protecting itself against a great deal of pathogenic viruses, bacteria, and fungi (Pieterse et al. 2014). Therefore, it seems that plants might have evolved their own language that makes them capable of interacting with their allied microbiota by secreting a greater diversity of chemical compounds via their leaves as well as roots. This phenomenon aids the plant systems to attract and select precise microbial population in the rhizospheric as well as phyllospheric environments that can offer explicit assistances that are desirable for the plant systems (Vorholt 2012).

## 3.6 Biofortification of Crops Using Microbes

The monster of undernourishment is deliberated among the supreme exalted universal challenges to mankind. It bothers almost a billion or more of the global people in both advanced and emerging nations. The disgrace of malnutrition takes account of food associated prolonged diseases in addition to the explicit nutrient paucities which further are held responsible for morbidity as well as abridged physical and psychological development. The occurrence of micronutrient paucity in human beings is largely reported, and the members of emerging nations mainly rely on the food from the staple crops which are further described by abridged bio-obtainability of vital micronutrients. The pervasiveness of such micronutrient paucities further intensifies the menace of widespread encumbrance of illness in different low- as well as middle-income nations (Black 2014), and the members of poor families are not able to manage expensive foods enriched with nutrients. As per the reports of United Nations System Standing Committee on Nutrition, the micronutrient hunger is deeply allied with over and above 50% of the total child mortality, and it further exhibits the leading risk factors for maternal death cases. Iron (Fe), zinc (Zn), and selenium (Se) are deliberated to be the micronutrients of utmost importance, and their intake is apposite for sustaining numerous life processes (UNSSCN 2004). The scarcity of any of the micronutrients has various deleterious effects on human health which are conveyed via numerous diseases. Zinc is an indispensable micronutrient for almost all the living entities comprising human beings, and its structural role in different proteins also makes it a significant micronutrient. Its deficiency is at utmost prevalence which results in abundant health-associated concerns, for instance, growth weakening, increased vulnerability toward different infections, diarrhea, retarded growth, delayed recovery of wounds, skeletal aberrations, and amplified danger of abortion (Salgueiro et al. 2000). On the other hand, the paucity of iron provokes nutritional anemia and also results in damaged working of immune system among the children along with the diminished neurocognitive growth (Murray-Kolb 2013). Selenium has numerous indispensable roles to play in a vast array of metabolic paths. Its deficiency is largely responsible for several heart diseases, reduced male fertility, hypothyroidism, weakened immune system, and high risk of infections, cancer, oxidative stress-related conditions, and epilepsy (Hatfield et al. 2014). The deficiency of different micronutrients also results in a weakened DNA molecule prone to easy damage. Therefore, there are ever-increasing endeavors targeting an enhanced assimilation of different micronutrients in plant systems.

Conversely, numerous approaches to augment the mineral elements uptake as well as food fortification haven't been fruitful every time. The approach of biofortification is only allied with entrusting the nutrient accumulation in the plant cells which makes it different from the "standard fortification" which encompasses practice of additives with the foods (Khan et al. 2019). The microbes having benefit effects on plant growth promotion are acknowledged for fortification of micro- as well as macronutrient concentrations in the essential food crops. There is a large number of mechanisms responsible for the fortification process, for instance, phosphate solubilization, zinc solubilization, siderophores production, and nitrogen fixation. The introduction of potent microbes in consort with the mineral fertilizers have the capability of enhancing the mineral uptake in turn growth as well as yield. Consequently, the biofortification of different food crops as an outcome of plant growth-promoting microbiota has been encouraged as a unique stratagem not only for upsurging the level of micronutrients in eatable food crops but also to elevate the output on soils having lesser fertility (Khan et al. 2019). Microbes are unseen soil engineers that are responsible for maintaining the status of a healthy soil and for the construction of a center for diverse biogeochemical cycles (Gadd 2010). A large number of microbes inhabiting soil ecosystems, for instance, bacteria, cyanobacteria, actinomycetes, and mycorrhiza, provide an environmental pleasant attitude for upgraded nutrient uptake along with the enhancement in plant growth. Microbes, more precisely the plant growth-promoting rhizobacteria (PGPR), are settled in the portion of rhizosphere and are capable of competently colonizing the plant roots, and they award the plants with many unique and beneficial attributes (Prasad et al. 2015). These microbes adopt diverse mechanisms for playing their central part in upsurging the nutrient consumption efficacy of the plants. These microbes are uniquely capable of playing an imperative role in the mineralization of organic matter coupled with the biotransformation of several inorganic nutrients which imparts these microbes with the innovative ability of biofortification. There are several other characteristics of these microbes, for instance, solubilization, chelation, and oxidation/reduction which have the aptitude of directly influencing nutrient accessibility (Khan 2005; Bonfante and Genre 2015). Besides obtaining improved crop yields, the present-day agricultural systems are mainly centered for producing nutritious safe food crops endowed with enriched micronutrient level especially in the edible part of the plant. Since the human population is predominantly reliant on diets grounded on staple food crops, therefore, consumption of foods with deprived or reduced concentration of micronutrients is greatly responsible for severe health concerns in human beings. The paucity of these micronutrients (selenium, zinc, iron, manganese, copper, and vitamins) in plant systems as well as in human beings is described as "hidden hunger" (Sharma et al. 2016) and imparts the risk of malnutrition among the global mankind. Consequently, the execution of the attitude of biofortification seems to be an imperative as well as lucrative method for endowment of distinguished and dominant solution for the production of crops having higher concentrations of required micronutrients. The plant growth-endorsing microbes are reportedly known for the biofortification of micronutrient components in different food harvests above and beyond the improvement in the soil productiveness as well as in the crop yield (Rana et al. 2012).

Iron, an important micronutrient, is present in the oxidized states owing to the prevalence of oxygen-rich environments. Different bacterial as well as fungal inhabitants of the soil ecosystems have evolved various unique mechanism for iron sequestration, for instance, the production of some lower-molecular-mass compounds called as siderophores which are immensely compassionate concerning $Fe^{3+}$ ions. These siderophores are eventually up taken by the plant systems via roots for conveyance of the sequestered iron to the plants. Hence, the siderophores of microbial origin are enormously capable of augmenting plant growth as a result of improved uptake of iron accompanied with inhibition of the pathogenic microbes by dint of competitive advantages (Srivastava et al. 2013). A diverse array of bacterial soil inhabitants have been largely acknowledged for the siderophore fabrication, viz., *Rhodospirillum, Bacillus, Azospirillum, Burkholderia, Pseudomonas, Azotobacter, Serratia, Arthrobacter, Enterobacter,* and *Rhizobium*, along with different fungal partners like *Syncephalastrum*, *Aspergillus*, *Rhizopus*, and *Penicillium* (Khan et al. 2019). There are numerous kinds of siderophores which are often secreted by the plant-allied microbiota, for instance, catecholate, carboxylate, and hydroxymate, and their exudation diverges across different species. Additionally, the siderophore production of mixed type has also been reported by many microbial species (Wandersman and Delepelaire 2004). Thus, the iron biofortification in different food crops is greatly supported by the secretion of these siderophore molecules. The microbes allied with plant systems also implement a number of tools for zinc solubilization, for instance, chelation (Whiting et al. 2001), dropping down the pH of soil (Subramanian et al. 2009), and by means of enhancing the root progression as well as the effective absorption area of the root (Bürkert and Robson 1994). The chelation of metallic ion zinc by diverse microorganisms coupled with enhancement in its availability for plant systems is a well-acknowledged occurrence. The microbial production of zinc-chelating compounds, metallophores, which sequester the zinc, is later released at the root surfaces and thus becomes available for the plants for its biofortification in the plant parts (Saravanan et al. 2004). Diverse microbial allies of plants, such as *Enterobacter, Pseudomonas, Bacillus, Microbacterium*, and different arbuscular mycorrhizae, are unbelievably proficient for zinc solubilization from composite mixtures which subsequently improves quality as well as nutrient content of the plants and their produces. Therefore, it can be said that the microbial approach of increasing nutrient content which further improves the quality of food crops is an economically viable and sustainability-advocating practice of superior crop production.

## 3.7 Enhancement of Abiotic Stress Tolerance Ability of Plants

The most discouraging phenomenon in the production of food crops is the appearance of several kinds of abiotic stresses which appear as an outcome of the intrinsic edaphic aspects as well as due to the anthropogenic activities. These stresses may be persuaded by the soil salinity, heavy metal pollution, and the prevalence of different organic pollutants. The developing nations has to face the problems of such abiotic stresses in a major proportion of their agricultural land which expressively condense the production of different crops (Lal 2000). The different abiotic stresses allied with the edaphic elements are comprised by soil-associated constrictions, such as reduced fertility, elevated salt content, low pH, and the occurrence of toxic heavy metals in the plant rhizosphere. In addition to these chemical factors, numerous physical restraints of the soil like poor texture, compaction, rockiness (Lal 1987), and slope abruptness (Lal 1998) also strongly affect the crop production. All these factors are held responsible for determination of the water-holding capability of soil, the cation-exchange capability, and the level of root communication with the soil and thus strongly effect the plant nutrient acquirement and crop harvests (Marchner 1995). The other abiotic stresses appear as a result of increased anthropogenic activities, such as superfluous irrigation patterns, and using saline water for irrigation purposes has backed the salt content in the agricultural soils. The employment of sewage water for irrigating fields and using sewage sludge as fertilizer have led to the accretion of poisonous heavy metals in various areas of the global agricultural sector. Another omnipresent difficulty arises as a result of the haphazard usage of organic pollutants and chemicals which persist in the soil environment for a longer duration of time and potentially interrupt the symmetry of the soft soil (Selvakumar et al. 2012). With the continuously deteriorating qualitative as well as quantitative aspects of freshwater assets, and underprivileged irrigation groundwork, the drought management especially in emerging nations seems to be imperfect. The insufficiency of moisture contents in agricultural soils is not just having a direct effect on the crop yields, but it also decreases the yields by influencing the obtainability as well as the nutrient conveyance. The drought stress also affects the crop yields by altering the hormonal balance of the plant systems as it is responsible for a reduction in the levels of cytokinin as well as a hike in the abscisic acid content in the leaves, which further leads to closing of stomatal pores (Figueiredo et al. 2008). Since water acts as a medium for carrying the nutrients from the plant roots, therefore, the low moisture level is responsible for a decreased diffusion of nutrients over small distances and the mass flow of water-soluble nutrients such as nitrate, sulfate, Ca, Mg, and Si over longer distances (Barber 1995). The drought is also responsible for a reduction in the obtainability of carbon dioxide for photosynthesis that further promotes the construction of reactive oxygen species that are highly potent for damaging the DNA and may also affect the functioning of the cell membranes (Sgherri et al. 2000). The salinity stress also affects the productivity of agro-ecosystems as it does not allow the plants to extract water from the soil. Additionally, some salts penetrate the plant systems and disturb various physiological processes of the plants which may lead to plant death (Tester and Davenport 2003).

Microbes that assist plants in alleviating certain kinds of abiotic stresses are awarded with several specified purposeful attributes. The plant systems are known for accretion of ethylene hormone under the conditions of stress (Jackson 1997). The microbial production of the enzyme 1-aminocyclopropane-1-carboxylate (ACC) deaminase is responsible for the cleavage of the ACC which acts as the precursor particle of ethylene, and the production of this enzyme significantly reduces the concentration of this gaseous hormone in the plants under stress (Abeles 1973). Some PGPR strains are endowed with the capability of removing stress by the possession of certain attributes. One such is the induced systemic tolerance (IST) which is meant to house the physical as well as chemical variations induced in the plant systems as a result of microbial activities which further aids plant systems in tolerating various abiotic stresses (Yang et al. 2009). The inoculation of the plants with the

PGPR *Paenibacillus polymyxa* is reported to have an improved tolerance of the drought stress (Timmusk and Wagner 1999). Plant systems suffer a lot due to the oxidative damages under abiotic stress conditions. The production of various antioxidant enzymes by numerous rhizobacterial species helps the plants in tolerating drought stress (Bowler et al. 1992; Scandalios 1994). The drought stress also leads to variations in the soil structures which disturbs the agronomic processes and affects the crop yields in a considerable way. In this case, the microbes which are known for the secretion of exopolysaccharide (EPS) seem to be of utmost importance. The EPS fashioned by such microbes defends these microorganisms from the hostile environmental conditions and thus permits their existence. The EPS also endows these microbes with a unique property of desiccation tolerance (Hartel and Alexander 1986; Konnova et al. 2001). The EPS secretion also allows the microbes to attach and inhabit the roots in an irreversible manner owing to engrossment of a system of fibrillary material which aids in the formation of a permanent connection amid bacteria and the root surface (Sandhya et al. 2009). The plant systems receiving treatments of EPS-constructing microbes exhibit an improved confrontation to the water stress (Bensalim et al. 1998). The EPS matrix is supposed to offer a microenvironment which clutches water and retains it for a longer interval of time as compared to the surrounding atmosphere and thus confers the bacteria as well as the protection of plant roots against desiccation (Hepper 1975). The EPS fabrication is also known to enhance the permeability by augmenting the soil aggregation along with the maintenance of developed water potential nearby the roots, thus allowing an increased uptake of nutrients by the plant systems, which further results in improved plant growth apart from the defense against drought stress (Alami et al. 2000).

## 3.8 Conclusion and Future Prospects

The products of microbial origin have the capability of enhancing crop yields and are equally potent to bring out the replacement of agrochemicals as well as chemical fertilizers. They can also improve the quality of the soil polluted by the explicit usage of chemicals for enhancing productivity of agricultural systems. Microbes are auspicious tools and can also augment the qualitative aspects of food crops and thus can be of immense importance for upgrading global health status of human beings. This potential of microbes is being utilized by several industries in which products of microbial origin are being formulated as biocontrol agents and effective biofertilizer products. The lab work is often carried out to unveil unique possessions of the microbial world, but at various times, the deeds of microbial technologies often fail to prove themselves at the field conditions. The belongings of microbial merchandises are often found to be unpredictable and unreliable between diverse studies and are found to be inconsistent under variable atmospheric conditions. Several times the microbial population fails to compete with the autochthonous microbiota which is a major logjam in the large-scale implementation of the technology.

Consequently, there is a burning requirement for the improvement of selection methods and implementation practices. Apart from this, a better understanding of the communications amid inoculated microbial strains and innate microbiota is strongly required under field conditions.

## References


Ab Rahman SF, Singh E, Pieterse CM, Schenk PM (2018) Emerging microbial biocontrol strategies for plant pathogens. Plant Sci 267:102–111

Abeles FB (1973) Ethylene in plant biology. Academic, New York

Ahmad Z, Imran M, Qadeer S, Hussain S, Kausar R, Dawson L, Khalid A (2018) Biosurfactants for sustainable soil management. Adv Agron:81–130. https://doi.org/10.1016/bs.agron.2018.02.002

Alami Y, Achouak W, Marol C, Heulin T (2000) Rhizosphere soil aggregation and plant growth promotion of sunflowers by exopolysaccharide producing Rhizobium sp. strain isolated from sunflower roots. Appl Environ Microbiol 66:3393–3398

Azcón-Aguilar C, Barea JM (1997) Arbuscular mycorrhizas and biological control of soil-borne plant pathogens–an overview of the mechanisms involved. Mycorrhiza 6(6):457–464

Banerjee S, Palit R, Sengupta C, Standing D (2010) Stress induced phosphate solubilization by 'Arthrobacter' Sp. and 'Bacillus' sp. isolated from tomato rhizosphere. Aust J Crop Sci 4(6):378

Barber SA (1995) Soil nutrient bioavailability: a mechanistic approach, 2nd edn. Wiley, New York

Bell CW, Asao S, Calderon F, Wolk B, Wallenstein MD (2015) Plant nitrogen uptake drives rhizosphere bacterial community assembly during plant growth. Soil Biol Biochem 85:170–182

Bensalim S, Nowak J, Asiedu SK (1998) A plant growth promoting rhizobacterium and temperature effects on performance of 18 clones of potato. Am J Potato Res 75:145–152

Berendsen RL, Pieterse CM, Bakker PA (2012) The rhizosphere microbiome and plant health. Trends Plant Sci 17(8):478–486

Bilali HE, Allahyari MS (2018) Transition towards sustainability in agriculture and food systems: role of information and communication technologies. Inf Process Agric 5:456–464. https://doi.org/10.1016/j.inpa.2018.06.006

Black RE (2014) Global distribution and disease burden related to micronutrient deficiencies. In: International nutrition: achieving millennium goals and beyond, vol 78. Karger Publishers, pp 21–28

Bonfante P, Genre A (2015) Arbuscular mycorrhizal dialogues: do you speak 'plantish' or 'fungish'? Trends Plant Sci 20(3):150–154

Bowler C, van Montagu M, Inze D (1992) Superoxide dismutase and stress tolerance. Annu Rev Plant Physiol Plant Mol Biol 43:83–116

Brunori G, Barjolle D, Dockes A-C, Helmle S, Ingram J, Klerkx L, Moschitz H, Nemes G, Tisenkopfs T (2013) CAP reform and innovation: the role of learning and innovation networks. EuroChoices 12:27–33. https://doi.org/10.1111/1746-692x.12025

Bürkert B, Robson A (1994) Zn uptake in subterranean clover (Trifolium subterraneum L.) by three vesicular-arbuscular mycorrhizal fungi in a root-free sandy soil. Soil Biol Biochem 26(9):1117–1124

Chang CH, Yang SS (2009) Thermo-tolerant phosphate-solubilizing microbes for multi-functional biofertilizer preparation. Bioresour Technol 100(4):1648–1658

Chaparro JM, Sheflin AM, Manter DK, Vivanco JM (2012) Manipulating the soil microbiome to increase soil health and plant fertility. Biol Fertil Soils 48(5):489–499

Chen YP, Rekha PD, Arun AB, Shen FT, Lai WA, Young CC (2006) Phosphate solubilizing bacteria from subtropical soil and their tricalcium phosphate solubilizing abilities. Appl Soil Ecol 34(1):33–41

Chen S, Sun D, Chung JS (2007) Treatment of pesticide wastewater by moving-bed biofilm reactor combined with Fenton-coagulation pretreatment. J Hazard Mater 144(1–2):577–584
Cho KM, Math RK, Islam SMA, Lim WJ, Hong SY, Kim JM, Yun MG, Cho JJ, Yun HD (2009) Biodegradation of chlorpyrifos by lactic acid bacteria during kimchi fermentation. J Agric Food Chem 57:1882–1889
Choudhary DK, Prakash A, Johri BN (2007) Induced systemic resistance (ISR) in plants: mechanism of action. Indian J Microbiol 47(4):289–297
Compant S, Duffy B, Nowak J, Clément C, Barka EA (2005) Use of plant growth-promoting bacteria for biocontrol of plant diseases: principles, mechanisms of action, and future prospects. Appl Environ Microbiol 71(9):4951–4959
Conant J (2005) Pesticides are poison, a community guide to environmental health. Hesperian Foundation, CA, USA
Cook RJ, Baker KF (1983) The nature and practice of biological control of plant pathogens. American Phytopathological Society
De Silva NI, Brooks S, Lumyong S, Hyde KD (2019) Use of endophytes as biocontrol agents. Fungal Biol Rev 33:133–148
DeFries RS, Ellis EC, Chapin FS III, Matson PA, Turner BL II, Agrawal A, Crutzen PJ, Field C, Gleick P, Kareiva PM, Lambin E (2012) Planetary opportunities: a social contract for global change science to contribute to a sustainable future. Bioscience 62:603–606
Donnelly JS (2002) Great Irish Potato Famine. The History Press
Doran JW, Zeiss MR (2000) Soil health and sustainability: managing the biotic component of soil quality. Appl Soil Ecol 15(1):3–11
Duong MH, Penrod SL, Grant SB (1997) Kinetics of p-nitrophenol degradation by Pseudomonas sp.: an experiment illustrating bioremediation. J Chem Educ 74(12):1451
Elliot LF, Lynch JM (1994) Biodiversity and soil resilience. In: Greenland DJ, Szabolos I (eds) Soil resilience and sustainable land use, vol 31. CAB International, Wallingford, pp 353–364
Evers D, Bonnechère S, Hoffmann L, Hausman JF (2007) Physiological aspects of abiotic stress response in potato. Belg J Bot 1:141–150
FAO (Food and Agriculture Organization of the United Nations) (2018) Available online: http://www.fao.org/faostat/en/#data/QC
FAO (Food and Agriculture Organization of the United Nations), IFAD (International Fund for Agricultural Development), UNICEF, WFP (World Food Programme), and WHO (World Health Organization) (2017) The State of Food Security and Nutrition in the World 2017: building resilience for peace and food security. FAO, Rome
Figueiredo MVB, Burity HA, Martinez CR, Chanway CP (2008) Alleviation of drought stress in common bean (Phaseolus vulgaris L.) by co-inoculation of Paenibacillus polymyxa and Rhizobium tropici. Appl Soil Ecol 40:182–188
Frey SD, Knorr M, Parrent JL, Simpson RT (2004) Chronic nitrogen enrichment affects the structure and function of the soil microbial community in temperate hardwood and pine forests. For Ecol Manage 196:159–171
Gadd GM (2010) Metals, minerals and microbes: geo-microbiology and bioremediation. Microbiology 156:609–643
Gardener BB, Fravel DR (2002) Biological control of plant pathogens: research, commercialization, and application in the USA. Plant Health Progress 3(1):17
Gertsson UE, Alsanius BW (2001) Plant response of hydroponically grown tomato to bacterization. In: International symposium on growing media and hydroponics, vol 644, pp 583–588
Giri B, Prasad R, Wu Q-S, Varma A (2019) Biofertilizers for sustainable agriculture and environment. Springer International Publishing, Cham. ISBN 978-3-030-18932-7. https://www.springer.com/gp/book/9783030189327
Glick BR (1995) The enhancement of plant growth by free-living bacteria. Can J Microbiol 41(2):109–117
Goshu D, Kassa B, Ketema M (2013) Measuring diet quantity and quality dimensions of food security in rural Ethiopia. J Dev Agric Econ 5:174–185

Gothwal RK, Nigam VK, Mohan MK, Sasmal D, Ghosh P (2007) Screening of nitrogen fixers from rhizospheric bacterial isolates associated with important desert plants. Appl Ecol Environ Res 6(2):101–109

Gray E, Smith D (2005) Intracellular and extracellular PGPR: commonalities and distinctions in the plant-bacterium signaling processes. Soil Biol Biochem 37:395–412

Gupta AK (2004) The complete technology book on biofertilizers and organic farming. National Institute of Industrial Research Press, India

Hartel PG, Alexander M (1986) Role of extracellular polysaccharide production and clays in the desiccation tolerance of cowpea Bradyrhizobia. Soil Sci Soc Am J 50:1193–1198

Hatfield DL, Tsuji PA, Carlson BA, Gladyshev VN (2014) Selenium and selenocysteine: roles in cancer, health, and development. Trends Biochem Sci 39(3):112–120

Heckman DS, Geiser DM, Eidell BR, Stauffer RL, Kardos NL, Hedges SB (2001) Molecular evidence for the early colonization of land by fungi and plants. Science 293(5532):1129–1133

Hepper CM (1975) Extracellular polysaccharides of soil bacteria. In: Walker N (ed) Soil microbiology, a critical review. Wiley, New York, pp 93–111

Igual J, Valverde A, Cervantes E, Velázquez E (2001) Phosphate-solubilizing bacteria as inoculants for agriculture: use of updated molecular techniques in their study

International Food Policy Research Institute (2018) 2018 Global Food Policy Report. International Food Policy Research Institute, Washington, DC. https://doi.org/10.2499/9780896292970

Jackson MB (1997) Hormones from roots as signal for the shoots of stressed plants. Trends Plant Sci 2:22–28

Kannaiyan S (ed) (2002) Biotechnology of biofertilizers. Alpha Science Int'l Ltd

Kaymak HC (2011) Plant growth and health promoting bacteria. In: Maheshwari DK (ed) Microbiology monographs, vol 18. Springer-Verlag, Berlin, pp 45–79. https://doi.org/10.1007/978-3-642-13612-2_3

Kazan K, Lyons R (2015) The link between flowering time and stress tolerance. J Exp Bot 67(1):47–60

Khan AG (2005) Role of soil microbes in the rhizospheres of plants growing on trace metal contaminated soils in phytoremediation. J Trace Elem Med Biol 18(4):355–364

Khan MS, Zaidi A, Musarrat J (eds) (2009) Microbial strategies for crop improvement. Springer, Berlin

Khan A, Singh J, Upadhayay VK, Singh AV, Shah S (2019) Microbial biofortification: a green technology through plant growth promoting microorganisms. In: Sustainable Green Technologies for Environmental Management. Springer, Singapore, pp 255–269. https://doi.org/10.1007/978-981-13-2772-8_13

Kiely PD, Haynes JM, Higgins CH, Franks A, Mark GL, Morrissey JP, O'gara F (2006) Exploiting new systems-based strategies to elucidate plant-bacterial interactions in the rhizosphere. Microb Ecol 51(3):257–266

Konnova SA, Brykova OS, Sachkova OA, Egorenkova IV, Ignatov VV (2001) Protective role of the polysaccharide containing capsular components of Azospirillum brasilense. Microbiology 70:436–440

Kumar A, Bohra B (2006) Green technology in relation to sustainable agriculture. In: Green technologies for sustainable agriculture. Daya Publishing House, Delhi

Kumar S, Kaushik G, Dar MA, Nimesh S, Lopez-Chuken UJ, Villarreal-Chiu JF (2018) Microbial degradation of organophosphate pesticides: a review. Pedosphere 28(2):190–208

Lal R (1987) Managing the soils of sub-Saharan Africa. Science 236:1069–1076

Lal R (1993) Tillage effects on soil degradation, soil resilience, soil quality and sustainability. Soil Tillage Res 27:1–7

Lal R (1997) Degradation and resilience of soils. Philos Trans R Soc Lond (B) 352:997–1010

Lal R (1998) Soil erosion impact on agronomic productivity and environment quality. Crit Rev Plant Sci 17:319–464

Lal R (2000) Soil management in the developing countries. Soil Sci 165:57–72

Lal R (2009) Soil degradation as a reason for inadequate human nutrition. Food Sec 1(1):45–57

Lal R, Greenland DJ (1979) Soil physical properties and crop production in the tropics. Wiley
Liu S, da Cunha AP, Rezende RM, Cialic R, Wei Z, Bry L, Comstock LE, Gandhi R, Weiner HL (2016) The host shapes the gut microbiota via fecal microRNA. Cell Host Microbe 19(1):32–43
Lucas JA (2011) Advances in plant disease and pest management. J Agric Sci 149:91–114. https://doi.org/10.1017/S0021859610000997
Lugtenberg B, Kamilova F (2009) Plant-growth-promoting rhizobacteria. Annu Rev Microbiol 63:541–556
Mansfield JW (2000) Antimicrobial compounds and resistance. In: Mechanisms of resistance to plant diseases. Springer, Dordrecht, pp 325–370
Marchner H (1995) Mineral nutrition of higher plants. Academic, San Diego
Mirzaee H, Shuey L, Schenk PM (2015) Transcriptomics of plants interacting with pathogens and beneficial microbes. In: Genomics, proteomics and metabolomics in nutraceuticals and functional foods, vol 4, pp 525–536
Mohammadi K, Sohrabi Y (2012) Bacterial biofertilizers for sustainable crop production: a review. J Agric Biol Sci 7:307–316
Mueller UG, Sachs JL (2015) Engineering microbiomes to improve plant and animal health. Trends Microbiol 23(10):606–617
Murray-Kolb LE (2013) Iron and brain functions. Curr Opin Clin Nutr Metab Care 16(6):703–707
Nannipieri P, Ascher J, Ceccherini MT, Landi L, Pietramellara G, Renella G (2003) Microbial diversity and soil functions. Eur J Soil Sci 54:655–670
Nkomoki W, Bavorová M, Banout J (2018) Adoption of sustainable agricultural practices and food security threats: effects of land tenure in Zambia. Land Use Policy 78:532–538. https://doi.org/10.1016/j.landusepol.2018.07.021
Oerke EC, Dehne HW (2004) Safeguarding production—losses in major crops and the role of crop protection. Crop Prot 23(4):275–285
Padmanabhan SY (1973) The great Bengal famine. Annu Rev Phytopathol 11(1):11–24
Pal KK, Gardener BM (2006) Biological control of plant pathogens. The Plant Health Instructor. https://doi.org/10.1094/PHI-A-2006-1117-02
Pandey G (2018) Challenges and future prospects of agri-nanotechnology for sustainable agriculture in India. Environ Technol Innov 11:299–307. https://doi.org/10.1016/j.eti.2018.06.012
Pieterse CM, Zamioudis C, Berendsen RL, Weller DM, Van Wees SC, Bakker PA (2014) Induced systemic resistance by beneficial microbes. Annu Rev Phytopathol 52:347–375
Pimentel D, McNair S, Janecka J, Wightman J, Simmonds C, O'connell C, Wong E, Russel L, Zern J, Aquino T, Tsomondo T (2001) Economic and environmental threats of alien plant, animal, and microbe invasions. Agr Ecosyst Environ 84(1):1–20
Prasad R, Kumar M, Varma A (2015) Role of PGPR in soil fertility and plant health. In: Plant-growth-promoting rhizobacteria (PGPR) and medicinal plants. Springer, Cham, pp 247–260
Prashar P, Kapoor N, Sachdeva S (2014) Rhizosphere: its structure, bacterial diversity and significance. Rev Environ Sci Biotechnol 13:63–77
Rana A, Joshi M, Prasanna R, Shivay YS, Nain L (2012) Biofortification of wheat through inoculation of plant growth promoting rhizobacteria and cyanobacteria. Eur J Soil Biol 50:118–126
Rangasamy K, Athiappan M, Devarajan N, Samykannu G, Parray JA, Aruljothi KN, Shameem N, Alqarawi AA, Hashem A, Abd_Allah EF (2018) Pesticide degrading natural multidrug resistance bacterial flora. Microb Pathog 114:304–310
Rasmann S, Turlings TC (2016) Root signals that mediate mutualistic interactions in the rhizosphere. Curr Opin Plant Biol 32:62–68
Roberts MJ, Schimmelpfennig DE, Ashley E, Livingston MJ, Ash MS, Vasavada U (2006) The value of plant disease early-warning systems: a case study of USDA's soybean rust coordinated framework
Rousk J, Baath E, Brookes PC, Lauber CL, Lozupone C, Caporaso JG, Knight R, Fierer N (2010) Soil bacterial and fungal communities across a pH gradient in an arable soil. ISME J 4:1340–1351
Sahu PK, Singh DP, Prabha R, Meena KK, Abhilash PC (2018) Connecting microbial capabilities with the soil and plant health: options for agricultural sustainability. Ecol Indic. https://doi.org/10.1016/j.ecolind.2018.05.084

Salgueiro MJ, Zubillaga M, Lysionek A, Cremaschi G, Goldman CG, Caro R, De Paoli T, Hager A, Weill R, Boccio J (2000) Zinc status and immune system relationship. Biol Trace Elem Res 76(3):193–205

Sandhya V, Ali SZ, Grover M, Reddy G, Venkateswarlu B (2009) Alleviation of drought stress effects in sunflower seedlings by the exopolysaccharides producing Pseudomonas putida strain GAP-P45. Biol Fertil Soils 46:17–26

Saravanan VS, Subramoniam SR, Raj SA (2004) Assessing in vitro solubilization potential of different zinc solubilizing bacterial (ZSB) isolates. Braz J Microbiol 35(1–2):121–125

Savary S, Ficke A, Aubertot JN, Hollier C (2012) Crop losses due to diseases and their implications for global food production losses and food security. Food Sec 4(4):519–537

Scandalios JG (1994) Regulation and properties of plant catalases. In: Foyer CH, Mullineaux PM (eds) Causes of photo-oxidative stress and amelioration of defense systems in plants. CRC, Boca Raton, pp 275–316

Schloss PD, Handelsman J (2006) Toward a census of bacteria in soil. PLoS Comput Biol 2:e92

Selvakumar G, Panneerselvam P, Ganeshamurthy AN (2012) Bacterial mediated alleviation of abiotic stress in crops. In: Bacteria in agrobiology: stress management. Springer, Berlin, Heidelberg, pp 205–224

Sgherri CLM, Maffei M, Navari-Izzo F (2000) Antioxidative enzymes in wheat subjected to increasing water deficit and rewatering. J Plant Physiol 157:273–279

Sharma SK, Ramesh A, Sharma MP, Joshi OP, Govaerts B, Steenwerth KL, Karlen DL (2010) Microbial community structure and diversity as indicators for evaluating soil quality. In: Lichtfouse E (ed) Biodiversity, biofuels, agroforestry and conservation agriculture, vol 5. Sustainable Agriculture Reviews, Springer, Netherlands, pp 317–358

Sharma P, Aggarwal P, Kaur A (2016) Biofortification: a new approach to eradicate hidden hunger. Food Rev Int 33(1):1–21

Shaw LJ, Morris P, Hooker JE (2006) Perception and modification of plant flavonoid signals by rhizosphere microorganisms. Environ Microbiol 8(11):1867–1880

Shoda M (2000) Bacterial control of plant diseases. J Biosci Bioeng 89(6):515–521

Singh DK (2008) Biodegradation and bioremediation of pesticide in soil: concept, method and recent developments. Indian J Microbiol 48(1):35–40

Singh AV, Prasad B, Goel R (2018) Plant growth promoting efficiency of phosphate solubilizing Chryseobacterium sp. PSR 10 with different doses of N and P fertilizers on Lentil (Lens culinaris var. PL-5) growth and yield. Int J Curr Microbiol App Sci 7(05):2280–2289

Srivastava MP, Tewari R, Sharma N (2013) Effect of different cultural variables on siderophores produced by Trichoderma spp. Int J Adv Res 1:1–6

Sturz AV, Christie BR (2003) Beneficial microbial allelopathies in the root zone: the management of soil quality and plant disease with rhizobacteria. Soil Tillage Res 72(2):107–123

Subbarao NS (1988) Phosphate solubilizing microorganism. In: Biofertilizer in agriculture and forestry. Regional Biofert. Dev. Centre, Hissar, India, pp 133–142

Subramanian KS, Tenshia V, Jayalakshmi K, Ramach V (2009) Role of arbuscular mycorrhizal fungus (Glomus intraradices)(fungus aided) in zinc nutrition of maize. J Agric Biotechnol Sustain Dev 1(1):029–038

Tang C, Komai K, Huang R (1989) Allelopathy and the chemistry of the rhizosphere. Phytochem Ecol 9

Tang J, Liu LX, Hu SF, Chen YP, Chen J (2009) Improved degradation of organophosphate dichlorvos by Trichoderma atroviride transformants generated by restriction enzyme mediated integration (REMI). Bioresour Technol 100:480–483

Tatum L (1971) The southern corn leaf blight epidemic. Science 171:1113–1116

Tester M, Davenport R (2003) Na+ tolerance and Na+ transport in higher plants. Ann Bot 91(5):503–527

Timmusk S, Wagner GH (1999) The plant-growth-promoting rhizobacterium Paenibacillus polymyxa induces changes in Arabidopsis thaliana gene expression: a possible connection between biotic and abiotic stress responses. Mol Plant Microbe Interact 12:951–959

Trivedi P, Schenk PM, Wallenstein MD, Singh BK (2017) Tiny microbes, big yields: enhancing food crop production with biological solutions. J Microbial Biotechnol 10(5):999–1003

Ullstrup A (1972) The impacts of the southern corn leaf blight epidemics of 1970–1971. Annu Rev Phytopathol 10:37–50
United Nations System Standing Committee on Nutrition (UNSSCN) (2004) 5th report on the world nutrition situation nutrition for improved development outcomes. SCN, Geneva
Valverde A, De Maayer P, Oberholster T, Henschel J, Louw MK, Cowan D (2016) Specific microbial communities associate with the rhizosphere of Welwitschia mirabilis, a living fossil. PLoS One 11(4):e0153353
Vorholt JA (2012) Microbial life in the phyllosphere. Nat Rev Microbiol 10(12):828
Walter GH, Chandrasekaran S, Collins PJ, Jagadeesan R, Mohankumar S, Alagusundaram K, Ebert PR, Daglish GJ, Nayak MK, Mohan S, Srivastava C (2016) The grand challenge of food security: general lessons from a comprehensive approach to protecting stored grain from insect pests in Australia and India. Indian J Entomol 78:7–16
Wandersman C, Delepelaire P (2004) Bacterial iron sources: from siderophores to hemophores. Annu Rev Microbiol 58:611–647
Wani SP (1990) Inoculation with associative nitrogen fixing bacteria: role in cereal grain production improvement. Indian J Microbiol 30:363–393
Welbaum GE, Sturz AV, Dong Z, Nowak J (2004) Managing soil microorganisms to improve productivity of agro-ecosystems. Crit Rev Plant Sci 23(2):175–193
Whitelaw MA (2000) Growth promotion of plants inoculated with phosphate solubilizing fungi. Adv Agron 69
Whiting SN, de Souza MP, Terry N (2001) Rhizosphere bacteria mobilize Zn for hyperaccumulation by Thlaspi caerulescens. Environ Sci Technol 35(15):3144–3150
Yang J, Kloepper JW, Ryu CM (2009) Rhizosphere bacteria help plants tolerate abiotic stress. Trends Plant Sci 14:1–4
Yin B, Crowley D, Sparovek G, De Melo WJ, Borneman J (2000) Bacterial functional redundancy along a soil reclamation gradient. Appl Environ Microbiol 66:4361–4365

# Chapter 4
# Beneficial Microbes as Alternative Food Flavour Ingredients for Achieving Sustainability

**Priyanka Roy, Intelli Kaur, Simranjeet Singh, and Vijay Kumar**

**Abstract** Microbial fermentation is an ancient concept, as old as human history, and flavour production by fermentation is also an old concept, as old as the practice of wine production. Wine production is one of the oldest arts of fermentation to be concerned with delicacy of food flavour and aroma compounds. Aromatic compounds have multipurpose uses in the food, feeds, cosmetics, pharmaceutical, and toiletries industries. The liquid and solid by-products can be used as a substrate in fermentation industries, which in turn manages the environmental waste. Industrial solid waste materials such as apple pomace, coffee husks, sugarcane bagasse, and sugar beet pulp are used as solid substrates in solid-state fermentation. Similarly, industrial liquid wastes such as corn steep liquor, blackstrap molasses, and potato and beet molasses are used as substrates for liquid-state fermentation. In both types of fermentation, the waste material as the substrate is the sole source of carbohydrates. In addition, however, that waste also contains amino acids, minerals, and vitamins, leading to variation in flavour production although the microbial culture remains the same. This chapter focuses on the gap area of microbial flavour production using food industry by-products as a substrate for solid-state fermentation, providing future opportunities for the food, pharmaceutical, cosmetics, and toiletry industries to produce high-scale industrial bioflavours that promise sustainability for future generations.

---

P. Roy · V. Kumar (✉)
Department of Basic and Applied Sciences, National Institute of Food Technology Entrepreneurship and Management, Sonipat, Haryana, India
e-mail: vijay.kumar@niftem.ac.in

I. Kaur
Department of Nutrition & Dietetics, National Health Services, Edinburgh, UK

S. Singh
Department of Biotechnology, Lovely Professional University, Phagwara, Punjab, India

Punjab Biotechnology Incubators, Mohali, Punjab, India

Regional Advanced Water Testing Laboratory, Mohali, Punjab, India



J. Singh et al. (eds.), *Microbial Biotechnology: Basic Research and Applications*, Environmental and Microbial Biotechnology,
https://doi.org/10.1007/978-981-15-2817-0_4

**Keywords** Sustainable flavour · Aromatic compounds · Solid-state fermentation · Liquid-state fermentation · Industrial by-products

## 4.1 Introduction

With the continuous increase in the human population, various countermeasures have been developed and proposed to enhance food production. To overcome this growing challenge we need efficient agricultural management techniques, reduction in food wastes, and changes in dietary patterns (Springmann et al. 2018). Most industrially processed foods such as puddings, custards, fillings, and ice cream use chemically synthesized flavours to obtain unique blends of buttery, chocolatey, and fruity flavour. However, although chemical flavouring is economical, it is deficient in enantioselectivity. Chemical flavourings are incorporated with the metabolism of the human body, causing several mutagenic effects. Flavour compounds isolated from chemical or plant sources are well known, but flavour compound production by fermentation or biotechnological processes has also gained attention in recent decades. Studies have shown natural flavours have high specificity with the least or no toxic effect. Chemical flavourings nowadays are being replaced by flavours isolated from microorganisms. In microbial flavour production in situ, microbes are inoculated into dairy-based food items such as cheese, yoghurt, and buttermilk originated by microbial processes. In ex situ, the production of flavours such as nutty, flowery, and fruity is carried out by a particular microbial strain and then further processed for flavour extraction. The first flavour compound was benzaldehyde, with characteristics such as flavours, smell, and taste that are similar to those of almonds (Krings and Berger 1998). Microbial bioflavours have been categorized into three classes based on their origin: bacterial, fungal, and algal. Therefore, the microbial populations synthesize bioflavours either by gene editing methods or by de novo pathways. It was also documented that microbial bioflavours show activity against various types of tumours. Alcoholic terpenes (α-terpineol) show anticancer and antitumour activity by minimizing the expression of transcription factor NF-KB3 without any damage to bodily systems (Sales et al. 2018).

Thus, the choice of natural flavours is in demand by consumers. The sensorial properties of natural flavours vary according to the variation of different enantiomers or regioisomers. The microbial bioflavour is considered as a "Natural Flavouring Substance" by article 3.2(c) of the regulation (EC) no. 1334/2008 (Bosse et al. 2013). Moreover, microbial flavour production by solid-state fermentation imparts value addition and promotes sustainability: it imparts sustainability because the raw materials used are mainly industrial waste. Solid-state fermentation (SSF) is a heterogeneous process that comprises three phases: solid, liquid, and gas (Thomas et al. 2013). This process also offers various advantages for the bulk production of enzymes and chemicals for flavouring purposes. SSF is one of the best processes for the production of secondary metabolites such as hydrolytic enzymes (El-Bakry et al. 2015), biopesticides (Ballardo et al. 2017; Monlau et al. 2015), humic acid (Motta and Santana 2014), bioplastics (Castilho et al. 2009), bioaromatic compounds (Martínez et al. 2017), biosurfactants (Jiménez-Peñalver et al. 2016; Vishal and Aniruddha 2012), and many others (Cerda et al. 2019). Moreover, the increasing food processing industries also add huge amounts of by-products as waste that does not have any potential uses, even though that waste contains potential carbon and nitrogen sources for microbial growth. As the demand for natural flavouring agents in industry is constantly increasing (Bosse et al. 2013), bioflavours from microbial sources can fulfil that demand. Moreover, consumer demands for natural flavours in foods are the result of health consciousness.

The aim of this chapter is to discuss the scope of utilization of solid wastes for developing low-cost fermentation processes for producing flavour compounds using microbial cultures. As microbial populations have a high capability to degrade lignocelluloses to simple sugars by enzymatic processes, such can be used by them for normal metabolism. SSF also promotes the growth of desirable microorganism, as well as production of secondary metabolites. Here, we consider flavour compounds as imparting sustainability by value addition using different approaches.

## 4.2 Solid-State Fermentation Is a Sustainable Approach for Microbial Bioflavouring Production

Various data have been published on the application of SSF for the production of metabolites, enzymes, flavouring agents, and spores. Agro-industry and the food industry provide various waste products in both solid and liquid form that further combine with various microorganisms for fermentation processes (Panesar et al. 2016). Enzyme production by the solid-state fermentation technique has greater benefits than commercial methods as this technique induces high productivity, generates concentrated products, and consists of simple processing system units. Another advantage of SSF is the use of agro- and food wastes as they are a rich source of nutrients and carbohydrates that act as substrate for production of bulk enzymes and chemicals (Saithi and Tongta 2016). Besides these fermentation processes, SSF is also helpful in minimizing the pollution problems by avoiding waste disposal into landfills, causing more ecosystem pollution (Behera and Ray 2016).

Every year a large amount of waste is generated by the food industries. Most of this waste is used as animal feed or simply burned, being mostly fruit pulp, peel, pomace, and bagasse, although after juice extraction from fruit the solid waste also contains some nutrients that can be used as substrate for SSF for production of valuable flavour compounds, which is a good approach for sustainable flavour production. Food industrial waste can be a better choice for fermentation as it provides carbon and nutrient sources combined with amino acids, salts, and minerals, as well as providing a solid support for microbial growth. The sugar industries in a region generated sugarcane bagasse at about a thousand tons per year that does not have any potential usage. Similarly, agro-industrial waste such as stems, stalks, leaves, straw, husks, shells, peel, seed/stones, pulp, or stubble from fruits, legumes, or cereals is a solid waste that contains sugars, fibres, proteins, and minerals which can be used as a solid substrate for SSF. Most of the solid wastes are lignocelluloses, mainly composed of cellulose, hemicelluloses, and lignin, that can be broken down to simple sugars by the enzymatic action of certain fungi during SSF. SSF, and understanding the biology for desirable products and to optimize conditions, are current priorities of research application (Barrios-González 2012).

Solid-state fermentation (SSF) is the utilization of carbohydrate or carbon content of the solid substrate by cultivating microorganisms. In most cases these cultivated microorganisms are fungi (Motta and Santana 2014). As fungi generally have the potential to produce enzymes that can dissolve the β-1,4-glycosidic linkages present in lignocelluloses, SSF is defined as the microorganism develops on the solid surface after utilizing the nutrients present in the solid matrix and in the absence of free water (Barrios-González 2012). SSF can be classified into two types: (1) SSF on natural solid substrates and SSF on impregnated inert supports (Barrios-González and Mejia 2007). The advantage of SSF is that the quantification of the biomass and the constituent can be easily determined at any time during the fermentation process; also, the production yield is quite higher than that of submerged fermentation (Diaz-Godinez et al. 2001). (2) SSF has a broad spectrum of application in the production of various enzymes, antibiotics, biocolours, and bioflavours. The added advantage is high-level production and product extraction that is simple as compared to submerged fermentation. It was reported that the yield of chymosin (heterologous protein) increases 500 fold when produced by SSF by *Aspergillus oryzae* as compared with submerged fermentation (SmF) (Barrios-González 2012). It was also proved that SSF is time efficient and a promising technology (Robinson et al. 2001) when considering solid waste from food industry that focuses on bioconversion (Kosseva 2013). Another use of the mixture of food industrial wastes is in a biorefinery as a sustainable approach for energy production (Aggelopoulos et al. 2013). Environmental sustainability is attained when value addition occurs to the solid waste generated by the food industry that also diminishes the amount of the waste (Motta and Santana 2014).

## 4.3 Bioflavour Synthesis in Solid-State Fermentation Depends on Substrate Composition

Bioflavour is a type of secondary metabolite that a microorganism produces during idiophase. Thus, all the factors that are responsible for microbial growth and aid in normal microbial metabolism influence the flavour production. The use of traditional starter culture on fermentation of by-products does not always gives promising results because the formation of flavour products depends on various factors, such as

(1) substrate composition, (2) type of starter culture, (3) amount of inocula, (4) fermentation temperature, and (5) moisture content of the substrate. This review focuses on solid food industrial waste; thus, production of flavours by solid-state fermentation is listed. Researchers have observed that when the substrate changes, the flavour compounds also change, and the concentration of the aromatic volatile compounds changes as well, even though the microorganism remains the same. The flavour-producing capability of a microorganism also changes among species and strains even when the genus of the organism is the same. Fungi are the most investigated microorganisms that produce natural flavour and aroma compounds as secondary metabolites during their normal metabolism. The fungus most studied for the production of fruity aroma compounds is *Ceratocystis fimbriata* on SSF by utilization of various food and agro-industrial wastes. A few yeasts, such as *Kluyveromyces marxianus*, *Saccharomyces cerevisiae*, and *Candida tropicalis*, also are being explored for the production of various aroma compounds. In most of the studies the researcher used agro-industrial waste such as sugarcane bagasse, cassava bagasse, wheat bran, and coffee husks in SSF for the biosynthesis of aroma compounds. In many studies fruit juice industrial waste, such as orange peel and apple pomace, was also used in SSF for flavour compound production. In Table 4.1, a few substrates are listed with the name of the microorganism used for the flavour production.

**Table 4.1** Substrates and microbial culture used in solid-state fermentation (SSF) for bioflavour production

| Substrate | Microorganism for fermentation | References |
|---|---|---|
| Cassava bagasse | *Ceratocystis fimbriata* | Felip et al. (2017) |
| Wheat bran | *Ceratocystis fimbriata* | Felip et al. (2017) |
| Soybean, soybean meal | *Ceratocystis fimbriata* | Rossi et al. (2015) |
| Sugarcane bagasse | *Trichoderma viride* | Fadel et al. (2015) |
| Coffee husks | *Ceratocystis fimbriata* | Soares et al. (2000) |
| Brewer's spent grains (BSG), Malt spent rootlets (MSR) | *Saccharomyces cerevisiae*, *Kluyveromyces marxianus* | Aggelopoulos et al. (2014) |
| Orange peel & pulp | *Saccharomyces cerevisiae* | Mantzouridou et al. (2015) |
| Apple pomace | *Saccharomyces cerevisiae*, *Hanseniaspora valbyensis*, *Hanseniaspora uvarum* | Madrera et al. (2015) |
| Plum bran, cassava bagasse | *Kluyveromyces marxianus* | Medeiros et al. (2000) |
| Olive mill waste | *Rhizopus oryzae*, *Candida tropicalis* | Guneser et al. (2017) |
| Orange pulp | *Saccharomyces cerevisiae* | (Mantzouridou and Paraskevopoulou 2013) |
| Sugarcane bagasse, Sugar beet molasses | *Kluyveromyces marxianus* | Martínez et al. (2017) |
| Sugarcane bagasse | *Ceratocystis fimbriata* | Christen et al. (1997) |

## 4.4 Achieving Environmental Sustainability by Utilization of Agro-industrial Waste

To decrease environmental pollution, alternative methods of waste management are presently being encouraged. Valorization of solid and liquid waste is an innovative approach in this regard that reduces environmental waste and can be a source of cheap raw material for the fermentation industries. Every year tons of solid waste are generated from the food processing industries, including food peels, pulp, aqueous residues, and bagasse, that cause a major disposal challenge because many of the wastes contain non-nutritional components and have a high phenolic content. Therefore, to overcome this challenge solid waste can be used as a solid substrate for SSF. Following are listed a few cheap raw materials with their potential contribution in the fermented flavour industries. Liquid waste such as wastewater is also used for biotechnological flavour production by submerged fermentation, but such use is very limited.

### *4.4.1 Sugarcane Bagasse*

Sugarcane bagasse, the main waste of the sugar industry, has application in bioenergy production by fermentation. It has also been used as a substrate for biosynthesis of antibiotics, enzymes, and organic acids. However, sugarcane bagasse can also be used in the production of a coconut-like aroma by SSF using *Trichoderma viride* (Fadel et al. 2015). *Trichoderma harzianum* is also known to extract a coconut-like flavour from the SSF of sugarcane bagasse (Ladeira et al. 2010). In a previous study, sugarcane bagasse was used as a substrate for the production of a fruity aroma by *Ceratocystis fimbriata* (Christen et al. 1997). The esters group of aroma compounds such as ethyl acetate, ethyl propionate, ethyl butyrate, ethyl hexanoate, and 3-methylbutyl acetate give a variety of fruity aromas after fermentation. The esters group of aroma compounds has high demand in the confectionery, bakery, and ice cream industries for its sweet and pleasurable fruity aroma, such as butyl butyrate, which gives an aroma similar to apple, pineapple, and apricot (Barros et al. 2012). Some research has used amino acid supplements such as leucine with sugarcane bagasse for the intense production of fruity flavours (Christen et al. 1997). However, not only esters but also ketones, aldehyde, and alcoholic derivatives are produced during the microbial fermentation of these substrates. Ketones, especially methyl ketones, are known to enhance a cheesy flavour (Gupta et al. 2015).

### 4.4.2 Orange Peel and Pulp

Orange peel and pulp are generated in huge amounts by the food processing industries, generally the fruit juice processing and packaging industry. Orange peel waste is a major citric waste as oranges are produced, processed, and consumed in large quantities. According to the juice-making industry, waste generated by the fruit after processing is 50% of total fruit weight. Every year millions of tons of citric waste are generated that cost nearly $10.00/ton for management for environmental concerns (Mantzouridou and Paraskevopoulou 2013). However, several alternatives for direct flavour extraction and isolation of phenolic compounds are also carried out for value addition and the utilization of solid orange peel waste. Orange peel contains good amounts of fermentable sugar such as glucose, fructose, cellulose, pectin, and minerals that can support microbial growth in SSF and thus can be used for the production of microbial secondary metabolites. In many studies it was used for ethanol production by fermentation. Moreover, orange peel can also be used for aroma compound production by fermentation when *Saccharomyces cerevisiae* is used as the fermentative organism. A diversity of flavour compounds is generated, such as isomayl acetate, which gives a strong banana-like aroma; additionally, a group of ethyl esters such as hexanoate, actanoate, decanoate, and dodecanoate gives a variety of fruity aromas similar to green apple, pineapple, peach, pear, etc. Both SSF and submerged fermentation of orange peel and pulp produce similar kinds of aroma compounds when fermented with the yeast *S. cerevisiae* (Mantzouridou et al. 2015). Researchers have also obtained a similar kind of aroma compound with the fruity aroma-producing microorganism *Ceratocystis fimbriata* (Rossi et al. 2015). Solid-state fermentation of the peel from a special variety of orange known as the ‘Valencia’ orange produces valencene, used to enhance the flavour of fruity drinks (Gupta et al. 2015).

### 4.4.3 Apple Pomace

The apple is a common source of fruit everywhere in the world. More than 68.3 million tons of apples was produced in 2014 (Madrera et al. 2015). Apple pomace is the main waste after apple juice extraction for processing and cider formation. Apple pomace contains 20–30% of the initial apple weight and does not have any potential use. As it contains non-nutritional factors such as pectin and tannins, pomace thus cannot be used as animal feed, although some studies conducted for the isolation of pectin and phenolic compounds from apple pomace also were successful (Dineiro Garcia et al. 2009). Using apple pomace as a substrate for SSF using the three different yeast cultures *Saccharomyces cerevisiae*, *Hanseniaspora valbyensis*, and *Hanseniaspora uvarum* gives a high level of acetic acid and esters after 7 to 14 days of incubation. A variety of esters such as isoamyl acetate (banana flavour), ethyl hexanoate (apple flavour), methyl octanoate (orange flavour), decyl acetate (floral), ethyl decanoate (grape), etc., along with aldehyde and ketones, are synthesized after fermentation (Madrera et al. 2015).

### 4.4.4 Coffee Husk

During coffee powder manufacturing, after dry processing of the coffee cherries coffee husk is generated. Coffee husk contains non-nutritional components such as tannin and caffeine and thus does not have any use, and its disposal also creates an environmental problem in many countries. Using coffee husk as a substrate for SSF can be a good alternative for fruity flavour production by using the fungal species *Ceratocystis fimbriata* (Soares et al. 2000). After fermentation, the volatile compounds generated are ethyl isobutyrate, isobutyl acetate, and ethyl-3-hexonate, thus producing a strong pineapple aroma. As the coffee husk contains very little sugar, a glucose supplement is needed to carry out this fermentation process.

### 4.4.5 Cassava Bagasse

Cassava (*Manihot esculenta* Crantz) is a main source of starch in many countries, and it is the world's sixth most consumed food crop. Cassava fibrous matter contains a high amount of starch that is used in the manufacture of refined sugar. After making sugar, the by-product, the bagasse, also contains an ample amount of starch that can be utilized by microorganisms for fermentation and the production of valuable secondary metabolites (Pandey et al. 2000). Felipe and coworkers in 2017 reported the use of cassava bagasse for flavour compound production by using the fungal mould *Ceratocystis fimbriata.* After fermentation it produces fruity aroma compounds such as isoamyl acetate and ethyl acetate. Supplementation of cassava bagasse with a higher concentration of glucose and fermentation using *Kluyveromyces marxianus* produces ethyl acetate, which is widely used in the food industry to enhance flavour (Medeiros et al. 2000).

### 4.4.6 Olive Mill Waste

Olive mill wastes are the major waste from the olive oil industry. In the Andalusian region, a Mediterranean region of Spain, during olive season about 2.5 million tons of olive oil is produced (Alvarez de la Pueute et al. 2010). The olive oil industry generates both olive mill waste and wastewater as its by-products. The solid by-product olive mill waste obtained after extraction of olive oil by press centrifugation contains about 25–55% water, 25–50% fibre, 5–8% residual oil, 2–6% ash, and 6–10% nitrogen in the insoluble fibre fraction (Gogus and Maskan 2006). Olive mill waste has also proved to be a good substrate for microbial growth and the production of desirable secondary metabolites. When *Rhizopus oryzae* and *Candida tropicalis* are allowed to ferment olive mill waste, a variety of sweet aromatic volatile compounds are detected after fermentation by gas chromatography in the medium (Guneser et al. 2017). Many of these volatile compounds, such as diactyl and (*E,E*)-2,4-decadienal, produce a buttery aroma. Some compounds such as benzene, acetaldehyde, and 2-phenylethanol give a flower aroma similar to that of the rose, and thus are proven to produce aromatic compounds. Solid-state fermentation of olive mill waste at the optimal pH 5 can produce vanillin (Cabrera et al. 2019), which is one of the most preferred flavouring agents used to give vanilla flavour to ice cream, cakes, and biscuits.

### 4.4.7 Green Coconut Husks

Solid-state fermentation of dried green coconut husks with the action of *Phanerochaete chrysosporium* produces vanillin (dos Santos et al. 2008). Vanillin can be quite an expensive flavouring agent if produced synthetically, whereas if it is produced with SSF of food waste, this is a cheap method of production that is also environmentally sustainable.

### 4.4.8 Maize-based Waste

Solid-state fermentation of waste from maize during the production of maize flour or maize products can produce strong buttery flavours, giving a taste similar to dairy products in the presence of microbes such as *Lactobacillus*. *Pediococcus* is also known to act on maize starch to produce a dairy flavour by releasing diacetyl (Escamilla-Hurtado et al. 2005).

## 4.5 Bioflavour Synthesis in Submerged Fermentation

Biosynthesis of flavour compounds can also be accomplished by submerged fermentation techniques. However, having restrictions, this method is not used frequently. In Brazil a group of researchers studied fruit aroma production by submerged fermentation by using cassava wastewater as substrate and *Geotrichum fragrans* as the microbial culture (Damasceno et al. 2003). They used cassava wastewater, which contains cyanogenic glycoside; thus, the isolated organism *Geotrichum fragrans* shows cyanide resistance.

## 4.6 Conclusion

The increasing consumer demands for natural flavours can be fulfilled by fermented bioflavours, which can be easily produced by microbial fermentation. When comparing the two fermentation processes, that is, solid-state and submerged fermentation, solid state fermentation (SSF) seems to be the best technique for secondary metabolite production. This technique is economically viable and generates less waste material or none. Moreover, the cost of flavour production is much lower as the raw material is solid waste from the food industry. Thus, this method of food flavour production can be a revolutionary approach in the food, feeds, cosmetics, pharmaceutical, and toiletries industries for bioflavour and bioaromatic products that also promise sustainability.

## References

Aggelopoulos T, Bekatorou A, Pandey A, Kanellaki M, Koutinas AA (2013) Discarded oranges and brewer's spent grains as promoting ingredients for microbial growth by submerged and solid state fermentation of agro-industrial waste mixtures. Applied Biochem Biotechnol 170:1885–1895

Aggelopoulos T, Katsieris K, Bekatorou A, Pandey A, Banat IM, Koutinas AA (2014) Solid state fermentation of food waste mixtures for single cell protein, aroma volatiles and fat production. Food Chem 145:710–716

Alvarez de la Pueute JM, Arana JJ, Garcia-Ruiz R (2010) Composting olive mill pomace: the Andalusian experience. Biocycle 51:31–32

Ballardo C, Barrena R, Artola A, Sánchez A (2017) A novel strategy for producing compost with enhanced biopesticide properties through solid-state fermentation of biowaste and inoculation with *Bacillus thuringiensis*. Waste Manag 70:53–58

Barrios-González J (2012) Solid-state fermentation: physiology of solid medium, its molecular basis and applications. Process Biochem 47:175–185

Barrios-González J, Mejia A (2007) Production of antibiotics and other commercially valuable secondary metabolites. In: Pandey A, Larroche C, Soccol CR, Rodríguez-León JA (eds) Current developments in solid-state fermentation. New York/New Delhi, Springer Science/Asiatech, pp 262–296

Barros DPC, Azevedo AM, Cabral JMS, Fonseca LP (2012) Optimization of flavour esters synthesis by *Fusarium solani pisi* cutinase. J Food Biochem 36:275–284

Behera SS, Ray RC (2016) Solid state fermentation for production of microbial cellulases: recent advances and improvement strategies. Int J Biol Macromol 86:656–669

Bosse AK, Fraatz MA, Zorn H (2013) Formation of complex natural flavours by biotransformation of apple pomace with basidiomycetes. Food Chem 141:2952–2959

Cabrera F, Serrano A, Torres Á, Rodriguez-Gutierrez G, Jeison D, Fermoso FG (2019) The accumulation of volatile fatty acids and phenols through a pH-controlled fermentation of olive mill solid waste. Sci Total Environ 20(657):1501–1507

Castilho LR, Mitchell DA, Freire DM (2009) Production of polyhydroxyalkanoates (PHAs) from waste materials and by-products by submerged and solid-state fermentation. Bioresour Technol 100(23):5996–6009

Cerda A, Mejias L, Rodríguez P, Rodríguez A, Artola A, Font X, Gea T, Sánchez A (2019) Valorisation of digestate from biowaste through solid-state fermentation to obtain value added bioproducts: a first approach. Bioresour Technol 271:409–416

Christen P, Meza JC, Revah S (1997) Fruity aroma production in solid state fermentation by *Ceratocystis fimbriata*: influence of the substrate type and the presence of precursor. Mycol Res 101:911–919

Damasceno S, Cereda MP, Pastore GM, Oliveira JG (2003) Production of volatile compounds by *Geotrichum fragrans* using cassava wastewater as substrate. Process Biochem 39:411–414

Diaz-Godinez G, Soriano-Santos J, Augur C, Viniegra-Gonzalez G (2001) Exopectinases produced by *Aspergillus niger* in solid-state and submerged fermentation: a comparative study. J Ind Microbiol Biotechnol 26:271–275

Dineiro Garcia Y, Suarez Valles B, Picinelli Lobo A (2009) Phenolic and anti-oxidant composition of by-products from the cider industry. Food Chem 117:731–738

El-Bakry M, Abraham J, Cerda A, Barrena R, Ponsá S, Gea T, Sánchez A (2015) From wastes to high value added products: novel aspects of SSF in the production of enzymes. Crit Rev Environ Sci Technol 45(18):1999–2042

Escamilla-Hurtado ML, Valdes-Martinez SE, Soriano-Santos J, Gomez-Pliego R, Verde-Calvo JR, Reyes-Dorantes A, Tomasini-Campocosio A (2005) Effect of culture conditions on production of butter flavor compounds by *Pediococcus pentosaceus* and *Lactobacillus acidophilus* in semisolid maize-based cultures. Int J Food Microbiol 105(3):305–316

Fadel HHM, Mahmoud MG, Asker MMS, Lotfy SN (2015) Characterization and evaluation of coconut aroma produced by *Tricoderma viride* EMCC-107 in solid state fermentation on sugarcane bagasse. Electron J Biotechnol 18:5–9

Felip LO, Oliveira AM, Bicas JL (2017) Bioaromas: perspectives for sustainable development. Trends Food Sci Technol 62:141–153

Gogus F, Maskan M (2006) Air drying characteristics of solid waste (pomace) of olive oil processing. J Food Eng 72(4):378–382

Guneser O, Demirkol A, Yuceer YK, Togay SO, Hosoglu MI, Elibol M (2017) Production of flavour compounds from olive mill waste by *Rhizopus oryzae* and *Candida tropicalis*. Braz J Microbiol 48:275–285

Gupta C, Prakash D, Gupta S (2015) A biotechnological approach to microbial based perfumes and flavours. J Microbiol Exp 3(1): 1–8

Jiménez-Peñalver P, Gea T, Sánchez A, Font X (2016) Production of sophorolipids from winterization oil cake by solid-state fermentation: optimization, monitoring and effect of mixing. Biochem Eng J 115:93–100

Kosseva MR (2013) Recovery of commodities from food wastes using solid-state fermentation. In: Kosseva MR, Webb C (eds) Food industry wastes. Elsevier, Amsterdam, pp 77–102

Krings U, Berger RG (1998) Biotechnological production of flavour and fragrances. Appl Microbiol Biotechnol 49:1

Ladeira NC, Peixoto VJ, Penha MP, de Paula Barros EB, Leite SG (2010) Optimization of 6-pentyl-alpha-pyrone production by solid state fermentation using sugarcane bagasse as residue. Bioresources 5(4):2297–306

Madrera RR, Bedrinana RP, Valles BS (2015) Production and characterization of aroma compounds from apple pomace by solid-state fermentation with selected yeasts. LWT Food Sci Technol 64:1342–1353

Mantzouridou FT, Paraskevopoulou A (2013) Volatile bio-ester production from orange pulp-containing medium using *Saccharomyces cerevisiae*. Food Bioprocess Technol 6:3326–3334

Mantzouridou FT, Paraskevopoulou A, Lalou S (2015) Yeast flavour production by solid state fermentation of orange peel waste. Biochem Eng J 101:1–8

Martínez O, Sánchez A, Font X, Barrena R (2017) Valorization of sugarcane bagasse and sugar beet molasses using *Kluyveromyces marxianus* for producing value-added aroma compounds via solid-state fermentation. J Clean Prod 158:8–17

Medeiros ABP, Pandey A, Freitas RJS, Christen P, Soccol CR (2000) Optimization of the production of aroma compounds by *Kluyveromyces marxianus* in solid-state fermentation using factorial design and response surface methodology. Biochem Eng J 6(1):33–39

Monlau F, Sambusiti C, Ficara E, Aboulkas A, Barakat A, Carrere H (2015) New opportunities for agricultural digestate valorization: current situation and perspectives. Energy Environ Sci 8(9):2600–2621

Motta FL, Santana MHA (2014) Solid-state fermentation for humic acid production by a *Trichoderma reesei* strain using an oil palm empty fruit bunch as the substrate. Appl Biochem Biotechnol 172:2205–2217

Pandey A, Soccol CR, Nigam P, Soccol VT, Vandenberghe LPS, Mohan R (2000) Biotechnological potential of agro-industrial residues. II. Cassava bagasse. Bioresour Technol 74:81–87

Panesar PS, Kaur R, Singla G, Sangwan RS (2016) Bio-processing of agro-industrial wastes for production of food-grade enzymes: progress and prospects. Appl Food Biotechnol 3(4):208–277

Robinson T, Singh D, Nigam P (2001) Solid-state fermentation: a promising microbial technology for secondary metabolite production. Appl Microbiol Biotechnol 55:284–289

Rossi A, Bene MC, Carlesimo M, Garelli V, Fortuna MC, d'Alessio PA (2015) Efficacy of orange peel extract in psoriasis. Global J Dermatol Venereol 3:1–4

Saithi S, Tongta A (2016) Phytase production of *Aspergillus niger* on soybean meal by solid-state fermentation using rotating drum bioreactor. Agric Agric Sci Procedia 11:25–30

Sales A, Paulino BN, Pastore GM, Bicas JL (2018) Biogeneration of aroma compounds. Curr Opin Food Sci 19:77

dos Santos BE, Perrone D, do Amaral Vendramini AL, Leite SG (2008) Vanillin production by Phanerochaete chrysosporium grown on green coconut agroindustrial husk in solid state fermentation. BioResources 3(4):1042–1050

Soares M, Christen P, Pandey A, Soccol CR (2000) Fruity flavour production by *Ceratocystis fimbriata* grown on coffee husk in solid-state fermentation. Process Biochem 35:857–861

Springmann M, Clark M, Mason-D'Croz D, Wiebe K, Bodirsky BL, Lassaletta L (2018) Options for keeping the food system within environmental limits. Nature 562(7728):519–525

Thomas L, Larroche C, Pandey A (2013) Current developments in solid-state fermentation. Biochem Eng J 81:146–161

Vishal J, Aniruddha P (2012) Solid state fermentation (SSF) for the production of sophorolipids from *Starmerella bombicola* NRRLY-17069 using glucose, wheat bran and oleic acid. Curr Trends Biotechnol Pharm 6(4):418–424

# Chapter 5
# Microalgae as Nutraceutical for Achieving Sustainable Food Solution in Future

**Pardeep Kaur**

**Abstract** Of late, a spurt in the general awareness about the biological aspects of nutrition has been witnessed. The changing trend demands for high nutritional value products that can easily and rapidly be produced at large scales in a cost-effective manner. Microalgae constitutes a distinct group of unicellular photosynthetic organisms and a broad variety of eukaryotic algae containing a plethora of beneficial compounds such as carbohydrate, proteins, fatty acids, vitamins, carotenoids, phycobiliproteins, astaxanthin, and lutein. These compounds find application in the production of high-quality nutraceuticals that provide health benefits such as controlling blood pressure, boosting immune system, reducing coronary heart diseases, serving as anticancer agents, and acting as antioxidants. Besides, the benefits of using microalgae are its high productivity on arable and nonarable land, thus posing no threat to the agricultural crop production. Although the nutritional value and its commercialization is still in nascent stage, intense efforts are underway all over the world to explore untapped potential of microalgae that could lead to the solution of several problems through green technologies and open gateway to a multibillion dollar industry. This chapter gives an overview of microalgae and its diversity, nutritional value, and current challenges on its use as nutraceutical product.

**Keywords** Microalgae · Nutrition · Nutraceuticals · Green technology · Toxic metabolites

## 5.1 Introduction

The holistic development of any nation depends on the good health and well-being of its public. The frequent consumption of packaged and junk foods, fast eating food habitats, long-scheduled work, sedentary lifestyle, etc. has resulted in a spurt

P. Kaur (✉)
Department of Biotechnology, Sri Guru Granth Sahib World University, Fatehgarh Sahib, India



J. Singh et al. (eds.), *Microbial Biotechnology: Basic Research and Applications*, Environmental and Microbial Biotechnology,
https://doi.org/10.1007/978-981-15-2817-0_5

in the prevalence of various diseases like carcinomas and heart-related disorders (Subudhi 2017). Therefore, there is an increasing global concern on the implementation of healthy foods in diet that will help to maintain a healthy lifestyle (Ranga et al. 2017). However, huge investment and capacity building are required to affirm to the food and nutritional security that often requires huge amount of farming land, livestock, and high density of natural resources. However, the current health policies demand the consumers to fend for themselves and entice them to attain maximum health benefits by spending minimum money (Lenoir-Wijnkoop et al. 2010). Therefore, a paradigm shift from traditional agriculture toward more sustainable solutions is the need of hour. To reduce the possibility of health problems, consumers prefer to use foods rich in adequate nutrients or nutraceuticals. Nutraceutical is a combination of two terms "nutrition" and "pharmaceutical," a food stuff either in the form of fortified food or dietary supplement that has the capability to impart health benefits by strengthening the body's response to ward off the infections and to get rid of diseases. Conventionally, these products are available in varied formulations. Nutraceutical are way more advantageous than the medicines as they are devoid of side effect, natural sources, easily available, and affordable. The term "nutraceutical" comprises several products such as vitamins, minerals, enzymes, antioxidants, probiotics, prebiotics, polyunsaturated fats, polyphenols, and spices (Chauhan et al. 2013). Depending upon their functions, nutraceuticals are grouped as dietary supplements, medicinal food, functional foods, etc. (Dillard and German 2000). On the basis of source of their origin, these are classified as products obtained from plants (vitamins, phytochemicals), animal (polysaccharides), or microorganisms (poly amino acids) (Pattanaik et al. 2019). Among the microorganisms, microalgae present a sustainable powerful reserve of number of functional ingredients for wide food applications (Pulz and Gross 2004). In general, the term "algae" circumscribes photosynthetic eukaryotic organisms. On the basis of their size, they can be classified as micro- and macroalgae. Macroalgae or seaweeds are multicellular organisms that include benthic marine organisms with variable sizes extending up to several meters (Markou et al. 2012). In contrast, the term microalgae, or microphytes, are microscopic photosynthetic algae and bacteria or cyanobacteria. Microalgae also have variable sizes, which can be as small as picometers (Waterbury 2006; Barsanti and Gualtiery 2006). The ecology of microalgae is also diverse, which ranges from marine to freshwater forms, and their photosynthetic mechanism resembles that of land-based plants. The microalgae have received an appreciable interest as sources of natural food, polysaccharides, fatty acids, and biomass due to its potentially simple and cost-effective cultivation techniques (Borowitzka 2013; Leu and Boussiba 2014; Aziz et al. 2017). Another major thrust for commercialization of microalgae is their generally regarded as safe (GRAS) status that has important relevance for products intended to be used for consumption purposes (Gangl et al. 2015). In contrast to plants, microalgae have higher growth rates (e.g., one to three doublings per day) and do not pose any competition for resources as they can be grown easily in open environment (Ng et al. 2015). In recognition of innumerable advantages of microalgae, several efforts are underway to promote them at biotechnological platforms, especially from genetic engineering point of view.

## 5.2 Indian Microalgal Diversity

India owing to its rich biodiversity, diverse ecologies, agronomic practices, and multitude of soil types is considered to be ideal for microalgal growth. The biodiversity-rich parts of the country like Eastern Ghats (Jena et al. 2005, 2008; Prasanna and Kaushik 2005; Rath and Adhikary 2005; Samantraray et al. 2002) and Chilika Lake in Odisha are also home to diverse species of microalgae (Adhikary 2000; Ratha et al. 2003; Rath and Adhikary 2005) and have been widely investigated. The investigation by Ratha et al. (2012) on qualitative distribution of microalgae in diverse ecological habitats from major biodiversity hotspots of India reported *Cyanophyceae* and *Chlorophyceae* as the abundantly found algal groups. Suresh et al. (2012) studied the microalgal diversity in Western and Eastern Ghats of Tamil Nadu and reported 97 species of microalgae. Kharkongar and Ramanujam (2014) reported a total of 85 taxa, including cyanobacteria and algal species belonging to diverse classes of algae with highest subaerial algal biodiversity in sacred grove compared to those of plantation and open disturbed forest. It is well reported in literature that the subaerial microhabitats are predominantly habitated by members belonging to Trentepohliales and cyanobacteria since they can easily adapt to adverse environmental conditions by adopting different survival mechanisms, such as production of carotenoids in Trentepohliales and extracellular polymeric substances (EPS) sheath by cyanobacteria as a protective sheath (Urzi and Realini 1998; Tomaselli 2003). Singh and Sharma (2014) reported 19 microalgal taxa belonging to Cyanophyta (8 taxa), Chlorophyta (7 taxa), and Bacillariophyta (4 taxa) from Sheer Khad (stream), which is a tributary of Sutlej River, Himachal Pradesh. Severes et al. (2018) identified microalgae from Western Ghats regions of India. The panorama depicted by the microalgal biodiversity suggests their potential application in diverse fields, nutraceutical one of them, since they are proficient in producing high-value compounds even in the harsh environmental conditions.

## 5.3 Microalgae as Nutraceutical

The consumption of microalgae as a human food source or nutritional supplements is not new, but their use dates back to prehistoric times. The species of Nostoc find use in various parts across the world, where they are consumed in traditional ways. In Japan, cyanobacterium *Aphanothece sacrum* is consumed as a popular delicacy known as "suizenji-nori." The microalgal biomass has gained immense popularity for the production of health-based food products, where it is being used exclusively for this purpose for over past few decades. Depending upon the organic composition, an array of products under different names are present in the market as nutraceuticals like dietary supplements, herbal preparations/products, traditional medicine, food supplements, or botanical supplements (Nicoletti 2012; Rajasekaran et al. 2008). The first-generation food supplements consist of primary metabolites, whereas other substances, relevant to nutraceuticals, are secondary metabolites. The seaweeds contradict this simplified version of classification as they contain both the type of substances. Therefore, a new terminology "superfoods" was suggested (Bishop and Zubeck 2012). A third type of food supplement called as pharmafoods or functional foods is emerging because of their resemblance with ordinary foods in terms of physiological properties. These pharmafoods or functional foods probably can best harness the potential of microalgae in the future for a wide variety of different products after doing away with associated limitation such as poor aesthetics in the current utilizations (Nicoletti 2012).

## 5.4 Nutritional Components of Microalgae

Microalgae are repertoire of valuable compounds such as proteins, carbohydrates, polyunsaturated fatty acids (PUFAs), minerals, and vitamins that can not only heighten the nutritional content of food but also serve to provide benefits to the consumer (Table 5.1 and Fig. 5.1).

### *5.4.1 Microalgal Proteins*

The essential amino acids (EAA) composition of algae is in sync with FAO requirements. *Chlorella vulgaris*, highly rich in protein content and desirable EAA composition, is popularly used as a food supplement (Becker 2007; Chronakis and Madsen 2011). Microalgae-based proteins have several advantages over other conventional protein sources, and they have scant land utilization compared to animal-based proteins (de Vries and de Boer 2010; Van Krimpen et al. 2013; Smetana et al. 2017), low water requirement, and ability to grow in saline water, among others (FAO 2010). Many species of microalgae produce proteins on par with egg, meat and milk, etc. (Gouveia et al. 2008a). Notably, red species of algae contain low concentrations of leucine and isoleucine, while brown algae species are often limited in methionine, cysteine, and lysine (Dawczynski et al. 2007; Mišurcová et al. 2014).

### *5.4.2 Microalgal Peptides and Protease Inhibitors*

The bioactive peptides produced by the microalgae are known to possess anticancer, antiviral, antioxidant, and immunomodulatory effects. Proteases are a group of proteolytic enzymes with fine applications in food, detergents, and pharmaceutical industries. Their deregulation can result in serious health-related issues, and that is why both proteases and protease inhibitors find applications as therapeutic agents (Drag and Salvensen 2010).

**Table 5.1** Microalgal species and products

| Products | Producers | References |
|---|---|---|
| Food | *Chlorella*, *Spirulina maxima*, *Odontella aurita*, *Tetraselmis chuii*, *Aphanizomenon flos-aquae*, *Nostoc*, *Aphanothece sacrum*, *Spirogyra*, *Oedogonium*, *Haematococcus pluvialis*, *Isochrysis galbana*, *Porphyridium cruentum*, *Diacronema vlkianum*, *Scenedesmus* sp. | Valenzuela-Espinoza et al. (2002), Gantar and Svirčev (2008), Liu and Chen (2016), and Bleakley and Hayes (2017) |
| Feed | *Chlorella*, *Spirulina*, *Tetraselmis*, *Isochrysis*, *Pavlova* | Gouveia et al. (2008a), Yaakob et al. (2014), Mobin and Alam (2017) |
| *Amino acid*<br>Mycosporine-like amino acids (MAA) | *Aphanizomenon* sp., *Chlorella luteoviridis*, *Chlorella minutissima*, *Chlorella sorokiniana*, *Chlorella sphaerica*, *Scenedesmus* sp., *Stichococcus* sp. *Chlamydomonas nivalis* | Xiong et al. (1999), Duval et al. (2000), Karsten et al. (2007), Chu (2012) |
| *Polyunsaturated fatty acids (PUFAs)*<br>Eicosapentaenoic acid (EPA)<br>Docosahexaenoic acid (DHA)<br>α-Linolenic acid<br>Arachidonic acid (AA)<br>Linolenic acid | *Phaeodactylum*, *Nannochloropsis*, *Schizochytrium*<br>*Crypthecodinium cohnii*, *Isochrysis galbana*, *Pavlova salina*, *Schizochytrium*<br>*Botryococcus* sp., *Chlamydomonas moewusii*, *Chlorella vulgaris*, *Dunaliella* sp., *Micromonas pusilla*, *Muriellopsis* sp., *Nannochloris atomus*, *Pseudokirchneriella subcapitata*, *Scenedesmus acutus*, *Scenedesmus obliquus*, *Scenedesmus quadricauda*, *Tetraselmis suecica*<br>*Nannochloris atomus*, *Porphyridium boryanum*<br>*Botryococcus* sp., *Chlorella* sp., *D. bardawil*, *Tetraselmis primolecta*, *Tetraselmis tertiolecta*, *N. atomus*, *Neochloris oleoabundans*, *P. subcapitata*, *Scenedesmus obliquus*, *T. suecica* | Pereira et al. (2012), Adarme-Vega et al. (2014)<br>Fried et al. (1982), Piorreck et al. (1984), Reitan et al. (1994), D'Souza and Loneragan (1999), Arisz et al. (2000), Becker (2004), Chiang et al. (2004), Poerschmann et al. (2004), Martinez-Fernandez et al. (2006), Patil et al. (2007)<br>Reitan et al. (1994), Zhang et al. (2002)<br>Fried et al. (1982), Piorreck et al. (1984), Reitan et al. (1994), D'Souza and Loneragan (1999), Arisz et al. (2000), Zhang et al. (2002), Chiang et al. (2004), Day et al. (2009), Gouveia and Oliveira (2009), Patil et al. (2007) |
| *Vitamins* | *Chlamydomonas eugametos*, *Chlorella pyrenoidosa*, *C. vulgaris*, *Chlamydomonas reinhardtii*, *Scenedesmus acutus*, *Scenedesmus obliquus*, *Scenedesmus quadricauda*, *C. protothecoides*, *Dunaliella tertiolecta*, *Prototheca moriformis*, *T. suecica* | Uhlik and Gowans (1974), Borowitzka (1988), Vilchez et al. (1997), Carballo-Cardenas et al. (2003), Matsukawa et al. (2000), Becker (2004) |

(continued)

**Table 5.1** (continued)

| Products | Producers | References |
|---|---|---|
| *Toxins*<br>Anatoxin and saxitoxin<br>Microcystins<br>Okadaic acid | *Flos-aquae*<br>*Microcystis aeruginosa*<br>*Dinophysis* sp. | Katircioglu et al. (2004), He et al. (2005) |
| Sterols | *Pyramimonas* cf. *cordata*, *T. suecica*, *D. salina*, *D. tertiolecta* | Ponomarenko et al. (2004), Cardozo et al. (2007), Luo et al. (2015) |
| *Carotenoids*<br>Astaxanthin<br>β-carotene<br>Canthaxanthin<br>Lutein | *C. nivalis*, *Chlamydocapsa* sp., *C. nivalis*, *C. vulgaris*, *Chromochloris zofingiensis*, *Coelastrella striolata*, *Haematococcus* sp., *S. obliquus*, *Chlorella* sp., *Dunaliella* sp., *Pyramimonas* sp., *Tetraselmis* sp.<br>*Chlorococcum* sp., *Chlamydocapsa* sp., *C. emersonii*, *C. fusa*, *C. vulgaris*<br>*Neospongiococcum* sp., *Chlorococcum* sp., *Chlamydocapsa* sp. | Borowitzka (1988), Liu and Lee (2000), Lorenz and Cysewski (2000), Yuan et al. (2002), Kang et al. (2005), Remias et al. (2005), Zhekisheva et al. (2005), Jin et al. (2006), Abe et al. (2007), Chattopadhyay et al. (2008), Fujii et al. (2008), Leya et al. (2009), Chu (2012), Domínguez-Bocanegra et al. (2004)<br>Rabbani et al. (1998), Egeland et al. (1995, 1997), Matsukawa et al. (2000), Hejazi and Wijffels (2003), Barbosa et al. (2005), Abe et al. (2007), Coesel et al. (2008), and Chu (2012)<br>Sathasivam et al. (2012) and Wu et al. (2016)<br>Yuan et al. (2002), Mendes et al. (2003), Pelah et al. (2004), Bhosale and Bernstein (2005), Abe et al. (2007), Chattopadhyay et al. (2008), Coesel et al. (2008) and Leya et al. (2009)<br>Chen (1998), Egeland et al. (1995, 1997), Barbosa et al. (2005), Tukaj et al. (2003), Bhosale and Bernstein (2005), Blanco et al.(2007), Ceron et al. (2008), Cha et al. (2008), Matsukawa et al. (2000), Shi et al. (2006), Sanchez et al. (2008) |
| Polysaccharides<br>Sulphated | *Porphyridium* sp. | Delattre et al. (2016), Xiao and Zheng (2016) |
| *Phenolic and volatile compounds*<br>B-cyclostirol,<br>α- and β-ionone, neophytadiene, nopole, phytol<br>Pentadecane<br>Heptadecane | *Chlorella*, *Nostoc*, *Anabaena*, *Tolypothrix*, *Chlamydomonas*, *D. salina*, *Synechocystis* sp., *Spirulina* | Plaza et al. (2010), de Morais et al. (2015) |

(continued)

**Table 5.1** (continued)

| Products | Producers | References |
|---|---|---|
| *Phycobiliproteins*<br>Phycocyanin<br>Phycoerythrin<br>Porphyridium<br>Allophycocyanin<br>Chlorophyll A | *Spirulina*, *A. flos-aquae* | Bishop and Zubeck (2012), Sonani et al. (2016) |
| *Antioxidant* | *Nostoc ellipsosporum*, *C. nivalis*, *Phaeodactylum tricornutum* | Mendiola et al. (2005), Wang et al. (2007), Jaime et al. (2007), Li et al. (2007), Ibáñez et al. (2008), Rodriguez-Garcia and Guil-Guerrero (2008) |
| *Antibacterial*<br>Diterpenoid<br>Nostocycline A<br>Tenuecyclamides<br>*Antiviral*<br>Spirulina<br>Nostoflan<br>*Antifungal*<br>Nostodione<br>Nostocyclamide | *Nostoc commune*<br>*Nosto* sp.<br>*Nostoc spongiaeforme*<br>*Spirulina platensis*<br>*Nostoc flagelliforme*<br>*Nostodione*<br>*Nostoc commune* | Asthana et al. (2009)<br>Banker and Carmeli (1998), Ploutno and Carmeli (2000)<br>Hayashi et al. (1996, 2008)<br>Moore et al.(1988), Bhadury and Wright (2004) |

Protease inhibitors are small molecules, and usually, peptides mimic the structure of a substrate and bind to enzymes. *Chlorella vulgaris* produces pepsin-hydrolyzed peptide with strong antioxidant activity and is resistant against gastrointestinal enzymes (Sheih et al. 2009a), besides containing a peptide with antiproliferative activity (Sheih et al. 2009b). Cyanobacteria also produce a set of potent metabolites in minute quantities (Janssen 2019). *Microcystis* has been known to produce aeruginosins, which helps to curtail the thromboembolic disorders by efficiently binding to serine proteases (Ersmark et al. 2008; Wang et al. 2009). The brackish water cyanobacterium *Nodularia spumigena* (Mazur-Marzec et al. 2013) and *Anabaena compacta* (Anas et al. 2012) are known to produce Spumigins, which inhibits trypsin-like serine proteases. The anabaenopeptins-cyclic hexapeptides act against exopeptidase, carboxypeptidase A (Murakami et al. 2000). The microginins have been shown to inhibit various exopeptidases (Welker and von Dohren 2006). Microginins are considered to be potential candidates for the development of new drugs for cardiovascular diseases as they are known to inhibit angiotensin-converting enzyme (ACE-I) and leucyl aminopeptidase (LAP) (Bagchi et al. 2016). The aerucyclamides have been reported to show activity against cancer cells and parasitic infections (Ishida et al. 2000; Portmann et al. 2008). The cyanopeptolins and aeruginosins show inhibitory action against serine proteases (Gademann and Portmann 2008; Hanessian et al. 2006). Ribulose-1,5-bisphosphate carboxylase/oxygenase (RuBisCO), a heterohexadecamer structured protein, possesses functionally bioactive peptides known to cure cardiovascular diseases, diabetes, neurodegenerative disorders, and oxidative stress (Selvaraj et al. 2017).

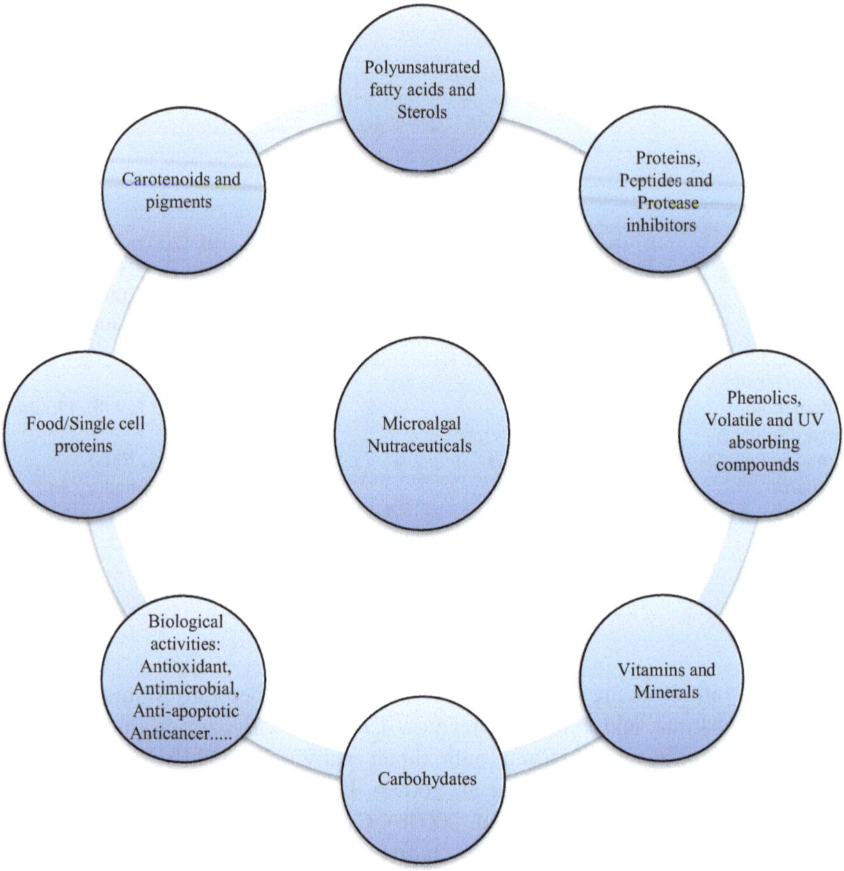


**Fig. 5.1** Important microalgal nutraceuticals

### *5.4.3 Microalgal Carbohydrates*

Carbohydrates constitute a wide variety of sugars or polysaccharides. Depending on the species, the microalgae produce discrete carbohydrates, e.g., glycogen, floridean starch, and amylopectin-like polysaccharides (Nakamura et al. 2005). Because of the absence of hemicelluloses and lignin, the algal biomass is amenable to easy digestion (Mussgnug et al. 2010). The cultivation and environmental conditions as well as the microlagal species determine the biomass carbohydrate content. Microalgal polysaccharides are recognized to promote gut microflora growth and regulation of blood glucose as they constitute a part of prebiotic supplements (Ibañez and Cifuentes 2013). Many cyanobacteria and green algae are surrounded by a special mucilaginous covering around their cells or filaments composed of exopolysaccharides (EPS) and are termed as slimes, sheaths, and/or capsules depending upon the species (Kumar and Adhikary 2018). EPS have received wide attention currently for their antimicrobial and anticarcinogenic roles (Mahendran et al. 2013, Bafanaa 2013). The cell wall polysaccharides from *Chlorella vulgaris* contain β-(1,3)-glucose, while the microalgal species contain heteropolysaccharides with different substituents (Raposo et al. 2014). The polysaccharides from *P. cruentum* are inhibitory against viruses, as well as bacteria (Huang et al. 2005; Raposo et al. 2014). The extracellular polysaccharides of *Rhodella reticulata* exhibit free radical scavenging and antioxidant activity (Chen et al. 2010a). Besides, EPS from red microalgae and *Arthrospira platensis* show antimicrobial and antioxidant activities (Rafika et al. 2011).

### 5.4.4 Microalgal Polyunsaturated Fatty Acids

Polyunsaturated fatty acids (PUFAs) which contain three or more double bonds play integral role in maintaining tissue integrity and imparting beneficial health effects, especially n-3 PUFAs, which are found to be effective in prevention or treatment of several ailments (e.g., heart-related disorders, various malignancies, and many more). Until recently, omega-3 and omega-6 fatty acids, the integral fatty acids, were mainly being derived from fish oil, but due to several concerns like overexploitation of marine sources, detection of toxic compounds in fishes, awful smell and taste, and oxidative instability, the interest has been deviated toward the exploitation of microalgae as an alternative source of PUFAs (Garcia et al. 2017). The three chief types of omega-3 fatty acids implicated in human health and physiology are α-linolenic acid (ALA), eicosapentaenoic acid (EPA), and docosahexaenoic acid (DHA). The microalgal species *Arthrospira* produce α-linolenic acid (ALA) and *Nannochloropsis*, *Phaeodactylum*, and *Nitzschia* produce EPA, while *Crypthecodinium* and *Schizochytrium* are the producers of DHA. The *Crypthecodinium cohnii* produces all the enzymes necessary for de novo synthesis of 22:6ω-3 (Henderson and Mackinlay 1999). *Nannochloropsis* is another promising candidate for pharmaceutical-based applications because it accumulates high levels of PUFA (Udayan et al. 2017). Arachidonic acid (AA), derivative of omega-6 fatty acid is considered as a precursor of prostaglandin and leucotriene synthesis, which has major role to play in circulatory and CNS functions (Medina et al. 1997).

### 5.4.5 Microalgal Sterols

Sterols, which are required to regulate the membrane fluidity, reduce the LDL-cholesterol levels and promote cardiovascular health (Piironen et al. 2000; Volkman 2003; Silvestro et al. 2013; Alsenani et al. 2015). Microalgae are recognized to produce both saturated and unsaturated sterols such as brassicasterol, sitosterol, and stigmasterol (Kohlhase and Pohl 1988; Volkman 2003). *Glaucocystophyte* has been reported to produce sitosterol, campesterol, and stigmasterol (Leblond et al. 2011), dinoflagellates produce 4α-methyl sterols and 24-propylidenecholesterol (Thomson et al. 2004; Giner et al. 2009), and *Dunaliella tertiolecta* and *D. salina* produce sterols with in vivo neuromodulatory activities (Francavilla et al. 2012).

### 5.4.6 Microalgal Pigments

Microalgal pigments like chlorophylls (a, b, and c), phycobiliproteins, phycocyanin, phycoerythrin, β-carotene, lutein, and astaxanthin are important bioactive compounds (Zhang et al. 2016). Chlorophyll or its derived products are widely been used for health benefit characteristics such as antioxidant, therapeutic properties, neuroprotective action, and protection against chronic diseases (Pangestuti and Kim 2011; Galasso et al. 2019). Chlorophyll a and its mixture with chlorophyll b exhibit chemopreventive effects, antioxidant activity, promotion of cell arrest, and apoptosis (Mishra et al. 2011). Some of the most important carotenoids with health benefits are ß-carotene, astaxanthin, lutein, and canthaxanthin. The antioxidant and therapeutic potentials of carotenoids include their role in prevention of diabetes, aging, cancer, obesity, and stroke, with higher provitamin A activity of ß-carotene and lipid peroxidation activity of astaxanthin (Chidambara-Murthy et al. 2005; Lin et al. 2016; Raposo et al. 2001, 2013). β-Carotene and astaxanthin strongly prevent oxidative stress through scavenging of free radicals and also exhibit anticancer effect (Cuvelier 2001; Demming-Adams and Adams 2002; Uttara et al. 2009; Lobo et al. 2010). β-Carotene produced by *D. salina* and *D. bardawil* lower plasma cholesterol and atherogenesis (García-González et al. 2005). The vitamins C and E, β-carotene, and zinc have proved to be effective against age-related macular degeneration, which is responsible for causing blindness (Taylor et al. 2002). Astaxanthin by *Haematococcus pluvialis* and *Chlorella zofingiensis* prevents obesity and fatty liver disease (Lorenz and Cysewski 2000; Ikeuchi et al. 2007). Astaxanthin has protective effects against many diseases, like cancers, and prevents against *Helicobacter pylori* (Olaizola 2005; Ikeuchi et al. 2007; Kamath et al. 2008; Satoh et al. 2009; Yuan et al. 2011). Dried biomass of *Haematococcus pluvialis* which is a rich source of astaxanthin has been commercialized owing to its strong antioxidant activity. Astaxanthin-rich *Haematococcus* available as dietary supplement (Lorenz and Cysewski 2000) has strong antioxidant activity (Higuera-Ciapara et al. 2006). Zeaxanthin by *Microcystis aeruginosa*, *Nannochloropsis*, and *D. salina* helps in vision and antioxidant protection of the body (Jin et al. 2003; Chen et al. 2005). Lutein produced by *Chlorella zofingiensis*, *Chlorella protothecoides*, and *Muriellopsis* sp. has protective antioxidative effect (Kleinegris et al. 2010). Studies found that chances of cataract decreases on consumption of lutein-/zeaxanthin-rich foods. Fucoxanthin is another microalgal pigment with nutraceutical abilities, which is associated with weight loss management (Abidov et al. 2010). The antioxidant and anti-inflammatory potential of phycocyanin and other pigments from red algae is also well documented (Kumar et al. 2014). Phycobiliproteins produced by cyanobacteria and rhodophyta (Eriksen 2008; Watanabe and Ikeuchi 2013; Mulders et al. 2014) reduce oxidative stress by neutralizing the reactive oxygen species (ROS), which is possible due to their chemical structures and chelating properties (Roy et al. 2007; Eriksen, 2008; Stengel et al. 2011; Rodriguez-Sanchez et al. 2012; de Jesus Raposo et al. 2013). Phycocyanin and phycoerythrin are two well-known commercially important phycobiliproteins produced by *Spirulina* sp. and *Porphyridium* sp., respectively (Plaza et al. 2009; Rodriguez-Sanchez et al. 2012; Borowitzka 2013). Microalgae produce chlorophylls a, b, c, d, and f with variable absorption spectra and tonality (Chen et al. 2010b; Roy et al. 2011). Studies have shown that the antimutagenic effect of microalgae is conferred by chlorophylls (Ferruzi and Blakeslee 2007; Gouveia et al. 2008a).

### *5.4.7 Microalgal Vitamins*

Microalgae form a part of human nutrition as it is a rich source of a number of vitamins. It has been reported that vitamin $B_{12}$ has a role to play in DNA repair and histone methylation, and it reduces the risk of breast cancer (Gruber 2016). *Nannochloropsis oculata*, *Chaetoceros calcitrans*, *Porphyridium cruentum*, and *Haslea ostrearia* are rich source of vitamin E (tocopherols) (Durmaz 2007; Bong and Loh, 2013; Santiago-Morales et al. 2018). Ascorbic acid is not only used as a food additive but also effective against diseases such as cancer and several infections (Boyera et al. 1998; Nunes-Alves et al. 2014). *Porphyridium cruentum* produces high quantities of vitamins E and C (ascorbic acid), as well as β-carotene (vitamin A) (Mus et al. 2013). The microalga *D. salina* produces vitamins A, $B_1$, $B_2$, $B_3$, $B_6$, $B_7$, and E (Hosseini Tafreshi and Shariati 2009). Microalgae is also known to contain vitamins $D_2$ and $D_3$ (Takeuchi et al. 1991; Rao and Raghuramulu 1996), together with provitamin $D_3$. Vitamin D and its metabolites are implicated in chemoprevention activities (Giammanco et al. 2015).

### *5.4.8 Microalgal Minerals*

The mineral composition of microalgae which is determined by geographical range and environmental conditions varies extensively among marine and freshwater microalgae on a strain, species, and generic basis (Fox and Zimba 2018). Minerals are present either in the form of compounds or elemental form such as Zn, Ca, and Mg and have important role to play (Nose 1972). The mineral content of microalgae is sufficient to fulfill the recommended daily intake for adults (Alsenani et al. 2015). Significant quantities of Zn, K, Na, Fe, P, Mg, and Mn is produced by *Isochrysis* sp., *Sochrisis* sp., *Chlorella* sp., and *Dunaliella* sp. (Fabregas and Herrero 1986; Tokuşoglu and Üunal 2003). The mineral content produced by the microalgae fulfills the Recommended Dietary Allowances (RDA), prescribed for various minerals (USDA 2002; Tokuşoglu and Üunal 2003).

### 5.4.9 Microalgal Phenolic and Volatile Compounds

Phenolic compounds (PCs) or polyphenolics are secondary metabolites produced by microalgae in response to stress conditions (Cabrita et al. 2010; La Barre et al. 2010). Phenolic compounds like caffeic acid, ferulic acid, and p-coumaric acid, which are used as dietary supplements as they represent important classes of natural antioxidants (Stengel et al. 2011). *Chlorella* and *Spirulina* are known to produce an array of phenolic compounds in appreciable quantities like phloroglucinol, p-coumaric acid, ferulic acid, and apigenin, which are produced by *Chlorella* at 74,000 ng/g, 540 ng/g, 0.63 ng/g, and 9.9 ng/g and by *Spirulina* at 51,000 ng/g, 920 ng/g, 0.67 ng/g, and 6.0 ng/g, respectively (Goiris et al. 2014). The other microalgal species known to produce phenolic compounds are *Nostoc* sp., *Chlorella* sp., *Anabaena* sp., *Tolypothrix* sp., and *Chlamydomonas* sp. (Li et al. 2007; Hajimahmoodi et al. 2010) with phenolic contents of par or more than fruits and vegetables (Ismail et al. 2004; Hassimotto et al. 2005; Lin and Tang 2007).

The volatile compounds are also the secondary metabolites that are responsible for imparting characteristic odors to the water (Abd El-Baky et al. 2002). They have diverse structures and biological activities such as antibacterial, antifungal, antiviral, and anticancer. The biologically active volatile compounds are aldehydes, ketones, fatty acids, and isoprenylated and brominated hydroquinones (Mathew et al. 1995; Morinoto et al. 1995; Borowitzka 1997). The heptadecane and tetradecane produced by *Spirulina* are known to have antibacterial capacity (Ozdemir et al. 2004). *Dunaliella salina* produces antimicrobial compounds like β-cyclocitral, α- and β-ionone, neophytadiene, and phytol (Herrero et al. 2006).

### 5.4.10 Microalgal UV-Absorbing Compounds

The sizeable loss of ozone layer and consequent increment in ultraviolet (UV) radiation have resulted in a host of skin problems like extrinsic skin aging and wrinkles, mottled hyperpigmentation, dilated blood vessels, and loss of skin tone. All these skin-related disorders have ameliorated the interest in quest for natural photoprotective compounds (Lee et al. 2015). In aquatic environments, where microalgae figure prominently, the presence and role of UV-absorbing compounds like sporopollenin, scytonemin, and mycosporine-like amino acids (MAAs) have been documented. Their presumed antioxidant and skin protective strategies raise the interest for possible medicinal and cosmetic applications (Dionisio-Sese 2010). Xiong et al. (1997) observed that UV-B-tolerant chlorophyte species of *Characium terrestre*, *Enallax coelastroides*, *Scenedesmus* sp., *Scotiella chlorelloidea*, and *Spongiochloris spongiosa contain* produce large amounts of sporopollenin. The biopolymer has been reported to be present also in *Dunaliella salina* zygotes (Komaristaya and Gorbulin 2006). Microalgal scytonemin appears restricted to cyanobacteria, specifically in the extracellular polysaccharide sheath of *Chlorogloeopsis* sp., *Scytonema* sp., and *Rivularia* sp. (Sinha et al. 1998). It is reportedly the most important UV-absorbing compound in *Lyngbya* cf. *aestuarii*, where its area content seems to follow the seasonal fluctuation of solar intensity (Karsten et al. 1998). MAAs are UV radiation-absorbing molecules and possess high molar extinction coefficients. Unlike sporopollenin and scytonemin, which are found in the cell wall or extracellular sheath of the microalgae, MAAs in microalgae are mostly intracellular. The presence of several MAAs in a species was observed in different chlorophytes, haptophytes, diatoms, and dinoflagellates grown in cultures or collected in a broad variety of aquatic habitats (Llewellyn and Airs 2010). MAAs also have been reported in several microalgae-invertebrate (sea anemone, coral, ascidian) symbiotic associations. Among the Chlorophytes, asterina and shinorine are produced by *Ankistrodesmus spiralis* and *Chlorella minutissima*; palythine, porphyra, and shinorine are produced by *Chlorella sorokiniana* and *Enallax coelastroides* (Xiong et al. 1999); and mycosporine-glycine is produced by *Pyramimonas parkeae* (Hannach and Sigleo 1998). Among the halophytes, mycosporine-glycine is produced by *Isochrysis* sp. and *Pavlova gyrans* (Hannach and Sigleo 1998); among the Diatoms, porphyra and shinorine are produced by *Chaetoceros* sp., *Corethron criophilumporphyra, Cosinodiscus centralisporphyra, Thalassiosira tumidaporphyra,* and Porosira glacialis (Helbling et al. 1996; Riegger and Robinson 1997). Among the Dinoflagellates, mycosporine-glycine, palythene, palythine, porphyra, and shinorine are produced by *Scrippsiella sweeneyae* (Taira et al. 2004); asterina, mycosporine-glycine, palythene, palythenic acid, palythine, palythinol, porphyra, shinorine, and usujirene are produced by *Alexandrium catenella, A. excavatum, A. minutum, and A. tamarense* (Carreto et al. 1990).

### *5.4.11 Microalgal Toxic Metabolites*

Microalgal species are known to produce toxins called as microcystins; homo- and anatoxin-*a*, also known as Very Fast Death Factor (VFDF); and saxitoxins. The most prevalent are the microcystins, produced by the blooming *Microcystis aeruginosa.* The most toxic amino acid known is microcystin LR produced by *M. aeruginosa*, which is lethal for both animals and human (Khan et al. 2018); its degree of toxicity is determined by the length of amino acid chain (Jungblut and Neilan 2006). Cytotoxins are pharmacologically active compounds with various biological activities such as anticancer, antimicrobial, antiplasmodial, and immunosuppressive (Abdo et al. 2012; Rath and Priyadarshani 2013; Malathi et al. 2014; Mukund and Sivasubramanian 2014; Semary and Fouda 2015; Shaieb et al. 2014). *Amphidinium* sp. produce an active antimycotic and antiprotozoal agent called as karatungiols (Washida et al. 2006). *Dinophysis* sp. is known to produce a potent neurotoxin called as okadaic acid for the treatment of cognitive disorders (He et al. 2005).

### 5.4.12 Microalgae as Food/Single Cell Protein

Microalgae has been consumed for its high protein content since thousands of the year. For example, *Nostoc* and *Arthrospira* or *Spirulina* have been used in Asia and Africa, respectively (Chacòn -Lee and Gonźalez-Maríno 2010). The most commonly consumed microalgae and with GRAS status are *Spirulina*, *Chlorella*, *Dunaliella*, *Haematococcus*, and *Schizochytrium* (Hayes et al. 2017). The microalgae-based foods are available as different formulations (Pulz and Gross 2004) or as fortified foods in the forms of confectionaries, refreshments, cereals, and beverages (Liang et al. 2004). Microalgae also serve as feed for animals (Gouveia et al. 2008a). Due to growing popularity of health foods, various food products are supplemented with microalgae. *Chlorella* and *Spirulina* are the popularly used microalgal species for the supplementation of pasta. Supplementation of pasta with *C. vulgaris* results in better nutritional value, enhanced sensorial properties, increased firmness, and improved swelling and water absorption (Fradique et al. 2010, Gouveia et al. 2007). *Isochrysis galbana* is another widely used species used for the supplementation of the food products. Supplementation of cookies with *I. galbana* is carried out to provide ω-3 PUFAs, besides exhibiting thermal resistance (Gouveia et al. 2008b). Phycocyanin extracts and whole *A. platensis* incorporation into cookies enhance the protein as well as fiber content (Singh et al. 2015). Supplementation of cookies with *Haematococcus pluvialis* provides antioxidant potential to the food product and also lowers the glycemic response (Hossain et al. 2017). *Spirulina platensis* is widely used in the supplementation of bread due to its antimicrobial activity, besides improving the protein content and mineral profile of the food product (Ak et al. 2016). *Arthrospira* is another popular supplement of the bread that helps to increase its protein content (Dinu et al. 2012; Achour et al. 2014; Ak et al. 2016). The techno-functional properties of microalgae are exploited as additives in food products (Caporgno and Mathys 2018). The property of microalgae to mimic fat molecules is utilized for its incorporation into emulsion resulting in reduction of percentage of oil as well as enhanced resistance to oxidation (Gouveia et al. 2006) and incorporation of microalgae into vegetarian desserts as coloring agents (Batista et al. 2008).

### 5.4.13 Microalgal Biological Activities

The microalgal metabolites have the following interesting biological activities:

#### 5.4.13.1 Antioxidant Activity

Antioxidants are compounds that defend the body from the damage caused by potentially harmful molecules called as free radicals. The free radicals such as reactive oxygen and nitrogen species attack biomolecules like DNA and proteins, causing several lethal diseases like carcinoma, cardiovascular, and brain-related disorders (Ngo et al. 2006). Microalgae produce several metabolites that have strong radical scavenging action, e.g., chlorophyll a (Cho et al. 2011); phycoerythrobilin (Yabuta et al. 2010); pigment fucoxanthin and its derivatives, auroxanthin, produced by *Undaria pinnatifida* (Sachindra et al. 2007); and carotenoids produced by *Dunaliella salina* (El-Baz et al. 2017).

#### 5.4.13.2 Anticancer Activity

Cancer, also called as malignancy, is caused by uncontrolled growth and division of cells. There are more than 100 different cancer types, a few of them lethal in nature due to their metastatic activity (Kevin et al. 2018). Microalgae produce different types of carotenoids with significant anticancer activities, β-carotene, lutein, astaxanthin, violaxanthin, and fucoxanthin. Significant inhibition of the growth of human colon cancer cell lines and LoVo colon carcinoma cells by β-carotene has been reported by Palozza et al. (2005) and Pham et al. (2013). The inhibitory effect of astaxanthin on the growth of cancer cell lines and colorectal cancer (CRC) cell lines has been studied by Palozza et al. (2009). The antiproliferative effects of lutein and violaxanthin on human colon cell line HCT-116 was studied by Shi and Chen (2002), Cha et al. (2008), Pasquet et al. (2011), Fu et al. (2013), and Talero et al. (2015). Kumar et al. (2013a, b) have reported the antiproliferative effects of fucoxanthin against different cells lines such as SK-Hep-1 and BNL CL.2. Depending on the culturing conditions, e.g., nutrients, temperature (Ingebrigtsen et al. 2016; Lauritano et al. 2016), and growth phase (Ribalet et al. 2007), the bioactivity of microalgae may vary. Lauritano et al. (2016) demonstrated that diatom *Skeletonema marinoi* shows anticancer activity under nitrogen starvation conditions only. Microalgae produce a repertoire of polysaccharides with anticancer activities (Raposo et al. 2015). Many microalgae present a high content of therapeutic proteins and peptides (Talero et al. 2015).

#### 5.4.13.3 Anti-Angiogenic Activity

Angiogenesis is completely a normal process, but under certain conditions, it may become pathological and cause malignancies, cardiovascular diseases, and many other lethal diseases (Cherrington et al. 2000; Armstrong et al. 2011). Fucoxanthin and fucoxanthinol suppress the angiogenesis process in rats (Sugawara et al. 2006) and significantly inhibit blood cell multiplication. Fucoxanthin also acts as therapeutic agent as it is effective against diabetes, impedes melanin formation (Shimoda et al. 2010), prevents photooxidation of DNA (Heo and Jeon 2009), and has also been reported to promote the synthesis of arachidonic acid and DHA content in mouse livers (Tsukui et al. 2009). Siphonaxanthin, an antiangiogenic agent, is also produced by some species of algae (Ganesan et al. 2010). Aerucyclamide, an antiplasmodial agent isolated from *M. aeruginosa*, is also a potential candidate to be used in pharmaceutical products (Pen et al. 2012).

#### 5.4.13.4 Anti-obesity Activity

Obesity, an abnormal or excessive fat accumulation, is a major epidemic. It is associated with a number of metabolic syndromes, type 2 diabetes, cardiovascular disease, cancer, and aging (Kopelman 2000). The underlying mechanism that causes obesity is overgrowth of adipose tissue (Wang et al. 2008). Microalgae are known to produce anti-hyperlidimic and fat-lowering agents. The fucoxanthin and fucoxanthinol predominantly produced by *Cylindrotheca closterium* and *Phaeodactylum tricornutum* not only inhibit differentiation of 3T3- L1 cells to adipocytes but also adipocyte differentiation (Hayato et al. 2006). The other biological activities fucoxanthin shows are anticancer, anti-oxidant, etc. (Maeda et al. 2007a; Kim et al. 2012). The fat in mouse feeds has been reported to be lowered by the activity of neoxanthin and fucoxanthin (Okada et al. 2008; Maeda et al. 2007b). The long-chain omega-3 fatty acid-rich oil obtained from *Aurantiochytrium* sp. KRS101 has been reported to decrease weight of obese animals treated with microalgal oil (Yook et al. 2015). Koo et al. (2019) documented that *Phaeodactylum* extract containing fucoxanthin exerts anti-obesity effects by promoting lipolysis and inhibiting lipogenesis.

#### 5.4.13.5 Antimicrobial Activity

The antimicrobial activity of *Chlorella* was first demonstrated by Pratt in 1944 (Pratt et al. 1944). Microalgae owes its antimicrobial activity to several compounds like indoles, phenols, and fatty acids (Mayer and Hamann 2005; Mendiola et al. 2007). Microalgae also produce antifungal compounds like okadaic acid, ciguatoxin, and karatungiols produced by *Prorocentrum lima*, *Gambierdiscus toxicus*, and dinoflagellate *Amphidinium*, respectively (Washida et al. 2006). *Microcystis aeruginosa* possess both antifungal and antibacterial activity (Khalid et al. 2010). *Dunaliella salina* and *Dunaliella primolecta* show inhibition to a large number of microorganisms like *Staphylococcus aureus*, *Escherichia coli*, *Candida albicans*, *Pseudomonas aeruginosa*, and *Klebsiella pneumonia* (Mendiola et al. 2008; Pane et al. 2015). Mudimu et al. (2014) observed an in vitro antifungal activity against *C. albicans* by *Heterochlorella luteoviridis* and *Porphyridium purpureum*. Ginsberg et al. (1947) first demonstrated the antiviral activity of algal polysaccharides against influenza B and mumps viruses. The cell wall sulfated polysaccharide of *Porphyridium* sp. is reported to act against *Herpes simplex* viruses types 1 and 2 (HSV 1, HSV 2) and *Varicella zoster* virus (VZV). *Scenedesmus obliquus* cells and protein extracts which were rich in Arg, Lys, Asp, Ala, and His have been found to be effective against Coxsackie $B_3$ virus (Afify et al. 2018).

#### 5.4.13.6 Antiapoptotic Activity

Apoptosis is programmed cell death that helps the body to get rid of damaged or infected cells. It is mediated by intracellular proteolytic cascade, besides the activation and downregulation of proapoptotic and antiapoptotic genes, respectively

(Nappo et al. 2012), and stated that diethyl ether extract from marine diatoms *Cocconeis scutellum* resulted in 89.2% apoptosis. Suh et al. (2017) reported significant inhibition of carcinogenesis and induction of cellular apoptosis through upregulation of apoptotic genes by the ethanol extract of *Botryidiopsidaceae* sp.

#### 5.4.13.7 Cytotoxic Activity

Microalgae produce several groups of compounds that act as cytotoxic agents. *Cyanophora paradoxa* (Cp) produce pigments that can efficiently inhibit various carcinomas. The Cp water and ethanol microalgal extracts significantly inhibited the growth of cancer cell lines in vitro at 100 μg $mL^{-1}$ (Baudelet et al. 2013). Ávila-Román (2019) reported the cytotoxic activity of oxylipins (OXLs) isolated from *Chlamydomonas debaryana* (13-HOTE) and *Nannochloropsis gaditana* (15-HEPE) against UACC-62 (melanoma).

## 5.5 Safety, Market Potential, and Current Challenges

Notwithstanding the nutritional and health aspects of microalgae, high content of nucleic acids associated with them is a major concern. The metabolic conversion of nucleic acid to uric acid may lead to the development of gout or renal calculus (Gantar and Svirčev 2008). The other important issue is that genetic characteristics and technological approaches adopted for biomass productions determine the digestibility and overall nutritional value of microalgae. For instance, toxin production, i.e., hepatotoxins and neurotoxins, by some cyanobacteria when grown in open conditions, poses a threat of microalgal biomass getting contaminated with toxins and other contaminants (Grobbelaar 2003).

Notwithstanding the acceptance of the use of microalgal biomass or myriad metabolites obtained from it, a small number of products can be seen in market. There are still a large number of avenue lying unexplored in this area and numerous challenges to be addressed to come up with innovative solutions. The most important challenge is to achieve high rate of microalgae production at economical scale. The commercial production of microalgae is being carried out in unsophisticated, low-productive man-made structures like artificial open ponds (Chisti 2007). Only a few species of microalgae such as *Spirulina*, *Chlorella*, and *Dunaliella* can be cultured in sustained open ponds. Despite the success of open systems, further advancements are required for closed system cultivation of microalgae. Closed photobioreactors such as tubular, flat panel, air lift, and bubble column which have been around for a long time are difficult to scale up and require high auxiliary energy, investment, and operation costs. The recovery and preservation of microalgae also require considerations. Microalgae cultures are usually very dilute suspensions with concentrations between less than 1 g/L (ponds) and 3–15 g/L (tubular or flat panel reactors) (Wijffels 2019). Thus, the recovery is a challenging task and is achieved by combining various operations such as sedimentation, flotation, filtration, and centrifugation followed by preservation of microalgal cells in order to maintain the protein quality of the biomass and the activity of other compounds of interest (Table 5.2).

**Table 5.2** Major microalgal products and producers

| Product | Microalgal source | Producer | Reference |
|---|---|---|---|
| Astaxanthin (dietary supplement) | *Haematococcus pluvialis* | Cyanotech (Hawaii, USA), EID Parry (India), Mera Pharma (USA), US Nutra (USA), Bioreal (Sweden), Parry Nutraceuticals (India) | Cyanotech (2019), EID Parry (2019), Mera Pharma (2019), Bioreal (2019), and US Nutra (2019) |
| Astaxanthin (food ingredient/ additive) | *Haematococcus pluvialis* | Algatech (Israel), Blue Biotech (Germany), Fuji Chemicals (Japan), Mera Pharma (USA), Bioreal (Sweden) | Algatech (2019), Bioreal (2019), Blue Biotech (2019), Fuji Chemical (2019), and Mera Pharma (2019) |
| *Spirulina* (dietary supplement) | *Spirulina* sp. | Cyanotech (Hawaii, USA), Earthrise (California, USA), Dainippon (Japan), EID Parry (India), Blue Biotech (Germany), Inner Mongolia Biomedical Eng (Mongolia), Panmol (Australia), Spirulina Mexicana (Mexico), Siam Alga Co (Thailand), Nippon Spirulina (Japan), Koor Foods Co (Israel), CBN Spirulina group Co., Ltd. (China), Beihai SBD Bio-Science Technology Co., Ltd. (China), Myanmar Spirulina (Myanmar), Blue Continent (NA) | Chacon-Lee and Gonzalez-Marino (2010), Tramroy (2011), Chen et al. (2015), Cyanotech (2019), Earthrise (2019), EID Parry (2019), US Nutra (2019), and Blue Biotech (2019) |
| *Chlorella* (dietary supplement) | *Chlorella* sp. | Blue Biotech (Germany), Earthrise (USA), Dainippon (Japan), Roquette Klotze (Germany), *Chlorella* Co. (Taiwan) | Chacon-Lee and Gonzalez-Marino (2010) and Blue Biotech (2019) |
| *Chlorella* (food supplement) | *Chlorella* sp. | Phycom (Netherlands), Dongying Diazen, Biological Engineering Co., Ltd. (China) | Chen et al. (2015) |
| EPA/DHA (omega-3) as dietary supplement | *Schizochytrium* | Flora Health (USA) | Flora Health (2019) |
| EPA/DHA (omega-3) (as food ingredient) | *Crypthecodinium*, *Nannochloropsis*, *Schizochytrium* | Cellena (USA), Martek/DSM (USA/NL) Blue Biotech (Germany), Xiamen Huison Biotech Co. (China) | Tramroy (2011) and Blue Biotech (2019) |

(continued)

**Table 5.2** (continued)

| Product | Microalgal source | Producer | Reference |
|---|---|---|---|
| $\beta$-Carotene (as additive/ vitamin) | *Dunaliella salina* | EID Parry (India), Cognis Australia (Australia), Natural Beta Technologies (Australia), Tianjin Lantai Laboratory (China), Nature Beta Technologies (Israel), Nikken Sohonsha (Japan), Aqua Carotene Ltd (Australia), Proalgen Biotech (India) | Carlsson et al. (2007), Chacon-Lee and Gonzalez-Marino et al. (2010), Tramroy (2011), Parry Nutraceuticals (2019), and Proalgen Biotech (2019) |

## 5.6 Future Direction of Research

Although many compounds of high biological value and health benefits have already been discovered, microalgae still remain one of the most unexplored groups of organisms in the world, as around 97% of marine microalgal compounds are yet to be isolated and characterized (Guedes et al. 2011). This lack of information demands for intensive research in the area of bioprospecting that involves isolation, identification, and growth optimization of new and locally available microalgal strains. The discovery of hyper-producing strains and novel metabolites with various health benefits helps to improve the economics of nutraceutical production. Among the around 10,000 algae species that are believed to exist, only a few thousand strains are kept in collections, a few hundred are investigated for chemical content, and just handful microalgae, e.g., *Chlorella*, *Dunaliella salina*, and *Haematococcus pluvialis*, are cultivated in industrial quantities (Spolaore et al. 2006). The development of a number of transgenic algal strains boasting recombinant protein expression, engineered photosynthesis, and enhanced metabolism encourages the prospects of engineered microalgae (Rosenberg et al. 2008). The other key area of further research is the complete genome sequencing of high nutraceutical-producing microalgae. Although *C. reinhardtii* and 30 other organellar and whole algal genomes have reportedly been sequenced (Guarnieri and Pienkos 2015), but in order to understand and characterize the genes and enzymes involved in the production of nutraceuticals and the mechanisms that trigger the production of these metabolites, such as nutrient deprivation and biotic and abiotic stress, information on whole genome sequences is integral. Major advancements are also required to construct bioreactors or open pond systems, which are made of inexpensive and environmentally friendly materials, and development of less energy-demanding and inexpensive harvesting and extraction techniques as extraction and purification of microalgal biomass and nutraceutical compound add up to the major costs of microalgae production plants (Garcia et al. 2017). Intensive research is also required for the identification of phenolic and volatile compounds from microalgae, and furthermore, advanced techniques are required for the isolation of such compounds, since this information is missing till now due to the complexity of their isolation. Despite being rich in proteins, the dried form of microalgae as food or food substitute has failed to attract the consumers due to the fishy smell and dark green color associated with these food products. To harness the benefits of nutritious proteins from microalgae, essential strategies are needed, which would be helpful to cover its smell or taste, for instance, microencapsulation techniques (Chacón-Lee and González-Mariño 2010).

## 5.7 Conclusion

Microalgae are known for their potential to produce food ingredients and bioactive compounds since ancient times. The microalgal biomasses are being widely cultivated to make the commercial formulation of functional foods and nutraceutical application. Microalgae contain many valuable compounds, which includes omega-3 fatty acids, vitamins, and pigment-protein complexes besides anticancer, antioxidative, cytotoxic, and anti-obesity activities. As research interests and investment in microalgae continue to grow in various parts of the world, their role in providing health benefits and nutrition will keep expanding, and these minuscule biofactories may bring about revolutionary changes in nutraceuticals in the future.

## References

Abd El-Baky HH, Shallan MA, El- Baroty G, El-Baz FK (2002) Volatile compounds of the microalga *Chlorella vulgaris* and their phytotoxic effect. Pak J Biol Sci 5(1):61–65

Abdo SM, Mona HH, Waleed ME, Rawheya ASED, Gamila HA (2012) Antiviral activity of freshwater algae. J Appl Pharm Sci 2:21–25

Abe K, Hattori H, Hirano M (2007) Accumulation and antioxidant activity of secondary carotenoids in the aerial microalga *Coelastrella striolata* var. *multistriata*. Food Chem 100:656–661

Abidov M, Ramazanov Z, Seifulla R, Grachev S (2010) The effects of xanthigen in the weight management of obese premenopausal women with non-alcoholic fatty liver disease and normal liver fat. Diabetes Obes Metab 12(1):72–81

Achour HY, Doumandji A, Sadi S, Saadi S (2014) Evaluation of nutritional and sensory properties of bread enriched with *Spirulina*. Ann Food Sci Technol 15:270–275

Adarme-Vega TC, Thomas-Hall SR, Schenk PM (2014) Towards sustainable sources for omega-3 fatty acids production. Curr Opin Biotechnol 26:14–18

Adhikary SP (2000) A preliminary survey of algae of estuaries and coastal areas in Orissa. Seaweed Res Util 22:1–5

Afify AEMR, El Bbaroty GS, El Baz FK, Abd El Baky HH, Murad AA (2018) Scendesmus obliquus: antioxidant and antiviral activity of proteins hydrolysed by three enzymes. J Genet Eng Biotechnol 16(2):399–408

Ak B, Avsaroglu E, Isık O, Ozyurt G, Kafkas E, Etyemez M, Uslu L (2016) Nutritional and physicochemical characteristics of bread enriched with microalgae *Spirulina platensis*. Int J Eng Res Appl 6:30–38

Algatech (2019) Corporate Website. http://www.algatech.com/. Accessed on 30 June 2019

Alsenani F, Ahmed F, Schenk PM (2015) Nutraceuticals from microalgae. In: Bagchi D, Preuss HG, Swaroop A (eds) Nutraceuticals and functional foods in human health and disease prevention. CRC Press, Boca Raton, pp 673–684

Anas AR, Kisingi T, Umezawa T, Matsuda F, Campiteli MR, Quinn RJ, Okino T (2012) Thrombin inhibitors from the freshwater cyanobacterium *Anabaena compacta*. J Nat Prod 75:1546–1552

Arisz SA, van Himbergen JA, Musgrave A, van den Ende H, Munnik T (2000) Polar glycerolipids of *Chlamydomonas moewusii*. Phytochemistry 53:265–270

Armstrong AW, Voyles SV, Armstrong EJ, Fuller EN, Rutledge JC (2011) Angiogenesis and oxidative stress: common mechanisms linking psoriasis with atherosclerosis. J Dermatol Sci 63:1–9

Asthana RK, Tripathi D, Srivastava MK, Singh A, Singh AP et al (2009) Isolation and identification of a new antibacterial entity from the Antarctic cyanobacterium *Nostoc* CCC 537. J Appl Phycol 21:81–88

Ávila-Román J, Taleroa E, de los Reyesb C, Zubíab E, Motilvaa V, García-Mauriñoc S (2019) Cytotoxic activity of microalgal-derived oxylipins against human cancer cell lines and their impact on ATP levels. Nat Prod Commun 11:1871–1875

Aziz N, Prasad R, Ibrahim AIM, Ahmed AIS (2017) Promising applications for the production of biofuels through algae. In: Patra JK, Vishnuprasad CN, Das G (eds) Microbial biotechnology. Springer Nature Singapore Pte Ltd, Singapore, pp 81–103

Bafanaa A (2013) Characterization and optimization of production of exopolysaccharide from *Chlamydomonas reinhardtii*. Carbohydr Polym 95:746–752

Bagchi SN, Sondhi S, Agrawal KM, Banerjee S (2016) An angiotensin-converting enzyme-inhibitory metabolite with partial structure of microginin in cyanobacterium *Anabaena fertilissima* CCC597, producing fibrinolytic protease. J Appl Phycol 28:177–180

Banker R, Carmeli S (1998) Tenuecyclamides A-D, cyclic hexapeptides from the cyanobacterium *Nostoc spongiaformae* var *tenue*. J Nat Prod 61:1248–1251

Barbosa MJ, Zijffers JW, Nisworo A, Vaes W, van Schoonhoven J, Wijffels RH (2005) Optimization of biomass, vitamins, and carotenoid yield on light energy in a flat-panel reactor using the A-stat technique. Biotechnol Bioeng 89:233–242

Barsanti L, Gualtiery P (2006) Algae: anatomy, biochemistry and biotechnology. CRC Taylor & Francis, New York

Batista AP, Gouveia L, Nunes MC, Franco JM, Raymundo A (2008) Microalgae biomass as a novel functional ingredient in mixed gel systems. In: Williams PA, Phillips GO (eds) Gums and stabilisers for the food industry. RSC Publishing, Cambridge, p 14

Baudelet PH, Gagez AL, Berard JB, Juin C, Bridian N, Kaas R, Thiery V, Cadoret JP, Picot L (2013) Antiproliferative activity of *Cyanophora paradoxa* pigments in melanoma, breast and lung cancer cells. Mar Drugs 11:4390–4406

Becker W (2004) Microalgae in human and animal nutrition. In: Richmond A (ed) Microalgal culture. Handbook. Blackwell, Oxford, pp 312–351

Becker EW (2007) Micro-algae as a source of protein. Biotechnol Adv 25(2):207–210

Bhadury P, Wright PC (2004) Exploitation of marine algae: biogenic compounds for potential antifouling applications. Planta 219:561–578

Bhosale P, Bernstein PS (2005) Microbial xanthophylls. Appl Microbiol Biotechnol 68:445–455

BioReal (2019) Corporate Website. http://www.bioreal.se/index.php?page=1&id=5. Accessed on 30 June 2019

Bishop WM, Zubeck HM (2012) Evaluation of microalgae for use as nutraceuticals and nutritional supplements. J Nutr Food Sci 2:147

Blanco AM, Moreno J, Del Campo JA, Rivas J, Guerrero MG (2007) Outdoor cultivation of lutein-rich cells of *Muriellopsis* sp. in open ponds. Appl Microbiol Biotechnol 73:1259–1266

Bleakley S, Hayes M (2017) Algal proteins: extraction, application, and challenges concerning production. Foods 6(5):33

Blue Biotech (2019) Blue Biotech Corporate Website. http://www.bluebiotech.de/. Accessed on 02 July 2019

Bong SC, Loh SP (2013) A study of fatty acid composition and tocopherol content of lipid extracted from marine microalgae, *Nannochloropsis oculata* and *Tetraselmis suecica*, using solvent extraction and supercritical fluid extraction. Int J Food Res 20:721–729

Borowitzka MA (1988) Vitamins and fine chemicals from micro-algae. In: Borowitzka MA, Borowitzka LJ (eds) Micro-algal biotechnology. Cambridge University Press, Cambridge, pp 153–196

Borowitzka MA (1997) Microalgae for aquaculture opportunities and constraints. J Appl Phycol 9:393–401

Borowitzka MA (2013) High-value products from microalgae – their development and commercialisation. J Appl Phycol 25:743 756

Boyera N, Galey I, Bernanrd BA (1998) Effect of vitamin C and its derivatives on collagen synthesis and cross-linking by normal human fibroblasts. Int J Cosmet Sci 20:151–158

Cabrita MT, Vale C, Rauter AP (2010) Halogenated compounds from marine algae. Mar Drugs 8(8):2301–2307

Caporgno MP, Mathys A (2018) Trends in microalgae incorporation into innovative food products with potential health benefits. Front Nutr 5:58. https://doi.org/10.3389/fnut.2018.00058

Carballo-Cardenas EC, Tuan PM, Janssen M, Wijffels RH (2003) Vitamin E (alpha-tocopherol) production by the marine microalgae *Dunaliella tertiolecta* and *Tetraselmis suecica* in batch cultivation. Biomol Eng 20:139–147

Cardozo KH, Guaratini T, Barros MP, Falcão VR, Tonon AP, Lopes NP, Campos S, Torres MA, Souza AO, Colepicolo P, Pinto E (2007) Metabolites from algae with economical impact. Comp Biochem Physiol C Toxicol Pharmacol 146:60–78

Carlsson AS, van Beilen JB, Möller R, Clayton D (2007) Micro- and macro-algae: utility for industrial applications. In: Bowles D (ed) Outputs from the EPOBIO: realising the economic potential of sustainable resources – bioproducts from non-food crops project. CNAP/CPL Press/University of York, York

Carreto JI, Carignan MO, Daleo G, de Marco SG (1990) Occurrence of mycosporine-like amino acids in the red tide dinoflagellate *Alexandrium excavatum*: UV-protective compounds. J Plankton Res 12:909–921

Ceron MC, Campos I, Sánchez JF, Acién FG, Molina E, Fernández-Sevilla JM (2008) Recovery of lutein from microalgae biomass: development of a process for *Scenedesmus almeriensis* biomass. J Agric Food Chem 56:11761–11766

Cha KH, Koo SY, Lee DU (2008) Antiproliferative effects of carotenoids extracted from *Chlorella ellipsoidea* and *Chlorella vulgaris* on human colon cancer cells. J Agric Food Chem 56:10521–10526

Chacòn -Lee TL, Gonźalez-Maríno GE (2010) Microalgae for "healthy" foods—possibilities and challenges. Compr Rev Food Sci Food Saf 9:655–675

Chattopadhyay P, Chatterjee S, Sen SK (2008) Biotechnological potential of natural food grade biocolorants. Afr J Biotechnol 7:2972–2985

Chauhan B, Gopal Kumar G, Kalam N, Ansari SH (2013) Current concepts and prospects of herbal nutraceutical: a review. J Adv Pharm Technol Res 4(1):4–8

Chen F (1998) Bioreactor technology for production of valuable algal products. Chin J Oceanol Limnol 16:84–90

Chen F, Li HB, Wong RNS, Ji B, Jiang Y (2005) Isolation and purification of the bioactive carotenoid zeaxanthin from the microalga *Microcystis aeruginosa* by high-speed counter-current chromatography. J Chromatogr 1064(2):183–186

Chen B, You W, Huang J, Yu Y, Chen W (2010a) Isolation and antioxidant property of the extracellular polysaccharide from *Rhodella reticulata*. World J Microbiol Biotechnol 26:833–840

Chen M, Schliep M, Willows RD, Cai ZL, Neilan BA, Scheer H (2010b) A red-shifted chlorophyll. Science 329:1318–1319

Chen J, Wang Y, Benemann JR, Zhang X, Hu H, Qin S (2015) Microalgal industry in China: challenges and prospects. J Appl Phycol 28(2):715–725. https://doi.org/10.1007/s10811-015-0720-4

Cherrington JM, Strawn LM, Shawver LK (2000) New paradigms for the treatment of cancer: the role of anti-angiogenesis agents. Adv Cancer Res 79:1–38

Chiang IZ, Huang WY, Wu JT (2004) Allelochemicals of *Botryococcus braunii* (Chlorophyceae). J Phycol 40:474–480

Chidambara-Murthy KN, Vanitha A, Rajesha J, Mahadeva-Swamy M, Sowmya PR, Ravishankar GA (2005) In vivo antioxidant activity of carotenoids from *Dunaliella salina*—a green microalga. Life Sci 76:1382–1390

Chisti Y (2007) Biodiesel from microalgae. Biotechnol Adv 25:294–306

Cho M, Lee H, Kang I, Won M, You S (2011) Antioxidant properties of extract and fractions from *Enteromorpha prolifera*, a type of green seaweed. Food Chem 27:999–1006

Chronakis IS, Madsen M (2011) Algal proteins. In: Phillips GO, Williams PA (eds) Handbook of food proteins. Woodhead Publishing Series in Food Sciences, Technology Nutrition, Sawston, pp 353–394

Chu WL (2012) Biotechnological applications of microalgae. Int e-J Sci Med Educ 6:24–37

Coesel SN, Baumgartner AC, Teles LM, Ramos AA, Henriques NM, Cancela L, Varela JC (2008) Nutrient limitation is the main regulatory factor for carotenoid accumulation and for Psy and Pds steady state transcript levels in *Dunaliella salina* (Chlorophyta) exposed to high light and salt stress. Mar Biotechnol 10:602–611

Cuvelier M-E (2001) Antioxidants. In: Morais R (ed) Functional foods: an introductory course. Escola Superior de Biotecnologia/UCP, Porto, pp 97–108

Cyanotech (2019) Cyanotech annual report. www.cyanotech.com. Accessed on 30 June 2019

D'Souza FML, Loneragan NR (1999) Effects of monospecific and mixed algae diets on survival, development and fatty acid composition of penaeid prawn (*Penaeus* spp.) larvae. Mar Biol 133:621–633

Dawczynski C, Schubert R, Jahreis G (2007) Amino acids, fatty acids, and dietary fibre in edible seaweed products. Food Chem 103:891–899

Day AG, Brinkmann D, Franklin S, Espina K, Rudenko G, Roberts A, Howse KS (2009) Safety evaluation of a high-lipid algal biomass from *Chlorella protothecoides*. Regul Toxicol Pharmacol 55:166–180

de Jesus Raposo MF, de Morais RM, De Morais AM (2013) Health applications of bioactive compounds from marine microalgae. Life Sci 93(15):479–486

de Morais MG, da Silva Vaz B, de Morais EG, Vieira Costa JA (2015) Biologically active metabolites synthesized by microalgae. Biomed Res Int 2015:835761. https://doi.org/10.1155/2015/835761

de Vries M, de Boer IJM (2010) Comparing environmental impacts for livestock products: a review of life cycle assessments. Livest Sci 128:1–11

Delattre C, Pierre G, Laroche C, Michaud P (2016) Production, extraction and characterization of microalgal and cyanobacterial exopolysaccharides. Biotechnol Adv 34:1159–1179

Demming-Adams B, Adams WW (2002) Antioxidants in photosynthesis and human nutrition. Science 298:2149–2153

Dillard CJ, German JB (2000) Phytochemicals: nutraceuticals and human health. J Sci Food Agric 80:1744–1756

Dinu M, Vlasceanu G, Dune A, Rotaru G (2012) Researches concerning the growth of nutritive value of the bread products through the *Spirulina* adding. J Environ Prot Ecol 13:660–665

Dionisio-Sese ML (2010) Aquatic microalgae as potential sources of UV-screening compounds. Philipp J Sci 139(1):5–16

Domínguez-Bocanegra AR, Guerrero Legarreta I, Martinez Jeronimo F, Tomasini Campocosio A (2004) Influence of environmental and nutritional factors in the production of astaxanthin from *Haematococcus pluvialis*. Bioresour Technol 92:209–214

Drag M, Salvensen GS (2010) Emerging principles in protease-based drug discovery. Nat Rev Drug Discov Perspect 9:690–701

Durmaz Y (2007) Vitamin E (α-tocopherol) production by the marine microalgae *Nannochloropsis oculata* (Eustigmatophyceae) in nitrogen limitation. Aquaculture 272:717–722

Duval B, Shetty K, Thomas WH (2000) Phenolic compounds and antioxidant properties in the snow alga *Chlamydomonas nivalis* after exposure to UV light. J Appl Phycol 11:559–566

Earthrise (2019) Earthrise company website. www.earthrise.com. Accessed on 30 June 2019

Egeland ES, Eikrem W, Throndsen J, Wilhelm C, Zapata M, Liaaen-Jensen S (1995) Carotenoids from further prasinophytes. Biochem Syst Ecol 23:747–755

Egeland ES, Guillard RRL, Liaaen-Jensen S (1997) Additional carotenoid prototype representatives and a general chemosystematic evaluation of carotenoids in *Prasinophyceae* (Chlorophyta). Phytochemistry 44:1087–1097

EID Parry (2019) EID Parry Annual Report. www.eidparry.com. Accessed on 30 June 2019

EL-Baz FK, Aly HF, Khalil WKB, Booles H, Ali GH (2017) Antineurodegenerative activity of microalgae *Dunaliella salina* in rats with Alzheimer's disease. Asian J Pharm Clin Res 10(1):134–139

Eriksen NT (2008) Production of phycocyanin a pigment with applications in biology, biotechnology, foods and medicine. Appl Microbiol Biotechnol 80:1–14

Ersmark K, Del Valle JR, Hanessian S (2008) Chemistry and biology of the aeruginosin family of serine protease inhibitors. Angew Chem Int Ed 47:1202–1223

Fabregas J, Herrero C (1986) Marine microalgae as a potential source of minerals in fish diets. Aquaculture 51:237–243

Ferruzi MG, Blakeslee J (2007) Digestion, absorption, and cancer preventive activity of dietary chlorophyll derivatives. Nutr Res 27:1–12

Flora Health (2019) Corporate Website. http://www.florahealth.com/home_usa.cfm. Accessed on 30 June 2019

Food and Agriculture Organization of the United Nations (2010) Algae-based biofuels: applications and co-products. Food and Agriculture Organization of the United Nations, Rome

Fox JM, Zimba PV (2018) Minerals and trace elements in microalgae. In: Microalgae in health and disease prevention. Academic, Cambridge, pp 177–193

Fradique M, Batista AP, Nunes MC, Gouevia L, Bandarra NM, Raymundo A (2010) Incorporation of *Chlorella vulgaris* and *Spirulina maxima* biomass in pasta products. Part 1. Preparation and evaluation. J Sci Food Agric 90:1656–1664

Francavilla M, Colaianna M, Zotti MG, Morgese M, Trotta P, Tucci P, Schiavone S, Cuomo V, Trabace L (2012) Extraction, characterization and in vivo neuromodulatory activity of phytosterols from microalga *Dunaliella tertiolecta*. Curr Med Chem 19(18):3058–3067

Fried A, Tietz A, Benamotz A, Eichenberger W (1982) Lipid-composition of the halotolerant alga *Dunaliella bardawil*. Biochim Biophys Acta 713:419–426

Fu W, Gudmundsson O, Paglia G, Herjolfsson G, Andrésson OS, Palsson BØ, Brynjolfsson S (2013) Enhancement of carotenoid biosynthesis in the green microalga *Dunaliella salina* with light-emitting diodes and adaptive laboratory evolution. Appl Microbiol Biotechnol 97:2395–2403

Fuji Chemicals (2019) Corporate website. http://www.fujichemicals.co.jp/English/life_science_/about_astareal/index.htm. Accessed on 02 July 2019

Fujii K, Nakajima H, Ann Y (2008) Potential of *Monoraphidium* sp. GK12 for energy-saving astaxanthin production. J Chem Technol Biotechnol 83:1578–1584

Gademann K, Portmann C (2008) Secondary metabolites from cyanobacteria: complex structures and powerful bioactivities. Curr Org Chem 12:326–341

Galasso C, Gentile A, Orefice I, Ianora A, Bruno A, Noonan DM et al (2019) Microalgal derivatives as potential nutraceutical and food supplements for human health: a focus on cancer prevention and interception. Nutrients 11:1226. https://doi.org/10.3390/nu11061226

Ganesan P, Matsubara K, Ohkubo T, Tanaka Y, Noda K, Sugawara T, Hirata T (2010) Anti-angiogenic effect of siphonaxanthin from green alga, *Codium fragile*. Phytomedicine 17:1140–1144

Gangl D, Zedler JAZ, Rajkumar PD, Martinez EMR, Riseley A et al (2015) Biotechnological exploitation of microalgae. J Exp Bot 66(22):6975–6990

Gantar M, Svirčev Z (2008) Microalgae and cyanobacteria: food for thought. J Phycol 44:260–268

Garcia JL, de Vicenete M, Glan B (2017) Microalgae, old sustainable food and fashion nutraceuticals. Microbial biotechnology. John Wiley & Sons Ltd and Society for Applied Microbiology, Hoboken

García-González M, Moreno J, Manzano JC, Florencio FJ, Guerrero MG (2005) Production of *Dunaliella salina* biomass rich in 9-cis-b-carotene and lutein in a closed tubular photobioreactor. J Biotechnol 115(1):81–90

Giammanco M, Di Majo D, La Guardia M, Aiello S, Crescimannno M, Flandina C et al (2015) Vitamin D in cancer chemoprevention. Pharm Biol 53:1399–1434

Giner JL, Zhao H, Boyer GL, Satchwell MF, Andersen RA (2009) Sterol chemotaxonomy of marine pelagophyte algae. Chem Biodivers 26(7):1111–1130

Ginsberg RH, Goebel WF, Horsfall FL (1947) Inhibition of mumps virus multiplication by a polysaccharide. Proc Soc Exp Biol Med 66:99–100

Goiris K, Muylaert K, Voorspoels S, Noten B, Paepe DD et al (2014) Detection of flavonoids in microalgae from different evolutionary lineages. J Phycol 50(3):485–492

Gouveia L, Oliveira AC (2009) Microalgae as a raw material for biofuels production. J Ind Microbiol Biotechnol 36:269–274

Gouveia L, Raymundo A, Batista AP, Sousa I, Empis J (2006) *Chlorella vulgaris* and *Haematococcus pluvialis* biomass as colouring and antioxidant in food emulsions. Eur Food Res Technol 222:362–367

Gouveia L, Batista AP, Miranda A, Empis J, Raymundo A (2007) Chlorella vulgaris biomass used as coloring source in traditional butter cookies. Innovative Food Sci Emerg Technol 8:433–436

Gouveia L, Batista AP, Sousa I, Raymundo A, Bandarra NM (2008a) Microalgae in novel food products. In: Papadopoulos KN (ed) Food chemistry research developments. Nova Science Publishers Inc., Hauppauge, pp 75–112

Gouveia L, Coutinho C, Mendonça E, Batista AP, Sousa I, Bandarra NM et al (2008b) Functional biscuits with PUFA- ω 3 from *Isochrysis galbana*. J Sci Food Agric 88:891–896

Grobbelaar J (2003) Quality control and assurance: crucial for the sustainability of the applied phycology industry. J Appl Phycol 15:209–215

Gruber BM (2016) BMB-group vitamins: chemoprevention? Adv Clin Exp Med 25:561–568

Guarnieri MT, Pienkos PT (2015) Algal omics: unlocking bioproduct diversity in algae cell factories. Photosynth Res 123:255–263

Guedes AC, Amaro HM, Malcata FX (2011) Microalgae as sources of carotenoids. Mar Drugs 9(4):625–644

Hajimahmoodi M, Faramarzi MA, Mohammadi N, Soltani N, Oveisi MR, Nafissi-Varcheh N (2010) Evaluation of antioxidant properties and total phenolic contents of some strains of microalgae. J Appl Phycol 22(1):43–50

Hanessian S, Del Valle JR, Xue Y, Blomberg N (2006) Total synthesis and structural confirmation of chlorodysinosin A. J Am Chem Soc 128(32):10491–10495

Hannach G, Sigleo AC (1998) Photoinduction of UV-absorbing compounds in six species of marine phytoplankton. Mar Ecol Prog Ser 174:207–222

Hassimotto NMA, Genovese MI, Lajolo FM (2005) Antioxidant activity of dietary fruits, vegetables, and commercial frozen fruit pulps. J Agric Food Chem 53(8):2928–2935

Hayashi T, Hayashi K, Maeda M, Kojima I (1996) Calcium spirulan, an inhibitor of enveloped virus replication, from a blue-green alga *Spirulina platensis*. J Nat Prod 59:83–87

Hayashi K, Kanekiyo K, Ohta Y, Lee JB, Takenaka H, Hayashi T (2008) Anti influenza A virus activity of an acidic polysaccharide from a blue green alga *Nostoc flagelliforme*. Planta Med 74:PA34. https://doi.org/10.1055/s-0028-1084032

Hayato M, Masashi H, Tokutake S, Nobuyuk T, Teruo K, Kazuo M (2006) Fucoxanthin and its metabolite, fucoxanthinol, suppress adipocyte differentiation in 3T3-L1 cells. Int J Mol Med 18:147–152

Hayes M, Skomedal H, Skjanes K, Mazur-Marzec H, Torunska-Sitarz A, Catala M, Hosoglu MI, García-Vaquero M (2017) Microalgal proteins for feed, food and health. In: Gonzalez-Fernandez C, Munoz R (eds) Microalgae-based biofuels and bioproducts. Woodhead publishers Elsevier, Sawston, pp 348–368

He J, Yang Y, Xu H, Zhang X, Li X-M (2005) Olanzapine attenuates the okadaic acid-induced spatial memory impairment and hippocampal cell death in rats. Neuropsychopharmacology 30(8):1511–1520

Hejazi MA, Wijffels RH (2003) Effect of light intensity on beta-carotene production and extraction by *Dunaliella salina* in two-phase bioreactors. Biomol Eng 20:171–175

Helbling EW, Chalker BE, Dunlap WC, Osmund HH, Villafañe VE (1996) Photoacclimation of Antarctic diatoms to solar ultraviolet radiation. J Exp Mar Biol Ecol 204:85–101
Henderson RJ, Mackinlay EE (1999) Polyunsaturated fatty acid metabolism in the marine dinoflagellate *Crypthecodinium cohnii*. Phytochemistry 30(6):1781–1787
Heo SJ, Jeon YJ (2009) Protective effect of fucoxanthin isolated from *Sargassum siliquastrum* on UV-B induced cell damage. J Photochem Photobiol B 95:101–107
Herrero M, Ibáñez E, Cifuentes A, Reglero G, Santoyo S (2006) *Dunaliella salina* microalga pressurized extracts as potential antimicrobials. J Food Prot 69:2471–2477
Higuera-Ciapara I, Felix-Valenzuela L, Goycoolea FM (2006) Astaxanthin: a review of its chemistry and applications. Crit Rev Food Sci Nutr 46:185–196
Hossain AKM, Brennan MA, Mason SL, Guo X, Zeng XA, Brennan CS (2017) The effect of astaxanthin-rich microalgae *Haematococcus pluvialis* and wholemeal flours incorporation in improving the physical and functional properties of cookies. Foods 6(57):1–10
Hosseini Tafreshi A, Shariati M (2009) *Dunaliella* biotechnology: methods and applications. J Appl Microbiol 107(1):14–35
Huang J, Chen B, You W (2005) Studies on separation of extracellular polysaccharide from *Porphyridium cruentum* and its anti-HBV activity in vitro. Chin J Mar Drugs 24:18–21
Ibañez E, Cifuentes A (2013) Benefits of using algae as natural sources of functional ingredients. J Sci Food Agric 93(4):703–709
Ibáñez E, Mendiola JA, Rodríguez-Meizoso I, Señorans FS, Reglero G, Cifuentes A (2008) Antioxidants in plant foods and microalgae extracted using compressed fluids. Elec J Env Agricult Food Chem 7(8):3301–3309
Ikeuchi M, Koyama T, Takahashi J, Yazawa K (2007) Effects of astaxanthin in obese mice fed a high-fat diet. Biosci Biotechnol Biochem 71(4):893–899
Ingebrigtsen RA, Hansen E, Andersen JH, Eilertsen HC (2016) Light and temperature effects on bioactivity in diatoms. J Appl Phycol 28:939–950
Ishida K, Nakagawa H, Murakami M (2000) Microcyclamide, a cytotoxic cyclic hexapeptide from the cyanobacterium *Microcystis aeruginosa*. J Nat Prod 63(9):1315–1317
Ismail A, Marjan ZM, Foong CW (2004) Total antioxidant activity and phenolic content in selected vegetables. Food Chem 87(4):581–586
Jaime L, Mendiola JA, Ibañez E, Martin-Alvarez PJ, Cifuentes A, Reglero G, Señorans FJ (2007) *β*-carotene isomer composition of sub- and supercritical carbon dioxide extracts. Antioxidant activity measurement. J Agric Food Chem 55:10585–10590
Janssen MLE (2019) Cyanobacterial peptides beyond microcystins – a review on co-occurrence, toxicity, and challenges for risk assessment. Water Res 151:488–499
Jena M, Ratha SK, Adhikary SP (2005) Algal diversity changes in Kathajodi river after receiving sewage of Cuttack and its ecological implication. Indian Hydrobiol 8:67–74
Jena M, Ratha SK, Adhikary SP (2008) Algal diversity in Rushikulya river, Orissa from origin till confluence to the sea. Indian Hydrobiol 11:9–24
Jin E, Feth B, Melis A (2003) A mutant of the green alga *Dunaliella salina* constitutively accumulates zeaxanthin under all growth conditions. Biotechnol Bioeng 81(1):115–124
Jin E, Lee CG, Polle JEW (2006) Secondary carotenoid accumulation in *Haematococcus* (Chlorophyceae): biosynthesis, regulation, and biotechnology. J Microbiol Biotechnol 16:821–831
Jungblut AD, Neilan BA (2006) Molecular identification and evolution of the cyclic peptide hepatotoxins, microcystin and nodularin, synthetase genes in three orders of cyanobacteria. Arch Microbiol 185:107–114
Kamath BS, Srikanta BM, Dharmesh SM, Sarada R, Ravishankar CA (2008) Ulcer preventive and antioxidative properties of astaxanthin from *Haematococcus pluvialis*. Eur J Pharmacol 590:387–395
Kang CD, Lee JS, Park TH, Sim SJ (2005) Comparison of heterotrophic and photoautotrophic induction on astaxanthin production by *Haematococcus pluvialis*. Appl Microbiol Biotechnol 68:237–241

Karsten U, Maier J, Garcia-Pichel F (1998) Seasonality in UV-absorbing compounds of cyanobacterial mat communities from an intertidal mangrove flat. Aquat Microb Ecol 16:37–44

Karsten U, Lembcke S, Schumann R (2007) The effects of ultraviolet radiation on photosynthetic performance, growth and sunscreen compounds in aeroterrestrial biofilm algae isolated from building facades. Planta 225:991–1000

Katircioglu H, Akin BS, Atici T (2004) Microalgal toxin(s): characteristics and importance. Afr J Biotechnol 3(12):667–674

Kevin A, Andrade M, Lauritano C, Romano G, Ianora A (2018) Marine microalgae with anticancer properties. Mar Drugs 16(5):165. https://doi.org/10.3390/md16050165

Khalid MN, Shameel M, Ahmad VU, Shahzad S, Leghari SM (2010) Studies on the bioactivity and phycochemistry of *Microcystis aeruginosa* (Cyanophycota) from Sindh. Pak J Bot 42:2635–2646

Khan MI, Shin JH, Kim JD (2018) The promising future of microalgae: current status, challenges, and optimization of a sustainable and renewable industry for biofuels, feed, and other products. Microb Cell Factories 17:36. https://doi.org/10.1186/s12934-018-0879-x

Kharkongar D, Ramanujam P (2014) Diversity and species composition of subaerial algal communities in forested area of Meghalaya, India. Int J Biodiversity 2014:1–10. https://doi.org/10.1155/2014/456202

Kim SM, Jung YH, Kwon O, Cha KH, Um BH (2012) A potential commercial source of fucoxanthin extracted from the microalga *Phaeodactylum tricornutum*. Appl Biochem Biotechnol 166:1843–1855

Kleinegris DM, van Es MA, Janssen M, Brandenburg WA, Wijffels RH (2010) Carotenoid fluorescence in *Dunaliella salina*. J Appl Phycol 22(5):645–649

Kohlhase M, Pohl P (1988) Saturated and unsaturated sterols of nitrogen-fixing blue-green algae (cyanobacteria). Phytochemistry 27:1735–1740

Komaristaya VP, Gorbulin OS (2006) Sporopollenin in the composition of cell walls of *Dunalliela salina* Teod. (*Chlorophyta*) zygotes. Int J Algae 8:43–52

Koo SY, Hwang JH, Yang SH, Um JI, Hong KW, Kang K, Pan CH, Hwang KT, Kim SM (2019) Anti-obesity effect of standardized extract of microalga *Phaeodactylum tricornutum* containing fucoxanthin. Mar Drugs 17:311. https://doi.org/10.3390/md17050311

Kopelman PG (2000) Obesity as a medical problem. Nature 404:635–643

Kumar D, Adhikary SP (2018) Exopolysaccharides from cyanobacteria and microalgae and their commercial application. Curr Sci 115(2):234–231

Kumar M, Tripathi MK, Srivastava A, Gour JK, Singh RK, Tilak R, Asthana RK (2013a) Cyanobacteria, *Lyngya aestuarii* and *Aphanothece bullosa* as antifungal and antileishmanial drug resources. Asian Pac J Trop Biomed 3:458–463

Kumar SR, Hosokawa M, Miyashita K (2013b) Fucoxanthin: a marine carotenoid exerting anticancer effects by affecting multiple mechanisms. Mar Drugs 11:5130–5147

Kumar D, Dhar DW, Pabbi S, Kumar N, Walia S (2014) Extraction and purification of C-phycocyanin from *Spirulina platensis* (CCC540). Indian J Plant Physiol 19(2):184–188

La Barre S, Potin P, Leblanc C, Delage L (2010) The halogenated metabolism of brown algae (Phaeophyta), its biological importance and its environmental significance. Mar Drugs 8(4):988–1010

Lauritano C, Andersen JH, Hansen E, Albrigtsen M, Escalera L, Esposito F, Helland K, Hanssen KØ, Romano G, Ianora A (2016) Bioactivity screening of microalgae for antioxidant, anti-inflammatory, anticancer, anti-diabetes and antibacterial activities. Front Mar Sci 3:1–12. https://doi.org/10.3389/fmars.2016.00068

Leblond JD, Timofte HI, Roche SA, Porter NM (2011) Sterols of *glaucocystophytes*. Phycol Res 59:129–134

Lee JH, Kim HS, Seo HH, Song MY, Kulkarni A, Choi YH, Kim KW, Moh SH (2015) Antiaging effects of algae-derived mycosporine-like amino acids (MAAs) on skin. In: Farage M, Miller K, Maibach H (eds) Textbook of aging skin. Springer, Berlin

Lenoir-Wijnkoop I, Dapoigny M, Dubois D, van Ganse E, Gutiérrez-Ibarluzea I, Hutton J, Jones P, Mittendorf T, Poley MJ, Salminen S, Nuijten MJ (2010) Nutrition economics–characterising the economic and health impact of nutrition. Br J Nutr 105(1):157–166

Leu S, Boussiba S (2014) Advances in the production of high-value products by microalgae. Indus Biotechnol 10:169–183

Leya T, Rahn A, Lütz C, Remias D (2009) Response of arctic snow and permafrost algae to high light and nitrogen stress by changes in pigment composition and applied aspects for biotechnology. FEMS Microbiol Ecol 67:432–443

Li HB, Cheng KW, Wong CC, Fan KW, Chen F, Jiang Y (2007) Evaluation of antioxidant capacity and total phenolic content of different fractions of selected microalgae. Food Chem 102(3):771–776

Liang S, Liu X, Chen F, Chen Z (2004) Current microalgal health food R & D activities in China. In: Ang PO (ed) Asian Pacific Phycol. 21st century Prospect. Challenges. Springer, Dordrecht, pp 45–48

Lin JY, Tang CY (2007) Determination of total phenolic and flavonoid contents in selected fruits and vegetables, as well as their stimulatory effects on mouse splenocyte proliferation. Food Chem 101(1):140–147

Lin J, Huang L, Yu J, Xiang S, Wang J, Zhang J, Yan X, Cui W, He S, Wang Q (2016) Fucoxanthin, a marine carotenoid, reverses scopolamine-induced cognitive impairments in mice and inhibits acetylcholinesterase in vitro. Mar Drugs 14:67. https://doi.org/10.3390/md14040067

Liu J, Chen F (2016) Biology and industrial applications of *Chlorella*: advances and prospects. Adv Biochem Eng Biotechnol 153:1–35

Liu BH, Lee YK (2000) Secondary carotenoids formation by the green alga *Chlorococcum* sp. J Appl Phycol 12:301–307

Llewellyn CA, Airs RL (2010) Distribution and abundance of MAAs in 33 species of microalgae across 13 Classes. Mar Drugs 8(4):1273–1291

Lobo V, Patil A, Phatak A, Chandra N (2010) Free radicals, antioxidants and functional foods: impact on human health. Pharmacogn Rev 4(8):118–126

Lorenz RT, Cysewski GR (2000) Commercial potential for *Haematococcus* microalgae as a natural source of astaxanthin. Trends Biotechnol 18:160–167

Luo X, Su P, Zhang W (2015) Advances in microalgae- derived phytosterols for functional food and pharmaceutical applications. Mar Drugs 13:4231–4254

Maeda H, Hosokawa M, Sashima T, Funayama K, Miyashita K (2007a) Effect of medium-chain triacylglycerols on anti-obesity effect of fucoxanthin. J Oleo Sci 56(12):615–621

Maeda H, Hosokawa M, Sashima T, Miyashita K (2007b) Dietary combination of fucoxanthin and fish oil attenuates the weight gain of white adipose tissue and decreases blood glucose in obese/diabetic KK-Ay mice. J Agric Food Chem 55:7701–7706

Mahendran S, Saravanan S, Vijayabaskar P, Anandapandian KTK, Shankar T (2013) Antibacterial potential of microbial exopolysaccharide from *Ganoderma lucidum* and *Lysinibacillus fusiformis*. Int J Recent Sci Res 4:501–505

Malathi T, Babu MR, Mounika T, Snehalatha D, Rao BD (2014) Screening of cyanobacterial strains for antibacterial activity. Phykos 44:6–11

Markou G, Angelidaki I, Georgakakis D (2012) Microalgal carbohydrates: an overview of the factors influencing carbohydrates production, and of main bioconversion technologies for production of biofuels. Appl Microbiol Biotechnol 95:631–645. https://doi.org/10.1007/s00253-012-4398-0

Martinez-Fernandez E, Acosta-Salmon H, Southgate PC (2006) The nutritional value of seven species of tropical microalgae for blacklip pearl oyster (*Pinctada margaritifera*, L.) larvae. Aquaculture 257:491–503

Mathew BR, Sankaranarayanan P, Nair C, Varghese T, Smanathon B et al (1995) Evaluation of chemoprevention of oral cancer with *Spirulina fusiformis*. Nutr Cancer 24:197–202

Matsukawa R, Hotta M, Masuda Y, Chihara M, Karube I (2000) Antioxidants from carbon dioxide fixing *Chlorella sorokiniana*. J Appl Phycol 12:263–267

Mayer AMS, Hamann MT (2005) Marine pharmacology in 2001–2002: marine compounds with anthelmintic, antibacterial, anticoagulant, antidiabetic, antifungal, anti-inflammatory, antimalarial, antiplatelet, antiprotozoal, antituberculosis, and antiviral activities; affecting the cardiovascular, immune and nervous systems and other miscellaneous mechanisms of action. Comp Biochem Physiol Part C Toxicol Pharmacol 140:265–286

Mazur-Marzec H, Kaczkowska MJ, Blaszczyk A, Ackaalan R, Spoof L, Meriluoto J (2013) Diversity of peptides produced by *Nodularia spumigena* from various geographical regions. Mar Drugs 11:1–19

Medina AR, Grima EM, Gimĕnez AG, González MJI (1997) Lipase-catalyzed esterification of glycerol and polyunsaturated fatty acids from fish and microalgae oils. Biotechnol Adv 16:517–580

Mendes RL, Nobre BP, Cardoso MT, Pereira AP, Palavra AF (2003) Supercritical carbon dioxide extraction of compounds with pharmaceutical importance from microalgae. Inorg Chim Acta 356:328–334

Mendiola JA, Mańn FR, Herńandez SF, Arredondo BO, Sĕnořans FJ, Ibánez E, Reglero G (2005) Characterization via liquid chromatography coupled to diode array detector and tandem mass spectrometry of supercritical fluid antioxidant extracts of *Spirulina platensis* microalga. J Sep Sci 28:1031–1038

Mendiola JA, Torres CF, Martín-Alvarez PJ, Santoyo S, Toré A, Arredondo BO, Señoráns FJ, Cifuentes A, Ibáñez E (2007) Use of supercritical $CO_2$ to obtain extracts with antimicrobial activity from *Chaetoceros muelleri* microalga. A correlation with their lipidic content. Eur Food Res Technol 224:505–510

Mendiola JA, Santoyo S, Cifuentes A, Reglero G, Ibanez E, Senorans FJ (2008) Antimicrobial activity of sub- and supercritical $CO_2$ extracts of the green alga *Dunaliella salina*. J Food Prot 71:2138–2143

Merapharma (2019) Corporate website. https://www.merapharma.com/about-mera.html. Accessed on 02 July 2019

Mishra VK, Bacheti RK, Husen A (2011) Medicinal uses of chlorophyll: a critical overview. In: Le H, Salcedo E (eds) Chlorophyll: structure, function and medicinal. Uses Publisher, Nova Science Publishers, Inc, Hauppauge

Mišurcová L, Buňka F, Vávra Ambrožová J, Machů L, Samek D, Kráčmar S (2014) Amino acid composition of algal products and its contribution to RDI. Food Chem 151:120–125

Mobin S, Alam F (2017) Some promising microalgal species for commercial applications: a review. Energy Procedia 110:510–517

Moore RE, Patterson GML, Carmichael WW (1988) New pharmaceuticals from cultured blue green algae. In: Fautin DG (ed) Biomedical importance of marine organisms. California Academy of Sciences, San Francisco, pp 143–150

Morinoto T, Nagatu A, Murakami N (1995) Anti-tumor promoting glyceroglycolipid from green alga *Chlorella vulgaris*. Phytochemistry 40:1433–1437

Mudimu O, Rybalka N, Bauersachs T, Born J, Friedl T, Schulz R (2014) Biotechnological screening of microalgal and cyanobacterial strains for biogas production and antibacterial and antifungal effects. Metabolites 4:373–393

Mukund S, Sivasubramanian V (2014) Anticancer activity of *Oscillatoria terebriformis*, cyanobacteria in human lung cancer cell line A549. Int J Appl Biol Pharm Technol 5:34–45

Mulders KJM, Lamers PP, Martens DE, Wijffels RH (2014) Phototropic pigment production with microalgae: biological constraints and opportunities. J Phycol 50:229–242

Murakami M, Suzuki S, Itou Y, Kodani S, Ishida K (2000) New anabaenopeptins, potent carboxypeptidase-A inhibitors from the cyanobacterium *Aphanizomenon flos-aquae*. J Nat Prod 63:1280–1282

Mus F, Toussaint JP, Cooksey KE, Fields MW, Gerlach R, Peyton BM et al (2013) Physiological and molecular analysis of carbon source supplementation and pH stress- induced lipid accumulation in the marine diatom *Phaeodactylum tricornutum*. Appl Microbiol Biotechnol 97:3625–3642

Mussgnug JH, Klassen V, Schlüter A, Kruse O (2010) Microalgae as substrates for fermentative biogas production in a combined biorefinery concept. J Biotechnol 150(1):51–56

Nakamura Y, Takahashi J, Sakurai A, Inaba Y, Suzuki E, Nihei S et al (2005) Some cyanobacteria synthesize semiamylopectin type α-Polyglucans instead of glycogen. Plant Cell Physiol 46(3):539–545

Nappo M, Berkov S, Massucco C, Di Maria V, Bastida J, Codina C et al (2012) Apoptotic activity of the marine diatom *Cocconeis scutellum* and eicosopentaenoic acid in BT20 cells. Pharm Biol 50:529–535

Ng DHP, Ng YK, Shen H, Lee YK (2015) Microalgal biotechnology: the way forward. In: Kim SK (ed) Handbook of marine microalgae. Elsevier, Amsterdam, pp 69–80

Ngo DN, Kim MM, Kim SK (2006) Chitin oligosaccharides inhibit oxidative stress in live cells. Carbohydr Polym 74:228–234

Nicoletti M (2012) Nutraceuticals and botanicals: overview and perspectives. Int J Food Sci Nutr 63:2–6

Nose T (1972) Mineral requirements. Suisan Zoshoku 20:289–300

Nunes-Alves C, Booty MG, Carpenter SM, Jayaraman P, Rothchild AC, Behar SM (2014) In search of a new paradigm for protective immunity to TB. Nat Rev Microbiol 12:289–299

Okada T, Nakai M, Maeda H, Hosokawa M, Sashima T, Miyashita K (2008) Suppressive effect of neoxanthin on the differentiation of 3T3-L1 adipose cells. J Oleo Sci 57(6):345–351

Olaizola M (2005) The health benefits of *Haematococcus* astaxanthin: cardiovascular health. Agro Food Ind Hi-Tech 16:35–37

Ozdemir G, Karabay NU, Dalay MC, Pazarbasi B (2004) Antibacterial activity of volatile component and various extracts of *Spirulina platensis*. Phytother Res 18(9):754–757

Palozza P, Serini S, Maggiano N, Tringali G, Navarra P, Ranelletti FO, Calviello G (2005) Beta-carotene downregulates the steady-state and heregulin-alpha-induced COX-2 pathways in colon cancer cells. J Nutr 135:129–136

Palozza P, Torelli C, Boninsegna A, Simone R, Catalano A, Mele MC, Picci N (2009) Growth-inhibitory effects of the astaxanthin-rich alga *Haematococcus pluvialis* in human colon cancer cells. Cancer Lett 283:108–117

Pane G, Cacciola G, Giacco E, Mariottini GL, Coppo E (2015) Assessment of the antimicrobial activity of algae extracts on bacteria responsible of external Otitis. Mar Drugs 13(10):6440–6452

Pangestuti R, Kim SK (2011) Biological activities and health benefit effects of natural pigments derived from marine algae. J Funct Foods 3(4):255–266

Parry Nutraceuticals (2019) Parry Nutraceuticals website. http://www.parrynutraceuticals.com/food-ingredients.Aspx

Pasquet V, Morisset P, Ihammouine S, Chepiedm A, Aumailley L, Berard JB et al (2011) Antiproliferative activity of violaxanthin isolated from bioguided fractionation of *Dunaliella tertiolecta* extracts. Mar Drugs 9:819–831

Patil V, Källqvist T, Olsen E, Vogt G, Gislerød HR (2007) Fatty acid composition of 12 microalgae for possible use in aquaculture feed. Aquac Int 15:1–9

Pattanaik A, Sukla LB, Pradhan D, Shukla V (2019) Artificial intelligence and virtual environment for microalgal source for production of nutraceuticals. Biomed J Sci Tech Res 13(5):4362–4367. https://doi.org/10.26717/BJSTR.2019.13.002459

Pelah D, Sintov A, Cohen E (2004) The effect of salt stress on the production of canthaxanthin and astaxanthin by *Chlorella zofingiensis* grown under limited light intensity. World J Microbiol Biotechnol 20:483–486

Pen S, Scarone L, Manta E, Stewart L, Yardley V et al (2012) Synthesis of a *Microcystis aeruginosa* predicted metabolite with antimalarial activity. Bioorg Med Chem Lett 22(15):4994–4997

Pereira H, Barreira L, Figueiredo F, Custodio L, Vizetto-Duarte C, Polo C et al (2012) Polyunsaturated fatty acids of marine macroalgae: potential for nutritional and pharmaceutical applications. Mar Drugs 10:1920–1935

Pham DNT, Leclerc D, Levesque N, Deng L, Rozen R (2013) β-Carotene 15,15′-monooxygenase and its substrate β-carotene modulate migration and invasion in colorectal carcinoma cells. Am J Clin Nutr 98:413–422

Piironen V, Lindsay DG, Miettinen TA, Toivo J, Lampi AM (2000) Plant sterols: biosynthesis, biological function and their importance to human nutrition. J Sci Food Agric 80:939–966
Piorreck M, Baasch KH, Pohl P (1984) Biomass production, total protein, chlorophylls, lipids and fatty acids of freshwater green and bluegreen algae under different nitrogen regimes. Phytochemistry 23:207–216
Plaza M, Herrero M, Cifuentes A, Ibanez E (2009) Innovative natural functional ingredients from microalgae. J Agric Food Chem 57:7159–7170
Plaza M, Santoyo S, Jaime L, Reina GG, Herrero M, Senorans FJ, Ibanez E (2010) Screening for bioactive compounds from algae. J Pharm Biomed Anal 51:450–455
Ploutno A, Carmeli S (2000) Nostocyclyne A. A novel antimicrobial cyclophane from the cyanobacterium *Nostoc* sp. J Nat Prod 63:1524–1526
Poerschmann J, Spijkerman E, Langer U (2004) Fatty acid patterns in *Chlamydomonas* sp. as a marker for nutritional regimes and temperature under extremely acidic conditions. Microb Ecol 48:78–89
Ponomarenko LP, Stonik IV, Aizdaicher NA, Orlova TY, Popovskaya GI, Pomazkina GV, Stonik VA (2004) Sterols of marine microalgae *Pyramimonas* cf. *cordata* (Prasinophyta), *Attheya ussurensis* sp. nov. (Bacillariophyta) and a spring diatom bloom from Lake Baikal. Comp Biochem Physiol B Biochem Mol Biol 138:65–70
Portmann C, Blom JF, Gademann K, Juttner F (2008) Aerucyclamides A and B: isolation and synthesis of toxic ribosomal heterocyclic peptides from the cyanobacterium *Microcystis aeruginosa* PCC 7806. J Nat Prod 71(7):1193–1196
Prasanna R, Kaushik BD (2005) Algal diversity from morphology to molecules. In: Satyanarayana T, Johri BN (eds) Microbial diversity: current perspectives and potential applications. IK International, New Delhi, pp 323–344
Pratt R, Daniels TC, Eiler JJ, Gunnison JB, Kumler WD, Oneto JF et al (1944) Chlorellin, an antibacterial substance from *Chlorella*. Science 99:351–352
Pro Aigen Biotech (2019) https://www.indiamart.com/proalgen-biotech-limited/. Accessed on 30 June 2019
Pulz O, Gross W (2004) Valuable products from biotechnology of microalgae. Appl Microbiol Biotechnol 65(6):635–648
Rabbani S, Beyer P, Lintig J, Hugueney P, Kleinig H (1998) Induced beta carotene synthesis driven by triacylglycerol deposition in the unicellular alga *Dunaliella bardawil*. Plant Physiol 116:1239–1248
Rafika C, Lamia T, Rym BD, Omeya EA, Ali Y, Khamissa G, Jihen BA, Hela O, Ouada HB (2011) Evaluation of cytotoxicity and biological activities in extracellular polysaccharides released by cyanobacterium *Arthrospira platensis*. Braz Arch Biol Technol 54:831–838
Rajasekaran A, Sivagnanam G, Xavier R (2008) Nutraceuticals as therapeutic agents: a review. Res J Pharm Technol 1:328–340
Ranga RA, Vijaya RD, Ravishankar GA (2017) Secondary metabolites from algae for nutraceutical application. Novel Tech Nutr Food Sci 1(1). https://doi.org/10.31031/NTNF.2017.01.000503
Rao SD, Raghuramulu N (1996) Food chain as origin of vitamin D in fish. Comp Biochem Physiol Part A Physiol 114:15–19
Raposo MFJ, Mendes-Pinto MM, Morais R (2001) Carotenoids, foodstuff and human health. In: Morais R (ed) Functional foods an introductory course. Universidade Católica Portuguesa—Escola Superior de Biotecnologia, Porto
Raposo MFJ, de Morais RMSC, Bernardo de Morais AMM (2013) Bioactivity and applications of sulphated polysaccharides from marine microalgae. Mar Drugs 11(1):233–252
Raposo MF, de Morais AM, de Morais RM (2014) Influence of sulphate on the composition and antibacterial and antiviral properties of the exopolysaccharide from *Porphyridium cruentum*. Life Sci 101:56–63
Raposo MFJ, Morais AMB, Morais RMS (2015) Marine polysaccharides from algae with potential biomedical applications. Mar Drugs 13:2967–3028
Rath J, Adhikary SP (2005) A check list of algae from Chilika Lake, Orissa. Nelumbo Bull Bot Surv India 47:101–114

Rath B, Priyadarshani I (2013) Antibacterial and antifungal activity of marine cyanobacteria from Odisha coast. Int J Curr Trends Res 2:248–251

Ratha SK, Naik K, Padhi SB (2003) Epiphytic algal diversity associated with different aquatic macrophytes of freshwater ponds in and around Berhampur University campus, Orissa. Nat Environ Pollut Technol 2:205–208

Ratha SC, Prasanna R, Gupta V, Dhar DW, Saxena AK (2012) Bioprospecting and indexing the microalgal diversity of different ecological habitats of India. World J Microbiol Biotechnol 28:1657–1667

Reitan KI, Rainuzzo JR, Olsen Y (1994) Effect of nutrient limitation on fatty acid and lipid content of marine microalgae. J Phycol 30:972–979

Remias D, Lütz-Meindl U, Lütz C (2005) Photosynthesis, pigments and ultrastructure of the alpine snow alga *Chlamydomonas nivalis*. Eur J Phycol 40:259–268

Ribalet F, Wichard T, Pohnert G, Ianora A, Miralto A, Casotti R (2007) Age and nutrient limitation enhance polyunsaturated aldehyde production in marine diatoms. Phytochemistry 68:2059–2067

Riegger L, Robinson D (1997) Photoinduction of UV-absorbing compounds in Antarctic diatoms and *Phaeocystis antarctica*. Mar Ecol Prog Ser 160:13–25

Rodriguez-Garcia I, Guil-Guerrero JL (2008) Evaluation of the antioxidant activity of three microalgal species for use as dietary supplements and in the preservation of foods. Food Chem 108:1023–1026

Rodriguez-Sanchez R, Ortiz-Butron R, Blas-Valdivia V, Hernandez-Garcia A, Cano-Europa E (2012) Phycobiliproteins or C-phycocyanin of *Arthrospira* (*Spirulina*) maxima protect against $HgCl_2$-caused oxidative stress and renal damage. Food Chem 135:2359–2365

Rosenberg JN, Oyler GA, Wilkinson L, Betenbaugh MJ (2008) A green light for engineered algae: redirecting metabolism to fuel a biotechnology revolution. Curr Opin Biotechnol 19:430–436

Roy KR, Arunasree KM, Dhoot A, Aparna R, Reddy GV, Vali S, Reddanna P (2007) C-phycocyanin inhibits 2-acetylaminofluorene-induced expression of MDR1 in mouse macrophage cells: ROS mediated pathway determined via combination of experimental and *in silico* analysis. Arch Biochem Biophys 459:169–177

Roy S, Llewellyn CA, Egeland ES, Johnson G (2011) Phytoplankton pigments: characterization, chemotaxonomy, and applications in oceanography. Cambridge University Press, New York

Sachindra N, Sato E, Maeda H, Hosokawa M, Niwano Y, Kohno M et al (2007) Radical scavenging and singlet oxygen quenching activity of marine carotenoid fucoxanthin and its metabolites. J Agric Food Chem 55:8516–8522

Samantraray SM, Mallick AK, Padhy SB (2002) Chilika environment and its algal vegetation. Seaweed Res Util 24:13–17

Sanchez JF, Fernández-Sevilla JM, Acién FG, Cerón MC, Pérez-Parra J, Molina-Grima E (2008) Biomass and lutein productivity of *Scenedesmus almeriensis*: influence of irradiance, dilution rate and temperature. Appl Microbiol Biotechnol 79:719–729

Santiago-Morales IS, Trujillo-Valle L, Márquez-Rocha FJ, López Hernández JF (2018) Tocopherols, phycocyanin and superoxide dismutase from microalgae as potential food antioxidants. Appl Food Biotechnol 5:19–27

Sathasivam R, Kermanee P, Roytrakul S, Juntawong N (2012) Isolation and molecular identification of β-carotene producing strains of *Dunaliella salina* and *Dunaliella bardawil* from salt soil samples by using species-specific primers and internal transcribed spacer (ITS) primers. Afr J Biotechnol 11(34):8425–8432

Satoh A, Tsuji S, Okada Y, Murakami N, Urami M, Nakagawa K, Ishikura M, Katagiri M, Koga Y, Shirasawa T (2009) Preliminary clinical evaluation of toxicity and efficacy of a new astaxanthin-rich *Haematococcus pluvialis* extract. J Clin Biochem Nutr 44:280–284

Selvaraj G, Kaliamurthi S, Cakmak ZE, Cakmak T (2017) RuBisCO of microalgae as potential targets for nutraceutical peptides: a computational study. Biotechnology 16(4–6):130–144

Semary NAE, Fouda M (2015) Anticancer activity of *Cyanothece* sp. strain extracts from Egypt: first record. Asian Pac J Trop Biomed 5:992–995

Severes A, Nivas S, D'Souza L, Hegde S (2018) Diversity study of freshwater microalgae of some unexplored water bodies of a rapidly developing industrial region in India. J Algal Biomass Util 9(2):31–40

Shaieb FA, Issa AA, Meragaa A (2014) Antimicrobial activity of crude extracts of cyanobacteria *Nostoc commune* and *Spirulina platensis*. Arch Biomed Sci 2:34–41

Sheih IC, Fang TJ, Wu TK (2009a) Isolation and characterisation of a novel angiotensin I-converting enzyme (ACE) inhibitory peptide from the algae protein waste. Food Chem 115(1):279–284

Sheih IC, Fang TJ, Wu TK, Lin PH (2009b) Anticancer and antioxidant activities of the peptide fraction from algae protein waste. J Agric Food Chem 58(2):1202–1207

Shi XM, Chen F (2002) High-yield production of lutein by the green microalga *Chlorella protothecoides* in heterotrophic fed-batch culture. Biotechnol Prog 18:723–727

Shi X, Wu Z, Chen F (2006) Kinetic modeling of lutein production by heterotrophic *Chlorella* at various pH and temperatures. Mol Nutr Food Res 50:763–768

Shimoda H, Tanaka J, Shan S, Maoka T (2010) Antipigmentary activity of fucoxanthin and its influence on skin mRNA expression of melanogenic molecules. J Pharm Pharmacol 62:1137–1145

Silvestro D, Andersen TG, Schaller H, Jensen PE (2013) Plant sterol metabolism. Δ7-sterol-C5-desaturase (STE1/DWARF7), Δ5,7-sterol-Δ7-reductase (DWARF5) and Δ24-sterol-Δ24-reductase (DIMINUTO/DWARF1) show multiple subcellular localizations in *Arabidopsis thaliana* (Heynh) L. PLoS One 8:e564529. https://doi.org/10.1371/journal.pone.0056429

Singh UB, Sharma C (2014) Microalgal diversity of Sheer Khad (stream): a tributary of Sutlej River, Himachal Pradesh, India. J Res Plant Sci 3(1):235–241

Singh P, Singh R, Jha A, Rasane P, Gautam AK (2015) Optimization of a process for high fibre and high protein biscuit. J Food Sci Technol 52:1394–1403

Sinha RP, Klisch M, Gröniger A, Häder DP (1998) Ultraviolet-absorbing/screening substances in cyanobacteria, phytoplankton and macroalgae. J Photochem Photobiol B Biol 47:83–94

Smetana S, Sandmann M, Rohn S, Pleissner D, Heinz V (2017) Autotrophic and heterotrophic microalgae and cyanobacteria cultivation for food and feed: life cycle assessment. Bioresour Technol 245(Pt A):162–170. https://doi.org/10.1016/j.biortech.2017.08.113

Sonani RR, Rastogi RP, Patel R, Madamwar D (2016) Recent advances in production, purification and applications of phycobiliproteins. World J Biol Chem 7:100–109

Spolaore P, Joannis-Cassan C, Duran E, Isambert A (2006) Commercial applications of microalgae. J Biosci Bioeng 101(2):87–96

Stengel DB, Connan S, Popper ZA (2011) Algal chemodiversity and bioactivity: sources of natural variability and implications for commercial application. Biotechnol Adv 29:483–501

Subudhi S (2017) Bioprospecting for algal based nutraceuticals and high value added compounds. J Pharm Pharm Sci 4(2):145–150. https://doi.org/10.15436/2377-1313.17.16511313.17.1651

Sugawara T, Matsubara K, Akagi R, Mori M, Hirata T (2006) Antiangiogenic activity of brown algae fucoxanthin and its deacetylated product, fucoxanthinol. Agric Food Chem 54:9805–9810

Suh SS, Kim SM, Kim JE, Hong JM, Lee SG, Youn UJ, Han SJ, Kim IC, Kim S (2017) Anticancer activities of ethanol extract from the Antarctic freshwater microalga, *Botryidiopsidaceae* sp. BMC Complement Altern Med 17:509. https://doi.org/10.1186/s12906-017-1991-x

Suresh A, Kumar RP, Dhanasekaran D, Thajuddin N (2012) Biodiversity of microalgae in Western a and Eastern Ghats, India. Pak J Biol Sci 15(19):919–928

Taira H, Aoki S, Yamanoha B, Taguchi S (2004) Daily variation in cellular content of UV-absorbing compounds mycosporine-like amino acids in the marine dinoflagellate *Scrippsiella sweeneyae*. J Photochem Photobiol B Biol 75:145–155

Takeuchi A, Okano T, Tanda M, Kobayashi T (1991) Possible origin of extremely high contents of vitamin D3 in some kinds of fish liver. Comp Biochem Physiol A Physiol 100:483–487

Talero E, García-Mauriño S, Ávila-Román J, Rodríguez-Luna A, Alcaide A, Motilva V (2015) Bioactive compounds isolated from microalgae in chronic inflammation and cancer. Mar Drugs 13(10):6152–6209

Taylor HR, Tikellis G, Robman LD, Mc Carty CA, Mc Neil JJ (2002) Vitamin E supplementation and macular degeneration: randomised controlled trial. BMJ 325(7354):11. https://doi.org/10.1136/bmj.325.7354.11

Thomson PG, Wright SW, Bolch CJS, Nichols PD, Skerratt JH, McMinn A (2004) Antarctic distribution, pigment and lipid composition, and molecular identification of the brine dinoflagellate *Polarella glacialis* (Dinophyceae). J Phycol 40:867–873
Tokuşoglu Ö, Üunal MK (2003) Biomass nutrient profiles of three microalgae: *Spirulina platensis*, *Chlorella vulgaris*, and *Isochrisis galbana*. J Food Sci 68:1144–1148
Tomaselli L (2003) Biodeterioration processes on inorganic substrata. Coalition 6:5–9
Tramroy P (2011) Microalgae market outlook report website. CBDM.T-Market and Business Intelligence. http://www.microalgae-market.com/
Tsukui T, Baba N, Hosokawa M, Sashima T, Miyashit K (2009) Enhancement of hepatic docosahexaenoic acid and arachidonic acid contents in C57BL/6J mice by dietary fucoxanthin. Fish Sci 75:261–263
Tukaj Z, Matusiak-Mikulin K, Lewandowska J, Szurkowski J (2003) Changes in the pigment patterns and the photosynthetic activity during a light-induced cell cycle of the green alga *Scenedesmus armatus*. Plant Physiol Biochem 41:337–344
Udayan A, Arumugam M, Pandey A (2017) Nutraceuticals from algae and cyanobacteria. In: Algal green chem. Elsevier, Amsterdam, pp 65–89
Uhlik DJ, Gowans CS (1974) Synthesis of nicotinic acid in *Chlamydomonas eugametos*. Int J Biochem 5:79–84
Urzi C, Realini M (1998) Colour changes of Noto's calcareous sandstone as related to its colonisation by microorganisms. Int Biodeterior Biodegrad 42(1):45–54
US Nutra (2019) Corporate website. http://www.usnutra.com/. Accessed on 21 July 2019
USDA (2002) USA: dietary reference intakes and recommended dietary allowances. Nat Acad of Sciences. Available at: http://arborcom.com/frame/nutrc.htm. Accessed on 21 July 2019
Uttara B, Singh AV, Zamboni P, Mahajan R (2009) Oxidative stress and neurodegenerative diseases: a review of upstream and downstream antioxidant therapeutic options. Curr Neuropharmacol 7(1):65–74. https://doi.org/10.2174/157015909787602823
Valenzuela-Espinoza E, Millan-Nuñez R, Núñez-Cebrero F (2002) Protein, carbohydrate, lipid and chlorophyll a content of *Isochrysis* aff. *galbana* (clone T-Iso) cultured with a low cost alternative to the f/2 medium. Aquac Eng 25(4):207–216
Van Krimpen M, Bikker P, Van der Meer I, Van der Peet-Schwering C, Vereijken J (2013) Cultivation, processing and nutritional aspects for pigs and poultry of European protein sources as alternatives for imported soybean products. Wageningen UR Livestock Research, Lelystad
Vilchez C, Garbayo I, Lobato MV, Vega JM (1997) Microalgae-mediated chemicals production and wastes removal. Enzyme Microb Technol 20:562–572
Volkman JK (2003) Sterols in microorganisms. Appl Microbiol Biotechnol 60(5):495–506
Wang L, Pan B, Sheng J, Xu J, Hu Q (2007) Antioxidant activity of *Spirulina platensis* extracts by supercritical carbon dioxide extraction. Food Chem 105:36–41
Wang H, Peiris TH, Mowery A, Le Lay J, Gao Y, Greenbaum LE (2008) CCAAT/enhancer binding protein-beta is a transcriptional regulator of peroxisome-proliferator-activated receptor-gamma coactivator-1alpha in the regenerating liver. Mol Endocrinol 22:1596–1605
Wang G, Goyal N, Hopkinson B (2009) Preparation of L-proline based aeruginosin 298-A analogs: optimisation of the P1-moiety. Bioorg Med Chem Lett 19:3798–3803
Washida K, Koyama T, Yamada K, Kitab M, Urmura D et al (2006) Karatungiols A and B two novel antimicrobial polyol compounds, from the symbiotic marine dinoflagellate *Amphidinium* sp. Tetrahedron Lett 47(15):2521–2525
Watanabe M, Ikeuchi M (2013) Phycobilisome: architecture of a light-harvesting supercomplex. Phytother Res 116:265–276
Waterbury JB (2006) The cyanobacteria—isolation, purification and identification. In: Dworkin M (ed) The prokaryotes—a handbook on the biology of bacteria, vol 4. Springer, New York, pp 1053–1073
Welker M, von Dohren H (2006) Cyanobacterial peptides—nature's own combinatorial biosynthesis. FEMS Microbiol Rev 30:530–563
Wijffels RH (2019) Algae for energy, http://www.knaw.nl/agenda/pdf/wijffels.pdf. Accessed on 02 July 2019

Wu Z, Dejtisakdi W, Kermanee P, Ma C, Arirob W, Sathasivam R, Juntawong N (2016) Outdoor cultivation of *Dunaliella salina* KU 11 using brine and saline lake water with raceway ponds in northeastern Thailand. Biotechnol Appl Biochem 64(6):938–943

Xiao R, Zheng Y (2016) Overview of microalgal extracellular polymeric substances (EPS) and their applications. Biotechnol Adv 34:1225–1244

Xiong F, Komenda J, Kopecky J, Nedbal L (1997) Strategies of ultraviolet-B protection in microscopic algae. Physiol Plant 100:378–388

Xiong FS, Kopecky J, Nedbal L (1999) The occurrence of UV-B absorbing mycosporine-like amino acids in freshwater and terrestrial microalgae (Chlorophyta). Aquat Bot 63:37–49

Yaakob Z, Ali E, Zainal A, Mohamad M, Takriff MS (2014) An overview: biomolecules from microalgae for animal feed and aquaculture. J Biol Res (Thessaloniki) 21(1):6. http://www.jbiolres.com/content/21/1/6/6

Yabuta Y, Fujimura H, Kwak CS, Enomoto T, Wata-nabe F (2010) Antioxidant activity of the phycoeryth-robilin compound formed from a dried Korean purple laver (*Porphyra* sp.) during in vitro digestion. Food Sci Technol Res 16:347–351

Yook JS, Kim KA, Park JE, Lee SH, Cha YS (2015) Microalgal oil supplementation has an anti-obesity effect in c57bl/6j mice fed a high fat diet. Prev Nutr Food Sci 20(4):230–237

Yuan JP, Chen F, Liu X, Li XZ (2002) Carotenoid composition in the green microalga *Chlorococcum*. Food Chem 76:319–325

Yuan JP, Peng J, Yin K, Wang JH (2011) Potential health-promoting effects of astaxanthin: a high-value carotenoid mostly from microalgae. Mol Nutr Food Res 55:150–165

Zhang CW, Cohen Z, Khozin-Goldberg I, Richmond A (2002) Characterization of growth and arachidonic acid production of *Parietochloris incisa* comb. Nov (Trebouxiophyceae, Chlorophyta). J Appl Phycol 14:453–460

Zhang D, Wan M, del Rio-Chanona EA, Huang J, Wang W, Li Y, Vassiliadis VS (2016) Dynamic modelling of *Haematococcus pluvialis* photoinduction for astaxanthin production in both attached and suspended photobioreactors. Algal Res 13:69–78

Zhekisheva M, Zarka A, Khozin-Goldberg I, Cohen Z, Boussiba S (2005) Inhibition of astaxanthin synthesis under high irradiance does not abolish triacylglycerol accumulation in the green alga *Haematococcus pluvialis* (Chlorophyceae). J Phycol 41:819–826

# Chapter 6
# Sustainable Approaches to Remove Heavy Metals from Water

**Andleeb Zehra, Mukesh Meena, Prashant Swapnil, Namita Anant Raytekar, and R. S. Upadhyay**

**Abstract** Water contamination by heavy metals is a worldwide issue undermining the whole biosphere and influencing the life of a huge number of individuals around the globe. Not exclusively is water contamination by metals one of the chief worldwide hazard factors for sickness, ailments and death, yet it likewise adds to the nonstop reduction of the accessible drinkable water around the world. These metals are discharged from an assortment of sources such as mining, urban sewage, smelters, tanneries, textile industry and chemical industry. Technologies utilized for their expulsion from aquatic bodies incorporate reverse-osmosis, ion-exchange, electrodialysis, adsorption, etc. Most of these technologies are quite costly, energy intensive and metal specific. These conventional technologies for the expulsion of the dangerous heavy metals are most certainly not practical and further create colossal amount of harmful chemical sludge. Delivering valuable solutions, which are easy to implement and affordable, often remains a challenge. Bioremediation is considered as one of the safer, cleaner, cost effective and promising sustainable approach for heavy metal removal from waste water. The objective of this chapter is to conduct a comprehensive review on different sustainable tools for treating heavy metals present in the water.

**Keywords** Heavy metals · Water pollution · Bioremediation · Eco-sustainable approach · Biotechnological techniques · Genetic engineering

---

A. Zehra (✉) · R. S. Upadhyay
Department of Botany, Banaras Hindu University, Varanasi, India

M. Meena
Department of Botany, Mohanlal Sukhadia University, Udaipur, India

P. Swapnil
International Centre for Genetic Engineering and Biotechnology, New Delhi, India

N. A. Raytekar
Department of Microbiology, Sandip University, Nasik, India



J. Singh et al. (eds.), *Microbial Biotechnology: Basic Research and Applications*, Environmental and Microbial Biotechnology,
https://doi.org/10.1007/978-981-15-2817-0_6

## 6.1 Introduction

According to recent estimates, more than 1.2 billion people worldwide do not have access to the most fundamental component of life i.e. clean drinking water (GWI 2018). There has been an extensive increment in the release of waste into the environment, particularly in water bodies, with the development of industries, and this has prompted the gathering of heavy metals, particularly in urban regions (Dixit et al. 2015, Musilova et al. 2016, Masindi and Muedi 2018). The discharge of untreated industrial waste into the water has turned into a preeminent interest in the developing nations and is seen as one of the most significant ecological issues (Burri et al. 2019). The unpredictable discharge of heavy metals into the water is a noteworthy wellbeing concern around the world, as they cannot be degraded into harmless forms and accordingly have enduring consequences for the biological systems (Mishra et al. 2018). Plants and animals require metals for their biological systems, but at raised levels, they meddle with metabolic responses of living beings. The decline in the plant development is attributed to the presence of various lethal heavy metals as it brings down the photosynthetic rate by reducing the enzymatic activity and also causes deprivation of essential mineral nutrients (Nematian and Kazemeini 2013). Heavy metals are potential carcinogenic agents since they have the ability to cause cancer in humans even at low concentrations (Dixit et al. 2015). Consumption of contaminated food causes collection of heavy metals through food chains and becomes a well-being danger to living organisms (Tak et al. 2013). Heavy metals cause free radical production by the process of oxidative stress (Chandra et al. 2015a, b; Mani 2015). Oxidative stress aids in the production of reactive oxygen species (ROS) (Chibuike and Obiora 2014) and hydrogen peroxide ($H_2O_2$), which causes breakage of cellular DNA and eventually results in cell damage (Chandra et al. 2015a, b; Mani 2015; Kapoor et al. 2019). Antioxidant system which protects the cells from reactive oxygen species (ROS) is suppressed by heavy metal toxicity which causes overproduction of ROS. In the event that this condition proceeds, the normal functioning of the living being is influenced, and this may constantly prompt cell death (Ojuederie and Babalola 2017). Thus, it is very important to remove or reduce the heavy metal contamination in water so as to prevent or reduce the contamination of environment and the possibility of uptake in the food web.

Bioremediation is a process being acknowledged as the standard practice for the reclamation of heavy metal-polluted sites since it is a more eco-accommodating, advantageous, and sustainable technique than the traditional chemical and physical strategies, which are frequently extravagant and inadequate (Igiri et al. 2018). Optimum temperature, pH, and moisture are important environmental factors which govern the ability of microbes to degrade the pollutants (Massoud et al. 2019). Bioremediation can possibly reestablish heavy metal-contaminated sites (Dowarah et al. 2009). However, an absence of data related to the elements controlling the microbial development and metabolism (Li et al. 2013) in contaminated conditions frequently confines its execution. Bioinformatics, in light of proteomics and genomics (Chauhan and Jain 2010; Poirier et al. 2013), offers momentous guarantee as tools for addressing long-standing inquiries with respect to the molecular components engaged in controlling mineralization pathways (Kim and Park 2013; Govarthanan et al. 2013; Achal et al. 2012). This chapter examines the sources of heavy metals in the aquatic environment and how they can be successfully remediated with the help of sustainable approaches, viz., bioremediation and phytoremediation as well as their mechanisms. The potential prospects and impediment of genetic engineering for bioremediation are also discussed.

## 6.2 Sources of Heavy Metals in the Aquatic Environment

Combustion of fossil fuels, forest fires, mining and smelting, weathering, municipal wastes, fertilizers, pesticides, and sewage are the basic sources of heavy metals in the environment (Rai 2009; Kabata-Pendias and Pendias 1989; Pillai 2010; Fig. 6.1). In developing countries like India, coal mining industries are also the major contributing source of heavy metal pollution (Sharma 2003; Rai 2012). Discharge of wastes containing heavy metals from different industries presents genuine dangers to water quality of rivers, lakes, and reservoirs and their biodiversity (Concas et al. 2006).

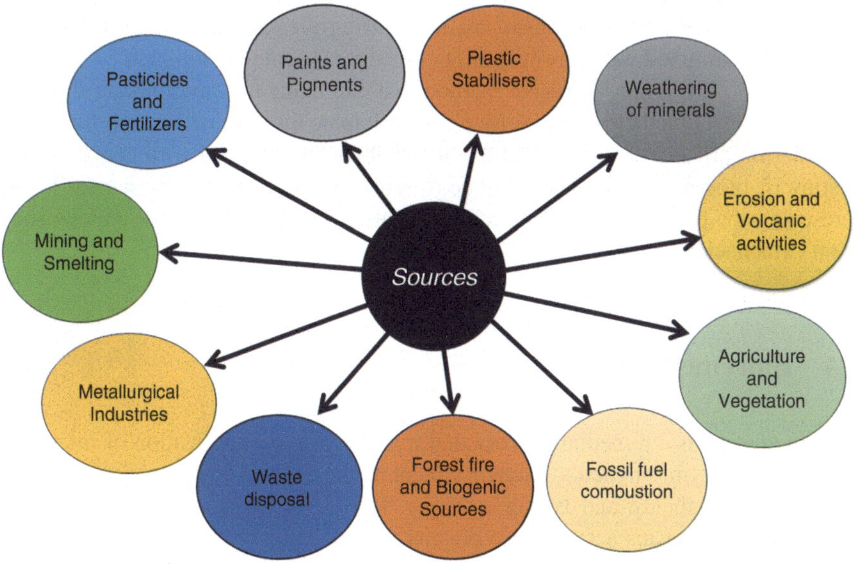


**Fig. 6.1** Sources of heavy metals in the aquatic environment

## 6.3 Prospects of Sustainable Approaches in Heavy Metal Management

This chapter reviews the prospect of sustainable approaches in heavy metal eradication in comparison to traditional chemical technologies. Different traditional technologies such as chemical precipitation method, ion exchange, sedimentation, microfiltration method, and reverse osmosis are not only cost-effective but also eco-friendly, as they represent a genuine danger to life in aquatic system because of different side effects and associatively pollute the aquatic environment (Igiri et al. 2018).

Bioremediation is an eco-accommodating, sustainable, and effective strategy for recovering sites polluted with different pollutants by utilizing the intrinsic biological systems of microbes and plants to degrade toxic contaminants (Kannabiran 2017; Prasad and Aranda 2018). Bioremediation is very cost-effective as compared to chemical technologies because the biosystems utilized in this process are prepared from the naturally available or waste biomass of bacteria, fungi, or algae which are clearly extremely modest (Kratochvil and Volesky 1998; Ayangbenro and Babalola 2017). Different plant wastes such as rice husks, spent grain, sawdust, sugarcane bagasse, fruit wastes, and weeds were used as adsorbents for different heavy metals, viz., Cd, Cu, Pb, Zn, and Ni (Ngah and Hanafiah 2008; Acharya et al. 2018). Bioremediation should be possible on location, in this way diminishing presentation dangers for cleanup faculty, or conceivably more extensive exposure because of transportation mishaps. Other than the above focal points, bioremediation is more affordable, disposes of waste enduringly, takes out long-term liability, and can be combined with physical or chemical treatment technologies.

Moreover, it is a non-obtrusive method that can make the environment flawless (Vidali 2001). Nonetheless, it is difficult to anticipate the pace of cleanup for a bioremediation practice as a few ecological variables are engaged with choosing the destiny of bioremediation, and till date, researchers are looking for standards for foreseeing the pace of removal of contaminants from various parts of the environment (Machackova et al. 2012).

## 6.4 Bioremediation

Bioremediation is a remedial process that mainly involves application of microbes and/or their enzymes to detoxify environmental contaminants for restoring its original form (Ayangbenro and Babalola 2017). This is a naturally occurring process where microorganisms act as major players that clean up pollutants of soil, water, and other environmental sources. In this process, the growth of certain microorganisms can be stimulated as they utilize these pollutants as a source of nutrition and energy (Ostrem Loss and Yu 2018). Through the metabolic activities of microbes, a variety of contaminants, especially heavy metals, can be degraded. Therefore, bioremediation can be very well used as a sustainable solution for heavy metal pollution which is constantly increasing due to industrialization and human activities (US EPA 2011; Masindi and Muedi 2018). Bioremediation process involves degradation, removal, alteration, immobilization, or detoxification of several heavy metals from the environment through the cellular processes of plants and microorganisms like bacteria and fungi (Artin 2010).

### *6.4.1 Mechanism of Bioremediation*

Bioremediation is the most economical way of heavy metal management. With the advent of technologies, it becomes possible to restore the heavy metal-polluted site by using different methodologies. These different techniques are employed based on various aspects such as aeration of area, characteristics of site, type of pollutant and its concentration, biosorption and bioavailability of pollutant, etc. (Smith et al. 2015; Azubuike et al. 2016; Ojuederie and Babalola 2017). However, no single technique can be helpful to achieve the complete restoration of heavy metal-contaminated environment. Therefore, different strategies are employed to clean up the environment (Verma and Jaiswal 2016). Broadly, bioremediation process can be carried out by two approaches as in situ and ex situ (Sharma 2012; Yuniati 2018).

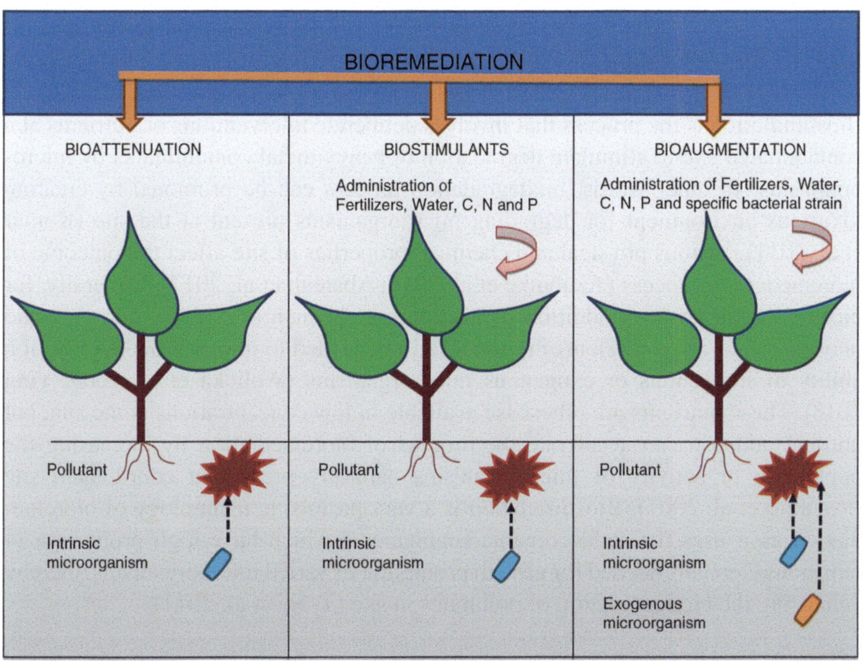


**Fig. 6.2** Types of bioremediation

Bioremediation can be achieved through three different technological processes (Adams et al. 2015; Fig. 6.2) as follows:

#### 6.4.1.1 Bioattenuation

Bioattenuation process is also known as natural attenuation as it relies on the natural way of heavy metal degradation which does not involve any intervention from humans (Mulligana and Yong 2004; Ying 2018). This process takes the advantage of metabolic diversity of intrinsic microorganisms present at the polluted site. Based on metabolic activities, these indigenous microbes can degrade, detoxify, neutralize, or transform the heavy metal (Abatenh et al. 2017). Bioattenuation comprises of various chemical, physical and biological processes in order to diminish the concentration and toxicity of recalcitrant. These processes encompass biodegradation by aerobic or anaerobic means, sorption, volatilization, stabilization, or transformation of pollutants (Mulligana and Yong 2004). The time required for natural attenuation of pollutants may vary from site to site depending upon the site conditions, type of contaminants, and degrading microbial flora of the site (Azubuike et al. 2016). This process is applied to the site where concentration of pollutants is minimal and no other bioremedial technique can work. Bioattenuation process is applicable to control soil as well as water heavy metal pollution that mainly relies upon the appropriate degrading microorganisms (Yu et al. 2005).

#### 6.4.1.2 Biostimulation

Biostimulation is the process that involves deliberate interventions of nutrients at a contaminated site to stimulate degradation of heavy metal contaminants by microorganisms. In other words, biodegradation process can be promoted by creating luxurious environment for degrading microorganisms present at the site (Kumar et al. 2011). Various physical and chemical properties of site affect the outcome of bioremediation process (Azubuike et al. 2016; Abatenh et al. 2017). Generally, for biostimulation process, addition of macronutrients such as carbon, nitrogen, and phosphorus or micronutrients in proper ratio is needed to improve the degradation ability of indigenous or exogenous microorganisms (Wolicka et al. 2009; Ying 2018). These nutrients are otherwise available in low concentrations at the site, but nutrient addition can accelerate the process of bioremediation by increasing the population or activity of microorganisms naturally present at coordinated site (Perfumo et al. 2007). Biostimulation is a very promising technology of bioremediation as it uses the stable organic supplements which have high proportion of nutritious elements needed for growth promotion of varied microorganisms, thereby enhancing the biodegradation of pollutants in site (Tyagi et al. 2011).

#### 6.4.1.3 Bioaugmentation

Bioaugmentation is one of the approaches of bioremediation that encompasses the induction of specific microbes with proficiency in degradation of heavy metal pollutants from contaminated site in order to improve the removal of contaminants through biodegradation (Goswami et al. 2018). Recently, bioaugmentation methods are gaining significant attention as a strategy of bioremediation. Basically, bioaugmentation is a one of the efficient ways of bioremediation where exogenous microorganisms with potential degradation ability are added to the site of contamination to speed up detoxification and decomposition of heavy metal pollutants. These altered microorganisms are either single strain of bacteria or consortia of microorganisms (Niu et al. 2009). These microorganisms can be isolated from natural environmental sources or genetically modified in the laboratory (Kulshreshtha 2013). The competency of bioaugmentation process relies upon many biotic and abiotic factors (Simon et al. 2004). The abiotic factors that affect bioaugmentation process include physiological and chemical properties of contaminated site, chemical structure, bioavailability of contaminants, and their concentration (Goswami et al. 2018). Biotic factors involve the selection of appropriate microorganisms that will have the ability to degrade heavy metal pollutants as well as to compete effectively with intrinsic microorganisms of the site (Abatenh et al. 2017).

Various approaches have been employed to make bioaugmentation as an efficacious remedial technique to recover contaminated site without destroying intrinsic microorganisms. Bioaugmentation process uses exogenous microorganisms which can be genetically modified in order to inherit desired catalytic capabilities among them. Therefore, genetically modified microorganisms exhibit enhanced decomposition ability covering numerous aromatic components (Abatenh et al. 2017). Moreover, bioaugmentation process can also be improvised by inoculating appropriate microorganism which is encapsulated using a variety of carriers like alginate (Mrozik and Piotrowska-Seget 2010). Certain newer approaches of bioaugmentation include gene augmentation technique where remediation gene is transferred to indigenous microorganisms. Rhizosphere bioaugmentation is another approach which includes introduction of microorganisms to the site along with plant to encourage microbial growth and degradation of pollutant (Kumar and Fulekar 2018). Phytoaugmentation approach does not involve the introduction of microbial inoculant; instead, it uses plants that are genetically engineered by transferring remediation genes (Gentry et al. 2004). It is believed that when bioattenuation and biostimulation process fail to work out, bioaugmentation technique should be employed (Mrozik and Piotrowska-Seget 2010).

Bioaugmentation approaches had been practiced to clean up undesirable compounds from a site that mainly include heavy metals. The selection of proper microorganism is very important for efficient bioaugmentation of the defected site. Several types of microbial species are used for bioaugmentation process. Many experiments exploit the efficiency of bacteria belonging to the genera *Pseudomonas*, *Bacillus*, *Sphingobium*, etc. The fungi belonging to the genera *Verticillium*, *Penicillium*, and *Aspergillus* had been experimented to remove undesirable heavy metals from polluted sites, especially to treat wastewater (Bahobil et al. 2017).

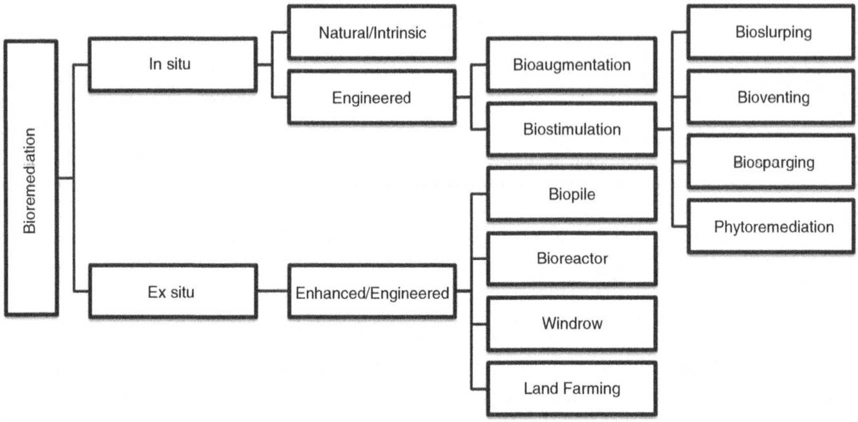


**Fig. 6.3** Strategies of bioremediation process

Bioremediation process is used to eliminate various contaminants like pesticides, heavy metals, hydrocarbons and chlorinated compounds, dyes, plastic waste, greenhouse gases, sewage, oil spills, and nuclear waste (Azubuike et al. 2016). Implementation of in situ or ex situ bioremediation strategy is decided according to the site of application (Igiri et al. 2018). Several in situ (on site) and ex situ (off site) approaches (Tomei and Daugulis 2013; Fig. 6.3) are useful in controlling environmental pollution.

Bioremediation techniques have been demonstrated as proficient and sustainable approaches in restoring polluted site containing a wide variety of pollutants (Gao et al. 2018). Microorganisms are the key players in bioremediation process; hence, it is important to consider variety in microbial species, microbial population, and their complexity present at contaminated environments for their appropriate exploitation and to decide the outcome of any bioremediation strategy. Additionally, environmental aspects that can affect microbial activities have to be maintained at the optimal level. Advanced microbial molecular detection methods such as genomics, proteomics (Singh 2006), metabolomics and transcriptomics (Chauhan and Jain 2010) have revolutionized the understanding of identification of microorganisms, their functions and metabolic pathways (Azubuike et al. 2016), which is required for developing microbial culture and its application. Less population of desired microorganisms, limitation of nutrients, low degradation capabilities, and bioavailability of pollutants are major governing factors that may affect the process of bioremediation. Therefore, optimization of such important parameters may determine the success of any bioremediation technique.

In a nutshell, bioremediation technology has been proven as an efficient, economical, and eco-friendly or sustainable approach for the restoration of contaminated site including soil and groundwater. Moreover, much attention should be paid to research specifically focusing toward development of more effective treatment design and performance. In addition to it, research should be also directed to develop newer strategies and approach for enhancement of bioavailability and mass transformation of contaminants, bioprocess optimization, and multidisciplinary integrative approach to reduce environmental pollution.

## 6.5 Phytoremediation

In the twenty-first century, the major challenge all over the world is rapid increase of industrialization and urbanization that has led to environmental pollution with several toxic and hazardous materials. Heavy metal contamination is the most distinguished concern since it directly affects the efficiency, growth rate, developmental stages, and productivity of plants (White et al. 2006). Several approaches (physiochemical and biological) have been used or established to restore heavy metal-polluted waters/soils including the landfill/damping locations (Das 2016; Ayangbenro and Babalola 2017). Thus, remediation approach is very critical to eliminate heavy metals from the water. These remediation techniques comprise several treatment methods for pollutant degradation, removal/separation (through accumulation or dissipation), or immobilization (Malik et al. 2017; Padmavathiamma and Li 2007). Phytoremediation approach takes account of soil microorganisms symbiotically associated with green plants to eliminate harmful contaminants from polluted soil and waters/wastewaters through degradation and detoxification mechanisms (Ali et al. 2013; Bharagava et al. 2017; Saxena et al. 2019). It can be efficient for the eco-restoration of locations mostly polluted with heavy metals, radioactive compounds, and several organic contaminants (Ali et al. 2013; Mahar et al. 2016). It is an environmentally friendly, non-invasive, and aesthetically attractive remediation technology that eliminates heavy metal contaminants from the polluted locations (Saxena et al. 2019). It consists of diverse phytoremediation techniques for the deterioration of numerous contaminants using altered mechanisms contingent on their applications. Based on the toxic contaminant source, field environments, required level of environmental clean-up, and plant nature, there are different phytoremediation techniques that can be used. These techniques include phytoextraction/phytovolatilization, phytodegradation, phytostabilization/phytoimmobilization, rhizodegradation, and rhizofiltration (Thangavel and Subbhuraam 2004; Saxena et al. 2019). Phytoremediation techniques consist of diverse plant-based technologies. The definition, mechanism, application, benefits, and restricted access of common and long-established phytoremediation practices are shown in Table 6.1.

Selected commonly recycled phytoremediation methodologies are as follows (Parmar and Singh 2015):

(1) Phytostabilization relies on either precipitation or immobilization of pollutants from groundwater and soil using plants, therefore reducing accessibility.

(2) Phytofiltration process uses roots and parts of plants to absorb pollutants from the water bodies.

**Table 6.1** Description of phytoremediation mechanisms and applications

| Phytoremediation processes | Definition | Mechanism | Pollutants | Applicability |
|---|---|---|---|---|
| Phytoextraction or phytoaccumulation or phytosequestration or phytoabsorption | Plants remove metal pollutants from contaminated sites via plant's root absorption and sequester/concentrate in aboveground harvestable plant parts | Hyperaccumulation | Pb, Cd, Zn, Ni, Cu, Pb, radionuclides, pentachlorophenol, aliphatic compounds (short chained) | Contaminated soil/sites, water, wastewaters |
| Phytofiltration or rhizofiltration | Plants concentrate and precipitate metal pollutants in low concentration from the aquatic environment in their roots | Rhizosphere accumulation | Pb, Cd, Zn, Ni, Cu, radionuclides (Cs, Sr, U), hydrophobic organics | Contaminated water and wastewaters |
| Phytostabilization or phytoimmobilization or phytotransformation | Plants immobilize or inactivate metal pollutants at their place involving absorption by roots, adsorption onto roots, and precipitation, complexation, and metal valence reduction in rhizosphere, e.g., reduction of $Cr^{6+}$ to $Cr^{3+}$ | Precipitation, complexation, and metal valence reduction | Pb, Cd, Zn, As, Cu, Cr, Se, U, polycyclic aromatic hydrocarbons (PAHs), polychlorinated biphenyl (PCBs), dioxins, furans, pentachlorophenol, DDT, dieldrin | Contaminated soil/sediments and sludge |
| Phytovolatilization or phytoevaporation | Plants take up metal pollutants through roots in low concentration, modify/transform them into less toxic form, and subsequently transpire/volatilize into the atmosphere through stomata | Volatilization or evaporation by leaves | Chlorinated solvents like carbon tetrachloride, trichloroethylene, methylene chloride, tetrachloroethylene, carbon tetrachloride, 1,1,1-trichloroethane, Hg (mercuric ion), Se | Contaminated wastewaters, soil, sediments, and sludges |
| Phytodegradation | Plants break down/convert highly toxic organic pollutants into less toxic forms through the action of enzymes secreted within plant tissues and released in the rhizosphere | Degradation in plant tissues | DDT, PAHs, bisphenol A, organophosphorus compounds | Contaminated soil, sediments, sludges, groundwater, surface water, and wastewaters |
| Rhizodegradation or rhizoremediation or phytostimulation | Plants break down/convert highly toxic organic pollutants into less toxic forms through enzymatic activity of rhizospheric microorganisms | Degradation in rhizosphere | Atrazine, ammunition wastes, petroleum hydrocarbon, PCBs, PAHs, TCE, diesel fuel | Contaminated soil, sediments, sludges, groundwater, and wastewaters |

Adapted from Yadav (2010), Ali et al. (2013), Jaishankar et al. (2014), Dixit et al. (2015), Sarwar et al. (2017), Saxena et al. (2019)

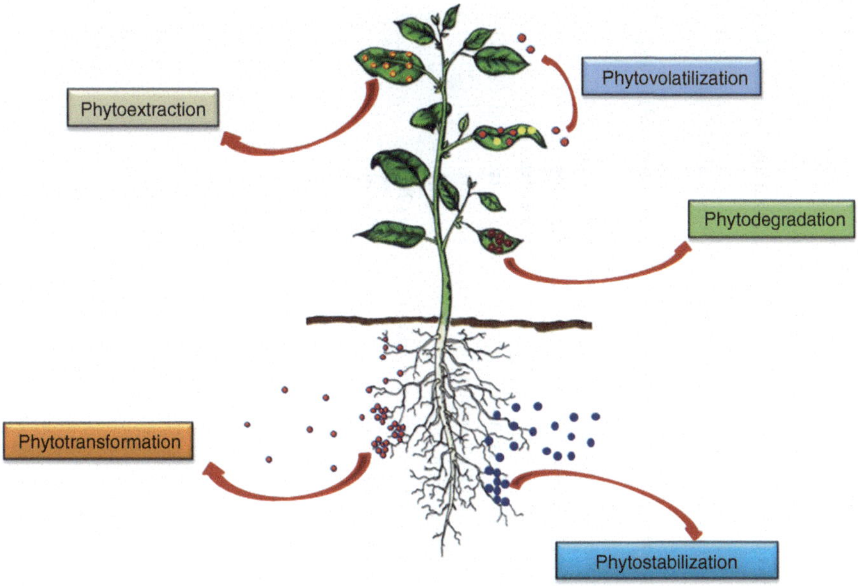


**Fig. 6.4** Schematic representation of phytoremediation techniques

(3) Phytovolatilization can clean up groundwater and soil by means of plants that can evapotranspirate pollutants like mercury (Hg), selenium (Se), and volatile hydrocarbons.

(4) Phytoextraction process involves uptake and absorption of metal pollutants from water and soil via plant tissues and consequent elimination of metal.

(5) Phytodegradation process utilizes plants or microbes for degradation of metals (or other contaminants) in groundwater as well as rhizospheric soil through metabolic activities.

(6) Phytotransformation implicates the uptake of organic pollutants from a polluted site and their transformation by plants into nontoxic or lesser toxic components.

(7) Evapotranspiration uses vegetative plants to prevent the leaching of pollutants.

Figure 6.4 shows the schematic presentation of the different phytoremediation approaches.

Uptake of metals is influenced by chemical speciation of metal and habitat characteristics of plants (aquatic, terrestrial, etc.). Thus, selection of plant is crucial for remediation of polluted sites. There are many plants exhibiting remediation characteristics that belong to diverse families, for example, Brassicaceae, Cyperaceae, Fabaceae, Poaceae, Lamiaceae, Caryophyllaceae, Euphorbiaceae, etc. (Sarma 2011). There are several plant species which are described as hyperaccumulators on the basis of their capability to sustain concentrations of toxic metal pollutants, as outlined in Table 6.2.

Of all the phytoremediation mechanisms, phytoextraction is the prominent technique to eliminate heavy metals from polluted locations. In phytoextraction

**Table 6.2** Some metal hyperaccumulator plants (with metal accumulation ability) used for phytoremediation

| Plant species | Family | Polluted medium | Metal | Metal accumulation capacity (mg $kg^{-1}$DW) | Phytoremediation mechanism and metal accumulation compartment | Reference |
|---|---|---|---|---|---|---|
| *Tagetes minuta* | Asteraceae | Water | As | 380.5 | Phytoextraction (shoots) | Salazar and Pignata (2014) |
| *Noccaea caerulescens* | Brassicaceae | Water | Pb | 1700–2300 | Rhizofiltration (aerial parts/root) | Dinh et al. (2018) |
| *Pteris vittata* | Pteridaceae | Water | As | 20,707 | Phytoextraction (shoots) | Kalve et al. (2011) and Oliveira et al. (2014) |
| | | | Cr | 35,303 | Phytoextraction (shoots) | Kalve et al. (2011) |
| | | | As | 20,707 | Phytoextraction (shoots) | Sakakibara et al. (2011) |
| *Eleocharis acicularis* | Cyperaceae | Water | Zn | 11,200 | Phytoextraction (shoots) | Sakakibara et al. (2011) |
| | | | Cu | 20,200 | Phytoextraction (shoots) | Sakakibara et al. (2011) |
| | | | Cd | 239 | Phytoextraction (shoots) | Sakakibara et al. (2011) |
| | | | As | 1470 | Phytoextraction (shoots) | Sakakibara et al. (2011) |
| *Azolla pinnata* | Salviniaceae | Water | Cd | 740 | Rhizofiltration (bioaccumulation) | Rai (2008) |
| *Lonicera japonica* | Caprifoliaceae | Water | Cd | – | Phytoextraction (shoots) | Liu et al. (2011a, b) |

technique, green plants are used to eliminate contaminants from polluted locations through root absorption and their sequestration (Saxena et al. 2019). This technique can have commercial applications since it is economically feasible to remove metals from polluted locations using plant biomass extracted and employed as "bio-metal" to get valuable, useful, and efficient metals; the procedure can be referred to as phytomining (Ali and Singh 2018). Therefore, it can create income and offer further employment opportunity for the people. Phytoextraction proficiency of green plants mainly depends on two factors, viz., bioconcentration factor (BCF) and translocation factor (TF). BCF characterizes concentration of metals in root/soil, representing metal accumulation, while TF describes concentration of metals in shoot/root that signifies metal translocation (Ali et al. 2013; Antoniadis et al. 2017).

Currently, there is an increasing curiosity for the exploitation of metal-accumulating roots and rhizomes of aquatic and semi-aquatic vascular plants for the elimination of heavy metals from polluted water bodies (Mémon et al. 2001). For instance, *Hydrocotyle umbellata*, *Eichhornia crassipes*, *Tagetes minuta*, *Lemna minor*, *Pteris vittata*, *Lonicera japonica*, *Eleocharis acicularis*, *Noccaea caerulescens*, and *Azolla pinnata* absorbed Cu, Cr, Hg, Pb, As, Cd, Fe, and Zn from polluted water bodies (Gallardo et al. 1999; Gustin et al. 2009; Oliveira et al. 2014). Furthermore, elimination of a widespread series of metal ions present in contaminated solutions using cell suspension cultures of *Datura innoxia* has been demonstrated (Waoo et al. 2017). Maximum eliminated metals were strongly chelated by unrevealed constituents of cell walls in a manner which did not involve metabolic activity. The hyperaccumulation of metals in several plant species has been broadly studied, and till date, extensive improvement has been accomplished. It has been clearly investigated that diverse mechanisms of metal accumulation, elimination, and compartmentation occur in many plant varieties. Now, the ever-increasing knowledge of biochemical pathways and metabolic processes of plants in relation to uptake/absorption of heavy metal, its accumulation, transport, and resistance will persuade enhancements of phytoremediation via recent genetic techniques. Therefore, phytoremediation proficiency of plants can be significantly enhanced by using these genetic engineering techniques.

## 6.6 Role of Advanced Biotechnological Techniques and Genetic Engineering in Bioremediation Process

Microbes are used in bioremediation in light of their capacity to deteriorate ecological pollutants because of their metabolism by means of biochemical pathways identified with the life form movement and development. Microbes have the capacity to degrade the harmful substances of polluted environment into harmless end products by the process of co-metabolism (Ojuederie and Babalola 2017). Degradation of hazardous products into harmless products utilizing inbred microbes did not yield much positive outcomes. Bioremediation of Hg from the polluted environment by indigenous bacteria has not been reported.

Nonetheless, recombinant DNA technology (RDT) has a noteworthy task to carry out bioremediation of heavy metals since it upgrades the remedial procedure (Kang et al. 2016). Introduction of genetic engineering techniques in bioremediation is intended to alter genotype of plants and microorganisms, thereby changing their functional proteins like enzymes, in order to use them as potential agents for deterioration of hazardous compounds (Wolejko et al. 2016). Azad et al. (2014) reported the use of genetically engineered bacteria for degradation of different heavy metals. Genetically engineered microorganisms (GEM) have been utilized to acquire skillful strains for bioremediation of polluted environment by having improved capacity to degrade an assortment of pollutants. Several studies have demonstrated the removal of Hg from polluted environments by using genetically modified *Escherichia coli* strain M109 and *Pseudomonas putida* with *merA* gene (Ojuederie and Babalola 2017). Different genes such as *merA* gene, *pheA*, *pheB*, *pheC*, *pheD*, and *pheR* genes (phenol catabolic genes), and ArsM gene have been extensively used for the removal of Hg, phenol, and As, respectively, with the help of genetic engineering (Liu et al. 2011a, b). Addition of *mer* genes into *Deinococcus geothermalis* bacterium from *Escherichia coli*, which are responsible for the degradation of Hg, enabled the bacterium for removal of Hg from polluted environments (Dixit et al. 2015). Sone et al. (2013) reported the addition of novel genes utilizing pMR68 plasmid for the synthesis of Hg-resistant strains of *Pseudomonas*. Thus, GEM can be used to assist remediation process to defeat the pollutants from the environment. It is also important to maintain the stability of these genetically modified microorganisms before applying them to the field, since the catabolic action of GEM is mainly related to the presence of stable recombinant plasmid in them (Ghosal et al. 2016).

Some modern techniques such as site-directed mutagenesis and rational designing have been used to engineer the microbes for the degradation of heavy metal contaminants (Kumar et al. 2013). Microbial biosensors are presently being utilized to set up the measure of heavy metal contaminants rapidly and accurately and are created utilizing genetic engineering. Dixit et al. (2015) detailed the utilization of biosensors to assess the degrees of different heavy metals in polluted environments. Usage of genetic engineering guarantees more prominent open doors for acquiring powerful pollutant-degrading microbes as they could have higher capability of ecological cleanup than the inbred microorganisms.

Transgenic plants can be obtained with insertion of specific genes in the genome of plants with enhanced phytoremediation capability using genetic engineering. Genetically engineered endophytes and plant growth-promoting microbes (PGPM) can adequately degrade the heavy metals in contaminated environment (Dixit et al. 2015; Ojuederie and Babalola 2017). Mani and Kumar (2014) reported that the expression of *merA* genes in transgenic rice and tobacco makes them ten times more resistant to Hg than those that do not express *merA* genes. Chen and Wilson (1997) observed that the different transgenic plants, such as *Arabidopsis thaliana*, *Nicotiana tabacum*, *Brassica juncea*, *Brassica oleracea* var. *botrytis*, and *Lycopersicon esculentum*, have been used for the degradation of heavy metals. Innovative research on the rapidly developing plants having capabilities of metal aggregation should be advanced. Additionally, microorganisms from different genera have to be explored for improving plants and rhizospheric microorganisms at genetic level, which can be eventually used for phytoremediation. Besides the *merA* genes, various genes ought to be investigated for their conceivable use in defeating a wide variety of heavy metals. Recombinant DNA technology is fundamental for the bioremediation procedure as it empowers analysts to examine, screen, and evaluate the execution of the procedure (Ojuederie and Babalola 2017). It ought to be utilized with alert and as per biosafety guidelines.

## 6.7 Conclusions

This chapter featured the heavy metal contamination sources and different mechanisms utilized by plants and microorganisms including their enzymes for the effective remediation of polluted environment, especially aquatic system. It uncovered the advantages of bioremediation as a superior alternate as well as sustainable approach in the expulsion of pollutants like heavy metals from the environment as compared to other existing physical and chemical strategies that are less effective and costly because of the measure of energy consumed. Microbes and plants have natural biological systems that empower them to make do under heavy metal pressure and expel the metals from the environment. Different processes such as precipitation, biosorption, enzymatic transformation of metals, and complexation are used during the bioremediation of heavy metals by the microbes for the removal of heavy metals from the polluted environments. Plants use phytoremediation techniques of which phytoextraction and phytostabilization have been very efficient. Environmental variables assume a noteworthy job in the achievement of bioremediation as the microorganisms utilized will be hampered if suitable ecological conditions are not accessible. Transgenic microorganisms and plants could successfully remediate polluted destinations of heavy metal and organic contaminations; however, its utilization ought to be liable to stringent biosafety techniques to guarantee that there is no well-being or ecological dangers.

Application of metagenomic approaches must be taken into consideration to understand the community structure of microorganisms present at the treatment site to explore metal-resistant genes for cleaning up various heavy metals by improving the degrading microbial strains.

Public impression of the utilization of different modern sustainable technologies for bioremediation will likewise have to change for its compelling usage; this determines necessity of collaboration among scientists and environmentalists.

## 6.8 Future Prospects

Rapid industrial development and innovation advancement pose negative effects on environment like water pollution where quality of water gets deteriorated. Because of the multifaceted nature engaged with the traditional approaches for biological treatment of water especially contaminated with heavy metals as a pollutant, the utilization of microorganisms has emerged as a help for bioremediation. Be that as it may, bioremediation innovation has impediments; few microorganisms cannot break down lethal metals into innocuous metabolites, and these affect microbial movement. Changes in the external layer proteins of microorganisms with potential bioremediation properties for improving metal restricting capacities are the possible method to upgrade their biotransformation ability of dangerous metals. Further investigations should concentrate on the variables engaged with improving in situ bioremediation methodologies utilizing GEMs and furthermore the applicability and flexibility of these GEMs in all the conceivable unfavorable/stressed conditions and environments polluted with different heavy metals. The hesitance among people in general to acknowledge GEM for bioremediation likewise needs to be taken into consideration as future investigations, and their non-harmfulness to the environment should be demonstrated.

## References

Abatenh E, Gizaw B, Tsegaye Z, Wassie M (2017) Application of microorganisms in bioremediation-review. J Environ Microbiol 1:02–09

Achal V, Pan X, Zhang D (2012) Bioremediation of strontium (Sr) contaminated aquifer quartz sand based on carbonate precipitation induced by Sr resistant *Halomonas* sp. Chemosphere 89:764–768

Acharya J, Kumar U, Rafi PM (2018) Removal of heavy metal ions from wastewater by chemically modified agricultural waste material as potential adsorbent-a review. Int J Curr Eng Technol 8:526–530

Adams GO, Fufeyin PT, Okoro SE, Ehinomen I (2015) Bioremediation, biostimulation and bioaugmentation: a review. Int J Environ Biorem Biodegrad 3:28–39

Ali S, Singh M (2018) An eco-friendly technique of phytoremediation. Int J Recent Trends Sci Technol ACAEE 2018:232–239

Ali H, Khan E, Sajad MA (2013) Phytoremediation of heavy metals-concepts and applications. Chemosphere 91:869–881

Antoniadis V, Levizou E, Shaheen SM, Ok YS, Sebastian A, Baum C, Prasad MNV, Wenzel WW, Rinklebe J (2017) Trace elements in the soil-plant interface: phytoavailability, translocation, and phytoremediation - a review. Earth Sci Rev 171:621–645

Artin H (2010) Principles of bioremediation processes. In: Grażyna P (ed) Trends in bioremediation and phytoremediation, vol 661. Research Signpost, Kerala, pp 23–54

Ayangbenro A, Babalola O (2017) A new strategy for heavy metal polluted environments: a review of microbial biosorbents. Int J Environ Res Pub Health 14(1):94

Azad MA, Amin L, Sidik NM (2014) Genetically engineered organisms for bioremediation of pollutants in contaminated sites. Chin Sci Bull 59(8):703–714

Azubuike CC, Chikere CB, Okpokwasili GC (2016) Bioremediation techniques-classification based on site of application: principles, advantages, limitations and prospects. World J Microb Biot 32:180

Bahobil A, Bayoumi RA, Atta HM, El-Sehrawey MM (2017) Fungal biosorption for cadmium and mercury heavy metal ions isolated from some polluted localities in KSA. Int J Curr Microbiol Appl Sci 6:2138–2154

Bharagava RN, Chowdhary P, Saxena G (2017) Bioremediation: an ecosustainable green technology: its applications and limitations. In: Bharagava RN (ed) Environmental pollutants and their bioremediation approaches, 1st edn. CRC Press, Boca Raton, pp 1–22. https://doi.org/10.1201/9781315173351-2

Burri NM, Weatherl R, Moeck C, Schirmer M (2019) A review of threats to groundwater quality in the anthropocene. Sci Total Environ 684:136–154

Chandra R, Saxena G, Kumar V (2015a) Phytoremediation of environmental pollutants: an ecosustainable green technology to environmental management. In: Chandra R (ed) Advances in biodegradation and bioremediation of industrial waste. CRC Press, Boca Raton, pp 1–30. https://doi.org/10.1201/b18218-2

Chandra K, Salman AS, Mohd A, Sweety R, Ali KN (2015b) Protection against FCA induced oxidative stress induced DNA damage as a model of arthritis and in vitro anti-arthritic potential of *costus speciosus* rhizome extract. Inter J Pharma Phyto Res 7(2):383–389

Chauhan A, Jain RK (2010) Biodegradation: gaining insight through proteomics. Biodegradation 21:861–879

Chen S, Wilson DB (1997) Genetic engineering of bacteria and their potential for $Hg^{2+}$ bioremediation. Biodegradation 8(2):97–103

Chibuike GU, Obiora SC (2014) Heavy metal polluted soils: effect on plants and bioremediation methods. Appl Environ Soil Sci 2014:12

Concas A, Ardau C, Cristini A, Zuddas P, Cao G (2006) Mobility of heavy metals from tailings to stream waters in a mining activity contaminated site. Chemosphere 63(2):244–253

Das DK (2016) Assessment of heavy metal accumulation in medicinal plants and possible remedial measures. J Pharm Res 15:63–66

Dinh N, van der Ent A, Mulligan DR, Nguyen AV (2018) Zinc and lead accumulation characteristics and in vivo distribution of $Zn^{2+}$ in the hyperaccumulator *Noccaea caerulescens* elucidated with fluorescent probes and laser confocal microscopy. Environ Exp Bot 147:1–12

Dixit R, Wasiullah MD, Pandiyan K, Singh UB, Sahu A, Shukla R, Singh BP, Rai JP, Sharma PK, Lade H, Paul D (2015) Bioremediation of heavy metals from soil and aquatic environment: an overview of principles and criteria of fundamental processes. Sustainability 7:2189–2212

Dowarah J, Boruah HD, Gogoi J, Pathak N, Saikia N, Handique AK (2009) Eco-restoration of a high-sulphur coal mine overburden dumping site in Northeast India: a case study. J Earth Sys Sci 118(5):597–608

Gallardo MT, Benson RF, Martin DF (1999) Lead accumulation by three aquatic plants. Symposia papers presented before the division of Environmental Chemistry, American Chemical Society 39, pp 46–47

Gao H, Xie Y, Hashim S, Khan AA, Wang X, Xu H (2018) Application of microbial technology used in bio remediation of urban polluted river: a case study of Chengnan River, China. Water 10:643

Gentry T, Rensing C, Pepper I (2004) New approaches for bioaugmentation as a remediation technology. Crit Rev Environ Sci Technol 34:447–494

Ghosal D, Ghosh S, Dutta TK, Ahn Y (2016) Current state of knowledge in microbial degradation of polycyclic aromatic hydrocarbons (PAHs): a review. Front Microbiol 7:1369

Global Water Intelligence (GWI) Report (2018)

Goswami M, Chakraborty P, Mukherjee K, Mitra G, Bhattacharyya P, Dey S, Tribedi P (2018) Bioaugmentation and biostimulation: a potential strategy for environmental remediation. J Microbiol Exp 6:223–231

Govarthanan M, Lee KJ, Cho M, Kim JS, Kamala-Kannan S, Oh BT (2013) Significance of autochthonous *Bacillus* sp. KK1 on biomineralization of lead in mine tailings. Chemosphere 8:2267–2272

Gustin JL, Loureiro ME, Kim D, Na G, Tikhonova M, Salt DE (2009) MTP1-dependent Zn sequestration into shoot vacuoles suggests dual roles in Zn tolerance and accumulation in Zn-hyperaccumulating plants. Plant J 57:1116–1127

Igiri BE, Okoduwa SI, Idoko GO, Akabuogu EP, Adeyi AO, Ejiogu IK (2018) Toxicity and bioremediation of heavy metals contaminated ecosystem from tannery wastewater: a review. J Toxicol 2018:1–16

Jaishankar M, Tseten T, Abbalagan N, Mathew BB, Beeregowda KN (2014) Toxicity, mechanism and health effects of some heavy metals. Interdiscip Toxicol 7:60–72

Kabata-Pendias A, Pendias H (1989) Trace elements in the soil and plants. CRC Press, Florida

Kalve S, Sarangi BK, Pandey RA, Chakrabarti T (2011) Arsenic and chromium hyperaccumulation by an ecotype of *Pteris vittata*-prospective for phytoextraction from contaminated water and soil. Curr Sci 100:888–894

Kang CH, Kwon YJ, So JS (2016) Bioremediation of heavy metals by using bacterial mixtures. Ecol Eng 89:64–69

Kannabiran K (2017) Actinobacteria are better bioremediating agents for removal of toxic heavy metals: an overview. Int J Environ Technol Manag 20(3–4):129–138

Kapoor D, Singh S, Kumar V, Romero R, Prasad R, Singh J (2019) Antioxidant enzymes regulation in plants in reference to reactive oxygen species (ROS) and reactive nitrogen species (RNS). Plant Gene. https://doi.org/10.1016/j.plgene.2019.100182

Kim S, Park CB (2013) Bio-inspired synthesis of minerals for energy, environment and medicinal applications. Adv Funct Mater 23:10–25

Kratochvil D, Volesky B (1998) Advances in the biosorption of heavy metals. Trends Biotechnol 16:291–300

Kulshreshtha S (2013) Genetically engineered microorganisms: a problem solving approach for bioremediation. J Bioremed Biodegr 4:1–2

Kumar P, Fulekar MH (2018) Rhizosphere bioremediation of heavy metals (Copper and Lead) by *Cenchrus ciliaris*. Res J Environ Sci 12:166–176

Kumar A, Bisht BS, Joshi VD, Dhewa T (2011) Review on bioremediation of polluted environment: a management tool. Int J Environ Sci 1:1079–1093

Kumar S, Dagar VK, Khasa YP, Kuhad RC (2013) Genetically modified microorganisms (GMOS) for bioremediation. In: Kuhad R, Singh A (eds) Biotechnology for environmental management and resource recovery. Springer, New Delhi, pp 191–218

Li M, Cheng X, Guo H (2013) Heavy metal removal by biomineralization of urease producing bacteria isolated from soil. Int Biodeterior Biodegrad 76:81–85

Liu S, Zhang F, Chen J, Sun G (2011a) Arsenic removal from contaminated soil via biovolatilization by genetically engineered bacteria under laboratory conditions. J Environ Sci 23(9):1544–1550

Liu ZL, Chen W, He XY (2011b) Cadmium-induced changes in growth and antioxidative mechanisms of a medicine plant (*Lonicera japonica* Thunb.). J Med Plants Res 5:141–1417

Machackova J, Wittlingerova Z, Vlk K, Zima J (2012) Major factors affecting in situ biodegradation rates of jet-fuel during large-scale biosparging project in sedimentary bedrock. J Environ Sci Health Part A 47(8):1152–1165

Mahar A, Wang P, Ali A, Awasthi MK, Lahori AH, Wang Q, Li R, Zhang Z (2016) Challenges and opportunities in the phytoremediation of heavy metals contaminated soils: a review. Ecotoxicol Environ Saf 26:111–121

Malik ZH, Ravindran KC, Sathiyaraj G (2017) Phytoremediation: a novel strategy and eco-friendly green technology for removal of toxic metals. Int J Agric Environ Res 3:1–18

Mani S (2015) Production of reactive oxygen species and its implication in human diseases. In: Vibha R, Umesh CSY (eds) Free radicals in human health and disease. Springer, New Delhi, pp 3–15

Mani D, Kumar C (2014) Biotechnological advances in bioremediation of heavy metals contaminated ecosystems: an overview with special reference to phytoremediation. Int J Environ Sci Technol 11(3):843–872

Masindi V, Muedi KL (2018) Environmental contamination by heavy metals. In: Saleh HE-DM, Aglan RF (eds) Heavy metals. IntechOpen, London. Available at https://www.intechopen.com/books/heavymetals/environmental-contamination-by-heavy-metals. Accessed 19 May 2019

Massoud R, Hadiani MR, Darani KK, Hamzehlou P (2019) Bioremediation of heavy metals in food industry: application of *Saccharomyces cerevisiae*. Electron J Biotechnol 37:56–60

Mémon AR, Aktoprakligil D, Ozdemir A, Vertill A (2001) Heavy metal accumulation and detoxification mechanisms in plants. Turk J Bot 25:111–121

Mishra S, Kumar A, Yadav S, Singhal MK (2018) Assessment of heavy metal contamination in water of Kali River using principle component and cluster analysis, India. Sustain Water Resour Manag 4(3):573–581

Mrozik A, Piotrowska-Seget Z (2010) Bioaugmentation as a strategy for cleaning up of soils contaminated with aromatic compounds. Microbiol Res 165:363–375

Mulligana CN, Yong RN (2004) Natural attenuation of contaminated soils. Environ Int 30:587–601

Musilova J, Arvay J, Vollmannova A, Toth T, Tomas J (2016) Environmental contamination by heavy metals in region with previous mining activity. Bull Environ Contam Toxicol 97:569–575

Nematian MA, Kazemeini F (2013) Accumulation of Pb, Zn, Cu and Fe in plants and hyperaccumulator choice in Galali iron mine area, Iran. Int J Agric Crop Sci 5(4):426

Ngah WW, Hanafiah MA (2008) Removal of heavy metal ions from wastewater by chemically modified plant wastes as adsorbents: a review. Bioresour Technol 99(10):3935–3948

Niu GL, Zhang JJ, Zhao S, Liu H, Boon N, Zhou NY (2009) Bioaugmentation of a 4-chloronitrobenzene contaminated soil with *Pseudomonas putida* ZWL73. Environ Pollut 57:763–771

Ojuederie O, Babalola O (2017) Microbial and plant-assisted bioremediation of heavy metal polluted environments: a review. Int J Environ Res Public Health 14(12):1504

Oliveira LM, Ma LQ, Santos JAG, Guilherme LRG, Lessl JT (2014) Effects of arsenate, chromate, and sulfate on arsenic and chromium uptake and translocation by arsenic hyperaccumulator *Pteris vittata* L. Environ Pollut 184:187–192

Ostrem Loss EM, Yu JH (2018) Bioremediation and microbial metabolism of benzo(a)pyrene. Mol Microbiol 109:433–444

Padmavathiamma PK, Li LY (2007) Phytoremediation technology: hyperaccumulation metals in plants. Water Air Soil Pollut 184:105–126

Parmar S, Singh V (2015) Phytoremediation approaches for heavy metal pollution: a review. J Plant Sci Res 2:139–146

Perfumo A, Banat IM, Marchant R, Vezzulli L (2007) Thermally enhanced approaches for bioremediation of hydrocarbon-contaminated soils. Chemosphere 66(1):179–184

Pillai P (2010) Accumulation of heavy metals from contaminated wastewater by aquatic plant Lemna minor and their biochemical effects on it. Nat Environ Pollut Technol 9:767–774

Poirier I, Hammann P, Kuhn L, Bertrand M (2013) Strategies developed by the marine bacterium *Pseudomonas fluorescens* BA3SM1 to resist metals: a proteome analysis. Aquat Toxicol 128–129:215–232

Prasad R, Aranda E (2018) Approaches in bioremediation: the new era of environmental microbiology and nanobiotechnology. Springer International Publishing, Cham. ISBN 978-3-030-02369-0. https://www.springer.com/gp/book/9783030023683

Rai PK (2008) Phytoremediation of Hg and Cd from industrial effluents using an aquatic free floating macrophyte *Azolla pinnata*. Int J Phytoremediation 10:430–439

Rai PK (2009) Heavy metal phytoremediation from aquatic ecosystems with special reference to macrophytes. Crit Rev Environ Sci Technol 39(9):697–753

Rai PK (2012) An eco-sustainable green approach for heavy metals management: two case studies of developing industrial region. Environ Monit Assess 184(1):421–448

Sakakibara M, Ohmori Y, Ha NTH, Sano S, Sera K (2011) Phytoremediation of heavy metal contaminated water and sediment by *Eleocharis acicularis*. Clean Soil Air Water 39:735–741

Salazar MJ, Pignata ML (2014) Lead accumulation in plants grown in polluted soils. Screening of native species for phytoremediation. J Geochem Explor 137:29–36

Sarma H (2011) Metal hyperaccumulation in plants: a review focusing on phytoremediation technology. J Environ Sci Technol 4:118–138

Sarwar N, Imran M, Shaheen MR, Ishaq W, Kamran A, Matloob A, Rehim A, Hussain S (2017) Phytoremediation strategies for soils contaminated with heavy metals: modifications and future perspectives. Chemosphere 171:710–721

Saxena G, Purchase D, Mulla SI, Saratale GD, Bharagava RN (2019) Phytoremediation of heavy metal-contaminated sites: eco-environmental concerns, field studies, sustainability issues, and future prospects. Rev Environ Contam Toxicol 249:71–131

Sharma DC (2003) Concern over mercury pollution in India. Lancet 362(9389):1050

Sharma S (2012) Bioremediation: features, strategies and applications. Asian J Pharmacy Life Sci 2:202–213

Simon MA, Bonner J, Page CA, Townsend RT, Mueller DC, Fuller C, Autenrieth RL (2004) Evaluation of two commercial bioaugmentation products for enhanced removal of petroleum from a wetland. Ecol Eng 22:263–277

Singh O (2006) Proteomics and metabolomics: the molecular make-up of toxic aromatic pollutant bioremediation. Proteomics 6:5481–5492

Smith E, Thavamani P, Ramadass K, Naidu R, Srivastava P, Megharaj M (2015) Remediation trials for hydrocarbon-contaminated soils in arid environments: evaluation of bioslurry and biopiling techniques. Int Biodeterior Biodegrad 101:56–65

Sone Y, Mochizuki Y, Koizawa K, Nakamura R, Pan-Hou H, Itoh T, Kiyono M (2013) Mercurial-resistance determinants in *Pseudomonas* strain K-62 plasmid pMR68. AMB Express 3(1):41

Tak HI, Ahmad F, Babalola OO (2013) Advances in the application of plant growth-promoting rhizobacteria in phytoremediation of heavy metals. In: Reviews of environmental contamination and toxicology. Springer, New York, pp 33–52

Thangavel P, Subbhuraam CV (2004) Phytoextraction: role of hyperaccumulators in metal contaminated soils. Proc Indian Natl Sci Acad B 70:109–130

Tomei MC, Daugulis AJ (2013) Ex situ bioremediation of contaminated soils: an overview of conventional and innovative technologies. Crit Rev Environ Sci Technol 43(20):2107–2139

Tyagi M, da Fonseca MM, de Carvalho CC (2011) Bioaugmentation and biostimulation strategies to improve the effectiveness of bioremediation processes. Biodegradation 22(2):231–241

US Environmental Protection Agency (2011) Green remediation best management practices: sites with leaking underground storage tank systems. EPA 542-F-11-008

Verma JP, Jaiswal DK (2016) Book review: advances in biodegradation and bioremediation of industrial waste. Front Microbiol 6:1–2

Vidali M (2001) Bioremediation. An overview. Pure Appl Chem 73(7):1163–1172

Waoo AA, Khare S, Ganguly S (2017) Evaluation of phytoremediation potential of *Datura innoxia* Mill., with reference to four heavy metals: Cd, Pb, Ni, and Cr. Curr Environ Eng 4(1):66–71. https://doi.org/10.2174/2212717804666161219171111

White PM, Wolf DC, Thoma GJ, Reynolds CM (2006) Phytoremediation of alkylated polycyclic aromatic hydrocarbons in a crude oil-contaminated soil. Water Air Soil Pollut 169:207–220

Wolejko E, Wydro U, Loboda T (2016) The ways to increase efficiency of soil bioremediation. Eco Chem Eng 23(1):155

Wolicka D, Suszek A, Borkowski A, Bielecka A (2009) Application of aerobic microorganisms in bioremediation in situ of soil contaminated by petroleum products. Bioresour Technol 100(13):3221–3227

Yadav SK (2010) Heavy metals toxicity in plants: an overview on the role of glutathione and phytochelatins in heavy metal stress tolerance of plants. S Afr J Bot 76:167–179

Ying G-G (2018) Chapter 14 - remediation and mitigation strategies. In: Maestroni B, Cannavan A (eds) Integrated analytical approaches for pesticide management. Academic, Cambridge, pp 207–217

Yu KSH, Wong AHY, Yau KWY, Wong YS, Tam NFY (2005) Natural attenuation, biostimulation and bioaugmentation on biodegradation of polycyclic aromatic hydrocarbons (PAHs) in mangrove sediments. Mar Pollut Bull 51:1071–1077

Yuniati MD (2018) Bioremediation of petroleum-contaminated soil: a review. In: IOP conference series: earth and environmental science, vol 118. IOP Publishing, Bristol, p 012063

# Chapter 7
# Microbial Synthesis of Nanoparticles and Their Applications for Wastewater Treatment

**Virendra Kumar Yadav, Samreen Heena Khan, Parth Malik, Anju Thappa, R. Suriyaprabha, Raman Kumar Ravi, Nisha Choudhary, Haresh Kalasariya, and G. Gnanamoorthy**

**Abstract** For the last few decades, with the emergence of nanoscience and nanotechnology, nanoparticles gained enormous attention due to their high surface/volume ratio and other novel, unique, and remarkable properties. Traditionally, nanoparticles are fabricated by either chemical or physical approaches which not only utilize toxic chemicals but also are energy-intensive and consequently costly. The microbe-based synthesis of nanoparticles is biocompatible, economical, eco-friendly, and energy-intensive. Metallic and nonmetallic, metal oxide, and sulfide nanoparticles are synthesized by bacteria, virus, fungi, and algae. These nanoparticles act as an adsorbent for the remediation of water and wastewater pollutants clearly due to their physicochemical properties, nano size, controlled growth, and surface modification. The carbohydrates, proteins, and enzymes present in such microbes act as surfactant and capping agents which reduce the use of harmful chemical surfactants. Nanoparticles find application in the remediation of dyes, heavy metals, microbial contaminants, and pesticides, in the area of wastewater treatment. The widely used adsorbents are iron oxide nanoparticle, zinc oxide,

V. K. Yadav (✉) · H. Kalasariya
Department of Microbiology, Nootan Science and Commerce College, Sankalchand Patel University, Visnagar, Gujarat, India

S. H. Khan · P. Malik · R. Suriyaprabha · N. Choudhary
School of Nanosciences, Central University of Gujarat, Gandhinagar, Gujarat, India

A. Thappa
School of Biotechnology, Shoolini University, Solan, HP, India

R. K. Ravi
School of Environment and Sustainable Development, Central University of Gujarat, Gandhinagar, Gujarat, India

G. Gnanamoorthy
Department of Inorganic Chemistry, University of Madras, Guindy campus, Chennai, Tamil Nadu, India



J. Singh et al. (eds.), *Microbial Biotechnology: Basic Research and Applications*, Environmental and Microbial Biotechnology,
https://doi.org/10.1007/978-981-15-2817-0_7

alumina, and titanium dioxide. The present chapter highlights the synthesis of nanoparticles from bacteria, algae, fungi, and viruses and their application for wastewater treatment.

**Keywords** Nanoparticles, bacteria · Metal oxide · Surface modification · Wastewater treatment

## 7.1 Introduction

Increasing population and advanced lifestyle have resulted in mounting pressure on environmental resources, deteriorating the process feasibilities of multiple natural environmental conservation pathways. The effects seem to be convincingly regulated, after the reports of environmental pollution assessment agencies, reflecting inescapable approaching threats. Though several efforts and resolving inroads are rapidly emerging to tackle this issue, nothing serious seems to shake the generating source, owing to which there is an urgent need to counteract the probable risk factors with powerful and stronger prediction technologies. For example, on a daily basis, more than half of the natural processes require conversion of harmful and complex materials into simpler and self-decomposable entities, through interventions that are increasingly energy sensitive, whereby stress levels and contaminants are bound to affect the environmental quality. So better and more powerful solutions, consuming little and effecting much higher problem domain, are urgently needed. In this context, the unique physical and chemical properties of nanoscale materials have been the focus of sheer interest ever since their inception. Nanotechnology and nanoparticles have drawn the attention of the scientific community towards them (Seqqat et al. 2019). Mostly nanoparticles due to their size tunable and fascinating properties find applications in the field of electronics, medicine, research, catalysis, and environmental cleanup (Khan et al. 2019), due to which their demand has increased drastically, and the load on the industries for their synthesis has increased manyfold. The commercial synthesis of nanoparticles is generally carried out by various chemical and physical routes. Among the chemical approaches, the most common methods are sol-gel, coprecipitation, hydrothermal, solvothermal (Ganachari et al. 2017), etc. On the other hand, the physical synthesis method includes laser sputtering, chemical vapor deposition (CVD), and ball milling. In addition to their advantages, these chemical and physical methods have several disadvantages as well. Chemical methods utilize numerous hazardous chemicals as surfactant reducing agents, or reactants, which are toxic to the environment. While the physical methods require sophisticated machinery for their synthesis, they are highly energy-intensive, which makes them expensive. So there is a need for another alternative source which should overcome the drawbacks of the abovementioned methods. A generalized method of nanoparticle synthesis is depicted in Fig. 7.1.

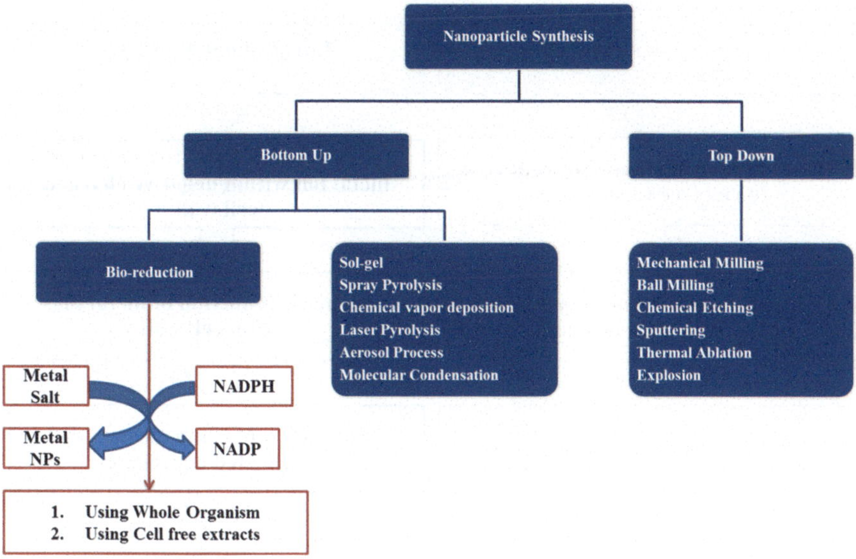


**Fig. 7.1** Generalized synthesis of nanoparticles

The biological synthesis of nanoparticles is the best alternative method which will overcome the drawback of the abovementioned problems. Among the biological methods, microbial-based synthesis of nanoparticles is of utmost importance due to its cost-effectiveness, ease of synthesis, and eco-friendly nature (Singh et al. 2016a; Prasad et al. 2016). Since the source of preparation is economically viable and renewable, interest in their scientific evaluation has attracted unconditional support from different parts of the globe. Besides this, the biosynthesized nanoparticles are also biocompatible which can be used directly in the field of medicine and drug delivery. Microorganisms are the organisms whose sizes fall in the range of microns. Generally, they can be classified into two classes: prokaryotic and eukaryotic. Prokaryotic involves bacteria and *Actinomycetes*, whereas eukaryotes involve algae, fungi, and yeast. There were several reports of using viruses as the templates for the synthesis of several nanoparticles. Microorganisms have numerous biomolecules such as enzymes, organic acids, polysaccharides, etc. known to play a key role in the synthesis process. Moreover, such biomolecules encapsulate the nanoparticle which further prevents the aggregation by acting as a capping agent. Microorganisms can synthesize all types of nanoparticles by reduction, for instance, single metallic nanoparticles (like gold, silver, copper, Se, Fe, or NZVI) (Das et al. 2017) or metal oxides (titanium oxide, zinc oxide, iron oxide, silica, and aluminum oxide) or metal sulfides (like PbS, CdS, PbSe, CdSe, ZnS). Metallic nanoparticles have almost replaced conventional catalysts in different biochemical and even industrial activities. The generalized flowchart for extracellular and intracellular biosynthesis is shown in Fig. 7.2.

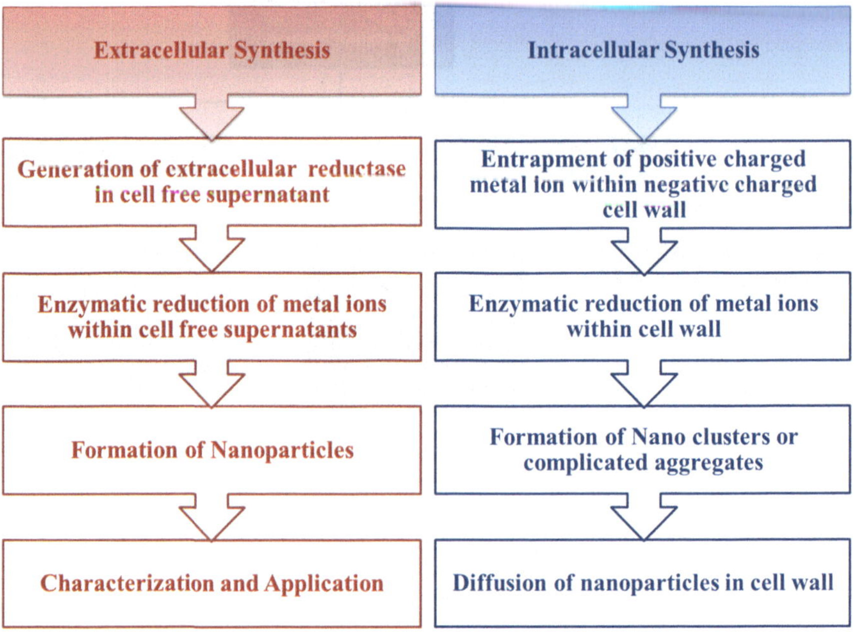


**Fig. 7.2** Schematic flowchart of extracellular and intracellular biosynthesis

The biologically synthesized biocompatible and eco-friendly nanoparticles have potential for wastewater treatment like remediation of water pollutants such as dyes, heavy metals, pesticides, and other organic/inorganic pollutants (Das et al. 2017). The nanoparticles being small in size have a high surface/volume ratio (SVR) and comparatively high surface energies have a special role in wastewater treatment (Shah et al. 2015a). Moreover, their microbial origin provides them special advantage like capping by biomolecules, which not only prevents their aggregation but also provides different functional active sites. These active sites help in the adsorption of pollutants from the wastewater.

The contamination of wastewater streams and its associated risks of environmental degradation is the source of polluting water bodies owing to which aquatic organisms are at risk (Bhateria and Jain 2016). The risk factors are enhanced significantly due to the heterogeneous nature of contaminants, requiring differential treatment measure for each specific kind of pollutant. The NPs with their high surface areas (SA) are able to initiate multiple interactions with chemically distinct pollutants at the same instant of time, owing to which in the same time interval, pollutant load is reduced to a greater extent as compared to conventional technologies (Lin et al. 2014). Keeping such characteristics of metal/metal oxide NPs in observance, the present chapter focuses on the advances in the field of microbial synthesis of single metals and metal oxide nanoparticles and their potential applications in the field of remediation of various organic and inorganic pollutants from the water and wastewater.

## 7.2 Advantages of Bacteria-Mediated Synthesis of Nanoparticles

Biological systems especially bacteria, due to their high growth rate, act as a unique candidate to achieve nonhazardous nanoparticle synthesis. In chemical methods, the growth of NPs is usually expensive and ends with toxic end products released into the environment. In biological systems, an individual bacterium can be a center and a matrix for the controlled growth on NPs.

Studies on bacterially synthesized NPs reveal interesting variations, with extracellular as well as intracellular formation sites, variability in the shapes and sizes, and numerous others. A key concern in all the cutting-edge applications of NPs (ranging from industrial, chemical, biological, and many others) is the extent for which these entities retain their low sizes (Nath and Banerjee 2013). The diverse chemical environments of application sites often lead to the aggregation of these moieties, so it is very important that these nanomaterials retain their native sizes for sufficiently longer durations. That is why physically prepared NPs are often ineffective in most applications and the problem of aggregation is frequently encountered. With biological sources, there is an advantage that an aggregation preventing agent does not need to be added separately, and the polysaccharides, protein fractions, and fatty acid conjugates of microbes serve as aggregation stabilizing agents (Siddiqi et al. 2018). Apart from this, culturing bacteria is relatively easy (since it is the largest active microbe among all varieties) and relatively less costly as compared to reducing agents of the chemical method. Furthermore, an incentive with the bacterial synthesis is that even if we have one bacterial strain, culturing it in different chemical environments (characterized by pH and temperature variation) could form different NPs at the same instant of time. For example, if the salt precursor for Fe is added in the culture medium of magnetotactic bacterium, we will get Fe-O NPs, while if a cobalt salt is added, the likely product would be Co (as such) or in some combined form as NPs. The reason is that both Co and Fe are magnetically sensitive and have unpaired electrons in their molecular orbitals. Similarly, the NPs of two metals belonging to the same group of the periodic table could be formed in a simplistic manner as they would have similar chemical characteristics. The only distinction is the presence of salt (for each metal type) in the culture medium of the chosen bacterial strain. So, low cost lesser synthesis time, variability, and better stability of the product (here NPs) are some of the reasons to choose a bacterial route for NP synthesis (Shah et al. 2015a).

The exact mechanism of microbial-based nanoparticle formation is not known and most of the studies project consensus for the metabolism of toxic compounds by bacteria after which a whole range of defensive genes is activated to combat the resultant toxicities. The activation of these genes is accompanied by the sequential modulations in oxidation and reduction chemical balances which ultimately result in the prevalence of a peculiar form that neutralizes the toxicity to a maximum extent. In this reference, some authentic claims are put forward by Ghashghaei et al. in their 2015 compilation, describing distinctive three-step nanoparticle formation by bacteria (Gahlawat and Choudhury 2019). The first discusses the bacterial self-assembled nanostructures such as pilus, flagellum, S-layer bipolyester, cyanophycin inclusions, phage, rhodopsin, and alginates. The second possibility projects the metal and metal oxide nanoparticle formation as bacterial metabolism by-products. In this mechanism, there seems to be an implicit role of bacterial enzymes. The third step of nanoparticle formation by bacteria relies on bacterial polymers (using polysaccharides and polysaccharide derivatives) which can be thereafter processed to NPs. Derivatives of plant products are highly reliable sources of preventing aggregation owing to their natural mode of existence, least energy requirements for operation, and much lower toxicity than chemical agents accomplishing the same. Discussing the diversity of mechanisms explaining nanoparticle formation, they concluded one common aspect in all approaches, involving the trapping of metal ions either on the surface of or inside microbial cells. Subsequently, these are reduced to NPs through enzymatic or nonenzymatic methods. Holding detoxification pathways as prime sources responsible for bacterial (and even other) nanoparticle formation, they illustrate a nonspecific uptake of toxic metal ions via the cationic membrane transport systems, which regulate the transport of metabolically important cations. Now since these uptaken metal ions are toxic, they are not tolerated by bacteria beyond a threshold concentration which is accomplished either via mutagenesis of metal-resistant bacterial strains or by complementation to identify the resistance restoring genes. The identification of genes responsible for causing toxic responses paves the way for their enzymatic inactivation of toxic ionic forms of the metal into nontoxic metal salts, modulation in metal ion efflux system activities, and diminishing membrane permeability. Figure 7.3 discusses the concentration-dependent enzymatic induction controls in a bacterium to counter the toxicity threats to the bacterium itself.

The higher the toxicity (caused by an increasing population of metal ions), the more pronounced is the activation of enzymes restoring the normal pH and chemical balance. Diversity of the product manifested by changes in the process parameters, such as pH, temperature, duration of culture, type of strain, culture conditions, and their mutual optimization, is a key factor affecting the nature of NPs formed. Interesting aspects within these predictors are the differences in the product's nature attained through differential regulation of comprising factors. For example, maintaining the temperature at 45 °C for half an hour is likely to confer different features compared to assuring the same for an hour or more. Similarly, the concentration of a characteristic protein source in the culture medium could be responsible for a peculiar nature of the product which may vary with the change in the protein source. The obvious reason is that proteins are composed of amino acids that play decisive roles in regulating 1:1 dispersion. Constituent amino acids may exhibit an altogether distinct response if they are dextro-rotatory instead if leave or vice versa. Likewise, composition of dispersion medium in the culture plays a critical role where if the water is the major source, polar activities are likely to be aggressive contrary to aqueous ethanol or some other additive (organic) source. On a similar note, if we change the pH, the dispersion trend is likely to be varied that is reflected in the $\zeta$-potential trend. It may be possible that the shape and size of NPs being

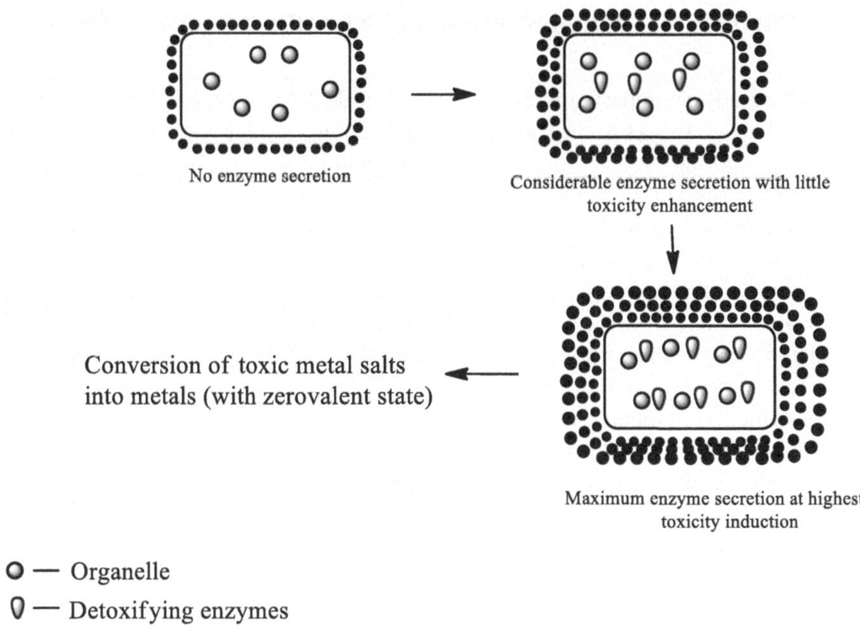


**Fig. 7.3** Correlative regulation of toxic metal accumulation and the intracellular enzyme activation in microbial cells

prepared are more regulatory and optimized at acidic pH which the alkaline pH is unlikely to support. The discussed factors present preliminary aspects, which, if replicated on a molecular scale, can have quite significant consequences. For instance, acidic pH may be supportive inactivating a particular bacterial gene/enzyme in a specific manner that the basic pH is unable to do. At the same point, it is imperious to elucidate the consequence of a peculiar pH on the expression of neighborhood genes and enzyme secretion, so that growth remains unaffected. An accurate generalization of the bacterial activities in different growth phases is the key to the choice of a bacterial species for a certain kind of nanoparticle synthesis. Generally, in the log phase, the most active nutrients are synthesized and the stationary phase is the domain where the functional activities could be stimulated.

Thus, exploiting the bacterial genetic makeup is the groundwork required for thoughtful and optimized regulation of nanoparticle formation. The mechanistic insights probably offer incentives that, with the variations in culture ingredients, even hybrid or core-shell NPs could also be prepared. Similar robustness is not attainable with fungal cultures since fungi require adequate growth of mycelia to exhibit the desired metabolic performance. Apart from this, the protein content of fungi is very much different from that of a bacterium, owing to which growth conditions remain less controlled with the former. Nevertheless, the logical estimation of obtaining still better results with consortium cultures suits well for the kind of NPs being formed.

## 7.3 Bacteria-Mediated Synthesis of Metallic Nanoparticles

Microbes used for preparing metal and metal oxide NPs comprise bacteria, fungi, virus, and algae, thanks to the robust chemical actions of microbial enzymes which remain operational even at extreme temperature and pH conditions. An interesting aspect of microbial biochemical activities is that the expression of enzymes could be modulated by changing the chemical compositions of their culture media. Optimization of culture parameters such as pH, light exposure, temperature, medium salt content, and mixing speed is a pivotal factor which can significantly modulate the enzyme activity (Sarker et al. 2013; Kirthi et al. 2011a).

Bacteria can significantly reduce heavy metal ions to produce nanoparticles; many researchers proved that bacteria facilitated interactive pathways accountable for the reduction of metal ions and their ability to precipitate metals to the nanometer scale (Kirthi et al. 2011a). One of the major benefits of bacteria-mediated nanoparticle synthesis is their large-scale production with negligible use of hazardous chemicals; however, there are firm boundaries like laborious bacterial culturing procedures and less control over their shape, size, and distribution.

So keeping track of media composition-dependent enzyme expressions easily serves as the lead for facilitating the requisite extent of chemical reduction. For example, the most common bacteria, *E. coli*, acts as a probiotic within the physiological environment, hinting that changing the pH and the chemical composition of the surrounding environment could activate another set of *E. coli* enzymes whereby functions other than the probiotic activities could be exercised (Fijan 2014). Another important feature responsible for regulating nanoparticle formation by bacteria is their ability to interact with their nearby environment arising due to the polarity of their lipidic membranes that catalyzes diverse oxidation-reduction processes (most of the nanoparticle preparation is enacted through a chemical reduction of metallic salts) (Singh et al. 2018). The distinction in cell wall polarities of gram-positive (GP) and gram-negative (GN) stains (dissimilar peptidoglycan extents) is, in itself, quite a significant prospect of bacterial structural diversity, which is missing in all other microbes. Another critical factor affecting NP formation by bacteria is the choice of a particular species (Singh et al. 2018; Wang et al. 2017). This is because all bacteria do not have a similar genetic makeup and that is why culturing two different species in the same media is likely to result in differential functional expressions of similar redox enzymes. To meet such challenges, a consortium of the bacterial population, comprising two or more species, generally performs better than a single bacterium. Furthermore, studies on the microbial synthesis of NPs have notified that only those biological moieties that have potential to accumulate the metallic precursors in the form of oxides, hydroxides, sulphates, etc. have the best chance of forming metal NPs (Zargar et al. 2011).

Synthesis of NPs using biological moieties (enzymes, sugar, and proteins secreted by a bacterial strain) is a bottom-up approach, producing less waste material and enabling the formation process to be stopped as per the requirement. This is particularly exciting, as applications of NPs are substantially expressed through their shape and size variations, which acutely affect their interaction abilities. Another advantage with bacterially synthesized NPs is that the enzymes responsible for initiating biochemical reduction of corresponding metallic precursors could be amplified in their expression through knowledge of their regulatory genes and pathways. Similarly, the introduction of these enzymes from the outside could facilitate an extracellular synthesis of both, i.e., pure metallic and their oxides NPs, the absence of which drives the synthesis at intracellular locations (Patra and Baek 2014; Malik et al. 2014).

## 7.4 Inception and Progression of Bacterially Formed NPs

### *7.4.1 Bacterial Synthesis of Metallic Nanoparticles*

Bacteria bear exceptional abilities for reducing metallic ions into their zerovalent forms (the nanoparticle state) and are perhaps the most befitting candidates for nanoparticle synthesis attributed to their ease of handling and robust culture medium requirement (Ruttkay-Nedecky et al. 2017a). In comparison to other microbes, it is very easy to mold and manipulate genes in bacteria for retrieval of metal ions. Generally, bacteria are exposed to harsh and increased heavy metal ion concentration in their surroundings. To combat these stressful conditions, bacteria have developed numerous defense processes, for instance, intracellular sequestration, efflux pumps, fluctuation in concentration ions of metals, and extracellular precipitation. These defense mechanisms form the basis of shape- and size-specific nanoparticle synthesis by bacteria. Table 7.1 comprises the various bacterial strains used for the synthesis of NPs along with their specific applications. The first incidence of bacterial synthesis of Au NPs was reported by *Beveridge and Murray*, who noted the extracellular deposition of Au NPs on *Bacillus subtilis* cell wall when $AuCl_4$ solution was exposed to its colonies (Beveridge and Murray 1980). Almost at the same time, *Klaus-Joerger* intracellularly synthesized AgNPs < 200 nm using an NADH-dependent reductase as an electron source and, in course, itself getting oxidized to $NAD^+$. It was noted that $e^-$ transfer from NADH results in a biological reduction of Ag + ions to their zerovalent form (the nanoparticle form) (You et al. 2013). The emergence of bacteria-mediated nanoparticle synthesis acquired attention in 2012 when Srivastava et al. intracellularly synthesized Pd, Ag, Rh, Ni, Fe, Co, Pt, and Li NPs (Srivastava and Constanti 2012). Interestingly, these synthesis procedures do not require an external stabilizing agent and electron donors and also did not need any pH modification in the course of biomineralization of different metal ions. Various metallic nanoparticles synthesized using various microbes are summarized in Table 7.1.

Some of the bacteria have shown exceptional phenomenon of nanoparticle synthesis, like synthesis at extremely high concentration of metal ions, acidic pH, and higher temperature. *Ps. stutzeri* and *Ps. aeruginosa* are bacteria that can grow at higher metallic concentration (Lu et al. 2016).

**Table 7.1** Various metallic nanoparticles synthesized using bacteria

| Sr. no. | Name of bacteria | Specific nanoparticle formed | Size, shape, and location | Reference |
|---|---|---|---|---|
| 1. | *Pseudomonas aeruginosa* | Au | 15–30 nm, spherical | Sarker et al. (2013) |
| 2. | *Pseudomonas stutzeri*, *Bacillus cereus* | Ag | Up to 200 nm, various shapes | Kirthi et al. (2011a) |
| 3. | *Alcaligenes faecalis* | Ag | 30–50 nm, spherical | Fijan (2014) |
| 4. | *Bacillus* sp. CS11 | Ag | 42–92 nm, spherical | Singh et al. (2018) |
| 5. | *Kocuria flava* | Cu | 5–30 nm, spherical | Wang et al. (2017) |
| 6. | *Ochrobactrum rhizosphaerae* | Ag | 10 nm, spherical | Zargar et al. (2011) |
| 7. | *Alteromonas macleodii* | Ag | 70 nm, spherical | Patra and Baek (2014) |
| 8. | *Deinococcus radiodurans* | Ag | 4–50 nm, spherical | Malik et al. (2014) |
| 9. | *Bacillus subtilis* | Au | 20–25 nm, spherical | Ruttkay-Nedecky et al. (2017a) |
| 10. | *Bacillus brevis* (NCIM 2533) | Ag | 41–68 nm, spherical | Beveridge and Murray (1980) |

Besides this, there are a few iron-reducing bacteria that can reduce ferric to ferrous form when grown on elemental sulfur, i.e., *Thiobacillus ferrooxidans*, *T. thiooxidans*, and *Sulfolobus acidocaldarius. This was reported by Brock and Gustafson, and the detailed process is given in IONP synthesis by non-magnetiotactic bacteria* (Brock and Gustafson 1976). Other biomineralization phenomena, such as the formation of tellurium (Te) in *Escherichia coli K12* (Iravani 2014), the direct enzymatic reduction of Tc (VII) by resting cells of *Shewanella putrefaciens* and *Geobacter metallireducens*, and the reduction of selenite to selenium by *Enterobacter cloacae*, *Desulfovibrio desulfuricans*, and *Rhodospirillum rubrum* (Kessi et al. 1999), have been reported as well.

Mullen et al. reported the leaching capability of $Ag^{+}$, $Cd^{2+}$, and $Cu^{2+}$ from solution by *Bacillus cereus*, *P. aeruginosa*, *B. subtilis*, and *E. coli*. Some of them can bind to large metallic ions, while some can form magnetic nanoparticles (Kessi et al. 1999) which are shown below in Table 7.1.

Multiple studies of the past 5 years have reported the quick and fast synthesis of NPs possessing varying size and shape from several bacterial strains, such as *E. coli*, *B. subtilis*, *B. cereus*, *B. megaterium*, *Ps. aeruginosa*, *Klebsiella pneumoniae*, *Alteromonas,* and *Ochrobactrum*. The idea of bacterial enzymatic functions can be well judged from the simplistic procedures, involving minimum time periods without any sophisticated pH and temperature optimization. For example, Das et al. reported that *B. cereus* could synthesize AgNPs extracellularly at ambient temperature within 24 h (Lakshmi Das et al. 2014). Kumari et al. reported enhanced antifungal activity of AgNPs prepared using biological route (fungus, *Trichoderma viride*) contrary to that of chemical method, although, in both modes, the sizes and shapes of produced NPs were similar. It was observed that biologically prepared particles effected (20 and 48.8)% higher decrements (compared to those prepared chemically) in the dry weight of fungal pathogens, namely, *Fusarium oxysporum* and *Alternaria brassicicola*. Staining assays using nitro blue tetrazolium and propidium iodide dyes revealed an enhanced generation of superoxide radicals, as the probable cause of death for the biologically prepared NPs. Scanning electron microscopy (SEM) analysis revealed altered osmotic balance and membrane disintegrity as the primary source of cell death (Kumari et al. 2019). This study, therefore, concluded that compared to the chemically produced NPs, the bacteria synthesized NPs effected comparatively higher ROS generation, antioxidant pathway downregulation, and disruption of osmotic balance and cellular integrity. From this research, the observations infer a higher toxic expression of biologically prepared NPs, which confers them a preference compared to the chemically prepared NPs in the treatment of wastewater streams. The hidden prospect of this study is the size of the synthesized NPs, which have an intricate effect on their functional properties. It is well known that smaller size at the nanoscale argues for stronger quantum confinement which has a direct correlation with the particle energies. Since this study observed higher toxicity for biologically synthesized NPs, it seems certain that biological sources could have produced a lower size as lower size could have increased the kinetic stability owing to which interactive potential of the formed NPs increases. Such size controls are attainable with chemical methods where the stabilizing or aggregation preventing agent needs to be added separately to prevent the size growth beyond a stage.

Further, AgNPs was also synthesized extracellularly by Kulkarni et al. using *Deinococcus radiodurans*. *D. radiodurans* are an extremely robust bacterium species which can survive cold, dehydration, vacuum, radiation, and acid exposures (Kulkarni et al. 2015). The NPs were synthesized by the reduction of AgCl solution, following which these were screened for their antibacterial and antibiotic responses towards GN and GP bacteria. The radiation resistance properties of *Deinococcus radiodurans* enabled its activity even in extreme environments, owing to which good results were obtained with regard to anticancer studies. Results of this study have encouraged the scientific community to optimize *Deinococcus radiodurans* growth kinetics (via changes in the culture medium composition) so that palladium, platinum, and several other NPs could also be produced with similar control and ease. An incentive with such microbial species is that they can be as such added to wastewater streams where they can feed and regulate their metabolism through processed consumption of organic contents, such as discarded proteins, carbohydrates, fats, nitrogen sources, and several others. The temperature-resistant functioning of *Deinococcus radiodurans* offers special benefits as it is likely to cause minimal structural changes of synthesized NPs owing to which the NPs retain an unaltered application potential to a longer extent. Digging for the mechanism of bacterial NP formation, Sneha et al. replicated the attempt of Liu et al. (synthesis of AuNP *Bacillus megaterium* dried cells) using *Corynebacterium* species to obtain NPs and proposed a nonenzymatic reduction mechanism to be the source of nanoparticle formation (Liu et al. 2000). They noted that the formation of NPs by bacteria depended on two major requirements, where the first was the presence of organic functional groups in the bacterial cell wall, while the second factor was the maintenance of optimum temperature and pH in the surrounding environment. Studies by He et al. provide the most significant justification of this fact where working on *Rhodopseudomonas capsulata* bacterium growth optimization, they noted that the peculiar morphology of the formed particle can be governed by concentration of metallic salts and pH of the medium. There was formation of spherical AuNPs within 10–20 nm at pH 6 using diluted $AuCl_4$ solutions. However, when the salt concentration was increased, the reaction formed Au nanowires (He et al. 2008). Interestingly, reduction of pH to 4 (enhancing the acidity), diluting salt concentrations, formed both spheres and triangular nanometer-scale plates. This study clearly established the role of pH and precursor concentration in the formation efficacy, size, and shape regime of NPs. So the same bacterium being fed on a similar metallic precursor could produce differently shaped NPs with the variation in medium pH. Similar attempts with *Lactobacillus* sp. A09 and *B. megaterium* D01 produced AgNPs through distinct reduction of silver ions (Gowramma et al. 2015). For details on the bacterial potential to synthesize NPs that are not discussed above (palladium, platinum, and some other metals), readers can refer to more specific literature sources (Shah et al. 2015b).

## 7.5 Bacterial Synthesis of Gold (Au) and Silver (Ag) Nanoparticles

Both Ag and AuNPs have drawn huge attention towards them due to their applications in medicine, electronics, cosmetics, coatings, packaging, and biotechnology. The production of nanoparticles emerges as an environment-friendly and exciting approach due to the use of microorganisms. Kushwaha et al. (2015) synthesized silver nanoparticles using *E. coli* and assessed their antibacterial potential against bacteria *Bacillus subtilis* and *Klebsiella pneumoniae* after confirming by UV-Vis and TEM (Sabri et al. 2016).

### *7.5.1 Gold (Au) Nanoparticles*

In the biotechnology and biomedical field, the gold and silver NPs are considered very essential and interesting for several applications. Gold and silver NPs mainly find applications in the biomedical field such as diagnosis and therapy. Gold nanoparticle synthesis gained a lot of interest from many researchers. *Rhodopseudomonas capsulate, Shewanella oneidensis, Escherichia coli, Yarrowia lipolytica*, and *Plectonema boryanum* have been applied for the formation of gold nanoparticles (Sehgal et al. 2018). Gold nanoparticles (AuNPs) are mainly applied in the field of drug delivery for cancer, diabetes mellitus, and heart-related diseases. It also finds application in medicine, mainly diagnosis and therapy of cancer and as anti-arthritic and malarial agents (Prasad et al. 2018). AuNPs are synthesized by *Klebsiella pneumoniae*, which is pathogenic bacteria, and their antimicrobial activity is assessed against pathogenic bacteria, i.e., *E. coli*, *Staphylococcus epidermidis*, *S. aureus*, *Pseudomonas aeruginosa*, and *Bacillus subtilis* (Prema et al. 2016). AuNPs of various shapes i.e. rectangular, square, cubic, and triangular shape with an average diameter of 60 nm were synthesized using *Rhodomonas capsulata*, which was later on applied for the treatment of human colon, lung, prostate, heart, and breast cancer (Menon et al. 2017). Pourrali et al. (2017) synthesized gold nanoparticles (AuNPs) using bacteria *Bacillus cereus* and fungus *Fusarium oxysporum* in vitro and their confirmation was done by UV-Vis, TEM, and XRD. Further, AuNPs were also used for the treatment of human fibroblast cell line CIRC-HLF (Pourali et al. 2017). Srinath et al. (2017) reported the synthesis of biocompatible gold nanoparticles using *Brevibacillus formosus* and assesed their antibacterial activity against *Staphylococcus aureus* and *Escherichia coli* (Srinath et al. 2017).

AuNPs were synthesized by using bacterium *Ps. fluorescens 417*, and properties confirmed by UV-Vis, FTIR, and TEM and finally showed their antimicrobial potential against *Ps. aeroginosa*, *Bacillus subtilis*, *E. coli*, *S. aureus*, and *Klebsiella pneumoniae* (Syed et al. 2016). Nangia et al. (2009) synthesized AuNPs using bacterium *Stenotrophomonas maltophilia (AuRed02)* and characterized them by electrophoresis, zeta potential, and FTIR (Nangia et al. 2009). Sharma et al. (2012) biosynthesized AuNPs using bacterium *Marinobacter pelagius* and analyzed them with TEM, UV-Vis, and dynamic light scattering (Sharma et al. 2012).

Besides bacteria, there are few reports in literature for the biosynthesis of AuNPs and AgNPs using microalgae (Agarwal et al. 2019). These microalgae are rich in proteins, flavonoids, and several other enzymes which are absent or present in very less content than the bacteria. Moreover, microalgae being a eukaryotic, have different modes of synthesis of both metallic and metal oxide nanoparticles. Sintubin et al. synthesized both Au and AgNPs using dried powders of *Spirogyra* spp. (Sintubin et al. 2011).

### 7.5.2 *Silver (Ag) Nanoparticles*

As we have already discussed above, AgNPs have various applications, especially in the field of medicine and drug delivery. They exhibit almost all applications shown by AuNPs. AgNPs are applied in food pacakaging in order to prevent their spoilage by foodborne pathogens. They are also used as antimicrobial agents for plant crop protection. All these applications of AgNPs are possible due to their economical and biocompatible nature (Hoseinnejad et al. 2017) (Siddiqi et al. 2018). In pharmaceutical products, AgNPs could be used as antimicrobials, agents for targeted drug delivery, and anti-biofilm agents. While in agriculture, it could be used for plant disease management. Further, it could also be used for environmental cleanup like water electrolysis, treatment of wastewater, degradation of dyes and pesticides, detection of pathogens and contaminants as a biosensor, etc. Silver nanoparticles showed antimicrobial potential against *B. cereus, B. subtilis, S. aureus, Corynebacterium rubrum, E. coli, Vibrio parahemolyticus, K. pneumoniae, Salmonella typhi, Ps. aeruginosa, Citobacter koseri,* and *Salmonella typhimurium* (Thomas et al. 2015). Gahlawat and Choudhury (2019) have already reported earlier the biosynthesis of AgNPs from various pathogenic and nonpathogenic bacteria (Gahlawat and Choudhury 2019).

In the field of research, Kumar and Ghosh (2016) studied that, in history, silver has been used as an antibiotic in human health care (Kumar and Ghosh 2016). For many centuries, the antimicrobial property of silver has been well known to cultures around the world. Nalenthiran et al. (2009) reported the synthesis of AgNPs from *Bacillus* sp. (*Brevibacillus borstelensis*_MTCC10642) by exposing them with $AgNO_3$ solution and further characterizing them with advanced instruments. The size of AgNPs was 5–15 nm, formed intracellularly in the periplasmic space of the bacteria (Nalenthiran et al. 2009). Suman et al. (2014) reported the synthesis of AgNPs from *Agrococcus* sp. and characterized them with TEM, UV-Vis, and FTIR. Silver nanoparticles were prepared using the bacterium *Escherichia coli* in liquid broth medium (Suman et al. 2014). Silver nanoparticles showed antimicrobial potential against *Lactobacillus fermentum, Streptomyces* sp., *Bacillus cereus, Brevibacterium casei, Staphylococcus aureus, Bacillus licheniromis, Enterobacteria,* and *Ureibacillus thermosphaerius* (Chintamani et al. 2018). Fang et al. (2019) synthesized AgNPs using bacteria *Bacillus amyloliquefaciens, Bacillus subtillis,* and *Stenotrophomonas* sp. *BHU-S7*, whereas gold nanoparticles were prepared using bacterium *Deinococcus radiodurans* (Fang et al. 2019).

Karthika et al. (2015) synthesized AgNPs from bacterium *Serratia marcescens*, whose characterization was done by sophisticated instruments. Here, the AgNPs exhibited antibacterial capability against *E. coli, Staphylococcus* sp., and *Pseudomonas* sp. (Karthika et al. 2015). El-Saadony et al. (2019) synthesized AgNPs from *Bacillus pseudomycoides* MT32 and showed antifungal properties against a group of Aspergillus spp. (*Asp. flavus, Asp niger,* and *Asp. tereus*), *Penicillum notatum, Rhizoctonias olina, Fusarium solani, Fusarium oxysporum, Trichoderma viride, Verticillium dahlia,* and *Pythium spinosum.* Characterization and analysis of silver nanoparticles were done using TEM, XRD, EDS, DLS, zeta potential, and UV-Vis (El-Saadony et al. 2019). Further, AgNPs synthesized from bacterium *Haemophilus influenzae* showed antimicrobial potential against various pathogenic bacteria like *Ps. aeruginosa, Klebsiella* spp., *Streptococcus* spp., *Serratia* spp., *S. aureus, E. coli,* and yeast (*Candida albicans*) using the agar well diffusion method. Characterization of AgNPs was confirmed using atomic force microscopy (Gahlawat and Choudhury 2019). Yadav and Fulekar (2018) reported the synthesis of AgNPs from bacterium *Pseudomonas* sp. ARS-22 and characterized them using various microscopic and spectroscopic instruments (Archana et al. 2015).

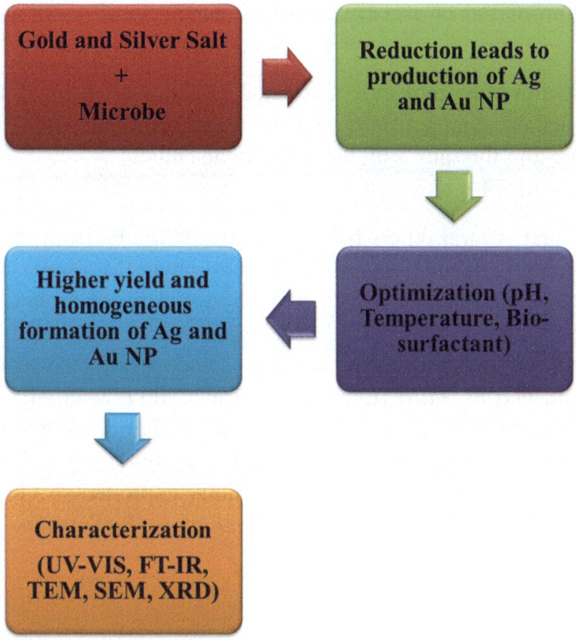


**Fig. 7.4** Flowchart for the biosynthesis of AuNPs and AgNPs from microorganisms

Further, Kumar and Ghosh (2016) reported the synthesis of AgNPs from bacterium *Brevibacillus centrosporus* DSM8445T, *B. levickii* LKG22481, *B. invocatus NCIMB13772T*, *B. panacihumi DCY35*, and *B. choshinensis* DSM8552T, and the confirmation of AgNPs was done by sophisticated instruments (Kumar and Ghosh 2016). The schematic representation of the biosynthesis of AgNPs and AuNPs is shown in Fig. 7.4.

## 7.6 Algal-Mediated Synthesis of Metallic Nanoparticles

Algae like fungi are eukaryotic organisms, which are widely utilized for the formation of both metal and metal oxide NPs, though their mode of synthesis of nanoparticles is more or less similar to prokaryotes. Several approaches have been made which are available in literature for the microalga-mediated formation of metal and metal oxide nanoparticles from their respective aqueous salt solutions (Kuppusamy et al. 2016). But the synthesis of nanoparticles by algae has numerous drawbacks as it is highly time taking, require more space for harvesting and tedious task (Agarwal et al. 2019). Simultaneously, there are several advantages also like photosynthetic property, ability to convert solar energy and $CO_2$ into biomass, and accumulation of nutrients in the form of C, N, and P. All these properties have direct impact on the morphological properties of nanocrystals. Microalgae could synthesize NPs in four different ways:

1. Direct utilization of biomolecules extracted after the breakage of the algae
2. Using supernatant obtained from microalgae culture media
3. Directly exposing the salts with the microalgae cells
4. Using living cells of microalgae

Merin et al. and several others have synthesized AgNPs from microalgae from the family of Haptophyta, Chlorophyta, and Ochrophyta (Mohseniazar et al. 2011). Mahdieha et al. reported synthesis of AgNPs by using living biomass of *Spirulina platensis* (Mahdieh et al. 2012), while Luangpipat et al. incubated gold chloride with *Chlorella vulgaris* under optimal conditions and reported the synthesis AuNPs, which was confirmed by TEM (Luangpipat et al. 2011). Microalgae due to their richness in various biomolecules like sulfated polysaccharides and proteins and pigments were used for the synthesis of Ag, Pd, Cd, and NPs, by either whole cells, supernatant, or dried biomass. Microalgae, mainly exploited so far for the above NPs, are cyanobacteria, *Chlorella* spp., *Lyngbya majuscula*, *Spirulina platensis*, and other *Chlorophyta* spp. (Patel et al. 2014).

## 7.7 Microbial-Mediated Synthesis of Metal Oxide Nanoparticles (MONPs)

### *7.7.1 Bacteria-Mediated Synthesis of Metal Oxide Nanoparticle*

Metal oxide NPs have changed the face of industries due to its application in research, medicine, ceramics, space, defense, wastewater treatment, and composite preparation (Patel et al. 2014). Some of the most commonly used metal oxide nanoparticles are iron oxide nanoparticles, silica nanoparticles, and alumina nanoparticles. These nanoparticles are mostly synthesized from a commercial precursor using chemical and physical approaches. The most common chemical approaches employ sol-gel, coprecipitation, hydrothermal and solvothermal decomposition (Bilton et al. 2012), while chemical vapor deposition and laser ablation are common examples of the physical method. Because it is highly energy-intensive, both of these approaches are not only expensive, but also hazardous due to applications of various chemicals in the synthesis. So the biosynthesis of metal oxide NPs is an economical, biocompatible, and green method (Jiang et al. 2018). Among the biological approaches, various photoproducts and microbes have also been used. Plant products like leaf, flower, fruit, bark, and stem have various proteins, phytochemicals, flavonoids, terpenoids, and enzymes that act as a reducing agent, while microbes like bacteria, fungi, yeast, and algae have numerous biomolecules (proteins and enzymes and organic acids) that act as both reducing and capping agent in addition to transforming the precursor into a biocompatible product (Shah et al. 2015a).

The mechanism of synthesis of metal oxides from the precursor material by microorganisms is similar to single metals. Generally, microorganisms can synthesize metal oxide NPs either intra- or extracellularly. In the previous process, metal precursors are taken up from the medium by the microbes and transform them into respective metal oxides using their machinery, whereas in the extracellular process, the metal precursors are present in the medium, where the microbes secrete their microbial products. The secreted microbial products interact with the metal precursors and get transformed into metal oxide nanoparticles (Mohd Yusof et al. 2019). Out of these two processes, extracellular mechanisms are considered advantageous, as the metal oxide nanoparticles form outside, so the recovery or downstream processing will be easier and economical in comparison to the intracellular process. Moreover, in the extracellular process, there are fewer chances of contamination.

Metal oxide nanoparticles are very effective in inhibiting the growth of various GP and GN bacteria and they have emerged as the most promising candidate as antimicrobial agents. The most common metal oxides include ZnO, $TiO_2$, CuO, $Fe_2O_3$, and MgO. The biological material used for the synthesis of NPs like bacteria, fungi, yeast, and plant extract lies on the principles of green chemistry and is compatible with the use of microorganisms (Mohd Yusof et al. 2019). The biologically synthesized NPs have various applications such as drug delivery, biolabeling, coating of medical products, and treatment of cancer (Yadav and Fulekar 2018). Opposite to physicochemical methods, biosynthesized NPs are nontoxic, making them suitable for biomedical applications. Moreover, the oxidized form of synthesized NPs is even more useful due to their physicochemical properties. Like metal NPs, bacteria could also synthesize metal oxide NPs by means of intracellular or extracellular process. Based on several works, it was found that among bacteria, NADH-dependent nitrate reductase (NDNR) enzyme activity plays a key role in the conversion of metallic ions to nanoparticles. The general synthesis methodology for the biosynthesis of nanoparticles using bacteria is shown in Fig. 7.5.

### *7.7.2 Iron Oxide Nanoparticles (IONPs): A Special Class*

The unique physicochemical properties of FeO nanoparticles have fueled their demand not only from research and industrial viewpoints, but perhaps these are one of the most preferred nanomaterials in biotechnology and microbiological domains (Gu et al. 2006; Tamer et al. 2009; Chang et al. 2008; Dong et al. 2011; Meng et al. 2011). The common link with hemoglobin present in the blood has recently fascinated research towards them in modified forms, whereby these entities could be prepared using facile approaches. On the basis of microbial stain and respective nanoparticle concentration, NPs of FeO could either stimulate or suppress microbial growth. The superparamagnetic attributes of these particles make them befitting heat coupling sources, forming the basis of their suitability in targeted drug delivery

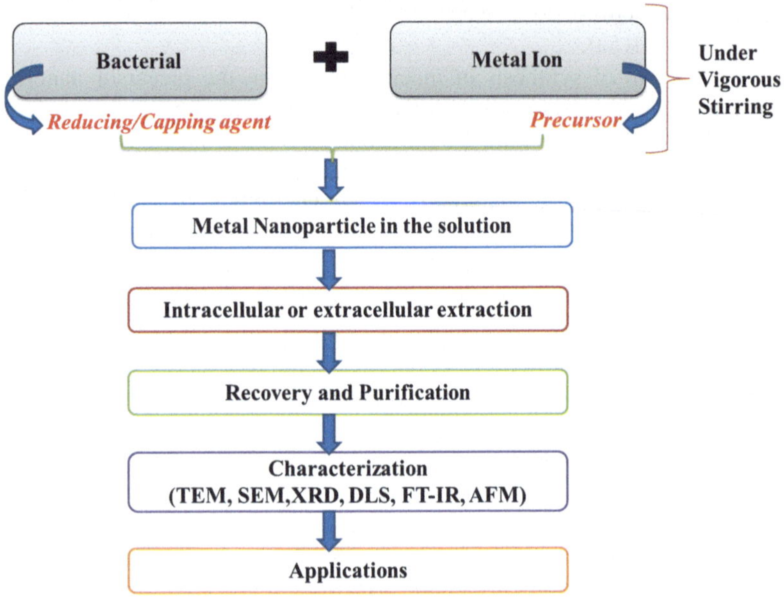


**Fig. 7.5** Generalized nanoparticle synthesis methodology using bacteria

to injurious or damaged cell or organ systems and improving the efficacy of modified drug delivery through hypothermic coupling (Khanna et al. 2018; Xue et al. 2018). The heat coupling efficacy of FeO NPs is already in robust consideration towards improving the antimicrobial efficacy of AgNPs (Senthil and Ramesh 2012). Similarly, the magnetically responsive ability of FeO NPs has rapidly increased their implication in biosensing (separation of pathogenic and nonpathogenic microbial strains), immobilization of select microbial population(s), and process intensification of biotechnological processes. Till date, FeO NPs are produced by several Fe-reducing bacterial species such as *Thiobacillus thiooxidans*, *T. ferrooxidans*, *Gallionella, Shewanella,* and magnetotactic bacterium, with existence in several phases, such as magnetite ($\gamma$-$Fe_3O_4$), hematite ($\beta$-$Fe_2O_3$), and maghemite ($\alpha$-$Fe_2O_3$) (Ali et al. 2016). Two broad mechanisms for synthesizing FeO NPs include either intervention of magnetotactic bacteria or through exposure of Fe precursors to non-magnetotactic bacteria. The two approaches are discussed ahead with their working distinctions.

#### 7.7.2.1 Magnetotactic Bacteria: Magnetite Nanoparticles

Magnetotactic bacteria comprise a peculiar category of bacteria, owing to their capability of intracellular generation of magnetite NPs in specialized organelles, called magnetosomes. These structures encode specialized globular protein, ferrin, which, in turn, comprises 24 distinctive subunits (Sakaguchi et al. 1993; Philipse and Maas 2002). This encapsulation aids in the storage of Fe in a nontoxic and soluble form, inside the magnetosomes, which could be extracted from bacteria through mechanical or chemical means (extracted magnetosomes could be ruptured to harvest magnetite NPs). The magnetite NPs are arranged in a crystalline manner within the magnetotactic bacteria with the entire population having equal size and shape. A characteristic/unique aspect of these magnetite NPs inside the magnetotactic bacteria is their ability to align them along the earth's magnetic field (Blakemore 1975). As harvested magnetic NPs are being used in diverse fields of cutting-edge, substantiating drug delivery, life sciences, MRI, CT scan, electronically equipped memory storage devices, ink printing, and several others. The most common bacterial species studied for the generation of magnetite are *Magnetospirillum magnetotacticum*, generating two distinct kinds of NPs, the first being magnetic ($Fe_3O_4$) NPs (having chain like structure), while the second being greigite ($Fe_3S_4$) NPs (Xie et al. 2009). Rarely, synthesis of both nanostructures is also reported from the same microbial species (but definitely under different culturing conditions). Popular species, among other bacteria well known for intracellular formation of magnetite ($Fe_3O_4$) NPs, include *Aquaspirillum magnetotacticum*, *Magnetospirillum candidatus*, *Magnetoglobus multicellularis*, and magnetotactic bacterium MV-1.

Magnetosomes are capable of forming robust crystalline as well as noncrystalline nanocrystal morphologies, having diverse shapes ranging from hexagonal to octahedral and faceted cuboctahedral shape and octahedral symmetry with either tied down or collected morphology within the phospholipid bilayers. It is pertinent to discuss some of the cahracteristic formations of magnetic NPs by bacterial stains, with the first being *Desulfovibrio magneticus* (RS-1), an anaerobic bacteria capable of reducing sulfur and intracellular accumulation of magnetite NPs, with sizes of most accumulated magnetite crystals remaining in the order of 30 nm (all particles being superparamagnetic in nature). In an interesting study, Klaus-Joerger demonstrated the possible replacement of Fe in biosynthesized magnetite NPs by Co, Cr, and Ni (all having an unpaired electron that contributes to spin-based characteristic magnetic sensitivity) in *Thermoanaerobacter ethanolicus*, a heat-enduring Fe-reducing bacterium. These substitutions produced octahedral magnetite NPs of <12-nm sizes in significant quantities, coexisting with poorly crystalline magnetite phase near the cell surface. Elblbesy and coworkers documented the formation of magnetic NPs from the *Magnetospirillum* strain AMB-1 to produce 47-nm-sized magnetite NPs, further working on the investigation of magnetic behavior as a function of varying incubation temperatures. Almost working on the same principle, *Philipse and Maas* reported the synthesis of magnetite crystals in single-domain, folded-chain, and flux-closure ring morphologies in *Magnetospirillum magnetotacticum* through regulation of bacterium locomotion via varying the externally applied magnetic field (Philipse and Maas 2002). Organisms living in Fe-rich surroundings harness their energy through a reduction of $mFe^{+3}$, with the partial reduction resulting in magnetosome having a peculiar protein-constituted phospholipid bilayer as its membrane. Several studies have screened the ability of diverse bacterial species isolated from Fe ore mining sites for their magnetic property conferring ability at laboratory conditions. In one such attempt, the *Thiobacillus thioparus* strain was identified through rybotyping, with microbial growth and magnetite production being optimized at varying pH, temperature, and substrate concentrations (Katayama and Kuraishi 1978). Moreover, magnetic NPs were purified via growth and lysis of bacteria and the magnetic properties were screened using empirical observations under the influence of magnetic field. With an interest to pursue their specific biological impact, the harvested NPs were monitored for their suitability in SDS-polyacrylamide gel electrophoresis (PAGE) in conjunction to those of coprecipitated magnetic NPs as well as particles coated with bacterial protein. The observations revealed that purified particles were synthesized using isolated bacterial strains having a protein coating, visualized on the stained polyacrylamide gel. The fluorescence property of the solution under magnetic field and aggregation of the particle along the edge of the wells in the absence of protein coating displayed the presence of monodispersive magnetic NPs in the preparation. Another attempt focused on the utilization of magnetically insensitive anaerobic bacterium GS-15, to reduce nonmagnetic brown disordered ferric oxide into a magnetic black solid particle, with the synthesized crystals residing outside the cell in a nonaligned hierarchy (Elcey et al. 2014).

#### 7.7.2.2 IONP Synthesis by Non-Magnetotactic Bacteria

Several studies have reported the synthesis of FeO NPs using non-magnetotactic bacteria, through culturing in peculiar environments at specific pH and temperature. One such attempt reports the biological synthesis of 73.30-nm-sized FeO NPs through *Lactobacillus rhamnosus* strain, followed by characterization using atomic force microscopy (AFM) and FTIR spectroscopy (Mohammed et al. 2016). Another attempt reports the utility of *P. islandicum* strain in getting rid of heavy metal manifested as pollution through a transformation of Fe, Cu, Cr, and U to their oxides at high temperatures (Tchounwou et al. 2012). Study results like these provide highly crucial information about the optimization of microbial species to attain optimum biological degradation as the genes coding for concerned biochemical conversions could be selected and engineered for a higher expression to maximize the detoxification efficacy. Many investigations report better efficiency of biodegradation using symbiotic cultures compared to single microbes, inferring a probable synergistic mode of functioning by microorganisms. These possibilities could be further optimized for enhancing the performance as requirement of distinctive nutrients from culture media would allow better growth of a bacterial and fungal consortium compared to bacteria or fungi alone.

##### Maghemite Nanoparticles ($\gamma$-$Fe_2O_3$)

Microbial species like *Actinobacter* (an aerobic bacterium) have been optimized to produce superparamagnetic $Fe_2O_3$ NPs at extracellular locations, using $FeCl_3$ and $FeSO_4$ under ambient culture conditions. Exploring the untapped potential of biologically abundant and environmentally safer technologies to utilize inorganic metals and their varying stoichiometry compounds has been on the rise over the past few years. The justified reason for this is the renewable nature of energy input needed to catalyze the bioconversions and catalysis processes. It seems quite overwhelming when the microbes residing in marine and terrestrial habitats are cultured for development into biologically altered forms that could selectively act on harmful synthetic dyes and poisonous heavy metals and their compounds, recalcitrants. These technologies reflect the untapped potential of natural biomaterials or bioresources to clean the environment and regulate the net balance of several useful metals and nonmetals, via precipitation, decomposition, and degradation. For instance, *Bacillus* species encode a significant population of enzymes and proteins aiding in the synthesis of metal oxide NPs, such as that of hexagonal and protein-capped crystalline $Fe_2O_3$ NPs by *Bacillus cereus* using $FeCl_2$ as a precursor. Capping with microbial proteins during synthesis ensures a thorough stability of as synthesized NPs using the same material as reducing and capping agents (Fang et al. 2019). So no separate capping agent needs to be added and NPs or nanomaterials made in this manner find astounding significance in drug delivery applications (where cytotoxicity has to be controlled).

The use of a thermophilic bacterium, *P. islandicum*, is immensely significant in this regard, through which amorphous Fe (III) oxyhydroxide is reduced to magnetically receptive iron oxides at 65 °C. Similarly, the enzyme extract of *Lactobacillus casei*, upon incubation with $FeSO_4$ solution (1 mM) at 5.6 pH, 37 °C, and in 5% $CO_2$-containing environment for 3 weeks, provided spherical NPs of 15 nm (evaluated using transmission electron microscopy) size. Preliminary formation of NPs was noted by the transparent to black color changes of the culture medium, after which confirmatory screening was made using electron microscopy and X-ray diffraction. Approaches like these offer benign methodology for FeO NP synthesis, with simple, efficient, and economic methodology, owing to which the as-harvested product could be readily utilized for drug delivery and pharmaceutical applications (Torabian et al. 2018). Similarly, *Bacillus subtilis* strains (isolated from rhizosphere-rich soil) are reported for the extracellular synthesis of $Fe_3O_4$ NPs, using a supernatant fraction of their culture medium (to write details). It would be interesting to note here that same yield of NPs is not possible with supernatant and pellet as the population of biological catalysts differs, thereby affecting the NP formation activity (Sundaram et al. 2012).

A peculiar aspect in FeO NP formation by magnetotactic bacteria involves the movement of Fe particles, leading to the formation of siderophores, which are the magnetosome vesicles facilitating Fe reduction from the +3 (ferric) to +2 (ferrous) state. After this biological reduction, controlled biomineralization of magnetite (in the last stage) happens. An important consideration here is the formation of siderophores, characteristic magnetic vesicles formed in magnetotactic bacteria, typically low-molecular-weight (0.5–1.5-kDa) $Fe^{+3}$-chelating molecules synthesized by most bacteria under Fe limiting conditions (Arakaki et al. 2008). In general, these siderophores are natively synthesized by magnetotactic bacteria.

**Table 7.2** Various metal oxide nanoparticles synthesized using bacteria

| | *Bacteria* | Size | Shape |
|---|---|---|---|
| ZnO | *Rhodococcus pyridinivorans* NT2 | 100–120 | Hexagonal |
| ZnO | *Bacillus licheniformis* MTCC 9555 | 100 | Hexagonal wurtzite |
| ZnO | *Staphylococcus aureus* and *Escherichia coli* | 3.9 ± 0.5 | Sphere |
| ZnO | *Klebsiella pneumoniae* | 20–25 | Hexagonal pyramids |
| FeO | *Actinobacteria* sp. | 10–50 | Spherical |
| $Fe_2O_3$ | *Shewanella oneidensis MR-1* | 30–40 | Irregular or rhombohedral |
| $Fe_3O_4$ | QH-2 | 80–90 | Rectangular |
| $Fe_3O_4$ | *Aeromonas hydrophila* | 50–60 | Spherical |
| $TiO_2$ | *Bacillus subtilis* | 66–77 | Spherical |
| $TiO_2$ | *Bacillus amyloliquefaciens* | 22.11–97.28 | Spherical |
| $TiO_2$ | *Lactobacillus sp.* | 8–35 | Spherical |
| $TiO_2$ | *Micrococcus lylae, Micrococcus aloeverae, Cellulosimicrobium* sp. | 14–17 | Spherical |
| $TiO_2$ | *Bacillus mycoides* | 40–60 | Spherical |
| $BaTiO_3$ | *Lactobacillus* sp. | 20–80 | Tetragonal |
| NiO | *Microbacterium* sp. MRS-*1* | 100–500 | Spherical flower |

Table 7.2 presents an account of different metal oxide NPs synthesized using bacterial species, with the characteristic shape, size, and species. Yet again, it is interesting to note the NPs of similar size, shape, and extent, which consequently form the basis of their distinct applications (owing to size- and shape-dependent NP functional responses).

#### 7.7.2.3 Fungus-Mediated Synthesis of FeO NPs

Recognized as eukaryotic organisms for their decomposition activities, fungi are ubiquitous in several ordinary lodgings and are one of the most well-known decomposer organisms. Biological diversity and robustness of living requirements for the fungi could be well judged from the supposedly 1.5 million species (reported on earth), even though only 70,000 of these are distinctly known. Digestion of extracellular food sources through secretion of specified enzymes to hydrolyze complex chemical compounds into simpler fractions that are finally absorbed and utilized as energy resources is a characteristic feature of these microorganisms. Latest data enabled through high-throughput screening methods estimate a prevalence of nearly 5.1 million fungal species (Wu et al. 2019). Higher tolerance of culture environment adversities (such as temperature and pH diversity) along with greater metal bioaccumulation abilities by the fungi renders them highly suitable compared to bacterial species (Archana et al. 2015). Apart from the higher culture fluctuation tolerance of fungi, the fascinating features for intense interest in fungal preparation of metal and metal oxide NPs are due to the simplicity of their scale-up process alongside an efficient extracellular secretion of enzymes (Boroumand Moghaddam et al. 2015; Prasad 2016, 2017; Prasad et al. 2018; Aziz et al. 2016). Utilization of fungi for making NPs is substantially preceded through thin solid substrate fermentation technique, manifesting the optimization of high wall-binding and intracellular metal uptake attributes, owing to their fast growth (Salvadori et al. 2015). Till date, several fungal species, including *Verticillium luteoalbum*, *Fusarium oxysporum*, *Asperigillus oryzae*, *Alternataalternata*, *Trichoderma viride*, etc., have been cultured and optimized to yield NPs of different shapes and sizes. An effort worth in this direction by *Gawande* et al., in 2016, reports the biomimetic mineralization of fungi to synthesize nano- or meso-sized NPs through activity of intra- and extracellularly secreted enzymes (responsible for a range of geometry and sizes of synthesized NPs). Fast growth and easier handling of fungal populations within the laboratory are the key advantages for their preferential usage in such applications (Gawande et al. 2016). The sole constraint behind the use of fungi in nanoparticle preparation is their difficulty in scaling up, hindering the commercial development and application of synthesized nanomaterials. This constraint arises due to higher biomass fraction of fungi (compared to bacteria), and for needful enzymatic activity, the biological activity of cultured microbe has to be adequately expressed in the absence of which the product synthesis becomes impaired or affected (with respect to product quality). Among all the fungal species, *Verticillium* and *Fusarium* are most utilized for metal oxide NP synthesis, owing to their possession of versatile enzymes, proteins, and other metabolites (Ovais et al. 2018). One study reports the synthesis of FeO NPs at room temperature using ferrous and ferric salt mixtures as precursors (Mazrouaa et al. 2019). Success has also been obtained regarding the formation of crystalline magnetite NPs through cationic protein enabled hydrolysis of anionic Fe complexes at the extracellular sites. The obtained NPs exhibited a characteristic ferromagnetic transition with an almost insignificant magnetization extent at low temperature. In processes like this, it is very pertinent to take note of the electrostatic proximity of cationic proteins with anionic Fe precursors, and it becomes imperative to have an aggregation preventing capping activities of cationic proteins, which in turn promotes interaction with negatively charged species/molecules (Fig. 7.6).

A rigorous investigation by *Latiffah* and coworkers reports the formation of spherical FeO NPs using three distinct manglicolous fungal species, namely, *Trichoderma asperellum*, *Phialemoniopsis ocularis*, and *Fusarium incarnatum*, isolated from mangroves. The NPs were, respectively, 25 ± 3.94, 13.13 ± 4.32, and 30.56 ± 8.68 nm in sizes, inferring highest monodispersion with *Phialemoniopsis ocularis*, a consideration which could be further screened with respect to the biological catalytic action of *Phialemoniopsis ocularis* (Latiffah et al. 2010).

Many research efforts have explored the bioreducing ability of the fungus *Aspergillus niger* for making magnetite ($Fe_3O_4$) NPs, in view of this fungi's ability to catalyze the $FeSO_4$ and $FeCl_3$ decomposition to FeS and $Fe_2O_3$, respectively. Upon exposure to ethanol-assisted supercritical conditions, the as-formed particles were stored at 300 °C and 850 PSI units of pressure. Subsequent analysis using phase structure and morphology aspects yielded spherical and pure Fe and $Fe_3O_4$ NPs having average sizes of 18 and 50 nanometers, respectively. Screening for

**Fig. 7.6** Advantages of fungus-based synthesis of nanoparticles

magnetic properties revealed ferromagnetic behavior with saturation magnetization extent of 68 emu per gram for $Fe_3O_4$. The analytical profile provided a novel paradigm shift to optimize biophysical methods for the large-scale synthesis of magnetic NPs optimized for biomedical applications (Abdeen et al. 2016).

Biosynthesis of Fe NPs is also reported using *Pleurotus* sp. in optimum culture media, where the fungus was allowed to grow in $2x10^{-4}$ M $FeSO_4$ solutions up to 72 h. UV-Vis absorbance peaks at 226 and 276 nanometers predicted the formation of NPs, while involvement of proteins in the NP formation (as component of culture medium) was ascertained through spectral analysis of broth containing culture medium at varying time intervals corresponding to 265 nm. Interactions of proteins with NPs were inferred through FTIR spectroscopy through a concomitant involvement of functional groups indicated by shift of functionalities of the treated cells. Morphology inspection through TEM revealed particle deposition at the inner as well as outer surface for completion of the synthesis process. The peculiar role of fungal proteins was confirmed by the absence of these depositions in control samples, where no *Pleurotus* sp. was furnished. This was also confirmed through X-ray fluorescence, which showed no Fe in control samples but prevailed in mycelium. The presence of few other elements was also noted, which could be due to randomized compositional fluctuations of culture medium. A characteristic aspect of FeO NP synthesis by fungi is the formation of Fe transport molecules like hydroxamates that bind the complex molecules and transport them inside the cells (Mazumdar and Haloi 2011). Similarly, the potential of *Aspergillus japonicus* isolate AJP01 is also reported for extracellular synthesis of 60–70-nm FeO NPs through hydrolysis of a mixture of Fe cyanide complexes (acting as precursor), under ambient growth conditions. Screening the mechanism, it was noted that hydrolysis of these complexes released $Fe^{+2}$ and $Fe^{+3}$, through involvement of protein facilitating coprecipitation and controlled nucleation, ultimately forming FeO NPs. Analysis using FTIR spectroscopy revealed the presence of proteins that confer stability to as-formed NPs, in agreement with the observations of an earlier preliminary investigation (Bhargava et al. 2011).

## 7.7.3 Microbially Mediated Synthesis of $TiO_2$, ZnO, CuO, $SiO_2$, and $Al_2O_3$ NPs

### 7.7.3.1 Titanium Dioxide ($TiO_2$) NPs

The significance of $TiO_2$ in multidimensional domains of electronics, optics, and environmental remediation is well known. In general, the nontoxic and biocompatible properties of $TiO_2$ impart its suitability for biomedical requirements such as bone or tissue engineering as well as in pharmaceutical industries. Majorly, $TiO_2$ is synthesized using chemical and biological reduction approaches such as sol-gel, hydrothermal, solvothermal, combustion, plant (extract), and microbially assisted biosynthesis. Microbes such as *Lactobacillus* (bacteria) and *Saccharomyces cerevisiae* (yeast) are used for $TiO_2$ NP synthesis. Few specific examples of $TiO_2$ NP synthesis using bacterial cultures are notified here. Firstly, *Jayaseelan* et al. and *Kirthi* et al. reported synthesis of $TiO_2$ NPs using $TiO(OH)_2$ as a precursor and *Bacillus subtilis* as a reducing agent. The morphology of the synthesized $TiO_2$ was found to be spherical/oval while the size was in the 66–77-nm range (Kirthi et al. 2011b; Jayaseelan et al. 2012). Secondly, Malarkodi et al. (2013) reported the synthesis of $TiO_2$ NPs of size 100–500 nm, irregularly shaped, using the biomass of *Planomicrobium* spp. obtained from melted ice. The synthesized NPs exhibited antimicrobial properties against *B. subtilis MTCC 3053* and *K. planticola MTCC 2727* in agar disk diffusion method (Malarkodi et al. 2013).

Thirdly, a report by Khan and Fulekar focuses on the biological synthesis of $TiO_2$ using *Bacillus amyloliquefaciens* as a capping agent. The synthesized $TiO_2$ exhibited a spherical morphology in the 22–97-nm size range. The synthesized $TiO_2$ were used further in the photocatalytic degradation of Reactive Red 31 (RR31, a poisonous dye) using platinum-doped $TiO_2$, developing the highest potential (90.98%) for RR31 degradation as compared to native $TiO_2$ (75.83%) (Khan and Fulekar 2016).

#### 7.7.3.2 Zinc Oxide (ZnO) NPs

Owing to its simplified synthesis approach as well as its increasing suitability in the fields of pharmaceuticals, cosmetics rubber industry, biosensor development, optics, solar cells, and environmental remediation, ZnO NPs have emerged as one of the most sought-after metal oxide NPs (Khan et al. 2016, 2018). The specific interest has been to explore the antimicrobial, antibacterial, and antioxidant activities through the inherent nontoxicity of nanoscale forms (Bhuyan et al. 2015). Zinc (Zn) is evidently the second most abundant metal in the earth's crust (after Fe) and it is the only metal present in all six enzyme classes (*lyases, transferases, oxidoreductases, hydrolases, isomerases, and ligases*). Similar to Zn, ZnO is the most researched and studied metal oxide after $TiO_2$. Compared to other synthesis methods, the microbial approach offers much advantages emerging helpful for NP synthesis at low pH, temperature, and pressure. In an interesting attempt, Kalpana et al. examined the antibacterial activity of ZnO NPs on *Escherichia coli, Salmonella typhi, Bacillus subtilis, Staphylococcus aureus*, and *Pseudomonas aeruginosa* (Kalpana et al. 2018). Further, Kundu et al. synthesized extracellular ZnO NPs of average diameter (100–120 nm) using *Rhodococcus pyridinivorans* NT2 and applied it for textile finishing and drug delivery in colon carcinoma (Kundu, Hazra) (Kundu et al. 2014). Jayseenlean et al. suggested a novel biological route for the synthesis of spherical, oval-shaped ZnO NPs of size 57.7 nm using bacterium *Aeromonas hydrophila*. The synthesized ZnO NPs were characterized by all the major sophisticated instruments and its potential was assessed against pathogenic bacteria and fungi (Jayaseelan et al. 2012). Taran et al. (2018) reported a simple, eco-friendly method for the synthesis of ZnO and $TiO_2$ NPs using bacterium *Halomonas elongata* IBRC-M 10214. The morphological confirmation of NPs by microscopy revealed that the particles were spherical shaped whose average diameter was $104.63 \pm 27.75$ for $TiO_2$ NPs and $18.11 \pm 8.93$ nm for ZnO NPs. Both of the NPs were assessed for their antibacterial activity against multidrug-resistant bacteria (MDRB), i.e., *E. coli* ATCC 25922 and *Staphylococcus aureus* ATCC 43300 (Taran et al. 2018).

#### 7.7.3.3 Copper Oxide (CuO) NPs

One of the most important transition metal oxides, CuO promises interesting applications in terms of its size-tunable and captivating properties, forming the basis of its inclusion in highly critical temperature gas sensors, superconductors, and catalytic, optical, photoconductive and photoelectronic, electrical, and energy storage purposes. It has been used as an antimicrobial agent recently against various bacterial species (Yadav et al. 2017).

#### 7.7.3.4 Silica ($SiO_2$) NPs

A green technique of $SiO_2$ NP formation, using a thermophilic bacterium (BKH1) as a biological template, is reported using an inorganic precursor (magnesium trisilicate) in an organic (tetraethyl orthosilicate) reducing agent (Show et al. 2015).

This novel method of bacterially synthesized $SiO_2$ NPs is eco-friendly and is workable at ambient temperature, thereby not requiring drastic energy inputs. This method avoids the complex protocol of multistep synthesis and is therefore much cost-effective. A notable aspect is the functional activity of thermophilic bacterium (BKH1) at high temperatures, pertaining to which identification of genes conferring survival and needed bioactivities at high temperatures could be an interesting framework for strengthening the impact of similar studies on other microbes. Such activities also create an interest in enzyme/protein stability against denaturation at high temperatures.

The synthesis of silicon/silica nanoparticle composites by the bacterium *Actinobacteria* sp. is reported by Marikani et al., by exposing the bacterium to $K_2SiF_6$ precursor under ambient conditions. The formation of silica NPs is due to the secretion of several reductases and oxidizing enzymes. Further, few scientists have also reported on the synthesis of silica NPs from *F. oxysporum* isolated from tomato wilt (Marikani et al. 2016). Silica NPs were synthesized from rice husk ash (RSA) by *F. oxysporum* in malt-glucose (MG) and malt-glucose-yeast-peptone (MGYP) media. The confirmation of size, shape, and purity was analyzed by sophisticated instruments. It was concluded from the results that solubilization of silica was not directly associated with their production of organic acids. The results showed that the production of organic acids was not directly related to the solubilization of silica. It was found that solubility and stability of silica are due to the extracellular proteins released into the medium during the exponential growth phase. Out of both the media, MG media were found more suitable for growth as well as for the formation of semicrystalline, quasi-spherical $SiO_2$ NPs of size 2–8 nm (Pineda-Vásquez et al. 2014).

### *7.7.4 Virus-Based Synthesis of Metal Oxide Nanoparticles*

Plant viruses have been used widely as templates for organic–inorganic hybrid synthesis. However, by simply adjusting the pH, fine-tuning of hybrid nanoparticle structures, especially the control of inorganic particle size and the location where silication occurs (i.e., outside and/or inside of the capsid), remains a challenge. By using the templating effect of the cowpea chlorotic mottle virus (CCMV) protein cage, it was shown that the silication at the outer or inner surface of the protein capsid, and the resulting structures of silica/virus hybrid nanoparticles, can be finely tuned by adjusting the pH (Liu et al. 2017). At pH 4.0, only small silica particles (2.5 nm in diameter) were formed inside the protein cages, whereas at pH 6.0, mainly silication occurred in the protein cages, resulting in monodisperse silica nanoparticles with 14-nm diameter. At pH 7.5, silica deposition was found on both the surfaces, i.e., inner and outer surfaces of the protein cage under aqueous conditions. In these reaction circumstances, multicomponent hybrid virus/nanoparticulate systems, such as CCMVAu/silica and Au/silica nanoparticles, were prepared stepwise. After removal of the CCMV template in thermal degradation, a single gold nanoparticle can be encapsulated within a hollow silica shell to simulate the structure of a baby rattle in which unattached solid particles are in the hollow particles. The Au/silica core-hollow shell nanoparticles can further be used as a stable catalyst. These synthetic methods are expected to provide a versatile method for preparing core-shell nanomaterials with well-designed structures and functions (Liu et al. 2017).

A sol-gel procedure has been developed to integrate bionanoparticles, such as cowpea mosaic virus, turnip yellow mosaic virus, tobacco mosaic virus, and ferritin into silica while maintaining the particle integrity and morphology. The structures of the resulting materials were characterized by TEM, small-angle X-ray scattering (SAXS), and N adsorption-desorption analysis. The results obtained show that the shape and surface morphology of the nanoparticles are largely retained after the incorporation of silica. After removal of the bionanoparticles by calcination, a mesoporous silica having monodisperse pores with well-defined shape and surface morphology of the bionanoparticles was replicated inside the silica (Niu et al. 2010).

## *7.7.5 Alga-Mediated Synthesis of Metal Oxide Nanoparticles*

### 7.7.5.1 Alumina Nanoparticles ($Al_2O_3$) by Algae

Algae are a good source of biomolecules among all other aquatic organisms; this is so because algae contain proteins, carbohydrates, fats, nucleic acids, pigments, and secondary metabolites such as alkaloids, some aromatic compounds, peptides, macrolides, and terpenes (Siddiqi and Husen 2016a). They can act as reducing agents that help in the preparation of nanoparticles from metal salts without producing any toxic by-product. Once the algal biomolecules are identified, the nanoparticles of the desired shape or size may be fabricated. The antimicrobial activity of the thus synthesized metal and metal oxide nanoparticles against several gram-positive and Gram-negative bacterial strains and fungi has been investigated (Shannon and Abu-Ghannam 2016).

The dimensions of synthesized alumina NPs depend on the pH, temperature, incubation time, and concentration of the solution. A new biological methodology is proposed for the production of ceramic α-aluminum oxide nanoparticles using an extract of the algae *Sargassum ilicifolium*. The algal extract works as a stabilizer as well as a bioreducing agent. The UV-Vis analysis shows the presence of an absorption peak at 227 nm, which confirmed the formation of the aluminum oxide nanoparticles. FTIR analysis has indicated that bioreduction of aluminum ions and stabilization of nanoparticles may be caused by interactions between aluminum and the biofunctional groups of the algal extract. The XRD pattern has confirmed that after calcination at ~1200 °C, the $Al_2O_3$ nanoparticles with 35-nm diameter were alpha crystalline in nature and have rhombohedral structure. TEM analysis showed that the alumina nanoparticles were well dispersed and spherical in shape with an average size of 20 ± 2.1 nm, while EDX spectroscopy has confirmed high purity of the alumina nanopowder, as sample contained only aluminum (46.31%) and oxygen (53.69%) (Koopi and Buazar 2018; Siddiqi and Husen 2016b).

## 7.8 Various Conventional Approaches for Wastewater Treatment

With the continuous advancement of technology, several methods have made huge impact on wastewater treatment. Although there are several conventional wastewater treatment methods, like coagulation, precipitation, electrolysis, filtration, absorption, adsorption (Azimi et al. 2017), etc., most of them are less efficient, expensive, and energy-intensive. Out of these, the most efficient and economical method is adsorption and the process can become cheaper by using waste materials derived from agriculture, industry, domestic, and poultry. The adsorbents developed from such materials have different groups on their surface that act as adsorption sites. These active sites have a role in the removal of multivariate pollutants from the wastewater.

Out of these adsorption is the most reliable technique by nanoparticles due to their economical nature and surface modification by microorganisms (Khan et al. 2019). Generally, nanoparticles adsorb inorganic pollutants on their surface. One more advantage with adsorption method is that the nanoparticles can be treated with 0.1 M NaOH for removal of all the surface-attached pollutants, mainly heavy metals. This helps in the reuse of nanoparticles, whereas if nanoparticles are IONPs, then it can be easily recovered after the reaction is complete, making the whole process economical.

## 7.9 Applications of Nanoparticles for Wastewater Treatment

### *7.9.1 Typical Composition of Wastewaters*

Before moving on to the mechanisms and efficacies of nanoparticle-enabled wastewater treatment technology, it is quite logical to know about the chemical diversity of wastewater streams in terms of its composition. Multiple research attempts on wastewater streams from different regions of the globe list six major deleterious components, namely, suspended solids, biodegradable organics, pathogens, nutrients, heavy metals, and soluble inorganic salts (Abdel-Raouf et al. 2012). All these constituents pose a significant risk to the pollution level of wastewaters and could be present either in native form or added at various stages of domestic water discharge. For example, the nutrients carbohydrates, fats, and proteins could be present from the source itself (may be due to excretion products of humans and animals), whereas heavy metals could be added to the wastewater stream if it has access to industries. Similarly, the organic constituents act as existence support towards the environmental microbial species, which continuously grow, propagate, and increase their metabolic activities while residing on it (Lareen et al. 2016). Consequently, the microbial load increases, and with waste material as dietary input, these organisms exhibit toxic responses and are easily the opportunistic carriers of communicable infections (Figs. 7.7 and 7.8).

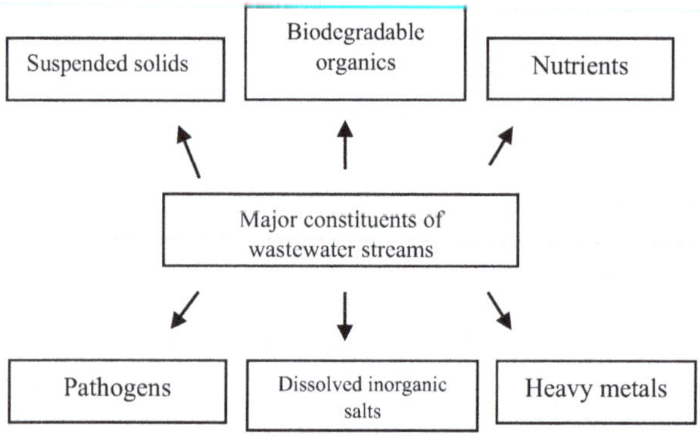


**Fig. 7.7** Major polluting constituents in a wastewater stream

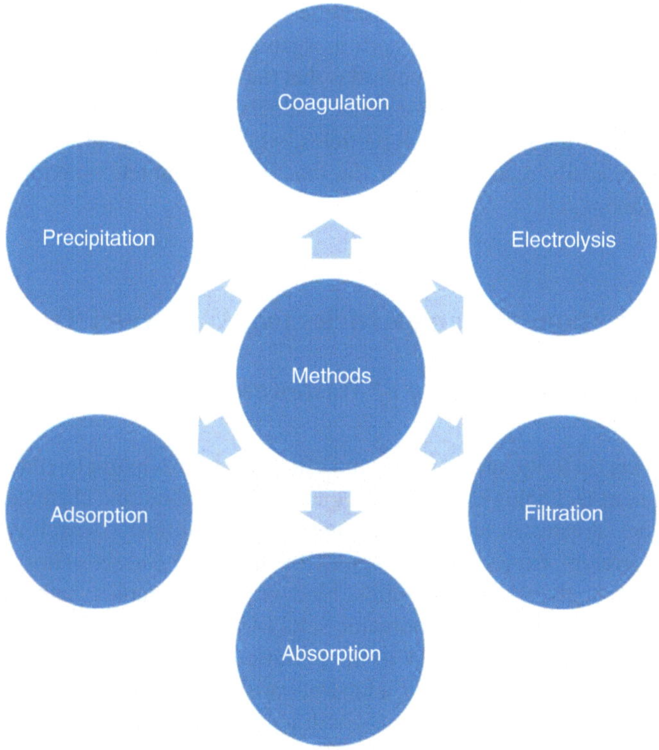


**Fig. 7.8** Different methods of wastewater treatment

The nitrogen in the discharged fats, proteins, and carbohydrates in the wastewater streams catalyzes the microbial activities through the eutrophication phenomenon, as a consequence of which the toxicity is enhanced (Puyol et al. 2017). The oxidative proximities of proteins, carbohydrates, and fats (summed as biological oxygen demand, BOD) deplete the environmental oxygen whereby serious loss to biodiversity could be driven. Like various other pollutants, inorganically dissolved impurities (such as calcium, sodium, and sulphate) could be added to wastewaters from household activities. These materials make the wastewater unfit for use owing to which its chemical load moderation is mandatory before final discharge as waste streams. Among the major organic materials (OM), fibers were found to prevail in the majority, followed by proteins and sugars, respectively accounting for 20.64, 12.38, and 10.65% of the total organic content (TOC). A certain percentage of endocrine-disrupting chemicals (mostly polyaromatic hydrocarbons [PAHs] and phthalates) were also found to contribute towards the pollutant load of wastewaters (Puyol et al. 2017). Together, volatile fatty acids, soluble proteins, and sugars contribute to nearly 30% of the total chemical oxygen demand (COD) of wastewaters, with the majority of proteins and sugars being >0.001 μm.

Several stages of treatment are employed to reduce the chemical load of wastewater streams, such as removal of impurities and solid materials through sedimentation, froth floatation, or salt-driven precipitation. Each successive stage is intended to reduce the toxic content through treatment procedures implicit towards the chemical diversity of impurities. The microbial decontamination of wastewaters has gained immense popularity in the past few years attributed to powerful chemical breakdown actions of microbial enzymes (Singh et al. 2016b). The advantage with this technology is that microbes can feed on the wastewater as their nutrient base and there is no requirement of maintaining their survival through external controls. The stage is now right to know about the specific bacterial species and strains for specific kinds of wastewater impurities.

Hyperthermophilic microorganisms can help in the migration of metal contaminants by reducing them to less mobile forms. Thermal radioactive or metal-containing industrial wastes may be treated potentially in bioreactors using microorganisms similar to the metabolism of *P. islandicum* or their enzymes (Kashefi and Lovley 2000). Many mesophilic microorganisms having the ability to use Fe (III) as a terminal electron acceptor can also reduce various metals and metalloids other than Fe (III) (Lovley 2013). Fe (III)-reducing bacteria and archaea were found capable to precipitate gold by reducing Au (III) to Au (0) (Kashefi et al. 2001). The reaction appeared to be enzymatically catalyzed, which was dependent on temperature and the presence of hydrogen (as a specific electron donor). Many Fe (III)-reducing microorganisms were able to reduce the forms of oxidized metals, including radionuclides such as uranium (VI) (Wu et al. 2006) and technetium (VII) and trace metals including arsenic (V), chromium (VI), cobalt (III), manganese (IV), and selenium (VI). Many of these metals and metalloids act as environmental pollutants. Thus, Fe (III)-reducing microorganisms can be used to remove metals from water and waste streams and to immobilize metals in the subsurface environments.

Microbial reduction of metals may play a major role in the formation of metal deposits, which may be especially important in hot environments containing metal-rich waters (Drewniak and Sklodowska 2013). Common products of bacterial iron-reduced nanosized magnetic particles enable early disease detection and accurate prognosis and personalized treatment, monitoring efficacy of prescribed therapy, or study of cellular interaction in a certain biological environment (Chircov et al. 2019). These particles might be used in different radionuclide therapies, drug delivery, magnetic resonance imaging (MRI) (Yadav and Fulekar 2018), diagnostics, immunoassays, molecular biology, DNA and RNA purification, cell separation and purification, cell adhesion studies, hyperthermia, and magnetic ferrofluids for magnetocaloric pumps.

## 7.10 Specific Metal and Metal Oxide NPs Appropriate for Wastewater Treatment

The suitability of metal or metal oxide NPs for wastewater treatment is manifested in their antimicrobial and antioxidant characteristics since the wastewater streams are rich in oxygen-consuming pathogenic microbes. The NPs are ideal candidates for reducing the chemical complexity of oxygenated aggregations of wastes, facilitating their smoother discharge and disposal (Chen et al. 2015). Among metals, mostly used NPs are Au, Ag, Cu, and Fe, while in oxides, $TiO_2$, ZnO, silica, and nano zerovalent iron (NZVI) NPs are preferred (Ruttkay-Nedecky et al. 2017b). To a smaller extent, MgO, CuO, and $Al_2O_3$ conjugates are also employed. The sole mandate of utilizing bacteria-driven NPs for wastewater treatment relies on the interaction enhancing the ability of NPs. In this reference, it is quite interesting to note that not only NPs can reduce the toxicity of wastewater streams but, sometimes, even the bacterial strains as such or in genetically altered forms are added to wastewater streams. Bacteria possess a number of enzymes which work as catalysts to break down the toxic wastes comprising the wastewater stream. With such abilities, sometimes the supplemented bacterial populations themselves become the sources to break down the toxic metal complexes into their salts which are further reduced to their zerovalent states. In the recent past, several interesting studies have reported the addition of genetically modified organisms (GMOs) into toxic waste streams to reduce their recalcitrant and heavy metal loads. This section selectively discusses bacterially synthesized metal and metal oxide NPs and their potential application in wastewater treatment.

In a very interesting 2013 contribution, *Yang* et al. have compared the wastewater treatment efficacies of nanoscale ZnO, $TiO_2$, zerovalent iron, and AgNPs. They discussed that even though a common toxicity reduction mechanism of all these nanomaterials relies on the generation of reactive oxygen species (ROS), the extent and manner of such activities differ on an individual basis. The important factors affecting the antimicrobial activities of these functionally distinct nanomaterials in wastewaters are their chemical properties and stability (Das et al. 2018). Interestingly, the researchers have confirmed that oxygen is essential for ROS generation by nanoscale Ag and iron particles, whereas for ZnO and $TiO_2$, illumination is needed to elicit the ROS expression. Interestingly, though having similar sizes, ZnO and $TiO_2$ exhibited greater toxicity than nanoscale Ag and Fe, owing to their oxidizing vulnerability and water solubility which result in the generation of toxicity-inducing metal ions. Regarding the toxicity of AgNPs, it is well reported that the toxic response is due to $Ag^+$ formation, which is very sensitively affected by the oxygen concentration of the surroundings.

Under aerobic conditions, the antimicrobial response of AgNPs is due to oxidative dissolution of AgNPs, concomitantly leading to gradual release of Ag ions and the production of ROS (Skandalis et al. 2017). The requirement of oxygen for the toxicity induction of AgNPs is further assured by no $Ag^+$ formation in the anaerobic environment, predicting a necessity of dissolved oxygen to facilitate oxidative dissolution of nanoscale Ag (Abdal Dayem et al. 2017).

Similarly, Fe has comparatively higher reactivity than Cu and Ag, owing to which nano zerovalent Fe (nZVI) can reduce several chemicals in wastewater streams, such as aromatic nitro compounds, chlorinated solvents, and oxidized heavy metals (Yang et al. 2019). In the absence of oxygen (solution phase), water could oxidize nano zerovalent Fe, forming $Fe^{+2}$ and $H_2$ gas. However, water causes rapid surface oxidation of nano zerovalent Fe under aerobic conditions, after which dissolved oxygen oxidize ferrous state ($Fe^{+2}$) to ferric ($Fe^{+3}$) (Yang et al. 2019). So, in totality, the pH and DO concentration regulated the reduction efficacy of nano zerovalent Fe (Park and Dempsey 2005).

Opposed to Fe and Ag, the nanoscale ZnO and $TiO_2$ generate ROS after photon absorption (Park and Dempsey 2005). Under illumination, electrons excited by photon absorption diffuse towards the surface and react with O to form superoxide anions ($O^{2--}$) and consequentially generated holes diffuse to the surface and react with water to form hydroxyl radicals ($OH^-$) (Liao and Reitberger 2013). A notable aspect is that nanoscale $TiO_2$ is nearly water-insoluble, whereas ZnO can dissolve in water to a higher extent. So unlike $TiO_2$, ZnO on being dissolved in water releases free Zn (as $Zn^{+2}$) and $Zn(OH)_2$ which are dominant zinc species at neutral pH conditions and exert a toxic response towards aquatic organisms (Reed et al. 2012). Thus, ZnO and $TiO_2$ mediate their toxicity-reducing responses in wastewater treatment through the additive association of sunlight, whereas nanoscale Ag and Fe express their toxicities through mutual oxidation and reduction. It seems reasonable here that a particular wastewater composition might be more specifically suited for treatment by ZnO and $TiO_2$ than Ag and Fe since ionic associations and redox existences are also affected by process temperature and pH. It is of further interest to note here that heterogeneous composition of wastewater streams complicates their detoxification, where one particular material may not be effective to a significant extent. In such situations, combinative approaches like adding microorganisms (preferably in consortium modes) to the wastewater reservoirs or adding two or more antimicrobial agents in simultaneous mode seem to be more effective than conventional technologies. Readers are suggested to refer to the 2016 contribution of *Lu* et al., where the various problems of nanoscale Ag, Fe, and Zn usage are discussed at length along with their probable resolutions. For example, nanoscale Ag alone often leads to fouling and clogging issues in the aqueous environment, while when AgNPs are attached to filter materials, not only the overall cost of the remediation process moderates but the disinfection efficacy also improves by significant extents (Mourdikoudis et al. 2018). Likewise, NZVI has been found effective in large-scale removal of contaminants (comprising nitroaromatic compounds, halogenated organic compounds, phenols, organic dyes, heavy metals, inorganic anions, nitrates, metalloids, and radio elements) through simultaneous attributes of reduction, oxidation, adsorption, and precipitation, but the complications of aggregation, oxidation, and separation difficulty restrict the long-range disinfection efficacy of nZVI. As a consequence, several modifications are implemented for better nZVI treatment efficacy that includes surface coating, doping with other metals, conjugation with supports, emulsification, and encapsulation in the matrix (Liang et al. 2014). Studies on these modifications suggest enhanced nZVI reactivity via doping with other metals. Similarly, surface coating and conjugation strategies forbid the aggregation and enhance the nZVI dispersion. The modifications of conjugations and matrix encapsulation ensure an optimum nZVI separation from degraded wastewaters, paving the way for their efficient removal from the treated wastewater streams (Ledakowicz et al. 2001). Similarly, the photocatalytic attributes of $TiO_2$ are the reasons for its extraordinary suitability in UV-radiation-driven particle separation. With its reasonable price and long-lasting photo-, chemical, and biological stability alongside a large energy band gap (3.2 eV), $TiO_2$ remains the most sustainable photocatalyst till date (Fujishima et al. 2000). Remarkable photocatalytic efficacy of this material facilitates ROS generation in reasonable time durations, facilitating an efficient removal of contaminants.

The prospect of nanomaterial recovery from the waste stream is definitely a concerning factor for their scale-up suitability in detoxification application (Shan et al. 2009). This challenge can be addressed by careful consideration of the source of these nanomaterials, which critically affects their shape, size, and aggregation stability. For instance, bottom-up approaches offers significantly higher control on their size and shape, owing to which the application stability and suitability would be higher. On the other hand, top-down approaches (mostly used in physical methods of nanomaterial synthesis) are elaborate, utilizing higher energy due to which the end products are less stable and, consequently, more vulnerable towards aggregation.

## 7.11 Conclusion

Microbial synthesis of nanoparticles has gained tremendous applications in wastewater treatment due to their high surface area to volume ratio, functionalization by biomolecules, etc. Microorganisms have various organic compounds like enzymes, acids, polysaccharides, etc. that act as a reducing agent for metallic nanoparticles, whereas some of biomolecules are also involved in the metal oxide synthesis of nanoparticles. These metallic and metal oxide nanoparticles have adsorption sites on their surface that help in the removal of pollutants from the wastewater. Nano adsorption has several advantages over other pollutant removal process. IONPs not only have diversity in the phases but also are economical, recyclable, and easily available. Microbially synthesized nanoparticles have huge potential for the green, economical, and biocompatible source of nanoparticles that find applications in the field of environmental cleanup.

## References

Abdal Dayem A, Hossain MK, Lee SB, Kim K, Saha SK, Yang G-M et al (2017) The role of reactive oxygen species (ROS) in the biological activities of metallic nanoparticles. Int J Mol Sci 18(1):120

Abdeen M, Sabry S, Ghozlan H, El-Gendy AA, Carpenter EE (2016) Microbial-physical synthesis of Fe and Fe3O4 magnetic nanoparticles using Aspergillus Niger YESM1 and supercritical condition of ethanol. J Nanomater 2016:1–7

Abdel-Raouf N, Al-Homaidan AA, Ibraheem IBM (2012) Microalgae and wastewater treatment. Saudi J Biol Sci 19(3):257–275

Agarwal P, Gupta R, Agarwal N (2019) Advances in synthesis and applications of microalgal nanoparticles for wastewater treatment. J Nanotechnol 2019:9

Ali A, Zafar H, Zia M, Ul Haq I, Phull AR, Ali JS et al (2016) Synthesis, characterization, applications, and challenges of iron oxide nanoparticles. Nanotechnol Sci Appl 9:49

Arakaki A, Nakazawa H, Nemoto M, Mori T, Matsunaga T (2008) Formation of magnetite by bacteria and its application. J R Soc Interface 6:977–999

Archana Y, Theivasanthi T, Paul PK, Upadhyay KC (2015) Extracellular biosynthesis of silver nanoparticles from plant growth promoting Rhizobacteria Pseudomonas sp. Int J Curr Microbiol App Sci 4(8):11

Azimi A, Azari A, Rezakazemi M, Ansarpour M (2017) Removal of heavy metals from industrial wastewaters: a review. ChemBioEng Rev 4:1–24

Aziz N, Pandey R, Barman I, Prasad R (2016) Leveraging the attributes of Mucor hiemalis-derived silver nanoparticles for a synergistic broad-spectrum antimicrobial platform. Front Microbiol 7:1984. https://doi.org/10.3389/fmicb.2016.01984

Beveridge TJ, Murray RG (1980) Sites of metal deposition in the cell wall of Bacillus subtilis. J Bacteriol 141(2):876–887

Bhargava A, Jain N, Panwar J (2011) Synthesis and application of magnetic nanoparticles: a biological perspective, pp 117–155

Bhateria R, Jain D (2016) Water quality assessment of lake water: a review. Sustain Water Res Manag 2(2):161–173

Bhuyan T, Mishra K, Khanuja M, Prasad R, Varma A (2015) Biosynthesis of zinc oxide nanoparticles from *Azadirachta indica* for antibacterial and photocatalytic applications. Mater Sci Semicond Process 32:55–61. https://doi.org/10.1016/j.mssp.2014.12.053

Bilton M, Milne S, Brown A (2012) Comparison of hydrothermal and sol-gel synthesis of nanoparticulate hydroxyapatite by characterisation at the bulk and particle level. Open J Inorg Nonmetal Mater 2:1–10

Blakemore R (1975) Magnetotactic bacteria. Science 190:377–379

Boroumand Moghaddam A, Namvar F, Moniri M, Tahir P, Azizi S, Mohamad R (2015) Nanoparticles biosynthesized by Fungi and yeast: a review of their preparation, properties, and medical applications. Molecules 20(9):16540–16565

Brock TD, Gustafson J (1976) Ferric iron reduction by sulfur- and iron-oxidizing bacteria. Appl Environ Microbiol 32:567–571

Chang JH, Kang KH, Choi J, Jeong YK (2008) High efficiency protein separation with organosilane assembled silica coated magnetic nanoparticles. Superlattice Microstruct 44(4–5):442–448

Chen D, Zhang X, Lee AF (2015) Synthetic strategies to nanostructured photocatalysts for CO2 reduction to solar fuels and chemicals. J Mater Chem A 3(28):14487–14516

Chintamani RB, Salunkhe KS, Chavan MJ (2018) Emerging use of green synthesis silver nanoparticle: an updated review. Int J Pharmaceut Sci Res 9(10)
Chircov C, Grumezescu AM, Holban AM (2019) Magnetic Particles for Advanced Molecular Diagnosis. Materials 12(13):2158
Das RK, Pachapur VL, Lonappan L, Naghdi M, Pulicharla R, Maiti S et al (2017) Biological synthesis of metallic nanoparticles: plants, animals and microbial aspects. Nanotechnol Environ Eng 2(1):18
Das S, Chakraborty J, Chatterjee S, Kumar H (2018) Prospects of biosynthesized nanomaterials for the remediation of organic and inorganic environmental contaminants. Environ Sci Nano 5(12):2784–2808
Dong H, Huang J, Koepsel RR, Ye P, Russell AJ, Matyjaszewski K (2011) Recyclable antibacterial magnetic nanoparticles grafted with quaternized poly(2-(dimethylamino)ethyl methacrylate) brushes. Biomacromolecules 12:1305–1311
Drewniak L, Sklodowska A (2013) Arsenic-transforming microbes and their role in biomining processes. Environ Sci Pollut Res 20(11):7728–7739
Elcey CD, Kuruvilla AT, Thomas D (2014) Synthesis of magnetite nanoparticles from optimized iron reducing bacteria isolated from iron ore mining sites. Int J Curr Microbiol App Sci 3(8):408–417
El-Saadony MT, El-Wafai N, El-Fattah HIA, Mahgoub S (2019) Biosynthesis, optimization and characterization of silver nanoparticles using a soil isolate of Bacillus pseudomycoides MT32 and their antifungal activity against some pathogenic Fungi. Adv Anim Vet Sci 7:238–249
Fang X, Wang Y, Wang Z, Jiang Z, Dong M (2019) Microorganism assisted synthesized nanoparticles for catalytic applications. Energies 12(1):190
Fijan S (2014) Microorganisms with claimed probiotic properties: an overview of recent literature. Int J Environ Res Public Health 11(5):4745–4767
Fujishima A, Rao T, Tryk D (2000) Titanium dioxide photocatalysis. J Photochem Photobiol C: Photochem Rev 1:1–21
Gahlawat G, Choudhury AR (2019) A review on the biosynthesis of metal and metal salt nanoparticles by microbes. RSC Adv 9(23):12944–12967
Ganachari SV, Banapurmath NR, Salimath B, Yaradoddi JS, Shettar AS, Hunashyal AM et al (2017) Synthesis techniques for preparation of nanomaterials. In: Martínez LMT, Kharissova OV, Kharisov BI (eds) Handbook of ecomaterials. Springer, Cham, pp 1–21
Gawande MB, Goswami A, Felpin F-X, Asefa T, Huang X, Silva R et al (2016) Cu and Cu-based nanoparticles: synthesis and applications in catalysis. Chem Rev 116(6):3722–3811
Gowramma B, Keerthi U, Rafi M, Muralidhara RD (2015) Biogenic silver nanoparticles production and characterization from native stain of Corynebacterium species and its antimicrobial activity. 3 Biotech 5(2):195–201
Gu H, Xu K, Xu C, Xu B (2006) Biofunctional magnetic nanoparticles for protein separation and pathogen detection. Chem Commun 9:941–949
He S, Zhang Y, Guo Z, Gu N (2008) Biological synthesis of gold nanowires using extract of Rhodopseudomonas capsulata. Biotechnol Prog 24(2):476–480
Hoseinnejad M, Mahdi Jafari S, Katouzian I (2017) Inorganic and metal nanoparticles and their antimicrobial activity in food packaging applications. Crit Rev Microbiol 44:1–21
Iravani S (2014) Bacteria in nanoparticle synthesis: current status and future prospects. Int Scholar Res Not 2014:359316
Jayaseelan C, Rahuman AA, Kirthi AV, Marimuthu S, Santhoshkumar T, Bagavan A, Gaurav K et al (2012) Novel microbial route to synthesize ZnO nanoparticles using Aeromonas hydrophila and their activity against pathogenic bacteria and fungi. Spectrochim Acta A Mol Biomol Spectrosc 90:7
Jiang J, Pi J, Cai J (2018) The advancing of zinc oxide nanoparticles for biomedical applications. Bioinorg Chem Appl 2018:18
Kalpana VN, Kataru BAS, Sravani N, Vigneshwari T, Panneerselvam A, Rajeswari VD (2018) Biosynthesis of zinc oxide nanoparticles using culture filtrates of Aspergillus Niger: antimicrobial textiles and dye degradation studies. OpenNano 3:48–55

Karthika D, Vadakkan K, Ashwini R, Shyamala A, Hemapriya J, Vijayanand S (2015) Prodigiosin mediated biosynthesis of silver nanoparticles (AgNPs) and evaluation of its antibacterial efficacy. Int J Curr Microbiol App Sci 4(11):7

Kashefi K, Lovley DR (2000) Reduction of Fe(III), Mn(IV), and toxic metals at 100 degrees C by Pyrobaculum islandicum. Appl Environ Microbiol 66(3):1050–1056

Kashefi K, Tor J, Nevin K, Lovley D (2001) Reductive precipitation of gold by dissimilatory Fe(III)-reducing bacteria and archaea. Appl Environ Microbiol 67:3275–3279

Katayama Y, Kuraishi H (1978) Characteristics of Thiobacillus thioparus and its thiocyanate assimilation. Can J Microbiol 24(7):804–810

Kessi J, Ramuz M, Wehrli E, Spycher M, Bachofen R (1999) Reduction of selenite and detoxification of elemental selenium by the phototrophic bacterium Rhodospirillum rubrum. Appl Environ Microbiol 65(11):4734–4740

Khan R, Fulekar MH (2016) Biosynthesis of titanium dioxide nanoparticles using Bacillus amyloliquefaciens culture and enhancement of its photocatalytic activity for the degradation of a sulfonated textile dye reactive red 31. J Colloid Interface Sci 475:184

Khan SH, Pathak B, Fulekar MH (2016) Development of zinc oxide nanoparticle by sonochemical method and study of their physical and optical properties. In: AIP conference proceedings vol 1724(1), p 020108

Khan S, Pathak B, Fulekar MH (2018) Synthesis, characterization and photocatalytic degradation of chlorpyrifos by novel Fe: ZnO nanocomposite material. Nanotechnol Environ Eng 3

Khan I, Saeed K, Khan I (2019) Nanoparticles: Properties, applications and toxicities. Arab J Chem 12:908–931

Khanna L, Verma NK, Tripathi SK (2018) Burgeoning tool of biomedical applications - Superparamagnetic nanoparticles. J Alloys Compd 752:332–353

Kirthi AV, Rahuman AA, Rajakumar G, Marimuthu S, Santhoshkumar T, Jayaseelan C et al (2011a) Biosynthesis of titanium dioxide nanoparticles using bacterium Bacillus subtilis. Mater Lett 65(17–18):2745–2747

Kirthi V, Rahuman A, Rajakumar G, Marimuthu S, Thirunavukkarasu S, Chidambaram J et al (2011b) Biosynthesis of titanium dioxide nanoparticles using bacterium Bacillus subtilis. Mater Lett 65:2745

Koopi H, Buazar F (2018) A novel one-pot biosynthesis of pure alpha aluminum oxide nanoparticles using the macroalgae Sargassum ilicifolium: a green marine approach. Ceram Int 44(8):8940–8945

Kulkarni RR, Shaiwale NS, Deobagkar DN, Deobagkar DD (2015) Synthesis and extracellular accumulation of silver nanoparticles by employing radiation-resistant Deinococcus radiodurans, their characterization, and determination of bioactivity. Int J Nanomedicine 10:963–974

Kumar A, Ghosh A (2016) Biosynthesis and characterization of silver nanoparticles with bacterial isolate from GangeticAlluvial soil. Int J Biotechnol Biochem 12(2):7

Kumari M, Giri VP, Pandey S, Kumar M, Katiyar R, Nautiyal CS et al (2019) An insight into the mechanism of antifungal activity of biogenic nanoparticles than their chemical counterparts. Pestic Biochem Physiol 157:45–52

Kundu D, Hazra C, Chatterjee A, Chaudhari A, Mishra S (2014) Extracellular biosynthesis of zinc oxide nanoparticles using Rhodococcus pyridinivorans NT2: multifunctional textile finishing, biosafety evaluation and in vitro drug delivery in colon carcinoma. J Photochem Photobiol B Biol 140:194–204

Kuppusamy P, Yusoff MM, Maniam GP, Govindan N (2016) Biosynthesis of metallic nanoparticles using plant derivatives and their new avenues in pharmacological applications – an updated report. Saudi Pharmaceut J 24(4):473–484

Lakshmi Das V, Thomas RT, Varghese R, Vasudevan S, Mathew JK, Radhakrishnan E (2014) Extracellular synthesis of silver nanoparticles by the Bacillus strain CS 11 isolated from industrialized area. 3 Biotech 4:121–126

Lareen A, Burton F, Schäfer P (2016) Plant root-microbe communication in shaping root microbiomes. Plant Mol Biol 90(6):575–587

Latiffah Z, Mah Kok F, Heng Mei H, Maziah Z, Baharuddin S (2010) Fusarium species isolated from mangrove soil in Kampung Pantai Acheh, Balik Pulau, Pulau Pinang, Malaysia. Trop Life Sci Res 21(1):21–29
Ledakowicz S, Solecka M, Zylla R (2001) Biodegradation, decolourisation and detoxification of textile wastewater enhanced by advanced oxidation processes. J Biotechnol 89(2):175–184
Liang D-W, Yang Y-H, Xu W-W, Peng S, Lu S-F, Xiang Y (2014) Nonionic surfactant greatly enhances the reductive debromination of polybrominated diphenyl ethers by nanoscale zerovalent iron: mechanism and kinetics. J Hazard Mater 278:592–596
Liao H, Reitberger T (2013) Generation of free OHaq radicals by black light illumination of Degussa (Evonik) P25 TiO2 aqueous suspensions. Catalysts 3(2):418–443
Lin P-C, Lin S, Wang PC, Sridhar R (2014) Techniques for physicochemical characterization of nanomaterials. Biotechnol Adv 32(4):711–726
Liu Y, Fu J, Chen P, Yu X, Yang P (2000) Studies on biosorption of Au3+ by Bacillus megaterium. Wei Sheng Wu Xue Bao 40(4):425–429
Liu A, Yang L, Verwegen M, Reardon D, Cornelissen JJLM (2017) Construction of core-shell hybrid nanoparticles templated by virus-like particles. RSC Adv 7(89):56328–56334
Lovley D (2013) Dissimilatory Fe(III)- and Mn(IV)-reducing prokaryotes. In: Rosenberg E, EF DL, Lory S, Stackebrandt E, Thompson F (eds) The prokaryotes: prokaryotic physiology and biochemistry. Springer, Berlin, pp 287–308
Lu H, Wang J, Stoller M, Wang T, Bao Y, Hao H (2016) An overview of Nanomaterials for water and wastewater treatment. Adv Mater Sci Eng 2016:10
Luangpipat T, Beattie IR, Chisti Y, Haverkamp R (2011) Gold nanoparticles produced in a microalga. J Nanopart Res 13:6439–6445
Mahdieh M, Zolanvari A, Azimee AS, Mahdieh M (2012) Green biosynthesis of silver nanoparticles by Spirulina platensis. Sci Iranica 19(3):926–929
Malarkodi C, Chitra K, Rajeshkumar S, Gnanajobitha G, Paulkumar K, Vanaja M et al (2013) Novel eco-friendly synthesis of titanium. Der Pharm Sinica 4(3):8
Malik P, Shankar R, Malik V, Sharma N, Mukherjee TK (2014) Green chemistry based benign routes for nanoparticle synthesis. J Nanopart 2014:14
Marikani DK, Uma Sangareswari K, Suganya P, Ganesan R, Rajarathinam K (2016) Biobased approach for the synthesis, characterization, optimization and application of silica nanoparticles by fungus Fusarium oxysporum. Pharmaceut Biol Eval 2:223–233
Mazrouaa AM, Mohamed MG, Fekry M (2019) Physical and magnetic properties of iron oxide nanoparticles with a different molar ratio of ferrous and ferric. Egypt J Pet 28(2):165–171
Mazumdar H, Haloi N (2011) A study on biosynthesis of iron nanoparticles by Pleurotus sp. J Microbiol Biotechnol Res 1(3):39–49
Meng X, Seton HC, Lu LT, Prior IA, Thanh NTK, Song B (2011) Magnetic CoPt nanoparticles as MRI contrast agent for transplanted neural stem cells detection. Nanoscale 3:977–984
Menon S, Shanmugam R, Kumar V (2017) A review on biogenic synthesis of gold nanoparticles, characterization, and its applications. Resource-Efficient Technol 3:516
Mohammed B, Abdulsattar J, Adnan W (2016) Biosynthesis of iron oxide nanoparticles using Lactobacillus rhamnosus and its effect against pathogenic bacteria
Mohd Yusof H, Mohamad R, Zaidan UH, Abdul Rahman NA (2019) Microbial synthesis of zinc oxide nanoparticles and their potential application as an antimicrobial agent and a feed supplement in animal industry: a review. J Anim Sci Biotechnol 10(1):57
Mohseniazar M, Barin M, Zarredar H, Alizadeh S, Shanehbandi D (2011) Potential of microalgae and lactobacilli in biosynthesis of silver nanoparticles. Bioimpacts 1:149–152
Mourdikoudis S, Pallares RM, Thanh NTK (2018) Characterization techniques for nanoparticles: comparison and complementarity upon studying nanoparticle properties. Nanoscale 10(27):12871–12934
Nalenthiran P, Sambandam A, Kathiravan G, Kannaian N, Prakash U, Crawford S, Muthupandian AK (2009) Microbial synthesis of silver nanoparticles by Bacillus sp. J Nanopart Res 11:1811–1815. https://doi.org/10.1007/s11051-009-9621-2

Nangia Y, Wangoo N, Goyal N, Shekhawat G, Suri CR (2009) A novel bacterial isolate Stenotrophomonas maltophilia as living factory for synthesis of gold nanoparticles. Microb Cell Fact 8(1):39
Nath D, Banerjee P (2013) Green nanotechnology – a new hope for medical biology. Environ Toxicol Pharmacol 36(3):997–1014
Niu Z, Kabisatpathy S, He J, Lee LA, Rong J, Yang L et al (2010) Synthesis and characterization of bionanoparticle—silica composites and mesoporous silica with large pores. Nano Res 2(6):474–483
Ovais M, Khalil AT, Ayaz M, Ahmad I, Nethi SK, Mukherjee S (2018) Biosynthesis of metal nanoparticles via microbial enzymes: a mechanistic approach. Int J Mol Sci 19(12):4100
Park B, Dempsey BA (2005) Heterogeneous oxidation of Fe(II) on ferric oxide at neutral pH and a low partial pressure of O2. Environ Sci Technol 39(17):6494–6500
Patel V, Berthold D, Puranik P, Gantar M (2014) Screening of cyanobacteria and microalgae for their ability to synthesize silver nanoparticles with antibacterial activity. Biotechnol Rep (Amst) 5:112–119
Patra JK, Baek K-H (2014) Green nanobiotechnology: factors affecting synthesis and characterization techniques. J Nanomater 2014:12
Philipse AP, Maas D (2002) Magnetic colloids from magnetotactic bacteria: chain formation and colloidal stability. Langmuir 18(25):9977–9984
Pineda-Vásquez TG, Casas-Botero AE, Ramírez-Carmona ME, Torres-Taborda MM, Soares CHL, Hotza D (2014) Biogeneration of silica Nanoparticles from Rice husk ash using Fusarium oxysporum in two different growth media. Ind Eng Chem Res 53(17):6959–6965
Pourali P, Badiee SH, Manafi S, Noorani T, Rezaei A, Yahyaei B (2017) Biosynthesis of gold nanoparticles by two bacterial and fungal strains, Bacillus cereus and Fusarium oxysporum, and assessment and comparison of their nanotoxicity in vitro by direct and indirect assays. Electron J Biotechnol 29:86–93
Prasad M, Lambe UP, Brar B, Shah I, Manimegalai J, Ranjan K et al (2018) Nanotherapeutics: an insight into healthcare and multi-dimensional applications in medical sector of the modern world. Biomed Pharmacother 97:1521–1537
Prasad R (2016) Advances and applications through fungal nanobiotechnology. Springer International Publishing, Cham. ISBN 978-3-319-42989-2. http://www.springer.com/us/book/9783319429892
Prasad R (2017) Fungal nanotechnology: applications in agriculture, industry, and medicine. Springer International Publishing, Cham. ISBN 978-3-319-68423-9. https://www.springer.com/gb/book/9783319684239
Prasad R, Kumar V, Kumar M, Shanquan W (2018) Fungal nanobionics: principles and applications. Springer Singapore, Singapore. ISBN 978-981-10-8666-3. https://www.springer.com/gb/book/9789811086656
Prasad R, Pandey R, Barman I (2016) Engineering tailored nanoparticles with microbes: quo vadis. WIREs Nanomed Nanobiotechnol 8:316–330. https://doi.org/10.1002/wnan.1363
Prema P, Iniya PA, Immanuel G (2016) Microbial mediated synthesis, characterization, antibacterial and synergistic effect of gold nanoparticles using Klebsiella pneumoniae (MTCC-4030). RSC Adv 6(6):4601–4607
Puyol D, Batstone DJ, Hülsen T, Astals S, Peces M, Krömer JO (2017) Resource recovery from wastewater by biological technologies: opportunities, challenges, and prospects. Front Microbiol 7:2106
Reed RB, Ladner DA, Higgins CP, Westerhoff P, Ranville JF (2012) Solubility of nano-zinc oxide in environmentally and biologically important matrices. Environ Toxicol Chem 31(1):93–99
Ruttkay-Nedecky B, Krystofova O, Nejdl L, Adam V (2017a) Nanoparticles based on essential metals and their phytotoxicity. J Nanobiotechnol 15(1):33
Ruttkay-Nedecky B, Krystofova O, Nejdl L, Adam V (2017b) Nanoparticles based on essential metals and their phytotoxicity. J Nanobiotechnol 15(1):33
Sabri MA, Umer A, Awan GH, Hassan MF, Hasnain A (2016) Selection of suitable biological method for the synthesis of silver nanoparticles. Nanomater Nanotechnol 6:29

Sakaguchi T, Burgess JG, Matsunaga T (1993) Magnetite formation by a sulphate-reducing bacterium. Nature 365(6441):47–49
Salvadori MR, Ando RA, Nascimento CAO, Corrêa B (2015) Extra and intracellular synthesis of nickel oxide nanoparticles mediated by dead fungal biomass. PLoS One 10(6):e0129799
Sarker PK, Talukdar SA, Deb P, Sayem SA, Mohsina K (2013) Optimization and partial characterization of culture conditions for the production of alkaline protease from Bacillus licheniformis P003. Springerplus 2:506
Sehgal N, Soni K, Gupta N, Kohli K (2018) Microorganism assisted synthesis of gold nanoparticles: a review. Asian J Biomed Pharmaceut Sci 8(64):–8
Senthil M, Ramesh C (2012) Biogenic synthesis of Fe3O4 nanoparticles using Tridax Procumbens leaf extract and its antibacterial activity on Pseudomonas aeruginosa. Dig J Nanomater Biostruct 7(3):6
Seqqat R, Blaney L, Quesada D, Kumar B, Cumbal L (2019) Nanoparticles for environment, engineering, and nanomedicine. J Nanotechnol 2019:2
Shah M, Fawcett D, Sharma S, Tripathy S, Poinern G (2015a) Green synthesis of metallic nanoparticles via biological entities. Materials 8(11):7278–7308
Shah M, Fawcett D, Sharma S, Tripathy SK, GEJ P (2015b) Green Synthesis of Metallic Nanoparticles via Biological Entities. Materials 8:7278–7308. (1996-1944 (Print))
Shan G, Yan S, Tyagi R, Surampalli R (2009) Applications of nanomaterials in environmental science and engineering: review. Pract Period Hazard Toxic Radioact Waste Manage 13:110
Shannon E, Abu-Ghannam N (2016) Antibacterial derivatives of marine algae: an overview of pharmacological mechanisms and applications. Mar Drugs 14(4):81
Sharma N, Pinnaka AK, Raje M, Ashish A, Bhattacharyya M, Choudhury AR (2012) Exploitation of marine bacteria for production of gold nanoparticles. Microb Cell Fact 11:86
Show S, Tamang A, Chowdhury T, Mandal D, Chattopadhyay B (2015) Bacterial (BKH1) assisted silica nanoparticles from silica rich substrates: a facile and green approach for biotechnological applications. Colloids Surf B Biointerfaces 126:245–250
Siddiqi KS, Husen A (2016a) Fabrication of metal and metal oxide nanoparticles by algae and their toxic effects. Nanoscale Res Lett 11(1):363
Siddiqi KS, Husen A (2016b) Fabrication of metal and metal oxide nanoparticles by algae and their toxic effects. Nanoscale Res Lett 11(1):363
Siddiqi KS, Husen A, Rao RAK (2018) A review on biosynthesis of silver nanoparticles and their biocidal properties. J Nanobiotechnol 16(1):14
Singh P, Kim Y-J, Zhang D, Yang D-C (2016a) Biological synthesis of nanoparticles from plants and microorganisms. Trends Biotechnol 34:588
Singh R, Kumar M, Mittal A, Mehta PK (2016b) Microbial enzymes: industrial progress in 21st century. 3 Biotech 6(2):174
Singh J, Dutta T, Kim K-H, Rawat M, Samddar P, Kumar P (2018) 'Green' synthesis of metals and their oxide nanoparticles: applications for environmental remediation. J Nanobiotechnol 16(1):84
Sintubin L, de Gusseme B, van der Meeren P, Pycke BF, Verstraete W, Boon N (2011) The antibacterial activity of biogenic silver and its mode of action. Appl Microbiol Biotechnol 91(1):10
Skandalis N, Dimopoulou A, Georgopoulou A, Gallios N, Papadopoulos D, Tsipas D et al (2017) The effect of silver nanoparticles size, produced using plant extract from Arbutus unedo, on their antibacterial efficacy. Nano 7(7):178
Srinath BS, Keerthiraj DN, Byrappa K (2017) Eco-friendly synthesis of gold nanoparticles by gold mine bacteria Brevibacillus formosus and their antibacterial and biocompatible studies. J Pharm 7:53–60
Srivastava SK, Constanti M (2012) Room temperature biogenic synthesis of multiple nanoparticles (Ag, Pd, Fe, Rh, Ni, Ru, Pt, Co, and Li) by Pseudomonas aeruginosa SM1. J Nanopart Res 14:831
Suman J, Neeraj S, Rahul J, Sushila K (2014) Microbial synthesis of silver nanoparticles by actinotalea sp. MTCC 10637. Am J Phytomed Clin Therapeut 2(8):7

Sundaram PA, Augustine R, Kannan M (2012) Extracellular biosynthesis of iron oxide nanoparticles by Bacillus subtilis strains isolated from rhizosphere soil. Biotechnol Bioprocess Eng 17(4):835–840

Syed B, Prasad NMN, Satish S (2016) Endogenic mediated synthesis of gold nanoparticles bearing bactericidal activity. J Microscopy Ultrastruct 4(3):162–166

Tamer U, Gündoğdu Y, Boyacı İH, Pekmez K (2009) Synthesis of magnetic core–shell Fe3O4–Au nanoparticle for biomolecule immobilization and detection. J Nanopart Res 12(4):1187–1196

Taran M, Rad M, Alavi M (2018) Biosynthesis of TiO2 and ZnO nanoparticles by Halomonas elongata IBRC-M 10214 in different conditions of medium. Bioimpacts 8(2):9

Tchounwou PB, Yedjou CG, Patlolla AK, Sutton DJ (2012) Heavy metal toxicity and the environment. Experientia Suppl 101:133–164

Thomas R, Janardhanan A, Varghese RT, Soniya EV, Mathew J, Radhakrishnan EK (2015) Antibacterial properties of silver nanoparticles synthesized by marine Ochrobactrum sp. Braz J Microbiol 45(4):1221–1227

Torabian P, Ghandehari F, Fatemi M (2018) Biosynthesis of iron oxide nanoparticles by cytoplasmic extracts of bacteria lactobacillus casei. Asian J Green Chem 1(2):181–188

Wang L, Hu C, Shao L (2017) The antimicrobial activity of nanoparticles: present situation and prospects for the future. Int J Nanomed 12:1227–1249

Wu Q, Sanford RA, Löffler FE (2006) Uranium(VI) reduction by Anaeromyxobacter dehalogenans strain 2CP-C. Appl Environ Microbiol 72(5):3608–3614

Wu B, Hussain M, Zhang W, Stadler M, Liu X, Xiang M (2019) Current insights into fungal species diversity and perspective on naming the environmental DNA sequences of fungi. Mycology 10(3):127–140

Xie J, Chen K, Chen X (2009) Production, modification and bio-applications of magnetic nanoparticles gestated by magnetotactic bacteria. Nano Res 2(4):261–278

Xue W, Liu Y, Zhang N, Yao Y, Ma P, Wen H et al (2018) Effects of core size and PEG coating layer of iron oxide nanoparticles on the distribution and metabolism in mice. Int J Nanomedicine 13:5719–5731

Yadav L, Tripathi RM, Prasad R, Pudake RN, Mittal J (2017) Antibacterial activity of cu nanoparticles against *E. coli*, *Staphylococcus aureus* and *Pseudomonas aeruginosa*. Nano Biomed Eng 9(1):9–14. https://doi.org/10.5101/nbe.v9i1.p9-14

Yadav VK, Fulekar MH (2018) Biogenic synthesis of maghemite nanoparticles (γ-Fe2O3) using Tridax leaf extract and its application for removal of fly ash heavy metals (Pb, Cd). Mater Today Proc 5(9, Part 3):20704–20710

Yang J, Hou B, Wang J, Tian B, Bi J, Wang N et al (2019) Nanomaterials for the removal of heavy metals from wastewater. Nanomaterials 9(3):424

You H, Yang S, Ding B, Yang H (2013) Synthesis of colloidal metal and metal alloy nanoparticles for electrochemical energy applications. Chem Soc Rev 42(7):2880–2904

Zargar M, Hamid AA, Bakar FA, Shamsudin MN, Shameli K, Jahanshiri F et al (2011) Green synthesis and antibacterial effect of silver nanoparticles using Vitex negundo L. Molecules 16(8):6667–6676

# Chapter 8
# Microbial Strategies for Controlling Harmful Cyanobacterial Blooms

**Digvijay Singh, Gurleen Kaur, Joginder Singh, and Saurabh Satija**

**Abstract** Cyanobacteria are earth's oldest heterogeneous group of prokaryotes that are predominantly photosynthetic. The anthropogenic nutrient enrichment in aquatic system fuels the exponential cyanobacterial growth along with optimum environmental conditions leading to the propagation of harmful scums known as "cyanoblooms". The catastrophic effects of these cyanoblooms include prevalence of anoxic conditions, alteration of food webs, accumulation of organic materials, and adverse effects on animals, birds, and humans due to release of toxic secondary metabolites (cyanotoxins). The CyanoHABs (cyanobacterial harmful algal blooms) pose a serious threat to the usability and substantiality of freshwater resources. The applicability and feasibility of different control strategies depend upon the intensity, frequency, and magnitude of invasion of the cyanobacterial population along with other environmental limitations. Moreover, the insights about the type of ecostrategist, the area/zone inhabited, the type of aquatic system escalated, and the mechanism of formation of CyanoHABs are yet other critical factors. The physical (sonication, aeration) and chemical (algicide, oxidants) agents used for the modulation of CyanoHABs have proved to be transitory and spatially constricted to smaller eutrophying systems. Additionally, chemical agents have potential to exterminate aquatic biota and also require intermittent dosage for its utility. The effectiveness of future cyanobloom management approaches thus could be magnified by employing biocontrol strategies which include use of virus, bacteria, fungi, and protozoans and biomanipulation for impeding cyanobacterial growth.

**Keywords** Blooms · Cyanotoxins · Biomanipulations · Metabolites

D. Singh (✉) · G. Kaur
Department of Biotechnology, Lovely Professional University, Phagwara, Punjab, India

J. Singh
Department of Microbiology, Lovely Professional University, Phagwara, Punjab, India

S. Satija
Department of Pharmaceutical Sciences, Lovely Professional University, Phagwara, Punjab, India



J. Singh et al. (eds.), *Microbial Biotechnology: Basic Research and Applications*, Environmental and Microbial Biotechnology,
https://doi.org/10.1007/978-981-15-2817-0_8

## 8.1 Introduction

Cyanobacteria are a diverse group of photosynthetically active (Hamilton et al. 2016) unicellular and multicellular prokaryotes which are also referred to as "pond scum" (Bhatnagar et al. 2014). This potent oxygen-producing machinery exhibits a variety of photosynthetic pigments like phycocyanin, chlorophyll, carotenoids, and phycobiliproteins which provide contrasting colours to the cyanobacterium giving it the popular name "blue-green algae" (Saini et al. 2018). Unlike photosynthetic eukaryotes which have thylakoids arranged in stacked grana, the cyanobacterial thylakoids have more uniformly arranged sheet-like pattern (Fig. 8.1) (Liberton et al. 2013). The plastids located in phototrophic eukaryotes are thought to have their ancestry in cyanobacteria via endosymbiosis. A cyanobacterium contributes globally to water and soil fertility through nitrogen fixation (Rai 1990). They are primarily found in limnic and marine environments (including salty, brackish, or freshwater). For energy metabolism, they require only water, carbon dioxide, inorganic substances, and light. Some species are reported to sustain long dark periods. Furthermore, some species are discovered to be surprisingly heterotrophs (Fay 1965). They have specialised cells called heterocysts which provide anaerobic environment for nitrogen fixation. Majority of them reproduce (asexually) to form dense colonies making buoyant algal mats on the surface of water bodies which can be of various colours including green, blue, brown, or red. These cyanobacterial blooms lead to depletion of oxygen in water bodies and affect aquatic flora and fauna with release of toxins (Paerl et al. 2011).

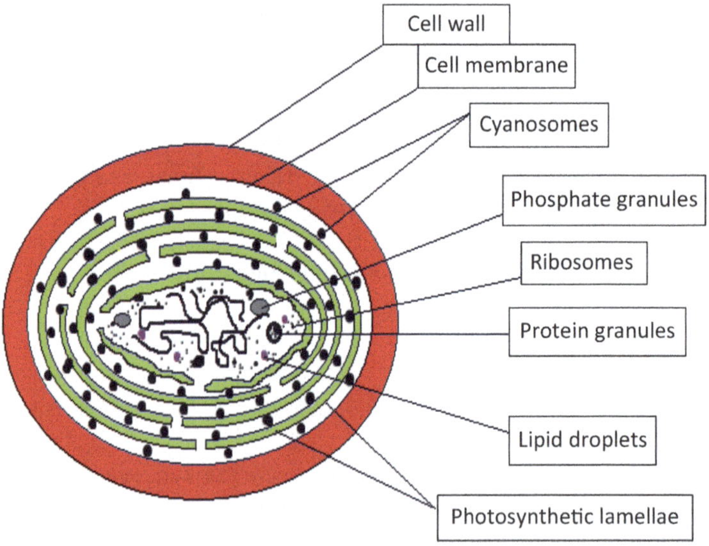


**Fig. 8.1** Structure of a Cyanobacterial Cell

## 8.2 Diseases Caused by Cyanobacteria

The diseases caused by cyanobacterial toxins or cyanotoxins respond in accordance with the type of toxin and the mode of water or water-related exposure (drinking, skin contact, etc.). When the cyanotoxins accumulate and reach a threshold point in an algal bloom, it causes harmful effects to not only the aquatic organisms but also to any other animal or bird which gets in contact with the infected water directly or indirectly. The three basic targets of cyanotoxins are the nervous system (neurotoxins), the liver (hepatotoxins), or the skin (dermatoxins). Some toxins are also reported to cause gastrointestinal symptoms and kidney-related ailments in humans (Christoffersen and Kaas 2000). People are generally exposed to these toxins by drinking contaminated water or bathing in it. Other reasons could be ingestion of algal tablets. Skin irritation, stomach cramps, vomiting, nausea, diarrhoea, fever, sore throat, headache, muscle and joint pain, blisters of the mouth, and liver damage are the major symptoms of cyanotoxins in humans. People involved in recreational activities in water containing cyanobacterial toxins may suffer allergic reactions, such as asthma, eye irritation, rashes, and blisters around the mouth and nose. Animals, birds, and fish can also be poisoned by high levels of toxin-producing cyanobacteria (Table 8.1).

**Table 8.1** General features of cyanotoxins, their targets, and sources

| Chemical structure | Toxin group | Target organ in mammals | Cyanobacterial member |
|---|---|---|---|
| Cyclic peptides | Microcystins | Liver | *Microcystis, Anabaena, Planktothrix (Oscillatoria) Nostoc, Hapalosiphon, Anabaenopsis* |
| | Nodularin | Liver | *Nodularia* |
| Alkaloids | Anatoxin-a | Nerve synapse | *Anabaena, Planktothrix (Oscillatoria), Aphanizomenon* |
| | Anatoxin-a(S) | Nerve synapse | *Anabaena* |
| | Aplysiatoxins | Skin | *Lyngbya, Schizothrix, Planktothrix (Oscillatoria)* |
| | Cylindrospermopsins | Liver | *Cylindrospermopsis, Aphanizomenon, Umezakia* |
| | Lyngbyatoxin-a | Skin, gastrointestinal tract | *Lyngbya* |
| | Saxitoxins | Nerve axons | *Anabaena, Aphanizomenon, Lyngbya, Cylindrospermopsis* |
| | Lipopolysaccharides | Skin irritant; affects | *All* |
| | (LPS) (endotoxins) | Any exposed tissue | |

## 8.3 Blooms and Its Types

In an aquatic ecosystem, an alga acts as a crucial organic source for supporting the food chains but relies upon optimum demand of nutrient supply for its growth. With increase in anthropogenic sources of nutrients, the accumulation of phytoplanktons in water bodies have increased at an alarming rate leading to discolouration of water bodies, which is termed as "algal bloom". Cyanobacteria being the most notorious agent for harmful algal blooms (HABs) formation is responsible not only for fluctuations of oxygen levels (hypoxia and anoxia), shellfish poisoning, and food web alterations but also for the release of toxins which cause hazardous health issues in animals as well as in humans. As depicted in Table 8.2, the diverse range of species from this phylum contributes to bloom formation but varies with parts of water bodies they inhabit or the type of nutrients they metabolise. These comprise of surface dwellers which form a scum layer over the water surface (e.g. *Anabaena, Aphanizomenon, Nodularia, Microcystis)*, subsurface bloom formers (*Cylindrospermopsis, Oscillatoria)*, $N_2$-fixing species (*Anabaena, Aphanizomenon, Cylindrospermopsis, Nodularia*), etc. Apart from these, some cyanobacterial species have potential to sustain in extreme environmental conditions which include varied light intensities, fluctuating temperatures, exposure to desiccation conditions, and low nutrient levels (Paerl et al. 2001). Cyanobacteria can be characterised into different "ecostrategists" on the basis of types of water bodies they inhabit (as described in Table 8.2). This classification is being done in accordance with eco-physiological laboratory work integrated with field observation (mostly in north-western Europe). This approach could help predict the presence of cyanobacteria under certain growth conditions (Mur et al. 1999).

### *8.3.1 Mechanism of Pathogenesis for Cyanobacteria to Cause Bloom*

The mechanism by which cyanobacteria form blooms is still debatable. Several inorganic/organic sources when present in excessive amount are thought to be responsible for fuelling the bloom formation in fresh- and marine water environments. Knowledge about the causative agents for the appearance of cyanobacterial blooms is still incomplete. The conventional approaches and investigations provide wisdom about the following facts:

- In early research experiments, the ratio of nitrogen to phosphorous (N/P) had been associated with the appearance of cyanoblooms (Glibert et al. 2004). Whereas other research models linked phosphorous concentration as a primary regulatory factor for growth of cyanobacteria and change in genotype, both these aspects were related to the temperature conditions of water (Joung et al. 2011).

**Table 8.2** Different types of cyanobacteria and their habitats

| Type | Water type or zone inhabited | Morphological features | Cyanobacterial genera | Description |
|---|---|---|---|---|
| Scum-forming ecostrategists | Euphotic zone | Coccoid/ filamentous, colonial, contain gas vesicles | *Microcystis*, *Anabaena*, *Aphanizomenon* | On the surface of water (euphotic zone), they undergo high photosynthesis. The carbohydrate thus concentrated makes them heavy and they migrate vertically and move to deep dark layers of water where they use their carbs to synthesise new gas vesicles which in turn provide them buoyant force to rise back to the euphotic zone |
| Homogenously dispersed ecostrategists | Eutrophic and hypertrophic water types; epilimnion layer | Filamentous; solitary | *Planktothrix (Oscillatoria) agardhii; Limnothrix (Oscillatoria) redekei* | Sensitive to high light intensities; no buoyancy regulation but entrained by passive water currents. Causes high turbidity and suppresses the proliferation of other phytoplanktons |
| Stratifying ecostrategists | Metalimnion layer of thermally stratified lakes | Single filamentous | *Planktothrix (Oscillatoria) rubescens* | These types of species have a phycoerythrin pigment which absorbs green light; hardly show any vertical migration but in late autumn season become buoyant and form red scums |
| Nitrogen-fixing ecostrategists | Deep as well as shallow systems | Can be colonial, possess gas vesicles | *Anabaena*, *Aphanizomenon*, *Cylindrospermopsis*, *Nodularia*, and *Nostoc* | Primarily present in the systems with low level of inorganic nitrogen; can regulate buoyancy because of the presence of gas vesicles and form scum along downwind shore |

(continued)

**Table 8.2** (continued)

| Type | Water type or zone inhabited | Morphological features | Cyanobacterial genera | Description |
|---|---|---|---|---|
| Small, colony-forming taxa | Small and intermittently flushed lakes | Small and colonial | *Aphanothece* | Not much information available about buoyant regulation and scum formation of these species. They are reported to have dominated after the decline of the *Planktothrix rubescens* population. But no evidential relationship supports this dominance |
| Benthic cyanobacteria | Benthic zone | Form coherent mats | *Oscillatoria limosa* | Grow on bottom sediments receiving ample amount of sunlight for photosynthesis. Sometimes, $O_2$ bubbles (by-product) may get entrapped within the mats making them buoyant and they cut loose and rise to the surface |

- The emergence of cyanobacteria through akinete (dormant cell) germination under oxygenic condition is light and temperature dependent (Cubillos-Ruiz et al. 2017).
- It is known that the N-fixing bacteria monopolise the phytoplankton community in nitrogen-limiting environment (Takamura et al. 2003). But the occurrence of non-N-fixing cyanobacteria (viz., *microcystis*) in ample N2 environment is still unknown.
- The population of cyanobacteria in oligotrophic environment (nutrient-poor) is less but dominates eukaryotic algae in mesotrophic and eutrophic conditions (nutrient-rich). This anomaly is explained with a hypothesis that in oligotrophic (low P) conditions, there is lower affinity for P transport by cyanobacteria than the eukaryotic algae thereby dominating them.
- On the other side, P sequestering ability of cyanobacteria is higher than eukaryotic algae in nutrient-rich (ample P) environment making them dominate in mesotrophic habitats.

- The most recent novel conceptual model (Molot et al. 2014) was proposed by Molot and co-workers in 2014, which investigated the reason or factor behind the dominance of cyanobacteria in mesotrophic environment and its absence in oligotrophic conditions. Clearly, it was something other than phosphorous which regulated this mechanism because the early field studies have stated that cyanobacteria have more affinity towards P than the eukaryotic algae; therefore, they must dominate in oligotrophic conditions also. In addition to the fact that P and N regulate the productivity of cyanobacteria with phosphorous concentration being very high, this model proposes the presence of $Fe^{2+}$ (ferrous ion) in enhancing the ability of cyanobacteria to compete and outnumber eukaryotic species. The evidence was also generated for the same by addition of oxine in field studies (P-enriched mesocosms in eutrophic Lake 227). Oxine oxidises $Fe^{2+}$ to $Fe^{3+}$ and chelates the latter; as a result, cyanobacterial growth was found to be inhibited. The cyanobacteria uptakes $Fe^{2+}$ but not $Fe^{3+}$ though $Fe^{3+}$ is more commonly found in aquatic environment. $Fe^{2+}$ is very soluble in anoxic environment but is present in very minimal concentration. The demand of $Fe^{2+}$ is higher for cyanobacteria than the eukaryotic algae. This model hypothesises that cyanobacteria combat the problem of low $Fe^{2+}$ concentration by internally loading ferrous ions. The anoxic sediments (low oxygen concentration) are rich in nutrients preferably $Fe^{2+}$ but the ions get rapidly re-oxidised to $Fe^{3+}$ when it comes in contact with upper oxygenated layers. The cyanobacteria tend to migrate with enough velocity to lower sediments for acquisition of these ferrous ions but the supporting data are limited. $Fe^{3+}$ and not $Fe^{2+}$ is required by eukaryotic systems. Some bacteria and fungi produce low-molecular-weight $Fe^{3+}$ chelators known as "siderophores". The presence of these factors positively supports the dominance of cyanobacteria in mesotrophic waters as siderophores scavenge $Fe^{3+}$ required for eukaryotic algae productivity. Siderophores are not necessarily responsible for initiating the bloom but maintaining its growth by cutting the supply of $Fe^{3+}$ for eukaryotic algae (Fig. 8.2).
- The primary factor for the bloom formation is thus the availability of $Fe^{2+}$ and factors which control the loading of these ions into the internal membranes of cyanobacteria. Sulphates when present in the system react with $Fe^{2+}$ to form ferrous sulphate which is further reduced to ferrous sulphide which is insoluble in nature. Therefore, the water bodies in which sulphide formation rate is higher tend to experience no cyanoblooms. This model states that once the light and temperature conditions are physiologically optimum, the cyanobloom formation is regulated by the availability of $Fe^{2+}$, with its internal loading rate which further depends on the anoxia conditions in sediments and sulphate reduction rate.
- This conceptual approach thus links anoxia, phosphorous (P), nitrogen (N), iron (Fe), and sulphate to the formation of cyanobacterial blooms.

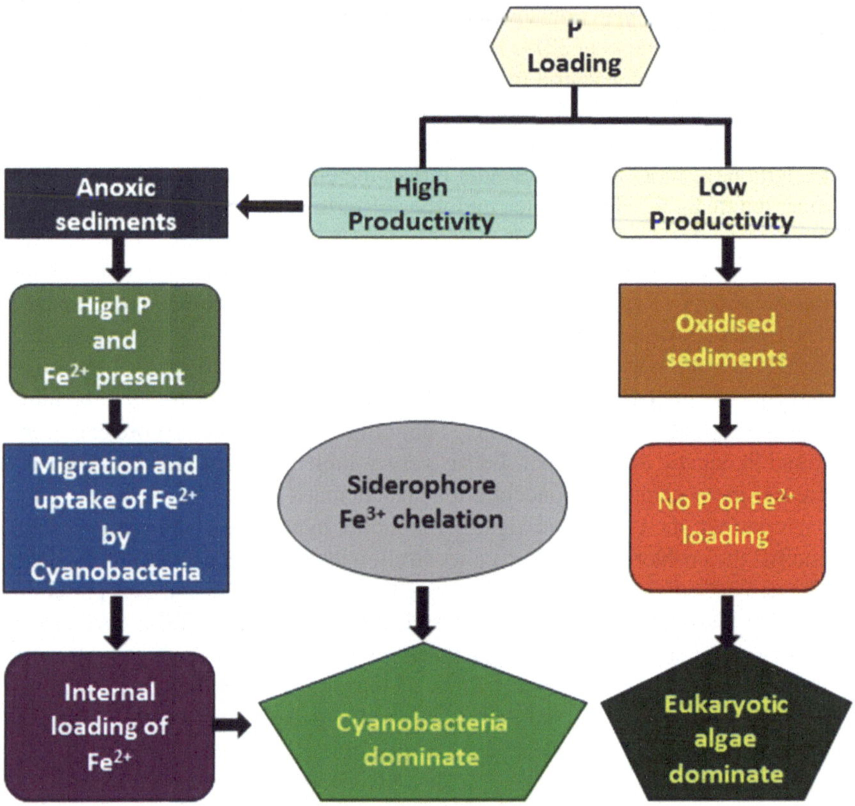


**Fig. 8.2** Simplified flow diagram of cyanobacterial bloom formation

## 8.4 Control Strategies for Cyanobloom Management

The cyanobacterial blooms are associated with many different factors for its formation and escalation. These factors include N, P, $Fe^{2+}$, anoxic sediments, sulphates, light intensity, temperature conditions, gas vesicles, etc. Different strategies are designed in accordance with the type and magnitude of a particular factor affecting specific species. These strategies are basically characterised into three groups—mechanical treatment, chemical treatment, and biological treatment. The mechanical treatment is employed to the blooms which are spatially constricted to a small area which includes pumping of surface scums, use of barriers, filters, etc., but this treatment method is transitory. The most commonly used chemical is algicide, whereas others include like copper compounds, barley straws, and certain oxidants such as chlorine, ozone, and peroxides. Though very robust controlling agents, these chemicals have the potential to kill other algal strains and aquatic biota. Therefore, an alternative approach of biological treatment has proved to be promising in mitigation of cyanoblooms.

## 8.4.1 *Different Types of Biocontrol Agents*

### 8.4.1.1 Viral Agents

The viruses which infect cyanobacteria are referred to as cyanogens and are considered as potential agents for mitigating cyanoblooms (Yau et al. 2011). Cyanophages were first isolated and partially purified by Safferman and Morris and it was named LPP-1 because it specifically lysed cyanobacterial species of *Lyngbya, Plectonema,* and *Phormidium*. Cyanogens are reported to be commonly present in aquatic as well as marine environment. The sudden downturns in cyanobacterium have been accompanied by the appearance of cyanophages as per the available data (Gons et al. 2002). The most attractive property of cyanogens is their rapid rate of propagation. For example, each contaminated cell of *Plectonema boryanum* filament by LPP-DUN1 has a generation time of 1 h and bursts out 100 phage particles (Barnet et al. 1984). The drawback of these agents is that they are specific to certain species and sometimes work differently for same species from two different water bodies and also this mechanism is temporary as the cyanobacteria becomes resistant to them. The host mutation results in alteration on the cell envelope leading to non-adsorption of phage particles on its surface. However, an alternative approach used mutant cyanophages (Rajasekhar et al. 2012) and it was found that the resistant cyanobacteria are susceptible to them. The use of mutant strains of cyanophages may be used as a better source for lysing a wide range of cyanobacteria (Table 8.3).

### 8.4.1.2 Bacterial Agents

Sudden drop in cyanobacterial population has been associated with the appearance of certain bacterial species. However, there is no such culmination about whether the bacteria act as a pathogen or are saprophytic in nature, i.e. it results in decay of dead cyanobacterial matter. There are three considerable ways by which these cyanobacteria mechanise—production of extracellular products, contact lysis, and entrapment lysis. Scientists had isolated filtrate of *Bacillus* sp. and it was reported to lyse seven different genera of cyanobacteria including *Anabaena* and *Microcystis* (Paerl et al. 2011). These bacillus species were found to release some volatile molecules which were low molecular weight and heat resistant and diffusible antibiotic molecules and were shown to be inhibitory towards filamentous cyanobacteria in particular. As per research, two species of *Flexibacter* were found to lyse *Oscillatoria williamsii* by disrupting electron transport, inhibiting glycolate dehydrogenase and nitrogenase activity. An extracellular metabolite which was later identified as lysozyme (by slab gel electrophoresis) caused growth inhibition (Table 8.3). Also, four different strains of *Myxobacteria* were found to lyse 40 different strains of cyanobacteria (from all orders) within a time frame of just 20 min (Gumbo et al. 2008). *Myxococci* was considered to be the best bacterial agent for cyanobacterial lyses on

**Table 8.3** Summary of different control methods for the mitigation of cyanoblooms

| Method | Advantages | Disadvantages |
|---|---|---|
| Algicide or other chemicals | • Effective<br>• Quick results | • Harmful to other aquatic organisms<br>• Short-term effect; frequent dosage required for long-term sustenance |
| Aeration | • Environment-friendly<br>• Controls the BOD of the system and maintains the optimum $O_2$ level in aquatic ecosystem | • Not cost-effective<br>• No direct effect on cyanobacteria |
| Mechanical mixing | • Eco-friendly<br>• Prevents stratification | • Maintenance cost is high<br>• Reduced efficiency in accordance with the water quality |
| Sonication (ultrasound) | • Application on large water surfaces possible<br>• Environment-friendly as well as cost-effective<br>• Cyanoblooms controlling efficiency up to 90% | • Results are vivid after a few weeks of experiment |
| Viral agents | • Present in both freshwater and marine environments<br>• Rapid growth rate | • Cultivation of cyanobacteria problematic<br>• Cyanobacteria becomes resistant to cyanophages after some encounters |
| Bacterial agents | • Effective through production of extracellular products; contact lysis and entrapment lysis<br>• Cause inhibition of cyanobacterial growth, photosynthesis, and metabolism | • Certain factors which effect bacterial invasion include its flexibility to changing physical conditional, wide host range, high growth rate, and adaptability to changes in host |
| Fungal agents | • Non-chytridiaceous fungus found effective against cyanoblooms<br>• Certain extracellular, heat-stable factors responsible for lyses of cyanobacteria | • Extracellular factors present in lower quantities and effective only when fungus is present in close proximity to cyanobacteria |
| Protozoan predators | • Acts as a natural predator of cyanobacteria by grazing and phagocytosing them | • Many factors influence the frequency of predation including grazing rates, predation specificity, growth rate of cyanobacteria and its predation by other higher organisms |
| Biomanipulation | • Acts as a sink for excess nutrients like P and N<br>• Natural method for limiting cyanobacterial population by addition of predators | • Applicable to small-sized water bodies<br>• Deliberate changes in the biodiversity of the system under consideration |

the basis of different peculiarities including flexibility towards changing physical conditions, ability to multiply, search for prey, broad-spectrum host range, adaptability towards the metabolic changes in the host, etc.

#### 8.4.1.3 Fungal Agents

Parasitic relationship was found between cyanobacteria and chytridiaceous fungus *Rhizophydium planktonicum* (Canter et al. 1990); however, later it was found to be of limited use as a biocontrol agent to control cyanoblooms because of its obligate nature and hardships in its large-scale production (Daft et al. 1985). Therefore, the interest of researchers shifted towards non-chytrid fungi. Out of 142 tested cultures of non-chytrid fungi, only 4.2% of the formed products have cyanobacterial inhibiting capacity. This antagonism research of fungi further led to isolation of isolate 62 out of 70 pure cultures of non-chytrid fungi of the genera *Acremonium*, *Emericellopsis*, and *Verticillium*, which were able to lyse cyanobacterial cells (Redhead and Wright 1978). These culture isolates showed antagonistic effect against Anabaena and other filamentous as well as unicellular cyanobacteria. *Acremonium* and *Emericellopsis* spp. of fungi lysed cyanobacteria through the formation of extracellular factors which were heat resistant and diffusible. Liquid cultures of the fungi *Emericellopsis salmosynnemata* and *Acremonium kiliense* partially purified the beta-lactam antibiotic cephalosporin C, and it was suggested that these organisms produced small amounts of these factors and were only antagonistic when present in close proximity of cyanobacteria. Also, these antibiotics were found to load inside the mucous sheath surrounding the cyanobacteria to reach a threshold level to cause an inhibitory effect.

#### 8.4.1.4 Protozoan Predators

Protozoans play a crucial role in regulating the cyanobacterial population in aquatic system by grazing them through phagocytosis (Sigee et al. 1999). Various cyanobacterial species act as a food source among ciliates like *Furgasonia*, *Nassula*, and *Pseudomicrothorax Stos*, *Amoeba*, and flagellate like *Monas guttula*. The rapacity of protozoans towards cyanobacterial species was reported in subjects directly obtained from natural in vitro experiments and in biocontrol field experiments which enlightened about the cyanobacterial-protozoa interactions and insights about the range and population being predated by them (Table 8.3). Many factors are reported to contribute to this interaction which includes growth and predation rate of protozoa, its specificity towards cyanobacterial species, growth rate of cyanobacteria, and its predation by other higher organisms.

#### 8.4.1.5 Biomanipulation

Biomanipulation involves construction of artificial wetlands comprising of floating mats which are further placed on water bodies. As the flora grows on these mats, they act as a sink for phosphorous, nitrogen, and other essential nutrients which are otherwise required by the cyanobacteria for its growth (Studer et al. 2017). However, nutrients being stored in these plants are systematically harvested to prevent its seepage in aquatic biota (Table 8.3), thus regulating the nutrient level in the water body and inhibiting the growth of cyanoblooms. The other approach could be used to increase the predation of cyanobacteria by promoting the growth of zooplanktons, benthic fauna, and other aquatic organisms which graze over cyanobacteria, thereby limiting its population. Though advantageous over chemical and physical methods, this method may cause modification of the conventional biodiversity of the aquatic ecosystem (Newcombe 2012).

## 8.5 Mechanism of Actions of Certain Biocontrol Agents

### *8.5.1 Cyanophages*

The mechanism begins when the LPP-1 cyanophage gets adsorbed on the cyanobacterial cell wall via the distal end of its tail (Padan and Shilo 1973). The tail penetrates into the host cell (depth of penetration not known) and the viral DNA is injected into the cell as suggested in bacteriophages. The mutant *Plectonema* does not adsorb cyanophage on its surface which suggests that there are specific receptors present on the susceptible cells. The first sign of infection is invasion of photosynthetic lamella and cessation of $CO_2$ as suggested by Smith and co-workers who studied the effect of LPP-1 on *Plectonema boryanum*. Later, the viral particles start developing in the space between the folded lamella and the plasma membrane which is termed as virogenic stroma (Suttle 2006). LPP-1 shows similar symptoms in other host species it infects. Before the lysis of cyanophage particles, no cytoplasmic changes are seen in the cell except for the host DNA. To complete its lytic cycle, the viral particles assemble together in the virogenic stroma which is separated from the nucleoplasm by the photosynthetic lamellae. After the cell lysis, the mature LPP-1G particles are released leaving behind vesicles of different sizes which include ruptured lamella, cell membrane, and wall material. It was precisely found that the proteins required for the invagination of the lamellae are synthesised in the first 3 h of infection and further the invagination occurs between 3 and 8 h. Furthermore, research showed that after the infection, the DNA of *Plectonema* is degraded into acid-soluble material after 3–7 h of infection. This acid-soluble product is incorporated into the viral DNA. Lastly, the protein synthesis machinery of the host cells is depressed soon after the infection (commonly at the fifth hour).

### 8.5.2 Cyanobacteriolytic Bacteria

The mechanism by which the bacterium mitigates the cyanobacterial population is broadly divided into three categories, viz., production of extracellular products, contact lysis, and entrapment lysis. In case of extracellular product formation, *Bacillus* sp. is reported to produce small molecular size, heat-stable, and diffusible factors which cause inhibition in seven different genera of cyanobacteria including *Anabaena* and *Microcystis*. Later, these volatile factors were found to inhibit filamentous cyanobacteria specifically (Nakamura et al. 2003a). Also, one of the volatile products was demonstrated as isoamyl alcohol (3-methyl-1-butanol). In other studies, lysozymes released by the bacteria were identified as another extracellular product which inhibited photosynthetic ETC reactions, glycolate dehydrogenase, and nitrogenase activity. The efficiency of these naturally released products was not considered significant due to their dilution and loss of activity with time. Therefore, two other mechanisms of contact lysis by bacteria and the entrapment lysis are studied in depth. Four different strains of *Myxobacteria* were reported to cause lysis of a wide range of cyanobacteria (40 genera) by contact lysis (Fraleigh and Burnham 1988). No formation of extracellular factors was found but presence of particular enzymes on the surface of the bacteria played an important role in the lysis of cyanobacteria causing disruption in just 20 min after the contact is maintained. The population levels of bacteria and cyanobacteria and nutrient status of the water body play an important role in this mechanism, whereas other factors like host density and inorganic nutrient concentration are secondary. Further in other studies, entrapment lysis was demonstrated in the *Myxococcus* group in which the cyanobacterium *Phormidium luridum* var. *Olivacea* was entrapped and lysed by the colonies of bacteria. These spherules acting as a predator entrap the cyanobacteria and lyse them by releasing lysozyme-like enzyme. It is also suggested that when the predator density is lower than the threshold required for its lysing property, the *Myxococci* tend to withdraw nutrition from the secretions produced by cyanobacteria or by preying on them without hampering host population (Nakamura et al. 2003b).

### 8.5.3 Fungal Pathogens

In fungal infection of cyanobacteria, the zoospore tends to attach itself on the part of the trichome which is wounded and on the apex. Simultaneously, many other zoospores encyst on the surface of the trichome and start developing into zoosporangia (Gerphagnon et al. 2013). The rhizoid, which is a tube-like structure emerging from the zoospore (may be branched), penetrates the trichome and kills the cells. The mature zoosporangium releases out the zoospores through one or more papillae. The infection cycle consists of different stages. In the first stage, germ tubes are produced soon after the zoospores penetrate mucilage sheath of the cyanobacteria which further penetrate the cell wall of the host. The second stage is marked by the release of the contents of the germ tube into the host cell and a globular structure known as prosporangium is formed. The third stage is the expansion stage in which the rhizoids protrude out of the prosporangium and start infecting the neighbouring cells. The next stage is marked by the formation of an epiphytic bud on the surface of the trichome which differentiates into flask shaped zoosporangium in which the development of zoospores occurs (Holfeld 1998). Lastly, the zoospores mature and are released through the papillae. Lysis of cyanobacteria by certain fungal species was reported to be linked with certain heat resistant extracellular factors. These were further purified from liquid cultures of the fungi Emericellopsis salmosynnemata and Acremoniumkiliense and identified to be β-lactam antibiotic cephalosporin C (Agha et al. 2016). It was suggested that the host cells must be present at close proximity to the fungus for the lysis because of the small levels of antibiotics being produced by the fungal cells. Antibiotics are found to load inside the mucilaginous sheath of the cyanobacteria and lyse it once they reach the threshold value.

## 8.6 Conclusion

Cyanobacteria are the only oxygenic photoautotrophs among the prokaryotes and have crucial properties through which they help in the regulation of the biosphere. They are responsible for fixation of nitrogen through specialised cells called heterocysts and also have an inherent property of surviving under harsh environmental conditions. Due to these properties, they get easily spread in any water system causing cyanobloom formation which scavenges nutrients from the aquatic ecosystem, causing anoxia, warming of system, and harming aquatic biodiversity. In addition, they form mutualistic and symbiotic relationships with the flora and fauna of the system which positively modulates their survival making them dominate the system. Also, a cyanobacterium produces certain toxins known as cyanotoxins which not only show lethal effect to aquatic life but also hamper the food web. In humans, the nervous system, liver, and skin are the three major targets of these cyanotoxins. The mechanism by which the cyanobacteria proliferate in a particular system is associated with anoxia, phosphorous (P), nitrogen (N), iron (Fe), and sulphate. Cyanoblooms can be controlled by the alteration in P, N, Fe, and sulphate levels, light, temperature, and other biotic interactions. Certain strategy-orientated approaches are being used for the mitigation of the cyanobacterial population which includes chemical, physical, and biological agents. Use of chemical agents (algicides, copper sulphate, oxidants) and physical agents (mechanical mixing, sonication) not only causes harmful side effects to the aquatic ecosystem but also is short-termed. A range of microbial approaches have proved to be efficient in mitigating the cyanoblooms which include viral agents (cyanophages), cyanobactericidal bacteria, fungal parasites, protozoans, etc. Field experiments are being done using microbial agents and have proved to be an asset in removing cyanoblooms. Cyanobloom formation is posing threats globally; therefore, its proper management is very important to circumvent its toxic threats.

## References

Agha R, Saebelfeld M, Manthey C, Rohrlack T, Wolinska J (2016) Chytrid parasitism facilitates trophic transfer between bloom-forming cyanobacteria and zooplankton. Sci Rep. https://doi.org/10.1038/srep35039

Barnet YM, Daft MJ, Stewart WDP (1984) The effect of suspended particulate material on cyanobacteria—cyanophage interactions in liquid culture. J Appl Bacteriol. https://doi.org/10.1111/j.1365-2672.1984.tb04701.x

Bhatnagar A, Kaczala F, Hogland W, Marques M, Paraskeva CA, Papadakis VG, Sillanpää M (2014) Valorization of solid waste products from olive oil industry as potential adsorbents for water pollution control--a review. Environ Sci Pollut Res Int. https://doi.org/10.1007/s11356-013-2135-6

Canter HM, Heaney SI, Lund JWG (1990) The ecological significance of grazing on planktonic populations of cyanobacteria by the ciliate Nassula. New Phytol. https://doi.org/10.1111/j.1469-8137.1990.tb00397.x

Christoffersen K, Kaas H (2000) Toxic cyanobacteria in water. A guide to their public health consequences, monitoring, and management. Limnol Oceanogr. https://doi.org/10.4319/lo.2000.45.5.1212

Cubillos-Ruiz A, Berta-Thompson JW, Becker JW, Van Der Donk WA, Chisholm SW (2017) Evolutionary radiation of lanthipeptides in marine cyanobacteria. Proc Natl Acad Sci U S A. https://doi.org/10.1073/pnas.1700990114

Daft MJ, Burnham JC, Yamamoto Y (1985) Algal blooms: consequences, and potential cures. J Appl Bacteriol. https://doi.org/10.1111/j.1365-2672.1985.tb04899.x

Fay P (1965) Heterotrophy and nitrogen fixation in Chlorogloea fritschii. J Gen Microbiol 39, 11–20

Fraleigh PC, Burnham JC (1988) Myxococcal predation on cyanobacterial populations: nutrient effects. Limnol Oceanogr. https://doi.org/10.4319/lo.1988.33.3.0476

Gerphagnon M, Latour D, Colombet J, Sime-Ngando T (2013) Fungal parasitism: life cycle, dynamics and impact on cyanobacterial blooms. PLoS One. https://doi.org/10.1371/journal.pone.0060894

Glibert PM, Heil CA, Hollander D, Revilla M, Hoare A, Alexander J, Murasko S (2004) Evidence for dissolved organic nitrogen and phosphorus uptake during a cyanobacterial bloom in Florida Bay. Mar Ecol Prog Ser. https://doi.org/10.3354/meps280073

Gons HJ, Ebert J, Hoogveld HL, Van Den Hove L, Pel R, Takkenberg W, Woldringh CJ (2002) Observations on cyanobacterial population collapse in eutrophic lake water. Antonie van Leeuwenhoek. https://doi.org/10.1023/A:1020595408169

Gumbo RJ, Ross G, Cloete ET (2008) Biological control of microcystis dominated harmful algal blooms. African J Biotechnol

Hamilton TL, Bryant DA, Macalady JL (2016) The role of biology in planetary evolution: cyanobacterial primary production in low-oxygen Proterozoic oceans. Environ Microbiol. https://doi.org/10.1111/1462-2920.13118

Holfeld H (1998) Fungal infections of the phytoplankton: seasonality, minimal host density, and specificity in a mesotrophic lake. New Phytol. https://doi.org/10.1046/j.1469-8137.1998.00126.x

Joung SH, Oh HM, Ko SR, Ahn CY (2011) Correlations between environmental factors and toxic and non-toxic microcystis dynamics during bloom in Daechung Reservoir, Korea. Harmful Algae. https://doi.org/10.1016/j.hal.2010.09.005

Liberton M, Page LE, O'Dell WB, O'Neill H, Mamontov E, Urban VS, Pakrasi HB (2013) Organization and flexibility of cyanobacterial thylakoid membranes examined by neutron scattering. J Biol Chem. https://doi.org/10.1074/jbc.M112.416933

Molot LA, Watson SB, Creed IF, Trick CG, Mccabe SK, Verschoor MJ, Sorichetti RJ, Powe C, Venkiteswaran JJ, Schiff SL (2014) A novel model for cyanobacteria bloom formation: the critical role of anoxia and ferrous iron. Freshw Biol. https://doi.org/10.1111/fwb.12334

Mur LR, Skulberg OM, Utkilen H (1999) Chapter 2. Cyanobacteria in the environment. Toxic cyanobacteria water: a guide to their public heal consequences. Monit Manag. doi: https://doi.org/10.1016/j.ecoleng.2012.12.089
Nakamura N, Nakano K, Sugiura N, Matsumura M (2003a) A novel cyanobacteriolytic bacterium, Bacillus cereus, isolated from a eutrophic lake. J Biosci Bioeng. https://doi.org/10.1263/jbb.95.179
Nakamura N, Nakano K, Matsumura M, Sugiura N (2003b) A novel control process of cyanobacterial bloom using cyanobacteriolytic bacteria immobilized in floating biodegradable plastic carriers. Environ Technol. https://doi.org/10.1080/09593330309385703
Newcombe G (2012) International Guidance Manual for the management of toxic cyanobacteria. Water Intell Online. doi: https://doi.org/10.2166/9781780401355
Padan E, Shilo M (1973) Cyanophages – viruses attacking blue green algae. Bacteriol Rev
Paerl HW, Fulton RS, Moisander PH, Dyble J (2001) Harmful freshwater algal blooms, with an emphasis on cyanobacteria. ScientificWorldJournal. https://doi.org/10.1100/tsw.2001.16
Paerl HW, Hall NS, Calandrino ES (2011) Controlling harmful cyanobacterial blooms in a world experiencing anthropogenic and climatic-induced change. Sci Total Environ. https://doi.org/10.1016/j.scitotenv.2011.02.001
Rajasekhar P, Fan L, Nguyen T, Roddick FA (2012) A review of the use of sonication to control cyanobacterial blooms. Water Res. https://doi.org/10.1016/j.watres.2012.05.054
Rai AN (1990). Handbook of symbiotic cyanobacteria, Boca Raton, CRC Press, London, United Kingdom pp 1–259. https://doi.org/10.1201/9781351071185
Redhead K, Wright SJL (1978) Isolation and properties of fungi that lyse blue green algae. Appl Environ Microbiol
Saini DK, Pabbi S, Shukla P (2018) Cyanobacterial pigments: perspectives and biotechnological approaches. Food Chem Toxicol. https://doi.org/10.1016/j.fct.2018.08.002
Sigee DC, Glenn R, Andrews MJ, Bellinger EG, Butler RD, Epton HAS, Hendry RD (1999) Biological control of cyanobacteria: principles and possibilities. Hydrobiologia
Studer I, Boeker C, Geist J (2017) Physicochemical and microbiological indicators of surface water body contamination with different sources of digestate from biogas plants. Ecol Indic. https://doi.org/10.1016/j.ecolind.2017.02.025
Suttle CA (2006) Cyanophages and their role in the ecology of cyanobacteria. Ecol Cyanobacteria. https://doi.org/10.1007/0-306-46855-7_20
Takamura N, Kadono Y, Fukushima M, Nakagawa M, Kim BHO (2003) Effects of aquatic macrophytes on water quality and phytoplankton communities in shallow lakes. Ecol Res. https://doi.org/10.1046/j.1440-1703.2003.00563.x
Yau S, Lauro FM, DeMaere MZ et al (2011) Virophage control of Antarctic algal host-virus dynamics. Proc Natl Acad Sci U S A. https://doi.org/10.1073/pnas.1018221108

# Chapter 9
# Biological Strategies Against Biofilms

**Ganga Sharma and Arun Karnwal**

**Abstract** Biofilms are microbial aggregates which consist of extracellular polymeric substances (EPSs) produced by the microorganism itself that adhere to biological environments such as in rivers, streams, and alimentary canal or living tissues of mammals or nonbiological surfaces like in wastewater treatment plant, tickling beds, indwelling medical devices (IMDs), and industrial or potable water system piping. Constituents of EPS are microorganism originated components of homologous proteins, polysaccharides, lipids, and DNA. The formation of biofilm involves the migration of microbial cells, the interaction between them through cell-to-cell signaling, synthesis of EPS, and in later stages, interaction between cell and EPS.

Biofilms have a unique biochemical profile rendering structural integrity to the microorganisms which the planktonic counterparts lack. This structural stability protects them from various troubles present in their environment such as antibiotics, the host's defense mechanism, harsh nutritive conditions, predators, etc. The survival of microorganisms in biofilms although beneficial to them gives rise to a significant amount of problems in humans in various essential fields including that of medicine and industries like pharmaceutical, food, and marine industries causing adverse health effects as well as economic loses. This resistance of microorganisms, therefore, is a major concern to handle in controlling biofilms. Various traditional strategies to control biofilms of pathogenic/spoilage bacterial species, which are either physical/mechanical removal of biofilms by cleaning, selection of appropriate bactericidal material, preconditioning of surfaces by methods like ultrasonication and plasma treatment, or chemical removal using antimicrobial agents such as disinfectants/sanitizers, are not always successful. In light of the above problems of biofilm control by conventional methods, in recent times, progress has been taking place in the field of fundamental biofilm research discovering novel methods of

G. Sharma
Central Research Institute, Kasauli, Himachal Pradesh, India

A. Karnwal (✉)
Department of Microbiology, School of Bioengineering and Biosciences, Lovely Professional University, Phagwara, Punjab, India

 
J. Singh et al. (eds.), *Microbial Biotechnology: Basic Research and Applications*, Environmental and Microbial Biotechnology,
https://doi.org/10.1007/978-981-15-2817-0_9

controlling biofilms. In the current chapter, we tend to discuss these recent and cutting-edge methods which are much more effective as an antibiofilm strategy focusing mainly on the use of biological components such as enzymes, phages, and antimicrobial molecules (AMPs, QS inhibitors) for the improvisation of areas of healthcare and food safety and in industrial processes.

**Keywords** Biofilm · Antimicrobial molecules · Quorum sensing inhibitors · Exo-poly Saccharides · Bacteriophage

## 9.1 Introduction

Biofilms are universal and found in a wide variety of environments, both natural, such as in rivers, streams, and alimentary canal or living tissues of mammals, and man-made like in wastewater treatment plant, tickling beds, indwelling medical devices (IMDs), and industrial or potable water system piping (Donlan 2002). Biofilms can evade host defense mechanisms that include both innate and adaptive immunity (Dunne Jr. 2002). It is the reason why biofilm formation is an increasing cause of concern throughout the world.

Bacterial biofilms not only contribute to hospital-acquired infections, but also are a leading cause of corrosion, fouling of water pipes, and food and pharmaceutical spoilage (Henderson 2010; Kumar and Anand 1998). Some of the health issues associated with biofilms are indirect such as in drinking water distribution system where biofilms corrode water pipes and weaken them and this loss of integrity weakens pipes aside from causing esthetic problems which may lead to a health concern. Microorganisms forming biofilms can cause infection in humans and animals and may be transmitted to each through cross-contamination. Biofilm-associated infection in animals can cause massive economic loss such as in livestock/poultry industry and others in terms of production (Chakraborty et al. 2018). Also, biofilms producing microorganisms contaminate foods and generate damage to the product, equipment, and consumers leading to economic loses.

In the food product manufacturing facilities, biofilm formation leads to deleterious hygiene issues due to adherence of a variety of microbes on food and degradation of equipment (Kabwanga et al. 2018). In the pharmaceutical industry, the development of biofilms and adherence of it into the production equipment and facilities are critical issues that need to be addressed (Kabwanga et al. 2018; Stewart 2015). Although most of the biofilm-forming microbes are harmful in many ways, some of them exhibit beneficial properties which have been put to use in several industrial processes (Morikawa 2006). The infections caused by biofilm-forming microbes are chronic, and for the treatment, antimicrobial agents need to be administered, but biofilms make the microbe resistant to antimicrobial agents compared to their planktonic counterparts (Costerton et al. 1999; Mah and Toole 2001; Stewart and Costerton 2001; Donlan and Costerton 2002).

Therefore, treatment of infections caused by biofilm-forming microbes is not resolved with the sole administration of antibiotics due to the problem of the development of resistance against them. Although highly sterile conditions and practices are fundamental to maintain a strategic distance from biofilm development, for proper resolution, some of the novel antibiofilm compounds should be explored as a potential antibiofilm agent in the near future. Some of them, which are already discovered or tested till date, are active herbal compounds such as essential oils, quorum-sensing inhibitors, antimicrobial peptide alone or in combination with antibiotics, and synthetic or genetically engineered compounds.

Out of these new control strategies which are continually emerging, most of the focus is on antibiofilm agents of biological origin such as enzymes, phages, AMPs, and QSIs. The present review will focus on describing in detail the various biocontrol agents explored till date for the eradication of biofilms from the site of its formation.

### 9.1.1 *Biofilms*

Biofilms are defined as the structural community of bacterial cells which are formed by a self-produced polymeric matrix known as exopolysaccharide (EPS), which takes around 85% of the volume of a biofilm. This community of cells adheres either to living or nonliving surfaces (Costerton et al. 1999) (Fig. 9.1).

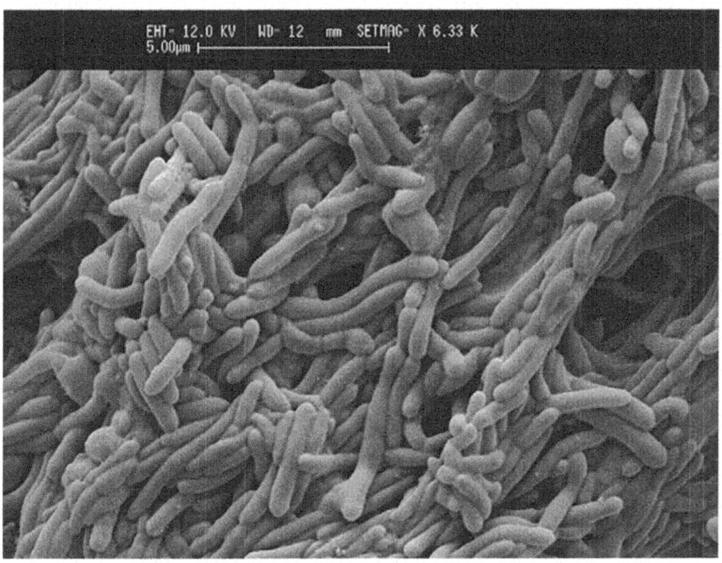


**Fig. 9.1** Scanning electron microscopy photomicrograph of a 6 days old *B. cereus* biofilm formed on a stainless steel surface. 6330 magnification; bar ¼ 5 mm (Simões et al. 2010)

The distinct levels in the process of biofilm formation can be divided into various steps (Crouzet et al. 2014). The following are the general stages for biofilm formation, though the precise details of the regulation of biofilm formation vary significantly from species to species.

1. Macromolecules in the liquid where biofilms are forming precondition the surface (living or nonliving) for adhesion.
2. Transportation of bacterial cells to the surface also occurs.
3. The cells transported are adsorbed to the surface.
4. Desorption of reversibly adsorbed cells and retention of irreversibly adsorbed cells occur.
5. Metabolism of the substrate by the biofilm-bound cells and then transportation of the by-products out of the biofilm.
6. The adsorbed cells produce cell-to-cell signaling molecules for monolayer/microcolony formation.
7. Maturation of biofilms occurs through the formation of extracellular matrix (EPS) and other cell materials. It forms a three-dimensional structure of cells known as a microcolony (O' Toole et al. 2000).
8. Detachment or dispersal of bacteria to migrate and then colonize in new areas (Landini et al. 2010).

The main composition of biofilms is the EPS matrix which is formed by retaining water and other bacterially originated substances released by bacterial cells which get embedded in this EPS matrix, and it provides the following advantages to the cells (Crouzet et al. 2014; Donlan and Costerton 2002; Jamal et al. 2018):

(a) Structural stability to the microbe due to aggregation and adhesion of cells to one another.
(b) Transportation of the necessary nutrients becomes easy in closely associated cells.
(c) Acts as an electron donor or receptor.
(d) Storage of most of the energy.
(e) Provides the binding or receptor site to enzymes.
(f) Protects from external factors such as antimicrobials and other environmental changes.
(g) Provides adaptation.

During biofilm formation, several species of bacteria communicate with one another through quorum sensing (Davies et al. 1998; Shirtliff et al. 2002). During biofilm formation, genetic information can be modified by horizontal gene transfer (HGT) within and between bacterial species and increase the adaptation in bacteria for changing environments. Moreover, this kind of higher gene transfer rates was observed more in biofilms than their counterparts. It confers protection and survival in adverse environmental conditions such as antibiotics (Costerton et al. 1999; Mah and Toole 2001), predators (Kadouri et al. 2007), and human immune system (Anderson and O'Toole 2008). This way biofilms enhance the virulence of microbes (Brooks et al. 2005).

HGT in biofilms is beneficial to microbes but are harmful to us because antimicrobial resistance and virulence genes get disseminated or new ones get emarginated, making multiple drug-resistant (MDR) strains which are known as multiresistant "superbugs." Moreover, biofilms' architecture is tuned under a specific environment with the help of different enzymes secreted by bacteria that modify its EPS composition when a change in nutrient availability occurs (Sauer et al. 2004; Ma et al. 2009).

In the natural environment, 99% of bacteria exist in biofilms. As per reports from the National Institutes of Health (NIH), up to 65% and 80% of all microbial and chronic infections, respectively, are related to biofilms which feature their immense clinical impact (Jamal et al. 2018). Biofilms are responsible for more than 65% of nosocomial infections (Böhme et al. 2009) and approximately 61% zoonotic human infections (García and Percival 2011). Not only human infections but most of the infections caused in animals like pneumonia, liver abscesses, enteritis, wound infections, and mastitis are caused by biofilm-forming microbes (Olson et al. 2002; Clutterbuck et al. 2007).

## 9.2 Biocontrol Agents Against Biofilms

### *9.2.1 Plant-Derived Essential Oils (EOs)*

Essential oils (EOs) are derivatives of the various parts of the plants such as flowers, roots, leaves, seeds, fruits, bark, herbs, twigs, and seeds. They are hydrophobic and aromatic liquids. From ancient times, herbs and spices are commonly used in our homes, as flavoring agents of food, as a food preservative for its long-time storage, or as a medicinal plant product. EOs perform a significant function in the defense of crops from various microorganisms, insects, and animals (Kerekes et al. 2015). They are obtained either traditionally by methods like extraction, steam distillation, cold press/expressing, and enfleurage or by modern techniques employing microwave or ultrasound waves for extraction or pressurized extractions. Among the 3000 EOs known, 300 EOs are commercially explored which comprise more than 60 individual compounds (Van de Braak and Leijten 1999; De Martino et al. 2009; Cowan 1999). The amount of extracted EOs from plants depends upon factors like the part of the plant used for the purpose, its age, and the extraction method used (Lemberkovics et al. 2004; Reyes-Jurado et al. 2015). Essential oils are classified as depicted below (Kerekes et al. 2015).

Their mechanism of action involves the following: they are lipophilic in nature and therefore are permeable in the cell membrane, and they may inhibit ATP production and ATPase activity and bring about the outward flow of ions or other cellular content (Bakkali et al. 2008), disrupting the genetic (de Oliveira et al. 2010, 2012) as well as cellular material of the microorganisms (Perricone et al. 2015). It was found that aldehyde and phenolic EOs are the most effective in fighting against microbes such as cinnamaldehyde, carvacrol, eugenol, or thymol (Bakkali et al. 2008; Perricone et al. 2015). Gram-positive microorganisms showed more sensitivity to EOs when compared to their Gram-negative counterparts (Burt 2004; Lambert et al. 2001).

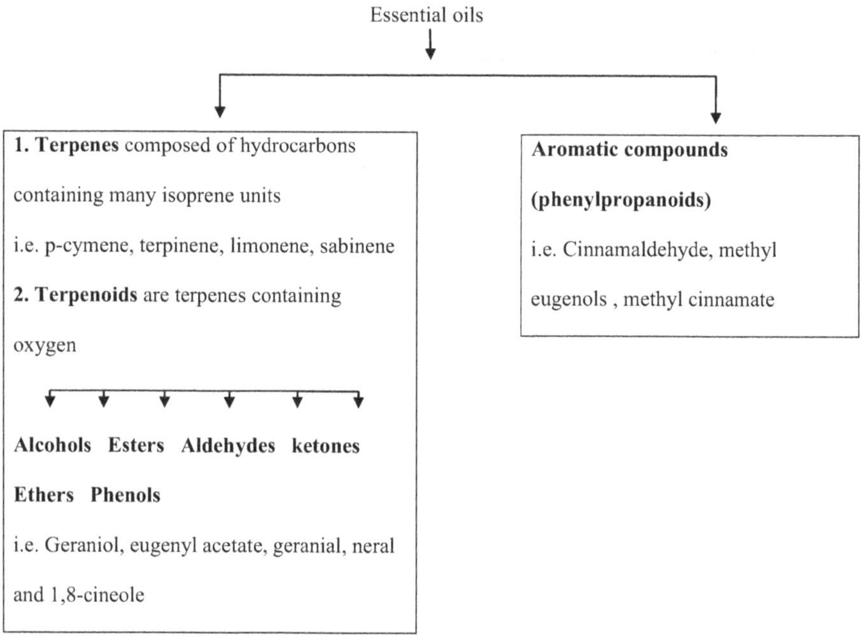


Some EOs can also act as quorum-sensing inhibitors (interfering with the communication and regulation of quorum-sensing genes) which leads to the reduced activity of biofilm formation and other virulence-related factors (Nazzaro et al. 2013). Some of the features which make EOs as future therapeutic agents are: they are easily extracted, are nontoxic to the tissue culture cell lines, are rapidly degraded when mixed in water, and have no side effects to health (Fabian et al. 2006; Warnke et al. 2006; Isman 2000). It has been observed that the presence of EOs modifies the antibiotic tolerance ability of the bacterial cell (Yap et al. 2014), and when the two antimicrobials, which target two different components of the bacterial cell, are combined, it changes the tolerance of the microorganism (Rosato et al. 2007; Cox et al. 1998; Langeveld et al. 2014; Longbottom et al. 2004; Cirino et al. 2014) (Table 9.1).

### 9.2.2 *Quorum-Sensing Inhibitors (QSIs)*

Quorum sensing (QS) is an interaction strategy in the microbial community that is chemical in nature and is used to regulate various behaviors such as virulence and biofilm formation (Uroz et al. 2009). As soon as the population of bacteria becomes dense, QS compounds start accumulating for the recognition of the population

**Table 9.1** Essential oil (EO) associated studies effective against biofilms

| Essential oils | Target biofilm organism in the study | Reference |
|---|---|---|
| Oregano essential oils, carvacrol, and thymol | *S. aureus* | Nostro et al. (2007) |
| Cassia, Peru balsam, and red thyme | *Pseudomonas* spp. and *S. aureus* | Kavanaugh and Ribbeck (2012) |
| 5% tea tree oil (TTO) | Coagulase-negative *Staphylococci* (CoNS)<br>1. Five out of nine of their biofilms are completely eradicated<br>2. 100% eradication after 1-h treatment to methicillin-susceptible *S. aureus* (MSSA) | Brady et al. (2006) |
| *Pelargonium graveolens* essential oil in combination with norfloxacin | Biofilms of two strains of *S. aureus* | Rosato et al. (2007) |
| Eugenol, cinnamaldehyde, citral, and geraniol | Clinical strains of *Staphylococcus aureus* | Jafri et al. (2014) |
| Cinnamon (*Cinnamomum zeylanicum*), TTO (*Melaleuca alternifolia*), and palmarosa (*Cymbopogon martini*), combined with ciprofloxacin | *P. aeruginosa* biofilm | Coelho and Pereira (2013) |

density to activate a corresponding response. Quorum-sensing inhibitors target the QS molecules to reduce the formation of biofilms, and this disruption reduces the growth, virulence, and dispersion of microorganisms (Papenfort and Bassler 2016).

It was proposed that quorum-sensing inhibitors mainly target the following:

1. The signal generator
2. The quorum-sensing molecule
3. The signal receptor

The QS signal receptor mediates the pharmacological action. One of the modes of action that often facilitates the transformation of biofilm pathogenicity is reducing the biofilm's resistance to conventional antimicrobial treatment. Rasamiravaka et al. (2015) reported several QS-inhibiting compounds, including penicillic acid, solenopsin A, catechin, ellagic acid derivatives, and curcumin. QSIs can be obtained from various sources, but their antibiofilm activity should be explored in future studies.

Most of the plant-derived QSIs have shown to exhibit remarkable antibiofilm activity. Several studies were performed related to QC-mediated inhibition of biofilm formation as shown in Table 9.2. These studies showed that the QSI when used alone or in synergism with various other antimicrobial agents can be used to control biofilms. Christensen et al. (2012) showed that antibiotic tobramycin, when combined with QS compounds including furanone and horseradish juice extract, disrupted the biofilms of *Pseudomonas aeruginosa* in mouse as experimental organism. The synergic effect of QS molecules and availability of QS inhibitors increased the

**Table 9.2** QSI associated with biofilm control

| QS inhibitor/QSI and antimicrobial agent combination | Synergized antibiotic if any | Target organism | Reference |
|---|---|---|---|
| RNAIII-inhibiting peptide (RIP) | Nil | *Staphylococcus* | Balaban et al. (2007) |
| Usnic acid (obtained from lichens) | Nil | *S. aureus* and *P. aeruginosa* | Francolini et al. (2004) |
| Pungent oil of fresh ginger (6-gingerol) | Nil | *P. aeruginosa* | Kim et al. (2015) |
| Lactonase from *Bacillus* spp. synergize | Ciprofloxacin gentamicin | *P. aeruginosa* | Kiran et al. (2011) |
| Patulin and penicillic acid obtained from *Penicillium* species | Nil | *P. aeruginosa* | Rasmussen et al. (2005) |
| Phenyl-DPD (phenyl-4,5-dihydroxy-2,3-pentanedione) | Gentamicin | *P. aeruginosa* | Roy et al. (2013) |
| Baicalin hydrate, cinnamaldehyde, hamamelitannin | Tobramycin, clindamycin, and vancomycin | *P. aeruginosa and S. aureus* | Brackman et al. (2011) |
| Chinese medicine baicalein | Nil | *P. aeruginosa* | Zeng et al. (2008) |
| 14-Alpha-lipoyl andrographolide (AL-1) obtained from green chiretta (*Andrographis paniculata*) | Nil | *P. aeruginosa* | Zeng et al. (2011) |
| LSFE | Tobramycin | *P. aeruginosa* | Jakobsen et al. (2012) |
| Ajoene synergized | Tobramycin | *P. aeruginosa* | Yang et al. (2006) Christensen et al. (2012) |

susceptibility of the *P. aeruginosa* biofilm to tobramycin. Such methods create a less favorable surface for biofilms to reside on, and they reduce biofilm pathogenicity using QS inhibitors, demonstrating a promising and exciting potential avenue for further exploration. However, more work needs to be done to incorporate these ideas into an in vivo environment, particularly in the case of biofilm formation, as in vitro biofilm models may not mimic complex in vivo conditions.

### 9.2.3 Antimicrobial Peptide (AMP)

Antimicrobial peptides (AMPs) are also known as "host defense peptides." In higher eukaryotic organisms, AMPs are "L"-shaped cationic molecules containing 15–50 amino acids having molecular weights between 1 and 5 KDa and are produced as part of an innate immune defense mechanism by eukaryotes and prokaryotes. They usually contain arginine and lysine residues in excess (Izadpanah and Gallo 2005;

Rossi et al. 2008; de la Fuente-Núñz et al. 2012). They act on a wide variety of organisms like bacteria, yeasts, fungi, viruses, and even cancer cells to directly kill them. They show specific and diverse activities related to normal immune homeostasis, which includes a variety of cytokine and growth factor-like effects. They mainly target cell membranes because the peptides with a positive charge and cell membranes/biofilm surfaces of microbes with a negative charge attract each other, killing the active and slow-growing bacteria in biofilms (Melo et al. 2009; Jorge et al. 2012). However, AMPs at deficient concentrations change their activity from bactericidal to bacteriostatic (Beloin et al. 2014). Cationic peptides induce gene expression in microorganisms by binding to their DNA because they can pass through the cell membrane.

As per the literature review done by Yasir et al. (2018), the following mode of actions of *antimicrobial peptides* worked for biofilm removal (Table 9.3):

Various studies (Table 9.4) reported that AMPs are more effective when combined with various conventionally used antibiotics. Also, it was found that by changing the amino acid composition of AMPs, antimicrobial activity can be increased (Ma et al. 2012; Xu et al. 2014; Tiwari et al. 2015). One such example of genetic manipulation is the replacement of functional "defective" sequence RR7 in one of the AMP R-FV-I16 by inserting the antibiofilm sequence FV7 (Xu et al. 2014). Another way in which the manipulation of AMPs can be done is by designing STAMPs (specifically targeted AMPs). The benefit of these STAMPs is that they harm pathogenic bacteria but not nonpathogenic ones (Li et al. 2010; He et al. 2009). These AMPs rupture the cell membrane or act as membrane perturbers (Wimley and Hristova 2011). Genetically engineered peptide such as peptide RN3 (5-17P22-36) of eosinophil granules can also be explored as a potential antibiofilm agent (Venge 1999; Acharya and Ackerman 2014).

**Table 9.3** Mode of action of AMP (Yasir et al. 2018)

| S. no. | Mode of action | Examples |
|---|---|---|
| 1. | The membrane potential of cells in biofilms is either disrupted or degraded | Nisin A, lacticin Q, and nukacin ISK-1, an engineered peptide RN3 (5-17P22-36), esculentin (CSA)-13 c |
| 2. | Quorum sensing is interrupted | Human cathelicidin LL-37 and indolicidin |
| 3. | Biofilm EPS matrix is degraded | Peptide PI, AMP derived from *Calliphora vicina*, hepcidin 20, peptide S4(1–16) M4Ka, piscidin-3 |
| 4. | Alarmone system is inhibited in both gram-positive and gram-negative bacteria to avoid the bacterial stringent response | Guanosine 50-diphosphate 30diphosphate (ppGpp) (p)ppGpp, 1018, DJK-5, and DJK-6, 1018 |
| 5. | Genes which are responsible for biofilm formation are downregulated and transportation of binding proteins is interrupted | Human β-defensin 3 (hBD-3), peptide Nal-P-113 |

**Table 9.4** Antimicrobial peptides associated with biofilm control

| Antimicrobial peptides | Synergized antibiotic if any | Targeted organism biofilm | Reference |
|---|---|---|---|
| A 9-amino acid peptide AMP 1037 | Nil | *P. aeruginosa*<br>*B. Cenocepacia*<br>*Listeria monocytogenes* | de la Fuente-Núñz et al. (2012) |
| LL-37 | Nil | *P. aeruginosa* | Overhage et al. (2008) |
| | | Group A *Streptococcus* (GAS) | Johansson et al. (2008) |
| | | *S. epidermidis* | Vuong et al. (2004) |
| | | *S. epidermidis ATCC35984* | Hell et al. (2010) |
| Tachyplesin III | Piperacillin-tazobactam (TZP) | *P. aeruginosa* | Hirakura et al. (2002) |
| Colistin | Ciprofloxacin | *P. aeruginosa* | Herrmann et al. (2010) |
| Nisin | Daptomycin/ ciprofloxacin | Methicillin-resistant *S. aureus* (MRSA) | Mataraci and Dosler (2012) |
| Indolicidin | Teicoplanin | | Dosler and Mataraci (2013) |
| Cecropin(1–7)-melittin A(2–9) amide (CAMA) | Ciprofloxacin | | |
| Cathelicidin peptide BMAP-28 | Quinupristin/ dalfopristin (Q/D)<br>Linezolid (LZD)<br>Vancomycin | *S. aureus* | Cirioni et al. (2006) |
| Peptide IB-367<br>LZD | NIL | *S. aureus* | Ghiselli et al. (2007) |
| Pal-Lys-LysNH2<br>Pal-Lys-Lys | Vancomycin | *S. aureus* on vascular grafts | Cirioni et al. (2007) |
| Peptide 1018 | Nil | It blocks or degrades guanosine pentaphosphate [(p)ppGpp], which is essential for biofilm formation. At low concentration, inhibition of biofilm and higher concentration eradication occurred | de la Fuente-Núñez et al. (2014) |
| D-Enantiomeric | Nil | Study on in vivo and in vitro antibiofilm activity of this newly synthesized broad-spectrum AMP | Low and White (1989) |
| Nisin A<br>Lacticin Q<br>Nukacin ISK-1 | Nil | *S. aureus* (an MRSA strain) | Okuda et al. (2013) |

(continued)

**Table 9.4** (continued)

| Antimicrobial peptides | Synergized antibiotic if any | Targeted organism biofilm | Reference |
|---|---|---|---|
| Esculentin | Nil | *P. aeruginosa* PAO1 | Luca et al. (2013) |
| (CSA)-13 c | Nil | *P. aeruginosa* | Nagant et al. (2013) |
| LL-37 and indolicidin | Nil | *P. aeruginosa* | Overhage et al. (2008) |
| Peptide PI | Nil | *Streptococcus mutans* | Ansari et al. (2017) |
| AMP derived from maggots of the blowfly *Calliphora vicina* | Nil | *Escherichia coli*<br>*Staphylococcus aureus*<br>*Acinetobacter baumannii* | Gordya et al. (2017) |
| Hepcidin 20 (human liver derived) | Nil | *S. epidermidis* | Brancatisano et al. (2014) |
| S4(1–16) M4Ka, a derivative of S4 | Nil | *P. aeruginosa* | Quilès et al. (2016) |
| Piscidin-3 (fish derived) | Nil | *P. aeruginosa* | Libardo et al. (2017) |
| Signaling nucleotides guanosine 50-diphosphate 30-diphosphate (ppGpp) (p)ppGpp | Nil | They can regulate the expression of a plethora of genes | Libardo et al. (2017)<br>Potrykus and Cashel (2008) |
| 1018<br>DJK-5<br>DJK-6 | Nil | They can block the synthesis and trigger degradation of (p)ppGpp in both Gram-positive and gram-negative bacteria | De la Fuente-Núñez et al. (2014) |
| | | *P. aeruginosa* | Pletzer et al. (2017) |
| Human β-defensin 3 (hBD-3) | Nil | *Staphylococcus epidermidis* ATCC 35984 | Zhu et al. (2013) |
| Nal-P-113 | Nil | It can inhibit genes controlling the mobility of extrachromosomal elements and transport and binding proteins such as Porphyromon | Wang et al. (2017) |

## 9.2.4 *Biofilm-Degrading Enzymes*

Primarily, enzymes whose composition are proteins or RNAs are natural catalysts that either accelerate chemical reactions without being consumed or altered or increase reaction rates without changing the chemical equilibrium between the reactants and products. Based on their functional characteristics on the ENZYME database (https://www.expasy.org/enzyme/), there are mainly six classes of enzymes (Shen and Chou 2007) (Table 9.5):

**Table 9.5** Different classes of enzymes with their mode of action (Shen and Chou 2007)

| S. no | Name of class | Mode of action |
|---|---|---|
| 1. | Oxidoreductases | Targets the quorum-sensing molecules by acting on peptide bonds, in linkages of acid anhydride |
| 2. | Transferases | Catalyzes reactions of oxidation and reduction by electron transfer producing $H_2O_2$. This affects the bacterial growth |
| 3. | Hydrolases | Targets the EPS matrix and transfers atoms between compounds |
| 4. | Lyases | Cleavage of C-C, C-O, and C-N bonds in EPS occurs leading to elimination of atoms |
| 5. | Isomerases | Catalyze the formation of a substrate's isomer by transferring the specific functional groups within the molecule |
| 6. | Ligases or synthetases | Catalyzes the joining together of two molecules using energy derived from ATP |

The biofilms produce an extracellular polysaccharide substance (EPS). The main composition of EPS are proteins, polysaccharides, and nucleic acids (Low and White 1989, Bayles 2007). EPS adheres to surfaces and protects the associated microorganisms from various antimicrobials and other shearing stress due to its structural stability factors (Cooksey and Wigglesworth-Cooksey 1995; Ramasamy and Zhang 2005). Therefore, disorganization of EPS with certain classes of enzymes will lead to detachment of biofilm (Stewart 2015) and would expose the bacteria to these agents.

Various actions of enzymes involve biochemical breakdown of EPS, inhibition of QS signaling, degradation of the adhesive bonds between cells, and the toxic substance accumulation, the cumulative effect of which leads to lysis of affected cell and deactivation of necessary enzymes needed for cell development (Thallinger et al. 2013). Enzymes such as DNase I-amylase and dispersin B (DspB) minimize the exopolysaccharide layer of the microbe; thus, the number of biofilm cells is reduced (Eckhart et al. 2007; Whitchurch et al. 2002; Kalpana et al. 2012). Moreover, the specificity of enzymes and their activities are interfered or influenced by many environmental factors like availability or nonavailability of activators, cofactors, or inhibitors, temperature, substrate, and pH (Baidamshina et al. 2017). One of the critical characteristics of enzymes is that they are substrate-specific, i.e., they cleave the EPS at a specific site (Bridier et al. 2015).

EPS composition governs a significant role in deciding whether enzymes alone or a blend of enzymes in synergy with other treatment methods, physical (ultrasound, stress) or chemical (chelating agents, buffers, surfactants, and detergents), is required to remove the EPS altogether (Thallinger et al. 2013; Darouiche et al. 2009; Izano et al. 2007). Additionally, significant reduction of biofilm mass is obtained, when the active enzyme is immobilized by entrapping in substances like poly(ethylene-alt-maleic anhydride), ceramics, polycaprolactam, etc. (Regina et al. 2012). The resistance of biofilm-forming pathogens to enzymes is quite uncommon; however, there are few exceptions, like *L. monocytogenes* resistant to lysozyme

(Nguyen and Burrows 2014), *S. aureus* mutant to lysostaphin (Gründling et al. 2006), and *P. aeruginosa* to peroxidase (Lewis 2001).

Though enzymatic therapies also have some limitations, the first one is that they are costly compared to several other antimicrobials. In the natural environment, biofilms are a composition of variably diversified microbial species; therefore, the EPS is also diverse (Jahid and Ha 2014). This diverse biofilm matrix is difficult to treat with substrate-specific enzymes. It is known that wrong selection of enzymes and their combinations sometimes leads to attenuation instead of killing (Baidamshina et al. 2017), or sometimes it does the reverse of increasing virulence factors and biofilm formation, i.e., induction of biofilm formation occurred in *Pseudomonas aeruginosa* and *Enterococcus faecalis* that is generated by a protease enzyme (Ołdak and Trafny 2005; Xu et al. 2014) (Table 9.6).

### 9.2.5 Bacteriophage

Bacteriophages were discovered by Frederick Twort in 1915 and Félix Bd'Hérelle in 1917 independently. These are viruses, shorter in size, and survive on host prokaryotes (d'Herelle 1917, 1918). Taxonomically, they are divided into Myoviridae, Siphoviridae, and Podoviridae (Ackermann 2009). Bacteriophages are bacterial viruses that exhibit two kinds of life cycles: the first one is lytic and the other is lysogenic. They have the ability to lyse the host bacterial cell or grow generation by generation with bacterial cell (Twort 1936). Bacteriophages have been applied medically to take care of human microbial diseases from the last 80 years in former Soviet Union and European countries (Clark and March 2006).

Bacteriophages penetrate biofilms (Pires et al. 2011; Vilas Boas et al. 2016); therefore, phages are active against both planktonic and biofilm form of bacteria (Kim et al. 2011; Gutiérrez et al. 2016). Antiphage refuges are formed in bacteria in biofilms, which establishes bacteria phage coexistence (Heilmann et al. 2012). The phage takes advantage of high cell density in biofilm and spreads rapidly; this weakens the biofilm structural integrity of bacterial cells and causes its lysis. Phages and antibiofilm substances can be applied together to target host bacteria for complete removal of biofilms (Uppuluri and Lopez-Ribot 2016). Alternatively, another method to enhance the broad host range of bacteriophages is that they can be genetically engineered. Dispersin B from *Aggregatibacter actinomycetemcomitans* is a biofilm-degrading enzyme expressed from engineered phages (Lu and Collins 2007).

The phage therapy has many advantages over conventional antibiotic therapy (Matsuzaki et al. 2005): it attacks the targeted microbe and does not affect the healthy microbial flora, is effective against MDR and phage-resistant bacterial mutants, is cheaper compared to antibiotics, and has minimum/rare side effects (Matsuzaki et al. 2003). One of the critical factors determining the efficacy of phage therapy is attaining high phage "killing titers" (Abedon and Thomas-Abedon 2010). However, Defence mechanisms and other host-mediated responses should be considered before adapting any conventional therapeutics methods in mammals.

**Table 9.6** Enzymes associated with biofilm control

| Enzymes | Source of enzyme | Synergized antimicrobial agent if any | Targeted organism biofilm | Reference |
|---|---|---|---|---|
| Amylase | *Bacillus subtilis* S8-18 | Nil | Methicillin-resistant *S. aureus* (MRSA)<br>*V. cholerae*<br>*P. aeruginosa* ATCC10145 | Kalpana et al. (2012) |
| | *B. subtilis*-derived | Nil | *S. aureus*<br>*P. aeruginosa* | Craigen et al. (2011) |
| | | | *P. aeruginosa* | Singh et al. (2015) |
| | *Bacillus subtilis* S8-18 | 10% human plasma | Methicillin-sensitive strain<br>MRSA strains | Watters et al. (2016) |
| DNase I | NA | Nil | *L. monocytogenes* | Nguyen and Burrows (2014) |
| | | Nil | *E. faecalis* | Abedon and Thomas-Abedon (2010) |
| | | Nil | *C. jejuni*<br>*Campylobacter coli* | Kim et al. (2017) |
| DNase I derivative (DNase1L2) | Human stratum corneum | Nil | *P. aeruginosa*<br>*S. aureus* | Eckhart et al. (2007) |
| | NA | Chlorhexidine gluconate povidone iodine | *S. aureus biofilm* | Kaplan et al. (2012) |
| Recombinant human DNase I (rhDNase I) | NA | Triclosan | *S. aureus* | Darouiche et al. (2009) |
| | *Staphylococcus simulans* | Clarithromycin | MRSA | Aguinaga et al. (2011) |
| Enzyme complex (amylase, cellulase, protease) | *Penicillium janthinellum mutant* EU2D-21 | Nil | *Escherichia coli*<br>*Salmonella enteric*<br>*Pseudomonas aeruginosa*<br>*Staphylococcus aureus* | Nagraj and Gokhale (2018) |

| | | | | |
|---|---|---|---|---|
| Lysozyme | Microorganisms<br>Plant extracts<br>Animals | Nil | *S. aureus*<br>*P. aeruginosa* strains 10,145 and 3989 | Hukić et al. (2017) |
| Alginate lyase | NA | Glutaraldehyde | *P. aeruginosa* | Meshram et al. (2016) |
| Cellulase | NA | Nil | *P. aeruginosa*<br>*S. aureus* | Fleming et al. (2017) |
| Ficin | NA | Nil | *S. aureus*<br>*S. epidermidis* | Baidamshina et al. (2017) |
| Subtilisin | Strains of *Bacillus* sp. | Nil | *Pseudoalteromonas* sp.<br>*Serratia marcescens* | Leroy et al. (2008) |
| | | Immobilization of enzyme on poly(ethylene-alt-maleic anhydride), in ceramics, and on polycaprolactam | *P. fluorescence*<br>*S. aureus*<br>*E. coli*<br>*B. subtilis*<br>*S. typhimurium*<br>*Staphylococcus xylosus* | Regina et al. (2012) |
| Proteinase K broad-spectrum serine protease | *Tritirachium album* | Nil | *Lactobacillus plantarum*<br>*Staphylococcus lentus*<br>*Staphylococcus cohnii*<br>*Staphylococcus saprophyticus* | Fagerlund et al. (2016) |
| Trypsin | *Rhodococcus ruber* strain C208 | Nil | *P. aeruginosa* | Banar et al. (2016) |
| Serratiopeptidase (Spep) | *S. marcescens* Spep | Nil | *Staphylococcus* spp. | Artini et al. (2013) |
| Papain<br>Bromelain | Present in living organism | Nil | *L. monocytogenes* pathogen<br>*Klebsiella pneumoniae*<br>*Acinetobacter* sp.<br>*S. aureus* | Mohamed et al. (2018)<br>Nguyen and Burrows (2014) |

(continued)

**Table 9.6** (continued)

| Enzymes | Source of enzyme | Synergized antimicrobial agent if any | Targeted organism biofilm | Reference |
|---|---|---|---|---|
| Endolysins peptidoglycan hydrolases | Bacteriophage lysins | Nil | Gram-positive pathogens | Gutierrez et al. (2016) |
| Extracellular enzyme complex | *Penicillium janthinellum* EU2D-2 | Nil | *Escherichia coli*<br>*Staphylococcus aureus*<br>*Pseudomonas aeruginosa*<br>*Salmonella enterica* | Nagraj and Gokhale (2018) |
| Endolysin SAL-2 | NA | Nil | *Staphylococcus* sp. | Son et al. (2010) |
| Endolysin LysH5 | NA | Nil | *S. aureus* | Gutierrez et al. (2015) |
| Artilysins | Engineered endolysin based | Nil | *MDR P. aeruginosa* | Briers et al. (2014) |
| Exopolysaccharide depolymerase, Dpo7 | NA | Nil | *S. epidermidis*<br>*S. aureus* | Gutierrez et al. (2015) |
| Glucose oxidase (Gox) | NA | Nil | *Acinetobacter calcoaceticum*<br>*Hansenula polymorpha*<br>*Corynebacterium aquaticum* | Thallinger et al. (2013) |
| Cellobiose dehydrogenase (Cdh) | NA | Nil | *S. aureus* on polydimethylsiloxane (PDMS)-based catheter surface | Thallinger et al. (2016) |
| Naturally derived lipases | NA | Nil | *Listeria* sp.<br>*Serratia* sp.<br>*B. cereus* | Seghal Kiran et al. (2014) |
| Serine<br>Cysteine<br>Metalloproteases | NA | Nil | *E. coli*<br>*P. fluorescens*<br>*Vibrio parahaemolyticus* | Hou et al. (2017)<br>Kumar (2008) Watters et al. (2016) |

The phage therapeutics should be developed to have active, harmless, safe, and long-term treatment options (Szczaurska-Nowak et al. 2009). Phages also modulate the immune system. One of the primary example is respiratory burst induced by bacterial cell wall that is inhibited by phagocytes in human blood (Levin and Bull 2004). Another essential feature observed about phages is the normalization of cytokine production by blood cells isolated from patients (Weber-Dabrowska et al. 2000). All these studies showed that mammal–phage interactions should be explored in detail for their further use as a treatment option either alone or in synergism with antibiotics.

Many phage combinations can be applied to obtain broader activity, i.e., cocktails of phages (Chan et al. 2013). Alternatively, an excellent strategy to fight against older biofilms is the use of combinations of both bacteriophages and antibiotics. The combination of a bacteriophage with amoxicillin was much more effective in reducing a mature biofilm of *Klebsiella pneumoniae* B5055 than each of the agents alone. The advantage of using phage–antibiotic combinations are decreased with the emergence of resistant cells that would appear upon using phages or antibiotica lone (Chhibber et al. 2009a, b). The recent multidrug-resistant (MDR) strains found in clinical isolates of bacteria are emerging day by day, and it has become difficult to treat these infections causing endemics (Alisky et al. 1998; Carlton 1999).

The principal downside of the use of therapeutic phages in medical treatment is the introduction of resistance against phages by pathogenic bacteria. The resistance of bacteria to phage may be developed due to inactivation of phage by the immune system of the host, and it may occur when virulence genes get incorporated into the host bacterial genome (Dolan 2009). The bacteriophage therapy has limitations of specificity towards the host which limits the phage to have a narrow range of host bacteria except for some exceptions, e.g., *Staphylococcal* phage K, Sb-1, and Stau2 (Curtin and Donlan 2006; Sharma et al. 2005). Cross-infections in closely related species, for example, of *Staphylococcus* by polyvalent phage K, SK311, U16, ϕ131, and ϕ812 are also one of the problems while using phage therapy (Pantůček et al. 1998). However, if a phage uses a bacterial virulence factor as a receptor, it should target the "virulent" subpopulation only (Bedi et al. 2009). Some of the obstacles which come across in the commercial production of phage as therapeutic agents are their complex manufacturing and testing methodology, current regulations, patenting and efficacy problems, and costly clinical trials (Debarbieux et al. 2016; Vandenheuvel et al. 2015). Despite these limitations, it can be summarized that phages are quite safe and effective as a future antibiofilm agent. Table 9.7 summarizes some of the phage-associated biofilm control studies done in the past years.

## 9.3 Conclusion

The infections caused by biofilms are chronic, recurrent, and resistant to antibiotics. Also, the contamination caused by them in industrial systems is challenging to eradicate. As a result of strengthening antimicrobial drug resistance, conventionally

**Table 9.7** Bacteriophage associated with biofilm control

| Phages | Target organism biofilm | Reference |
|---|---|---|
| T4 | *E. coli* | Corbin et al. (2001) |
| 2307-B1 | *L. monocytogenes* | Hibma et al. (1997) |
| 53b SF153b | *E. agglomerans* | Hughes et al. (1998) |
| F116 | *P. aeruginosa* | Hanlon et al. (2001) |
| 11229, φEnt, φ1.15 | *E. cloacae* | Tait et al. (2002) |
| φS1 | *P. fluorescens* | Sillankorva et al. (2004) |
| KH1 | *E. coli O157* | Sharma et al. (2005) |
| 456 | *S. epidermidis* | Curtin and Donlan (2006) |
| φ11, φ12 | *S. aureus* | Sass and Bierbaum (2006) |
| K | *S. epidermidis* | Cerca et al. (2007) |
| TG1 T7 | *E. coli* | Lu and Collins (2007) |
| C2 | *S. maltophilia* | Briandet et al. (2008) |
| φS1 | *P. fluorescens* | Sillankorva et al. (2008) |
| B5055 phage synergizes with antibiotic | *K. pneumoniae* | Bedi et al. (2009) |
| SAP-2 | *S. aureus* | Son et al. (2010) |
| P100 | *L. monocytogenes* | Soni and Nannapaneni (2010) |
| IBB-PF7A, IBB-SL58B | *P. fluorescens, S. lentus* | Sillankorva et al. (2010) |
| M4 | *P. aeruginosa* | Fu et al. (2010) |
| Bacteriophage, from the Myoviridae family T4-like phage | NA | Yoon et al. (2010) |
| phiIBB-PAP21, phiIBB-PAA | *P. aeruginosa* | Pires et al. (2011) |
| Aab01, Aab01-1 | *Aggregatibacter actinomycetemcomitans* | Castillo-Ruiz et al. (2011) |
| BVPaP-3 | *P. aeruginosa* | Ahiwale et al. (2011) |
| λW60, PB-1 | *E. coli, P. aeruginosa* | Kay et al. (2011) |
| CP8, CP30 | *C. jejuni* | Siringan et al. (2011) |
| phi 15 | *P. putida* | Cornelissen et al. (2011) |

used antibiotic therapy alone is not sufficient to control biofilm-related infections. Hence, another category of molecules/remedies to treat biofilm-associated threats is an appealing area and still has to be explored by researchers. Each new novel molecule has some advantages and limitations. Although in AMPs, enzymes, and bacteriophages, QSIs have broad-spectrum antibacterial function and tend to be protected from the occurrence of microbial resistance and could work synergistically with antibiotics, extensive research is needed such as chemical studies of the EPS matrix of various microbes and complex immunomodulatory activities inside the host cells which can reduce/enhance their efficacy. However, the effectiveness of biological control strategies might be affected through a range of physical and chemical factors. These factors include temperature or time applied in the biocontrol method, treatment of single species or multiple species biofilm, development strategy used by an organism to develop a biofilm, and composition of the surface matrix. Therefore, strategically defined control methods or validation studies of new emerging biocontrol assays against microbial biofilms need to be done before the commercialization of these products.

## References


Abedon ST, Thomas-Abedon C (2010) Phage therapy pharmacology. Curr Pharm Biotechnol 11:28–47

Acharya KR, Ackerman SJ (2014) Eosinophil granule proteins: form and function. J Biol Chem 289(25):17406–17415

Ackermann HW (2009) Phage classification and characterization. Methods Mol Biol 501:127–140

Aguinaga A, Francés ML, Del Pozo JL, Alonso M, Serrera A, Lasa I, Leiva J (2011) Lysostaphin and clarithromycin: a promising combination for the eradication of *Staphylococcus aureus* biofilms. Int J Antimicrob Agents 37:585–587

Ahiwale S, Tamboli N, Thorat K, Kulkarni R, Ackermann HW, Kapadnis B (2011) In vitro management of hospital *Pseudomonas aeruginosa* biofilm using indigenous T7-like lytic phage. Curr Microbiol 62:335–340

Alisky J, Iczkowski K, Rapoport A, Troitsky N (1998) Bacteriophages show promise as antimicrobial agents. J Infect 36(1):5–15

Anderson GG, O'Toole GA (2008) Innate and induced resistance mechanisms of bacterial biofilms. Curr Top Microbiol Immunol 322:85–105

Ansari JM, Abraham NM, Massaro J, Murphy K, Smith-Carpenter J, Fikrig E (2017) Anti-biofilm activity of a self-aggregating peptide against *Streptococcus mutans*. Front Microb 8:488

Artini M, Papa R, Scoarughi GL, Galano E, Barbato G, Pucci P, Selan L (2013) Comparison of the action of different proteases on virulence properties related to the staphylococcal surface. J Appl Microbiol 114(1):266–277

Baidamshina DR, Trizna EY, Holyavka MG, Bogachev MI, Artyukhov VG, Akhatova Kayumov AR (2017) Targeting microbial biofilms using Ficin, a nonspecific plant protease. Sci Rep 7:46068

Bakkali F, Averbeck S, Averbeck D, Idaomar M (2008) Biological effects of essential oils - a review. Food Chem Toxicol 46:446–475

Balaban N, Cirioni O, Giacometti A, Ghiselli R, Braunstein JB, Silvestri C, Mocchegiani F, Saba V, Scalise G (2007) Treatment of Staphylococcus aureus biofilm infection by the quorum-sensing inhibitor RIP. Antimicrob Agents Chemother 51:2226–2229

Banar M, Emaneini M, Satarzadeh M, Abdellahi N, Beigverdi R, van Leeuwen WB, Jabalameli F (2016) Evaluation of mannosidase and trypsin enzymes effects on biofilm production of *Pseudomonas aeruginosa* isolated from burn wound infections. PLoS One 11(10):e0164622

Bayles KW (2007) The biological role of death and lysis in biofilm development. Nat Rev Microbiol 5:721–726

Bedi MS, Verma V, Chibber S (2009) Amoxicillin and specific bacteriophage can be used together for eradication of biofilm of *Klebsiella pneumoniae* B5055. World J Microb Biotechnol 25:1145–1151

Beloin C, Renard S, Ghigo JM, Lebeaux D (2014) Novel approaches to combat bacterial biofilms. Curr Opin Pharmacol 18:61–68

Böhme A, Risse-Buhl U, Küsel K (2009) Protists with different feeding modes change biofilm morphology. FEMS Microbiol Ecol 69(2):158–169

Brackman G, Cos P, Maes L, Nelis HJ, Coenye T (2011) Quorum sensing inhibitors increase the susceptibility of bacterial biofilms to antibiotics in vitro and in vivo. Antimicrob Agents Chemother 55:2655–2661

Brady A, Loughlin R, Gilpin D, Kearney P, Tunney M (2006) In vitro activity of tea tree oil against clinical skin isolates of methicillin-resistant and -sensitive *Staphylococcus aureus* and coagulase-negative staphylococci growing planktonically and as biofilms. J Med Microbiol 55:1375–1380

Brancatisano FL, Maisetta G, Di Luca M, Esin S, Bottai D, Bizzarri R, Campa M, Batoni G (2014) Inhibitory effect of the human liver-derived antimicrobial peptide hepcidin 20 on biofilms of polysaccharide intercellular adhesin (PIA)-positive and PIA-negative strains of *Staphylococcus epidermidis*. Biofouling 30:435–446

Briandet R, Lacroix-Gueu P, Renault M, Lecart S, Meylheuc T, Bidnenko E, Steenkeste K, Bellon-Fontaine MN, Fontaine-Aupart MP (2008) Fluorescence correlation spectroscopy to study diffusion and reaction of bacteriophages inside biofilms. Appl Environ Microbiol 74:2135–2143

Bridier A, Sanchez-Vizuete P, Guilbaud M, Piard JC, Naïtali M, Briandet R (2015) Biofilm-associated persistence of food-borne pathogens. Food Microbiol 45:167–178

Briers Y, Walmagh M, Van Puyenbroeck V, Cornelissen A, Cenens W, Aertsen A, Oliveira H, Azeredo J, Verween G, Pirnay JP, Miller S (2014) Engineered endolysin-based "Artilysins" to combat multidrug-resistant gram-negative pathogens. MBio 5(4):e01379–e01314

Brooks JT, Sowers EG, Wells JG, Greene KD, Griffin PM, Hoekstra RM, Strockbine NA (2005) Non-O157 Shiga toxin producing *Escherichia coli* infections in the United States,1983–2002. J Infect Dis 192:1422–1429

Burt SA (2004) Essential oils: their antibacterial properties and potential applications in foods: a review. Int J Food Microbiol 94:223–253

Carlton RM (1999) Phage therapy: past history and future prospects. Arch Immunol Ther Exp 47(5):267–274

Castillo-Ruiz M, Vinés ED, Montt C, Fernández J, Delgado JM, Hormazábal JC, Bittner M (2011) Isolation of a novel *Aggregatibacter actinomycetemcomitans* serotype b bacteriophage capable of lysing bacteria within a biofilm. Appl Environ Microbiol 77(9):3157–3159

Cerca N, Oliveira R, Azeredo J (2007) Susceptibility of *Staphylococcus epidermidis* planktonic cells and biofilms to the lytic action of staphylococcus bacteriophage K. Lett Appl Microbiol 45:313–317

Chakraborty S, Duttal TK, De A, Das M, Ghosh S (2018) Impact of bacterial biofilm in veterinary medicine: an overview. Int J Curr Microbiol App Sci 7(4):3228–3239

Chan BK, Abedon ST, Loc-Carrillo C (2013) Phage cocktails and the future of phage therapy. Future Microbiol 8(6):769–783

Chhibber S, Bedi MS, Verma V (2009a) Amoxicillin and specific bacteriophage can be used together for eradication of biofilm of *Klebsiella pneumoniae* B5055. World J Microbiol Biotechnol 25:1145–1151

Chhibber S, Verma V, Harjai K (2009b) Restricting ciprofloxacin-induced resistant variant formation in biofilm of *Klebsiella pneumoniae* B5055 by complementary bacteriophage treatment. J Antimicrob Chemother 64:1212–1218

Christensen LD, van Gennip M, Jakobsen TH, Alhede M, Hougen HP, Høiby N, Bjarnsholt T, Givskov M (2012) Synergistic antibacterial efficacy of early combination treatment with tobramycin and quorum-sensing inhibitors against *Pseudomonas aeruginosa* in an intraperitoneal foreign-body infection mouse model. J Antimicrob Chemother 67:1198–1206

Cirino IC, Menezes-Silva SM, Silva HT, de Souza EL, Siqueira-Júnior JP (2014) The essential oil from *Origanum vulgare* L. and its individual constituents carvacrol and thymol enhance the effect of tetracycline against *Staphylococcus aureus*. Chemotherapy 60:290–293

Cirioni O, Giacometti A, Ghiselli R, Bergnach C, Orlando F, Mocchegiani F, Silvestri C, Licci A, Skerlavaj B, Zanetti M, Saba V, Scalise G (2006) Pre-treatment of central venous catheters with the cathelicidin BMAP-28 enhances the efficacy of antistaphylococcal agents in the treatment of experimental catheter-related infection. Peptides 27:2104–2110

Cirioni O, Giacometti A, Ghiselli R, Kamysz W, Silvestri C, Orlando F, Mocchegiani F, Vittoria AD, Kamysz E, Saba V, Scalise G (2007) The lipopeptides Pal-Lys-Lys-NH2 and Pal-Lys-Lys soaking alone and in combination with intraperitoneal vancomycin prevent vascular graft biofilm in a subcutaneous rat pouch model of staphylococcal infection. Peptides 28:1299–1303

Clark JR, March JB (2006) Bacteriophages and biotechnology: vaccines, gene therapy and antibacterials. Trends Biotechnol 24(5):212–218

Clutterbuck AL, Woods EJ, Knottenbelt DC, Clegg PD, Cochrane CA, Percival SL (2007) Biofilms and their relevance to veterinary medicine. Vet Microbiol 121(1–2):1–17

Coelho FA, Pereira MO (2013) Exploring new treatment strategies for *Pseudomonas aeruginosa* biofilm infections based on plant essential oils. In: Méndez-Vilas A (ed) Microbial pathogens and strategies for combating them: science, technology and education, vol 1. Formatex Research Center, Badajoz, pp 83–89

Cooksey KE, Wigglesworth-Cooksey B (1995) Adhesion of bacteria and diatoms to surfaces in the sea: a review. Aquat Microb Ecol 9:87–96

Corbin BD, McLean RJC, Aron GM (2001) Bacteriophage T4 multiplication in a glucose-limited *Escherichia coli* biofilm. Can J Microbiol 47:680–684

Cornelissen A, Ceyssens PJ, T'Syen J, Van Praet H, Noben JP, Shaburova OV, Krylov VN, Volckaert G, Lavigne R (2011) The T7-related *Pseudomonas putida* phage phi 15 displays virion-associated biofilm degradation properties. PLoS One 6:e18597

Costerton JW, Stewart PS, Greenberg EP (1999) Bacterial biofilms: a common cause of persistent infections. Science 284:1318–1322

Cowan MM (1999) Plant products as antimicrobial agents. Clin Microbiol Rev 12:564–582

Cox SD, Gustafson JE, Mann CM, Markham JL, Liew YC, Hartland RP, Bell HC, Warmington JR, Wyllie SG (1998) Tea tree oil causes K leakage and inhibits respiration in *Escherichia coli*. Lett Appl Microbiol 26:335–358

Craigen B, Dashiff A, Kadouri DE (2011) The use of commercially available alpha-amylase compounds to inhibit and remove *Staphylococcus aureus* biofilms. Open Microbiol J 5:21–31

Crouzet M, Le Senechal C, Brözel VS, Costaglioli P, Barthe C, Bonneu M, Vilain S (2014) Exploring early steps in biofilm formation: set-up of an experimental system for molecular studies. BMC Microbiol 14:253

Curtin JJ, Donlan RM (2006) Using bacteriophages to reduce formation of catheter associated biofilms by Staphylococcus epidermidis. Antimicrob Agents Chemother 50:1268–1275

d'Herelle F (1917) Sur un microbe invisible antagoniste des bacilles dysentériques. Comptes Rendus de l'Académie des Sciences Paris 165:373–375

d'Herelle F (1918) Technique de la recherche du microbe filtrant bactériophage (*Bacteriophagum intestinale*). Comptes Rendus des Seances de la Societe de Biologie Paris 81:1060–1062

Darouiche RO, Mansouri MD, Gawande PV, Madhyastha S (2009) Antimicrobial and antibiofilm efficacy of triclosan and Dispersin B combination. J Antimicrob Chemother 64:88–93

Davies DG, Parsek MR, Pearson JP, Iglewski BH, Costerton JW, Greenberrg EP (1998) The involvement of cell-to-cell signals in the development of a bacterial biofilm. Science 280(5361):295–298

de la Fuente-Núñez C, Reffuveille F, Haney EF, Straus SK, Hancock REW (2014) Broad-spectrum anti-biofilm peptide that targets a cellular stress response. PLoS Pathol 10:e1004152

de la Fuente-Núñz C, Korolik V, Bains M, Nguyen U, Breidenstein EBM, Horsman S, Lewenza S, Burrows L, Hancock RE (2012) Inhibition of bacterial biofilm formation and swarming motility by a small synthetic cationic peptide. Antimicrob Agents Chemother 56:2696–2704

De Martino L, De Feo V, Formisano C, Mignola E, Senatore F (2009) Chemical composition and antimicrobial activity of the essential oils from three chemotypes of *Origanum vulgare* L. ssp. hirtum (link) Ietswaart growing wild in Campania (Southern Italy). Molecules 14:2735–2746

de Oliveira MMM, Brugnera DF, Cardoso MDG, Alves E, Piccoli RH (2010) Disinfectant action of *Cymbopogon* sp. essential oils in different phases of biofilm formation by *Listeria monocytogenes* on stainless steel surface. Food Control 21:549–553

de Oliveira MMM, Brugnera DF, Do Nascimento JA, Batista NN, Piccoli RH (2012) Cinnamon essential oil and cinnamaldehyde in the control of bacterial biofilms formed on stainless steel surfaces. Eur Food Res Technol 234:821–832

Debarbieux L, Pirnay JP, Verbeken G, De Vos D, Merabishvili M, Huys I, Patey O, Schoonjans D, Vaneechoutte M, Zizi M, Rohde C (2016) A bacteriophage journey at the European medicines agency. FEMS Microbiol Lett 363(2):fnv225. https://doi.org/10.1093/femsle/fnv225

Dolan RM (2009) Preventing biofilms of clinically relevant organisms using bacteriophage. Trends Microbiol 17:66–72

Donlan RM (2002) Biofilms: microbial life on surfaces. Emerg Infect Dis 8(9):881–890

Donlan RM, Costerton JW (2002) Biofilms: survival mechanisms of clinically relevant microorganisms. Clin Microbiol Rev 15(2):167–193

Dosler S, Mataraci E (2013) In vitro pharmacokinetics of antimicrobial cationic peptides alone and in combination with antibiotics against methicillin resistant *Staphylococcus aureus* biofilms. Peptides 49:53–58

Dunne WM Jr (2002) Bacterial adhesion: seen any good biofilms lately? Clin Microbiol Rev 15(2):155–166

Eckhart L, Fischer H, Barken KB, Tolker-Nielsen T, Tschachler E (2007) DNase1L2 suppresses biofilm formation by *Pseudomonas aeruginosa* and *Staphylococcus aureus*. Br J Dermatol 156:1342–1345

Fabian D, Sabol M, Domaracka K, Bujnakova D (2006) Essential oils–their antimicrobial activity against Escherichia coli and effect on intestinal cell viability. Toxicol In Vitro 20:1435–1445

Fagerlund A, Langsrud S, Schirmer BC, Moretro T, Heir E (2016) Genome analysis of Listeria monocytogenes sequence type 8 strains persisting in salmon and poultry processing environments and comparison with related strains. PLoS One 11(3):e0151117

Fleming D, Chahin L, Rumbaugh K (2017) Glycoside hydrolases degrade polymicrobial bacterial biofilms in wounds. Antimicrob Agents Chemother 61(2):e01998–e01916

Francolini I, Norris P, Piozzi A, Donelli G, Stoodley P (2004) Usnic acid, a natural antimicrobial agent able to inhibit bacterial biofilm formation on polymer surfaces. Antimicrob Agents Chemother 48:4360–4365

Fu W, Forster T, Mayer O, Curtin JJ, Lehman SM, Donlan RM (2010) Bacteriophage cocktail for the prevention of biofilm formation by Pseudomonas aeruginosa on catheters in an in vitro model system. Antimicrob Agents Chemother 54(1):397–404

García AB, Percival SL (2011) Zoonotic infections: the role of biofilms. In: Biofilms and veterinary medicine. Springer, Berlin, pp 69–110

Ghiselli R, Giacometti A, Cirioni O, Mocchegiani F, Silvestri C, Orlando F, Kamysz W, Licci A, Nadolski P, Della Vittoria A, Łukasiak J, Scalise G, Saba V (2007) Pretreatment with the protegrin IB-367 affects gram positive biofilm and enhances the therapeutic efficacy of linezolid in animal models of central venous catheter infection. J Parenter Enter Nutr 31:463–468

Gordya N, Yakovlev A, Kruglikova A, Tulin D, Potolitsina E, Suborova T, Bordo D, Rosano C, Chernysh S (2017) Natural antimicrobial peptide complexes in the fighting of antibiotic resistant biofilms: Calliphora vicina medicinal maggots. PLoS One 12:e0173559

Gründling A, Missiakas DM, Schneewind O (2006) Staphylococcus aureus mutants with increased lysostaphin resistance. J Bacteriol 188:6286–6297

Gutierrez D, Briers Y, Rodríguez-Rubio L, Martínez B, Rodríguez A, Lavigne R, García P (2015) Role of the pre-neck appendage protein (Dpo7) from phage vB SepiS-philPLA7 as an antibiofilm agent in staphylococcal species. Front Microbiol 6:1315

Gutiérrez D, Rodríguez-Rubio L, Martínez B, Rodríguez A, García P (2016) Bacteriophages as weapons against bacterial biofilms in the food industry. Front Microbiol 7:825

Hanlon GW, Denyer SP, Olliff SJ, Ibrahim LJ (2001) Reduction of exopolysaccharide viscosity as an aid to bacteriophage penetration through *Pseudomonas aeruginosa* biofilms. Appl Environ Microbiol 67:2746–2753

He J, Anderson MH, Shi W, Eckert R (2009) Design and activity of a 'dual-targeted' antimicrobial peptide. Int J Antimicrob Agents 33:532–537

Heilmann S, Sneppen K, Krishna S (2012) Coexistence of phage and bacteria on the boundary of self-organized refuges. Proc Natl Acad Sci U S A 109:12828–12833

Hell E, Giske CG, Nelson A, Römling U, Marchini G (2010) Human cathelicidin peptide LL37 inhibits both attachment capability and biofilm formation of *Staphylococcus epidermidis*. Lett Appl Microbiol 50(2):211–215

Henderson P (2010) Fouling and anti-fouling in other industries –power stations, desalination plants-drinking water supplies and sensors. In: Biofouling. Wiley-Blackwell, Chichester, pp 288–305

Herrmann G, Yang L, Wu H, Song Z, Wang H, Høiby N, Ulrich M, Molin S, Riethmüller J, Döring G (2010) Colistin-tobramycin combinations are superior to monotherapy concerning the killing of biofilm Pseudomonas aeruginosa. J Infect Dis 202:1585–1592

Hibma AM, Jassim SA, Griffiths MW (1997) Infection and removal of L-forms of listeria monocytogenes with bred bacteriophage. Int J Food Microbiol 34:197–207

Hirakura Y, Kobayashi S, Matsuzaki K (2002) Specific interactions of the antimicrobial peptides cyclic-sheet tachyplesin I with lipopolysaccharides. Biochim Biophys Acta 1562:32–36

Hou HM, Zhu YL, Wang JY, Jiang F, Qu WY, Zhang GL, Hao HS (2017) Characteristics of N-acylhomoserine lactones produced by *Hafnia alvei* H4 isolated from spoiled instant sea cucumber. Sensors 17(4):772

Hughes KA, Sutherland IW, Clark J, Jones MV (1998) Bacteriophage and associated polysaccharide depolymerases- novel tools for study of bacterial biofilms. J Appl Microbiol 85:583–590

Hukić M, Seljmo D, Ramovic A, Ibrišimović MA, Dogan S, Hukic J, Feric Bojic E (2017) The effect of lysozyme on reducing biofilms by *Staphylococcus aureus*, *Pseudomonas aeruginosa*, and *Gardnerella vaginalis*: an in vitro examination. Microb Drug Resist 24(4):353–358

Isman MB (2000) Plant essential oils for pest and disease management. Crop Prot 19:603–608

Izadpanah A, Gallo RL (2005) Antimicrobial peptides. J Am Acad Dermatol 52(3):381–390

Izano EA, Wang H, Ragunath C, Ramasubbu N, Kaplan JB (2007) Detachment and killing of *Aggregatibacter actinomycetemcomitans* biofilms by dispersin B and SDS. J Dent Res 86:618–622

Jafri H, Husain FM, Ahmad I (2014) Antibacterial and antibiofilm activity of some essential oils and compounds against clinical strains of *Staphylococcus aureus*. J Biomed Ther Sci 1:65–71

Jahid IK, Ha SD (2014) Inactivation kinetics of various chemical disinfectants on *Aeromonas hydrophila* planktonic cells and biofilms. Foodborne Pathog Dis 11:346–353

Jakobsen TH, van Gennip M, Phipps RK, Shanmugham MS, Christensen LD, Alhede M, Skindersoe ME, Rasmussen TB, Friedrich K, Uthe F, Jensen PØ, Moser C, Nielsen KF, Eberl L, Larsen TO, Tanner D, Høiby N, Bjarnsholt T, Givskov M (2012) Ajoene, a sulfur-rich molecule from garlic, inhibits genes controlled by quorum sensing. Antimicrob Agents Chemother 56:2314–2325

Jamal M, Ahmad W, Andleeb S, Jalil F, Imran M, Nawaz MA, Hussain T, Ali M, Rafiq M, Kamil MA (2018) Bacterial biofilm and associated infections. J Chin Med Assoc 81(1):7–11

Johansson L, Thulin P, Sendi P, Hertzén E, Linder A, Åkesson P, Low DE, Agerberth B, Norrby-Teglund A (2008) Cathelicidin LL-37 in severe *Streptococcus pyogenes* soft tissue infections in humans. Infect Immun 76(8):3399–3404

Jorge P, Lourenco A, Pereira MO (2012) New trends in peptide-based anti-biofilm strategies: a review of recent achievements and bioinformatic approaches. Biofouling 28:1033–1061

Kabwanga IT, Yetişemiyen A, Nankya S (2018) Dairy industrial hygiene: a review on biofilm challenges and control. Int J Res Granthaalayah 6(2):268–273. https://doi.org/10.5281/zenodo.1194694

Kadouri D, Venzon NC, O' Toole GA (2007) Vulnerability of pathogenic biofilms to *Micavibrio aeruginosavorus*. Appl Environ Microbiol 73(2):605–614

Kalpana BJ, Aarthy S, Pandian SK (2012) Antibiofilm activity of amylase from Bacillus subtilis S8-18 against biofilm forming human bacterial pathogens. Appl Biochem Biotechnol 167:1778–1794

Kaplan JB, LoVetri K, Cardona ST, Madhyastha S, Sadovskaya I, Jabbouri S, Izano EA (2012) Recombinant human DNase I decreases biofilm and increases antimicrobial susceptibility in staphylococci. J Antibiot 65:73–77

Kavanaugh NL, Ribbeck K (2012) Selected antimicrobial essential oils eradicate *Pseudomonas* spp. and *Staphylococcus aureus* biofilms. Appl Environ Microbiol 78:4057–4061

Kay MK, Erwin TC, McLean RJ, Aron GM (2011) Bacteriophage ecology in *Escherichia coli* and *Pseudomonas aeruginosa* mixed-biofilm communities. Appl Environ Microbiol 77(3):821–829

Kerekes EB, Vidács A, Török Jenei J, Gömöri C, Takó M, Chandrasekaran M, Kadaikunnan S, Alharbi NS, Krisch J, Vágvölgyi C (2015) Essential oils against bacterial biofilm formation and quorum sensing of food-borne pathogens and spoilage microorganisms. In: The Battle against microbial pathogens: basic science, technological advances and educational programs. Microbiology book series, vol 5 (5). Formatex Research Center, Badajoz, pp 429–437

Kim M, Park J-M, Um H-J, Lee K-H, Kim H, Min J, Kim Y-H (2011) The antifouling potentiality of galactosamine characterized from *Vibrio vulnificus* exopolysaccharide. Biofouling 27:851–857

Kim HS, Lee SH, Byun Y, Park HD (2015) 6-Gingerol reduces Pseudomonas aeruginosa biofilm formation and virulence via quorum sensing inhibition. Sci Rep 5:8656

Kim SH, Park C, Lee EJ, Bang WS, Kim YJ, Kim JS (2017) Biofilm formation of campylobacter strains isolated from raw chickens and its reduction with DNase I treatment. Food Control 71:94–100

Kiran S, Sharma P, Harjai K, Capalash N (2011) Enzymatic quorum quenching increases antibiotic susceptibility of multidrug resistant *Pseudomonas aeruginosa*. Iran J Microbiol 3:1–12

Kumar JK (2008) Lysostaphin: an antistaphylococcal agent. Appl Microbiol Biotechnol 80(4):555–561

Kumar CG, Anand SK (1998) Significance of microbial biofilms in food industry: a review. Int J Food Microbiol 42:9–27

Lambert RJW, Skandamis PN, Coote PJ, Nychas G-JE (2001) A study of the minimum inhibitory concentration and mode of action of oregano essential oil, thymol and carvacrol. J Appl Microbiol 91:453–462

Landini P, Antoniani D, Burgess JG, Nijland R (2010) Molecular mechanisms of compounds affecting bacterial biofilm formation and dispersal. Appl Microbiol Biotechnol 86(3):813–823

Langeveld WT, Veldhuizen EJA, Burt SA (2014) Synergy between essential oil components and antibiotics: a review. Crit Rev Microbiol 40:76–94

Lemberkovics E, Kéry A, Simándi B, Kakasy A, Balázs A, Héthelyi E, Szoke E (2004) Influence of extraction methods on the composition of essential oils. Acta Pharm Hung 74(3):166–170

Leroy C, Delbarre C, Ghillebaert F, Compere C, Combes D (2008) Influence of subtilisin on the adhesion of a marine bacterium which produces mainly proteins as extracellular polymers. J Appl Microbiol 105(3):791–799

Levin BR, Bull JJ (2004) Population and evolutionary dynamics of phage therapy. Nat Rev Microbiol 2(2):166–173

Lewis K (2001) Riddle of biofilm resistance. Antimicrob Agents Chemother 45:999–1007

Li L, Guo L, Lux R, Eckert R, Yarbrough D, He J, Anderson M, Shi WY (2010) Targeted antimicrobial therapy against Streptococcus mutans establishes protective non-cariogenic oral biofilms and reduces subsequent infection. Int J Oral Sci 2:66–73

Libardo MD, Bahar AA, Ma B, Fu R, McCormick LE, Zhao J, McCallum SA, Nussinov R, Ren D, Angeles-Boza AM, Cotten ML (2017) Nuclease activity gives an edge to host-defense peptide piscidin 3 over piscidin 1, rendering it more effective against persisters and biofilms. FEBS J 284(21):3662–3683

Longbottom CJ, Carson CF, Hammer KA, Mee BJ, Riley TV (2004) Tolerance of Pseudomonas aeruginosa to Melaleuca alternifolia (tea tree) oil is associated with the outer membrane and energy-dependent cellular processes. J Antimicrob Chemother 54:386–392

Low CS, White DC (1989) Regulation of external polymer production in benthic microbial communities. In: Cohen Y, Rosenberg E (eds) Microbial mats: physiological ecology of benthic microbial communities, pp 228–238

Lu TK, Collins JJ (2007) Dispersing biofilms with engineered enzymatic bacteriophage. Proc Natl Acad Sci 104:11197–11202

Luca V, Stringaro A, Colone M, Pini A, Mangoni ML (2013) Esculentin (1-21), an amphibian skin membrane-active peptide with potent activity on both planktonic and biofilm cells of the bacterial pathogen *Pseudomonas aeruginosa*. Cell Mol Life Sci 70(15):2773–2786

Ma L, Conover M, Lu H, Parsek MR, Bayles K, Wozniak DJ (2009) Assembly and development of the *Pseudomonas aeruginosa* biofilm matrix. PLoS Pathog 5(3):e1000354

Ma L, Liu X, Liang H, Che Y, Chen C, Dai H, Yu K, Liu M, Ma L, Yang CH, Song F (2012) Effects of 14-alpha-lipoyl andrographolide on quorum sensing in Pseudomonas aeruginosa. Antimicrob Agents Chemother 56(12):6088–6094

Mah TFC, Toole GA (2001) Mechanisms of biofilm resistance to antimicrobial agents. Trends Microbiol 9(1):34–39

Mataraci E, Dosler S (2012) In vitro activities of antibiotics and antimicrobial cationic peptides alone and in combination against methicillin resistant *Staphylococcus aureus* biofilms. Antimicrob Agents Chemother 56:6366–6371

Matsuzaki S, Yasuda M, Nishikawa H, Kuroda M, Ujihara T, Shuin T, Shen Y, Jin Z, Fujimoto S, Nasimuzzaman MD, Wakiguchi H (2003) Experimental protection of mice against lethal Staphylococcus aureus infection by novel bacteriophage ϕMR11. J Infect Dis 187(4):613–624

Matsuzaki S, Rashel M, Uchiyama J, Sakurai S, Ujihara T, Kuroda M, Ikeuchi M, Tani T, Fujieda M, Wakiguchi H, Imai S (2005) Bacteriophage therapy: a revitalized therapy against bacterial infectious diseases. J Infect Chemother 11(5):211–219

Melo MN, Ferre R, Castanho MA (2009) Antimicrobial peptides: linking partition, activity and high membrane bound concentrations. Nat Rev Microbiol 7:245–250

Meshram P, Dave R, Joshi H, Dharani G, Kirubagaran R, Venugopalan VP (2016) Biofouling control on ultrafiltration membrane through immobilization of polysaccharide-degrading enzyme: optimization of parameters. Desalin Water Treat 57(55):26861–26870

Mohamed SH, Mohamed MS, Khalil MS, Azmy M, Mabrouk MI (2018) Combination of essential oil and ciprofloxacin to inhibit/eradicate biofilms in multidrug-resistant Klebsiella pneumoniae. J Appl Microbiol 125(1):84–95

Morikawa M (2006) Beneficial biofilm formation by industrial bacteria *Bacillus subtilis* and related species. J Biosci Bioeng 101(1):1–8

Nagant C, Pitts B, Stewart PS, Feng Y, Savage PB, Dehaye JP (2013) Study of the effect of antimicrobial peptide mimic, CSA-13, on an established biofilm formed by *Pseudomonas aeruginosa*. Microbiol Open 2(2):318–325

Nagraj AK, Gokhale D (2018) Bacterial biofilm degradation using extracellular enzymes produced by *Penicillium janthinellum* EU2D-21 under submerged fermentation. Adv Microbiol 8(09):687

Nazzaro F, Fratianni F, De Martino L, Coppola R, De Feo V (2013) Effect of essential oils on pathogenic bacteria. Pharmaceuticals 6:1451–1474

Nguyen UT, Burrows LL (2014) DNase I and proteinase K impair *Listeria monocytogenes* biofilm formation and induce dispersal of pre-existing biofilms. Int J Food Microbiol 187:26–32

Nostro A, Roccaro AS, Bisignano G, Marino A, Cannatelli MA, Pizzimenti FC, Cioni PL, Procopio F, Blanco AR (2007) Effects of oregano, carvacrol and thymol on *Staphylococcus aureus* and *Staphylococcus epidermidis* biofilms. J Med Microbiol 56:519–523

O'Toole G, Kaplan HB, Kolter R (2000) Biofilm formation as microbial development. Annu Rev Microbiol 54:49–79

Okuda KI, Zendo T, Sugimoto S, Iwase T, Tajima A, Yamada S, Sonomoto K, Mizunoe Y (2013) Effects of bacteriocins on methicillin-resistant *Staphylococcus aureus* biofilm. Antimicrob Agents Chemother 57(11):5572–5579

Ołdak E, Trafny EA (2005) Secretion of proteases by Pseudomonas aeruginosa biofilms exposed to ciprofloxacin. Antimicrob Agents Chemother 49:3281–3288

Olson ME, Ceri H, Morck DW, Buret AG, Read RR (2002) Biofilm bacteria: formation and comparative susceptibility to antibiotics. Can J Vet Res 66(2):86

Overhage J, Campisano A, Bains M, Torfs ECW, Rehm BHA, Hancock REW (2008) Human host defense peptide LL-37 prevents bacterial biofilm formation. Infect Immun 76:4176–4182

Pantucek R, Rosypalová A, Doškař J, Kailerová J, Růžičková V, Borecká P, Snopková Š, Horváth R, GoÈtz F, Rosypal S (1998) The polyvalent staphylococcal phage φ812: its host-range mutants and related phages. Virology 246(2):241–252

Papenfort K, Bassler BL (2016) Quorum sensing signal–response systems in gram-negative bacteria. Nat Rev Microbiol 14(9):576–588

Perricone M, Arace E, Corbo MR, Sinigaglia M, Bevilacqua A (2015) Bioactivity of essential oils: a review on their interaction with food components. Front Microbiol 6:76

Pires D, Sillankorva S, Faustino A, Azeredo J (2011) Use of newly isolated phages for control of Pseudomonas aeruginosa PAO1 and ATCC 10145 biofilms. Res Microbiol 162(8):798–806

Pletzer D, Wolfmeier H, Bains M, Hancock RE (2017) Synthetic peptides to target stringent response-controlled virulence in a Pseudomonas aeruginosa murine cutaneous infection model. Front Microbiol 8:1867

Potrykus K, Cashel M (2008) (p) ppGpp: still magical? Annu Rev Microbiol 62:35–51

Quilès F, Saadi S, Francius G, Bacharouche J, Humbert F (2016) In situ and real time investigation of the evolution of a *Pseudomonas fluorescens* nascent biofilm in the presence of an antimicrobial peptide. Biochim Biophys Acta Biomembr 1858(1):75–84

Ramasamy P, Zhang X (2005) Effects of shear stress on the secretion of extracellular polymeric substances in biofilms. Water Sci Technol 52:217–223

Rasamiravaka T, Labtani Q, Duez P, El Jaziri M (2015) The formation of biofilms by *Pseudomonas aeruginosa*: a review of the natural and synthetic compounds interfering with control mechanisms. Biomed Res Int 2015:1–17. https://doi.org/10.1155/2015/759348

Rasmussen TB, Skindersoe ME, Bjarnsholt T, Phipps RK, Christensen KB, Jensen PO, Andersen JB, Koch B, Larsen TO, Hentzer M, Eberl L, Hoiby N, Givskov M (2005) Identity and effects of quorum-sensing inhibitors produced by *Penicillium* species. Microbiology 151:1325–1340

Regina VR, Søhoel H, Lokanathan AR, Bischoff C, Kingshott P, Revsbech NP, Meyer RL (2012) Entrapment of subtilisin in ceramic sol–gel coating for antifouling applications. ACS Appl Mater Interfaces 4(11):5915–5921

Reyes-Jurado F, Franco-Vega A, Ramirez-Corona N, Palou E, López-Malo A (2015) Essential oils: antimicrobial activities, extraction methods, and their modeling. Food Eng Rev 7(3):275–297

Rosato A, Vitali C, De Laurentis N, Armenise D, Milillo MA (2007) Antibacterial effect of some essential oils administered alone or in combination with norfloxacin. Phytomedicine 14:727–732

Rossi LM, Rangasamy P, Zhang J, Qiu X-Q, Wu GY (2008) Research advances in the development of peptide antibiotics. J Pharm Sci 97:1060–1070

Roy V, Meyer MT, Smith JA, Gamby S, Sintim HO, Ghodssi R, Bentley WE (2013) AI-2 analogs and antibiotics: a synergistic approach to reduce bacterial biofilms. Appl Microbiol Biotechnol 97:2627–2638

Sass P, Bierbaum G (2006) Lytic activity of recombinant bacteriophage phi11 and phi12 endolysins on whole cells and biofilms of *Staphylococcus aureus*. Appl Environ Microbiol 73:347–352

Sauer K, Cullen MC, Rickard AH, Davies DG, Gilbert P (2004) Characterization of nutrient – induced dispersion in *Pseudomonas aeruginosa* PAO1 biofilm. J Bacteriol 186(21):7312–7326

Seghal Kiran G, Nishanth Lipton A, Kennedy J, Dobson AD, Selvin J (2014) A halotolerant thermostable lipase from the marine bacterium *Oceanobacillus* sp. PUMB02 with an ability to disrupt bacterial biofilms. Bioengineered 5(5):305–318

Sharma M, Ryu JH, Beuchat LR (2005) Inactivation of *Escherichia coli* O157:H7 in biofilm on stainless steel by treatment with an alkaline cleaner and a bacteriophage. J Appl Microbiol 99:449–459

Shen HB, Chou KC (2007) EzyPred: a top-down approach for predicting enzyme functional classes and subclasses. Biochem Biophys Res Commun 364:53–59

Shirtliff ME, Mader JT, Camper AK (2002) Molecular interactions in biofilms. Chem Biol 9:859–871

Sillankorva S, Oliveira R, Vieira MJ, Sutherland I, Azeredo J (2004) Bacteriophage Φ S1 infection of *Pseudomonas fluorescens* planktonic cells versus biofilms. Biofouling 20(3):133–138

Sillankorva S, Oliveira R, Vieira MJ, Azeredo J (2008) Real-time quantification of *Pseudomonas fluorescens* cell removal from glass surfaces due to bacteriophage ϕS1 application. J Appl Microbiol 105(1):196–202

Sillankorva S, Neubauer P, Azeredo J (2010) Phage control of dual species biofilms of Pseudomonas fluorescens and Staphylococcus lentus. Biofouling 26(5):567–575

Simões M, Simões LC, Vieira MJ (2010) A review of current and emergent biofilm control strategies. LWT-Food Sci Technol 43(4):573–583

Singh V, Verma N, Banerjee B, Vibha K, Haque S, Tripathi CK (2015) Enzymatic degradation of bacterial biofilms using *Aspergillus clavatus* MTCC 1323. Microbiology 84:59–64

Siringan P, Connerton PL, Payne RJ, Connerton IF (2011) Bacteriophage-mediated dispersal of Campylobacter jejuni biofilms. Appl Environ Microbiol 77(10):3320–3326

Son JS, Lee SJ, Jun SY, Yoon SJ, Kang SH, Paik HR, Kang JO, Choi YJ (2010) Antibacterial and biofilm removal activity of a podoviridae *Staphylococcus aureus* bacteriophage SAP-2 and a derived recombinant cell-wall-degrading enzyme. Appl Microbiol Biotechnol 86(5):1439–1449

Soni KA, Nannapaneni R (2010) Removal of Listeria monocytogenes biofilms with bacteriophage P100. J Food Prot 73:1519–1524

Stewart PS (2015) Prospects for anti-biofilm pharmaceuticals. Pharmaceuticals 8(3):504–511

Stewart PS, Costerton JW (2001) Antibiotic resistance of bacteria in biofilms. Lancet 358:135–138

Szczaurska-Nowak K, Dąbrowska K, Celka M, Kurzępa A, Nevozhay D, Wietrzyk J, Świtala-Jeleń KI, Syper D, Poźniak G, Opolski A, Gorski A (2009) Antitumor effect of combined treatment of mice with cytostatic agents and bacteriophage T4. Anticancer Res 29(6):2361–2370

Tait K, Skillman LC, Sutherland IW (2002) The efficacy of bacteriophage as a method of biofilm eradication. Biofouling 18:305–311

Thallinger B, Prasetyo EN, Nyanhongo GS, Guebitz GM (2013) Antimicrobial enzymes: an emerging strategy to fight microbes and microbial biofilms. Biotechnol J 8(1):97–109

Thallinger B, Brandauer M, Burger P, Sygmund C, Ludwig R, Ivanova K, Kun J, Scaini D, Burnet M, Tzanov T, Nyanhongo GS (2016) Cellobiose dehydrogenase functionalized urinary catheter as novel antibiofilm system. J Biomed Mater Res B Appl Biomater 104(7):1448–1456

Tiwari SK, Noll KS, Cavera VL, Chikindas ML (2015) Improved antimicrobial activities of synthetic-hybrid bacteriocins designed from enterocin E50-52 and pediocin PA-1. Appl Environ Microbiol 81:1661–1667

Twort TW (1936) Further investigations on the nature of ultra-microscopic viruses and their cultivation. J Hyg 36:204–235

Uppuluri P, Lopez-Ribot JL (2016) Go forth and colonize: dispersal from clinically important microbial biofilms. PLoS Pathol 12(2):e1005397

Uroz S, Dessaux Y, Oger P (2009) Quorum sensing and quorum quenching: the yin and yang of bacterial communication. Chembiochem 10:205–216

Van de Braak SAAJ, Leijten GCJJ (1999) Essential oils and oleoresins: a survey in the Netherlands and other major markets in the European Union. CBI, Centre for the Promotion of Imports from Developing Countries, Rotterdam, p 116

Vandenheuvel D, Lavigne R, Brüssow H (2015) Bacteriophage therapy: advances in formulation strategies and human clinical trials. Ann Rev Virol 2:599–618

Venge P (1999) Eosinophil cationic protein (ECP): molecular and biological properties and the use of ECP as a marker of eosinophil activation in disease. Clin Exp Allergy 29:1172–1186

Vilas Boas D, Almeida C, Sillankorva S, Nicolau A, Azeredo J, Azevedo NF (2016) Discrimination of bacteriophage infected cells using locked nucleic acid fluorescent in situ hybridization (LNA-FISH). Biofouling 32:179–190

Vuong C, Voyich JM, Fischer ER, Braughton KR, Whitney AR, DeLeo FR, Otto M (2004) Polysaccharide intercellular adhesin (PIA) protects Staphylococcus epidermidis against major components of the human innate immune system. Cell Microbiol 6(3):269–275

Wang HY, Lin L, Tan LS, Yu HY, Cheng JW, Pan YP (2017) Molecular pathways underlying inhibitory effect of antimicrobial peptide Nal-P-113 on bacteria biofilms formation of Porphyromonas gingivalis W83 by DNA microarray. BMC Microbiol 17(1):37

Warnke PH, Sherry E, Russo PA, Acil Y, Wiltfang J, Sivananthan S, Sprengel M, Roldan JC, Schubert S, Bredee JP, Springer IN (2006) Antibacterial essential oils in malodorous cancer patients: clinical observations in 30 patients. Phytomedicine 13(7):463–467

Watters CM, Burton T, Kirui DK, Millenbaugh NJ (2016) Enzymatic degradation of in vitro Staphylococcus aureus biofilms supplemented with human plasma. Infect and Drug Resist 9:71

Weber-Dabrowska B, Zimecki M, Mulczyk M (2000) Effective phage therapy is associated with normalization of cytokine production by blood cell cultures. Arch Immunol Ther Exp 48(1):31–37

Whitchurch CB, Tolker-Nielsen T, Ragas PC, Mattick JS (2002) Extracellular DNA required for bacterial biofilm formation. Science 295:1487

Wimley W, Hristova K (2011) Antimicrobial peptides: successes, challenges and unanswered questions. J Membr Biol 239:27–34

Xu W, Zhu X, Tan T, Li W, Shan A (2014) Design of embedded-hybrid antimicrobial peptides with enhanced cell selectivity and anti-biofilm activity. PLoS One 9:e98935

Yang JY, Della-Fera MA, Nelson-Dooley C, Baile CA (2006) Molecular mechanisms of apoptosis induced by Ajoene in 3T3-L1 adipocytes. Obesity 14:388–397

Yap PSX, Yiap BC, Ping HC, Lim SHE (2014) Essential oils, a new horizon in combating bacterial antibiotic resistance. Open Microbiol J 8:6–14

Yasir M, Willcox M, Dutta D (2018) Action of antimicrobial peptides against bacterial biofilms. Materials 11(12):2468

Yoon S, Choi Y, Lee SJ, Son J, Jun S, Kang S (2010) Bacteriophage or lytic protein derived from the bacteriophage which effective for treatment of *Staphylococcus aureus* biofilm. US 2010/0254950-A1

Zeng Z, Qian L, Cao L, Tan H, Huang Y, Xue X, Shen Y, Zhou S (2008) Virtual screening for novel quorum sensing inhibitors to eradicate biofilm formation of *Pseudomonas aeruginosa*. Appl Microbiol Biotechnol 79:119–126

Zeng X, Liu X, Bian J, Pei G, Dai H, Polyak SW, Song F, Ma L, Wang Y, Zhang L (2011) Synergistic effect of 14-alpha-lipoyl andrographolide and various antibiotics on the formation of biofilms and production of exopolysaccharide and pyocyanin by *Pseudomonas aeruginosa*. Antimicrob Agents Chemother 55:3015–3017

Zhu C, Tan H, Cheng T, Shen H, Shao J, Guo Y, Shi S, Zhang X (2013) Human β-defensin 3 inhibits antibiotic-resistant Staphylococcus biofilm formation. J Surg Res 183(1):204–213

# Chapter 10
# Microbial Options Against Antibiotic-Resistant Bacteria

**Mahantesh M. Kurjogi, Ram S. Kaulgud, and Poondla Naresh**

**Abstract** Antibiotics are routinely used to treat human and animal infectious diseases since the time of invention. However, due to continuous and long-term usage of antibiotics, infectious organisms naturally developed resistance over time through genetic changes. Further indiscriminate use of antibiotics in public health care is accelerating the emergence of drug-resistant bacteria. The spread of antimicrobial resistant bacteria among people, animals, food, and environment is of growing concern that requires urgent attention to control the widespread occurrence of antibiotic-resistant bacteria. Transition from antibiotics to nontraditional treatments is one option to overcome this global challenge. Small peptides like bacteriocin, synthesized by certain bacteria, showed good antimicrobial activity against pathogenic bacteria. Use of microbial cell-free probiotic along with regular antibiotics has significantly increased the antibacterial activity against multidrug-resistant bacteria. The application of phage therapy and quorum sensing inhibitors are also well-known options against antibiotic-resistant bacteria. Recent developments in genome editing showed successful cleavage of specific target gene, coding for pathogenesis or re-sensitizing pathogenic organisms for antibiotics, this strategy proves their ability to kill specific pathogenic bacteria based on their sequence rather than targeting group of bacteria. Similarly, nanotechnology has attracted worldwide interest due to its promising results in drug delivery system, and the versatile characteristics

M. M. Kurjogi (✉)
Multi-Disciplinary Research Unit, Karnataka Institute of Medical Sciences, Hubballi, Karnataka, India

R. S. Kaulgud
Multi-Disciplinary Research Unit, Karnataka Institute of Medical Sciences, Hubballi, Karnataka, India

Department of General Medicine, Karnataka Institute of Medical Sciences, Hubballi, Karnataka, India

P. Naresh
Graduate School of Biomedical Science and Engineering, Hanyang University, Seoul, South Korea

 
J. Singh et al. (eds.), *Microbial Biotechnology: Basic Research and Applications*, Environmental and Microbial Biotechnology,
https://doi.org/10.1007/978-981-15-2817-0_10

like potent antimicrobial activity of nanoparticles make it extremely outstanding candidate for the management of infectious diseases.

**Keywords** Antibiotics · Antimicrobial resistance · Penicillin · Bacteriocins · Phage therapy · Quorum sensing

## 10.1 Introduction

Antibiotics are a class of compounds naturally produced by certain types of microorganisms for inhibition of other competitive microorganisms in their habitat. These compounds can be isolated or synthesized by various mechanisms which can be used as medications for the treatment of infectious disease caused by pathogenic bacteria/etiology. The invention of antibiotics has made a tremendous contribution in medical field; many diseases that once known to cause huge damage on human health have been successfully treated after the invention of antibiotics (CDC report 2013). However, over a period of time, some bacteria naturally developed resistance against antibiotics due to prolonged usage of these antibiotics (Kaliwal et al. 2011). Antibiotic resistance is the ability of bacteria to develop some mechanism to neutralize the action of antimicrobials against them. As a result, use of such antimicrobials against these bacteria becomes ineffective. Recently, the World Health Organization has announced the alarming level of resistance among various species of bacteria (Tacconelli et al. 2018). The development of resistance among bacteria goes beyond the available antibiotics and also exceeds the rate of antibiotic discovery at present situation. Studies suggest that around 444 million people could suffer infectious disease by 2050 (Gould and Bal 2013). High prevalence of antibiotic resistant bacteria has become one of the major public health problems and especially control of mortality due to nosocomial infections poses a biggest challenge to the health care professionals.

The evolution of multidrug-resistant bacteria in hospital environment is the result of prolonged exposure of bacteria to various antibiotics and also transfer of resistant bacteria between individuals; another important factor for evolution of resistant bacteria is transmission of resistance genes from resistant bacteria to susceptible bacteria (Guillemot et al. 2001). In the pre-antibiotic era, *S. aureus* infection resulted in 80% mortality (Smith and Vickers 1960). With the inventions of antibiotics, the organism was reported to be successfully controlled by the earliest antibiotics like penicillin. However, during the 1950s, the extensive applications of these antibiotics against infectious disease resulted in the emergence of beta-lactamase-producing bacterial strains (Fisher and Knowles 1978). Further beta-lactamase-resistant penicillins were developed to overcome the emergence of beta-lactamase-producing strains, but resistant bacteria developed against this new antibiotic were first reported in the 1960s from Europe and in the 1970s from the USA (Peacock et al. 1980). Furthermore, in the 1980s, *S. aureus* strains were also frequently reported to have developed resistance against potent antibiotics like methicillin in hospital environment (Hughes 1987). By the 1990s, studies reported the emergence of bacteria resistance to semi-synthetic penicillins like nafcillin and oxacillin. Moreover, several bacteria also developed resistance against antibiotics like macrolides, tetracyclines, and aminoglycosides, compromising the use of these drugs for empiric therapy for infectious diseases in a number of regions. This has led to the introduction of glycopeptide antibiotic known as vancomycin for the management of methicillin-resistant *S. aureus* (MRSA) infections (Hiramatsu et al. 2001). However, Hiramatsu and his coworker (1997) reported the first vancomycin-resistant *Staphylococcus aureus* from Japan. Similarly, other countries like the USA (Smith et al. 1999), Belgium (Denis et al. 2002), and India (Assadullah et al. 2003) also reported reduced susceptibility of *S. aureus* to vancomycin. Hence, it is necessary to understand the genetics and defense pathway mechanism of the resistant bacteria at individual level in order to develop effective therapy against infectious disease. Therefore, comprehensive efforts are required to discover novel molecules or new technologies for the control of multidrug-resistant bacteria. In recent times, several progressive approaches have been made with unique properties, which include use of probiotic, phage therapy, quorum sensing inhibition, genome editing technology, and nanoparticle therapeutics, and they prove their potential ability to control resistant bacteria. Though the alternative strategy for combating bacterial infection is in its infancy, its potential to re-sensitize or eliminate resistant bacteria cannot be underestimated. Therefore, in this chapter, we summarized novel strategies for handling the crisis of bacterial infections, and overview of new strategies for the management of multidrug-resistant bacteria is represented in Fig. 10.1.

## 10.2 Small Peptide as a Novel Antimicrobial Agent

As the prevalence of resistant bacteria increases, it is necessary to search for a new molecule that plays an important role in controlling the widespread occurrence of resistant bacteria. Bacteriocins are ribosomally produced small peptide molecules known for their potential antimicrobial agent. Bacteriocins are broadly classified into two classes based on the mode of their production. Class I bacteriocins are produced after modification during post-translation process. This class of bacteriocins is also identified by the presence of unusual amino acids where threonine and serine residues are dehydrated to dehydrobutyrine and dehydroalanine, respectively, during post-translation modification (Cotter et al. 2013). Class II bacteriocins are unmodified with cyclic structure and further divided into class IIa to class IIe (Cotter et al. 2013). Several class I bacteriocins are isolated, among which the most commonly studied bacteriocins include nisin, lacticin, staphylococcin, mersacidin, etc. (Brotz et al. 1995; McAuliffe et al. 1998; Navaratna et al. 1998; Xie et al. 2004; Field et al. 2008). Similarly, class II bacteriocins are also extensively studied which revealed their strong affinity with mannose phosphotransferase receptor suggesting their specificity in antimicrobial activity against pathogens (Oppegard et al. 2007). Bacteriocin usually recognizes either a general or specific receptor molecule on a

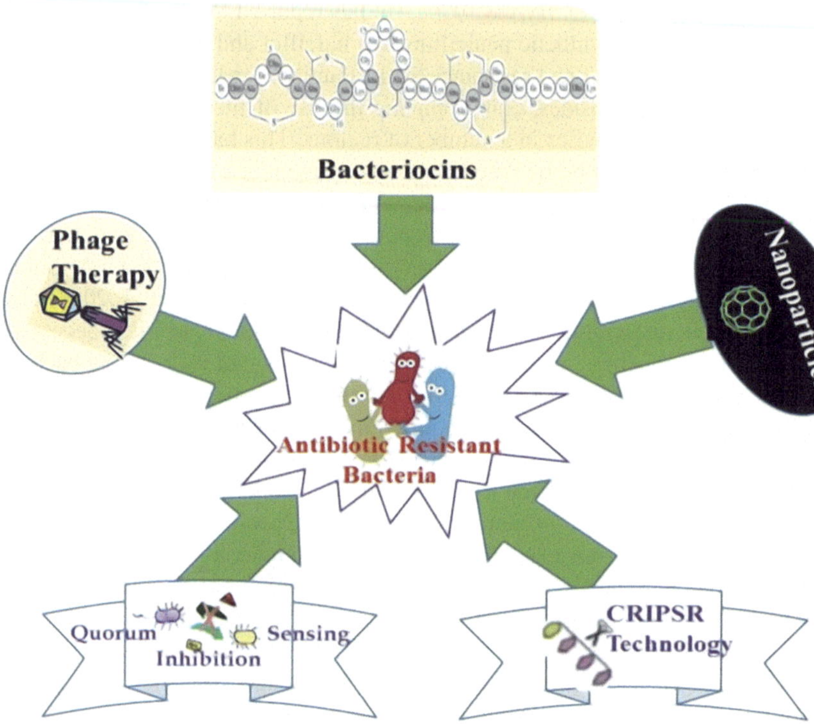


**Fig. 10.1** Overview of new strategies for the management of antibiotic-resistant bacteria

target cell to which it binds and disrupts membrane structure by pore formation (Brogden 2005). Bacteriocins produced by different types of bacteria vary in their structural and functional characteristics. Extensive studies on the mode of action of bacteriocins showed that bacteriocins produced by Gram-positive bacteria have broad-spectrum antimicrobial activity (Sang and Blecha 2008). Nisin is a common bacteriocin produced by *Lactobacillus* and *Lactococcus* species which is well known for its antimicrobial activity against foodborne pathogens (Dimitrieva-Moats and Unlu 2012). *Staphylococcus* species also produce different types of bacteriocins such as lysostaphin, aureocin, and nukacin. Lysostaphin is produced by coagulase-negative staphylococci and is proven to be an outstanding candidate for the control of human and animal infection caused by multidrug-resistant *S. aureus* (Bastos et al. 2010). Aureocin is produced by *S. aureus*; successful isolation and application of aureocin were demonstrated to control the udder infection in domestic cattle (2007). Similarly, nukacin is another bacteriocin produced by *S. simulans* also used in the therapeutic applications of animal disease (Ceotto et al. 2010). *Carnobacterium* spp. also produce bacteriocins such as carnobacteriocin X and carnocyclin A which have potential to kill *Listeria* spp. (Martin-Visscher et al. 2008; Tulini et al. 2014). Likewise, *Enterococcus* spp. produce different types of bacteriocin such as enterocin, enterocin X, and enterocin A. All these bacteriocins showed their efficacy against foodborne and also clinical pathogens (Gálvez et al. 2007; Hu et al. 2010; De la Fuente-Salcido et al. 2015).

Different types of bacteriocins are also produced by Gram-negative bacteria. It is important to note that the first reported bacteriocin in 1952 was isolated from Gram-negative bacteria. Among Gram-negative bacteria, *Escherichia coli* is the predominant producer of bacteriocins known as colicins, which are reported to have the ability to kill the target organism by the action of pore formation or nuclease activity (Bakkal et al. 2010). Similarly, Klebcins are proteinase bacteriocins produced by *Klebsiella pneumoniae* (Gillor et al. 2004). Pyocin is another small antimicrobial peptide produced by *Pseudomonas aeruginosa* (Gulluce et al. 2013). Further, bacteriocins with their potential antimicrobial ability are shown to inhibit the growth of multidrug-resistant bacteria such as vancomycin- and methicillin-resistant bacteria. This suggests that cell wall peptidoglycan acts as a receptor for bacteriocins.

## 10.3 Phage Therapy as a Complimentary Strategy for the Control of Antibiotic-Resistant Bacteria

In recent years, increased prevalence of antibiotic-resistant bacteria has emerged as a major threat to the global population. Presently available antibiotics fail to control the spread of infectious diseases caused by resistant bacteria. Therefore, rather than the old antibiotic therapy, a new strategy such as phage therapy is required to control the adverse effect of resistant bacteria on human health. Phages are considered as natural predators of bacteria; bacteria feeding viruses are generally known as bacteriophages. Bacteriophages are very specific in their host. Hence, they are effectively used for the biotyping of bacterial strains, since they are specific to their target bacteriophages. They have been considered as a promising strategy against bacterial infections. Several studies have shown the use of phage therapy in treating bacterial infections, but still, it has not gained much interest all around the world. Phage therapy has several advantages such as target specificity and does not harm the other normal microflora and replication takes place inside the infected cell. The mode of action of phage therapy involves adsorption of phages to their target bacteria and killing the host bacteria after making several copies of itself with the host DNA replication process. The newly formed phages, lysis the host bacterial cell and released into surrounding environment, further infect the nearby target bacteria. This process continues until the target bacteria get eliminated from the surrounding environment. Once all the bacteria are killed from the surrounding, bacteriophages are eliminated through natural cleansing process without affecting the human tissue. To date, clinical applications of phages in treating infectious disease have been extensively studied (Morello et al. 2011; Vieira et al. 2012; Waters et al. 2017). Around 137 different phages have been characterized for targeting *Pseudomonas* genus (Pires et al. 2015). Several institutes in Europe have carried out extensive research on application of phages on human trials to treat common bacterial infections caused by *E. coli*, *S. aureus*, *Streptococcus* spp., *Proteus* spp., *P. aeruginosa*, *S. dysenteriae*, *Salmonella* spp., and *Enterococcus* spp. (Kutateladze and Adamia 2008). Similarly, specific phages were successfully used to treat diabetic foot ulcers caused by multidrug-resistant *S. aureus* (Fish et al. 2016). In another study, patients administered with phage cocktail consisting of various phages targeting different types of bacteria responsible for dysentery such as *Salmonella typhi*, *Shigella*, *E. coli*, *Salmonella paratyphi*, *Proteus* spp., *P. aeruginosa*, *Shigella flexneri*, and *Staphylococcus* spp., *Streptococcus* spp., and *Enterococcus* spp. were found to recover from the symptoms within 24 h of phage cocktail treatment (Chanishvili and Sharp 2008). Recently, Forti et al. (2018) demonstrated the successful use of six different phages in treating *P. aeruginosa* infection in mice and *Galleria mellonella* models. It is observed that some phages also disrupt the *P. aeruginosa* biofilms. This ability of phages is one of the important contributions over traditional antibiotic treatment (Waters et al. 2017; Fong et al. 2017). Recent studies showed promising results in controlling *E. coli* infection in mice (Vahedi et al. 2018). Kumari and coworkers confirm that topical application of phage on burn wounds of mouse showed significant reduction in mortality of mice (Kumari et al. 2011). Apart from clinical application, phages have potential to control the growth of food pathogen and are considered to be safe; several commercial phages were used for biocontrol of bacterial pathogens, viz., *Pseudomonas syringae*, *Listeria monocytogenes*, *Salmonella* spp., *E. coli*, and *Campylobacter* spp. (El-Shibiny and El-Sahhar 2017). Another progress in phage therapy research is that phages and purified phage lytic proteins can be genetically engineered, thereby increasing the efficacy of treatment. A variety of phages are also used as a vehicle in drug delivery process. M13 phages were successful in delivering the coding sequence to the target cell resulting in the death of the target cell (Westwater et al. 2003). Correspondingly, a variety of bioengineered phages were constructed to control *E. coli* infection by destroying biofilms and interrupting DNA replication and delivery of RNA-guided virulence nucleases. Further, some phages also function as an effective adjuvant that increases the efficacy of antibiotics against bacterial infections (Citorik et al. 2014; Lu and Collins 2007, 2009).

## 10.4 Quorum Sensing Inhibition as a Novel Approach to Diminish Bacterial Resistance

For many years, the search for an effective treatment to fight against infectious diseases has been one of the biggest challenges to the scientific community. Use of plant-based bioactive molecules to control infectious agent was one of the well-known methods practiced before the invention of antibiotic. However, the use of such molecules to treat infectious disease was substituted by chemotherapy due to its broad spectrum activity and low toxicity. Prolonged usage of antibiotics has led to the development of resistant strains, and in recent years, emerging antibiotic-resistant strains have become a serious threat worldwide. Currently, revival of bioactive molecules that block quorum sensing in bacteria is necessary to mitigate infectious diseases. Quorum sensing is a unique mechanism of bacteria that regulates the expression of genes; most pathogenic bacteria acclimatize to their habitat by expressing virulence genes via quorum sensing system (Heilmann et al. 2015). For instance, in *Pseudomonas aeruginosa*, Las and Rhl are two different quorum sensing transcription factors responsible for production of biofilm and expression of multiple virulence genes (Rutherford and Bassler 2012). Similarly, in *Staphylococcus aureus*, quorum sensing system controls the expression of accessory gene regulator (AGR) genes that are responsible for the production of several toxins and exoenzymes (Martin et al. 2013). The survival of *S. aureus* in the host environment against the immunity of the host is attributed to the self-defense mechanism expressed by *S. aureus* (Kurjogi et al. 2010). Hence, quorum sensing system plays an important role in regulation of various virulence genes and biofilm formation leading the resistant bacteria. Therefore, quorum sensing inhibition approach is considered to be a promising strategy for the management of antimicrobial resistant bacteria. Available reports show that use of anti-quorum sensing agents can obstruct the quorum sensing signals among the bacteria; several bacterial enzymes like lactonase, acylase, oxidoreductases, and 3-hydroxy-2-methyl-4(1H)-quinolone 2, 4-dioxygenase have been reported to be potential quorum sensing inhibitors (Jiang et al. 2019). Similarly, bioengineered *E. coli* was successful to disrupt the proteolytic activity and pyocyanin production of *Pseudomonas aeruginosa* (Dong et al. 2018). Liu et al. reported that fishes supplemented with lactonase were found to be resistant to *Aeromonas hydrophila* infection (Liu et al. 2016). In addition, studies suggest that lactonase can also disrupt the biofilm formation by *Vibrio parahaemolyticus* in shrimps (Torres et al. 2018). Acylase is another bacterial enzyme known as quorum sensing inhibitor found to be successfully used for the control of *Pseudomonas aeruginosa* in healthcare sectors (Grover et al. 2016). Further it is also noted that oxidoreductases by bacteria can abolish the biofilm formation and inhibit the growth of *Klebsiella oxytoca* and *K. pneumoniae* (Wildschut et al. 2006; Zhang et al. 2018). Overall, the quorum sensing inhibitors are the promising alternative strategy to tradition antibiotic therapy. Use of anti-quorum sensing agent in health sectors not only kills the pathogenic bacteria but also controls the spread of antibiotic-resistant bacteria. However, further studies are needed to ensure the stability of quorum sensing inhibitors to convey the potential ability of quorum sensing therapy for management of infectious diseases.

## 10.5 Gene Editing Technique for Management of Infectious Diseases

Invention of antibiotics is one of the greatest discoveries that revolutionized the medical field. Antibiotics have saved millions of lives since their discovery. However, increased use of antibiotics to treat common infectious disease has led to the development of resistant strains. Therefore, in recent years, most pharmaceutical companies have stopped the production of several antibiotics due to declined use of such antibiotics. On the other hand, use of genetic engineering to edit the bacterial gene to re-sensitize the bacteria to antibiotic is providing a novel approach for management of infectious diseases.

CRISPR-Cas system not only protects bacteria against invaders but also controls endogenous transcription and the pathogenicity of bacteria. For example, *Francisella novicida*, which is known as intracellular parasite, can successfully replicate by surpassing the host immune system. This bacterium has several mechanisms to mitigate the defense mechanisms of the host. On macrophage engulfment, *F. novicida* enters the phagosome, where numerous antimicrobials and immune recognition receptors are present (Jones et al. 2012). Toll-like receptor 2 (TLR2) is one of those receptors that can detect bacterial lipoproteins (BLPs). TLR2 activation initiates a pro-inflammatory response and triggers immune cells, thereby eliminating pathogen. However, *F. novicida* uses cas9, sacRNA, and tracrRNA to inhibit BLP expression (Sampson and Weiss 2014). Therefore, by preventing TLR2 activation, this pathogen can survive within the host. *F. novicida* induces inflammatory response in the absence of these regulators, as it was stated that cas9, sacRNA, and tracrRNA deletion mutants induce stronger inflammatory immune response compared to wild type. In contrast, deletion mutants of sacRNA, cas9, and tracrRNA are not capable of causing lethal infection in mice, further emphasizing the importance of CRISPR-cas system as an *F. novicida* virulence regulator. In addition, cas9 is required for invasion and attachment of *Campylobacter jejuni* (Louwen et al. 2013). Nevertheless, *C. jejuni* attachment to host cells protects this bacterium from the inherent complementary mechanism of the host. A study recently confirmed that *C. jejuni* has a role to play in controlling CRISPR-cas9 associated virulence genes (Shabbir et al. 2018).

Resistance may evolve by inactivating CRISPR-Cas loci through mutations or deletions in target cleavage cas genes or by deleting targeting spacers (Bikard et al. 2012; Jiang et al. 2013). At present, more than 20 distinct acr gene families have been identified, both type I and II CRISPR-Cas systems (Pawluk et al. 2018; Borges et al. 2017). Many of the Acr protein families targeting type I CRISPR-Cas systems have been associated with *Pseudomonas aeruginosa* as well as other Proteobacteria species. While most of these Acr proteins tend to target only one CRISPR-Cas subtype, one Acr targeting both the type I-E and I-F CRISPR-Cas subtypes has been published. More recently, Acr proteins have been established as target type II systems—including the CRISPR-Cas9 systems used for gene editing—one of which is particularly wide in its target range (Pawluk et al. 2016). The massive sequence diversity and high specificity of Acrs indicate that they are likely ubiquitous and possibly carried by MGEs such as phages and plasmids to circumvent targeting by CRISPR-Cas (Harrington et al. 2017). The implications of CRISPR-Cas targeting resistant genes and their effect on other population need to be studied in detail, especially using clinical pathogens to understand the ecological and evolutionary risks.

Until CRISPR-Cas can be used to target antibacterial resistance, several hurdles remain to be overcome. Future research is needed to identify the effective method to explore CRISPR-Cas technology. However, the social and legislative challenges are to draft guidelines for regulation of CRISPR technology and to encourage the proper use of this technology to ensure its responsible and safe use (Makarova et al. 2015).

### *10.5.1 Green Nanotechnology to Combat Against Antibiotic-Resistant Bacteria*

For many years, the search for an effective treatment to fight against infectious diseases has been one of the biggest challenges to the scientific community. The concern of drug-resistant clinical pathogens is not only limited to humans but also reported in domestic animals. Several studies show that bacteria have developed resistance to many commonly used veterinary antibiotics (Kaliwal et al. 2011; Kurjogi and Kaliwal 2011). It is essential to explore novel antimicrobial agents with potent antimicrobial activity as an alternative to traditional antibiotics. In this context, nanotechnology has attracted worldwide interest due to its promising results in drug delivery system, and it is not surprising to see the antimicrobial activity of nanoparticles against clinical pathogens. The versatile characteristics of nanoparticles make them extremely outstanding candidate in several research fields like clinical, agricultural, and physical sciences (Chaudhuri and Paria 2012; Tran et al. 2013; Rauwel et al. 2015). Nanoparticles can be synthesized in different ways. Till date, several chemical and physical approaches have been made for the synthesis of nanoparticles. However, nanoparticles synthesized by chemical and physical processes are not suitable for clinical application since the reducing or stabilizing agent used in chemical or physical process is not biocompatible and hazardous to environment. Therefore, recently environmentally benign biological methods like green synthesis of nanoparticles have gathered global scientific attention (Lee et al. 2016; Cerda et al. 2017; Prasad 2014, 2016, 2019). Nanoparticles synthesized by biological approach are more advantageous in terms of safety, efficiency, and biocompatibility (Kumar et al. 2015; Quester et al. 2016; Prasad et al. 2017). Several microbes are considered as a novel source for green synthesis of nanoparticles since microbes are rich source of enzymes and other metabolites that act as a reducing or stabilizing agent in the process of nanoparticle synthesis (Prasad et al. 2016, 2018). Another advantage of using microbes is they can be easily cultured in a controlled condition. Several studies have proved the antimicrobial efficacy of microbe-based nanoparticles in different way. Nanoparticles can be used as adjuvant with available antibiotics that increases the efficiency of antibiotics against the target pathogen (Hassan and Hemeg 2017). On the other hand, several metal nanoparticles like silver, copper, gold, zinc, etc. are known to be used directly as an antimicrobial agent against pathogenic bacteria (Aziz et al. 2014, 2015, 2016). Recently, studies reported that silver nanoparticles synthesized by *Ganoderma applanatum* demonstrated in vitro antibacterial activity against clinical pathogens (Jogaiah et al. 2017). Similarly, silver nanoparticles were synthesized by edible mushrooms such as *Pleurotus pulmonarius* and *Pleurotus djamor* which showed high bactericidal activity against clinically important *Staphylococcus aureus* and *Pseudomonas aeruginosa* (Shivashankar et al. 2013). In the same way, silver nanoparticles synthesized using *Trichoderma viride* also showed good bactericidal activity against Gram-positive and Gram-negative bacteria (Chitra and Annadurai 2013). Pandey and colleagues prepared nano-capsulation for oral dosage of streptomycin and other antibiotics that are non-injectable (Pandey et al. 2003). Elechiguerra et al. (2005) demonstrated the inhibition of HIV from binding to host cells when treated with silver nanoparticles. Nanoparticles produced by *R. stolonifer* successfully inhibited the growth of antibiotic-resistant *P. aeruginosa* isolated from burn patients (Afreen and Ranganath 2011). Studies also revealed that *P. glomerata*-based nanoparticles performed synergistic antibacterial activity against *E. coli*, *P. aeruginosa*, and *S. aureus* when combined with standard antibiotics (Birla et al. 2009). The authors also used saprophytic fungi like *Nigrospora oryzae* for nanoparticle production and successfully demonstrated their efficacy against several clinical pathogens like *E. coli*, *B. cereus*, *Proteus vulgaris*, *P. aeruginosa*, and *Micrococcus luteus* (Saha et al. 2011). The mechanism involved in antimicrobial activity of nanoparticles is attributed to their size and shape. Further, surface charge of the nanoparticles is also an important factor of antibacterial activity where bacterial cell wall electrostatically attracts the oppositely charged nanoparticles, causing damage to the cell membrane leading to the death of the bacterial cell (Aziz et al. 2014, 2015, 2016, 2019).

## 10.6 Conclusions

Current studies on alternative strategies specifically against multidrug-resistant bacterial infections suggest that these novel therapies have all the potential ability to control antibiotic-resistant bacteria. Further research has to be carried out to show the application of these therapies in large population through clinical trials. However, ever remaining challenges in these therapies are purification of bacteriocins, development of phage bank that includes collection of various identified phages against the resistant bacteria, identification of quorum sensing inhibitors, implementation of CRISPR technology, and controlled synthesis of microbe-based nanoparticles.

**Acknowledgments** The authors thank Department of Health Research, Ministry of Family and Welfare, Government of India, New Delhi for supporting Multi-Disciplinary Research Unit at Karnataka Institute of Medical Sciences, Hubli.

## References

Afreen VR, Ranganath E (2011) Synthesis of monodispersed silver nanoparticles by *Rhizopus stolonifer* and its antibacterial activity against MDR strains of *Pseudomonas aeruginosa* from burnt patients. Int J Environ Sci 1:1830–1840

Assadullah S, Kakru DK, Thoker MA, Bhat FA, Hussain N, Shah A (2003) Emergence of low level vancomycin resistance in MRSA. Indian J Med Microbiol 21:196–198

Aziz N, Faraz M, Pandey R, Sakir M, Fatma T, Varma A, Barman I, Prasad R (2015) Facile algae-derived route to biogenic silver nanoparticles: synthesis, antibacterial and photocatalytic properties. Langmuir 31:11605–11612. https://doi.org/10.1021/acs.langmuir.5b03081

Aziz N, Fatma T, Varma A, Prasad R (2014) Biogenic synthesis of silver nanoparticles using Scenedesmus abundans and evaluation of their antibacterial activity. J Nanoparticles, 689419. https://doi.org/10.1155/2014/689419

Aziz N, Pandey R, Barman I, Prasad R (2016) Leveraging the attributes of Mucor hiemalis-derived silver nanoparticles for a synergistic broad-spectrum antimicrobial platform. Front Microbiol 7:1984. https://doi.org/10.3389/fmicb.2016.01984

Aziz N, Faraz M, Sherwani MA, Fatma T, Prasad R (2019) Illuminating the anticancerous efficacy of a new fungal chassis for silver nanoparticle synthesis. Front Chem 7:65. https://doi.org/10.3389/fchem.2019.00065

Bakkal S, Robinson SM, Ordonez CL, Waltz DA, Riley MA (2010) Role of bacteriocins in mediating interactions of bacterial isolates taken from cystic fibrosis patients. Microbiology 156:2058–2067

Bastos MCF, Coutinho BG, Coelho MLV (2010) Lysostaphin: a staphylococcal bacteriolysin with potential clinical applications. Pharmaceuticals (Basel) 3:1139–1161

Bikard D, Hatoum-Aslan A, Mucida D, Marraffini LA (2012) CRISPR interference can prevent natural transformation and virulence acquisition during in vivo bacterial infection. Cell Host Microbe 12(2):177–186

Birla SS, Tiwari VV, Gade AK, Ingle AP, Yadav AP, Rai MK (2009) Fabrication of silver nanoparticles by *Phoma glomerata* and its combined effect against *Escherichia coli*, *Pseudomonas aeruginosa* and *Staphylococcus aureus*. Lett Appl Microbiol 48:173–179

Borges AL, Davidson AR, Bondy-Denomy J (2017) The discovery, mechanisms, and evolutionary impact of anti-CRISPRs. Annu Rev Virol 4(1):37–59

Brogden KA (2005) Antimicrobial peptides: pore formers or metabolic inhibitors in bacteria? Nat Rev Microbiol 3:238–250

Brotz H, Bierbaum G, Markus A, Molitor E, Sahl HG (1995) Mode of action of the lantibiotic mersacidin: inhibition of peptidoglycan biosynthesis via a novel mechanism? Antimicrob Agents Chemother 39:714–719

Centers for Disease Control and Prevention, Office of Infectious Disease (2013) Antibiotic resistance threats in the United States. Available at: http://www.cdc.gov/drugresistance/threat-report

Ceotto H, Holo H, Silva da Costa KF, Nascimento JS, Salehian Z, Nes IF, Bastos MCF (2010) Nukacin 3299, An antibiotic produced by Staphylococcus simulans 3299 identical to nukacin ISK-1. Vet Microbiol 146:124–131

Cerda SJ, Gomez EH, Nunez AG, Rivero IA, Ponce GY, Lopez FLZ (2017) A green synthesis of copper nanoparticles using native cyclodextrins as stabilizing agents. J Saudi Chem Soc 21:341–348

Chanishvili N, Sharp R (2008) Bacteriophage therapy: experience from the Eliava Institute, Georgia. Microbiol Australia 29:96–101

Chaudhuri GR, Paria S (2012) Core/shell nanoparticles: classes, properties, synthesis mechanisms, characterization, and applications. Chem Rev 112:2373–2433

Chitra K, Annadurai G (2013) Bioengineered silver nanobowls using *Trichoderma viride* and its antibacterial activity against Gram-positive and Gram-negative bacteria. J Nanostruct Chem 3:9. https://doi.org/10.1186/2193-8865-3-9

Citorik RJ, Mimee M, Lu TK (2014) Sequence-specific antimicrobials using efficiently delivered RNA-guided nucleases. Nat Biotechnol 32:1141–1145

Coelho MLV, Nascimento JS, Fagundes PC, Madureira DJ, Oliveira SS, Brito MAP, Bastos MCF (2007) Activity of staphylococcal bacteriocins against *Staphylococcus aureus* and *Streptococcus agalactiae* involved in bovine mastitis. Res Microbiol 158:625–630

Cotter PD, Ross RP, Hill C (2013) Bacteriocins - a viable alternative to antibiotics? Nat Rev Microbiol 11:95–105

De la Fuente-Salcido NM, Cataneda RJC, Garc ABE (2015) Isolation and characterization of bacteriocinogenic lactic bacteria from Tuba and Tepache, two traditional fermented beverages in México. Food Sci Nutr 3(2):1–9

Denis O, Nonhoff C, Byl B, Knoop C, Bobin-Dubreux S, Struelens MJ (2002) Emergence of vancomycin-intermediate *Staphylococcus aureus* in a Belgian hospital: microbiological and clinical features. J Antimicrob Chemother 50:383–391

Dimitrieva-Moats GY, Unlu G (2012) Development of freeze-dried bacteriocin-containing preparations from lactic acid bacteria to inhibit *Listeria monocytogenes* and *Staphylococcus aureus*. Probiotics Antimicrob Proteins 4:27–38

Dong W, Zhu J, Guo X et al (2018) Characterization of AiiK, an AHL lactonase, from Kurthia huakui LAM0618T and its application in quorum quenching on Pseudomonas aeruginosa PAO1. Sci Rep 8(1):6013

Elechiguerra JL, Burt JL, Morones JR, Camacho-Bragado A, Gao X, Lara HH, Yacaman MJ (2005) Interaction of silver nanoparticles with HIV-1. J Nanobiotechnol 3(6)

El-Shibiny A, El-Sahhar S (2017) Bacteriophages: the possible solution to treat infections caused by pathogenic bacteria. Can J Microbiol 63:865–879

Field D, Connor PM, Cotter PD, Hill C, Ross RP (2008) The generation of nisin variants with enhanced activity against specific Gram-positive pathogens. Mol Microbiol 69:218–230. https://doi.org/10.1111/j.1365-2958.2008.06279.x

Fish R, Kutter E, Wheat G, Blasdel B, Kutateladze M, Kuhl S (2016) Bacteriophage treatment of intransigent diabetic toe ulcers: a case series. J Wound Care 25(7):S27–S33

Fisher JF, Knowles JR (1978) Bacterial resistance to β-lactams: the β-lactamases. Ann Rep Med Chem 13:239–248. https://doi.org/10.1016/S0065-7743(08)60628-4

Fong SA, Drilling A, Morales S, Cornet ME, Woodworth BA, Fokkens WJ et al (2017) Activity of bacteriophages in removing biofilms of *Pseudomonas aeruginosa* isolates from chronic rhinosinusitis patients. Front Cell Infect Microbiol 7:418

Forti F, Roach DR, Cafora M, Pasini ME, Horner DS, Fiscarelli EV et al (2018) Design of a broad-range bacteriophage cocktail that reduces *Pseudomonas aeruginosa* biofilms and treats acute infections in two animal models. Antimicrob Agents Chemother 62:e02573-17

Gálvez A, Abriouel H, López RL, Ben Omar N (2007) Bacteriocin-based strategies for food biopreservation. Int J Food Microbiol 120:51–70

Gillor O, Kirkup BC, Riley MA (2004) Colicins and microcins: the next generation antimicrobials. Adv Appl Microbiol 54(1):129–146

Gould IM, Bal AM (2013) New antibiotic agents in the pipeline and how they can help overcome microbial resistance. Virulence 4(2):185–191

Grover N, Plaks JG, Summers SR, Chado GR, Schurr MJ, Kaar JL (2016) Acylase-containing polyurethane coatings with anti-biofilm activity. Biotechnol Bioeng 113(12):2535–2543

Guillemot DP, Courvalin, and the French Working Party to Promote Research to Control Bacterial Resistance (2001) Better control of antibiotic resistance. CID 33:542–547

Gulluce M, Karaday M, Barıs O (2013) Bacteriocins: promising natural antimicrobials. In: Méndez-Vilas A (ed) Microbial pathogens and strategies for combating them: science, technology and education. Formatex, Extremadura, pp 1016–1027

Harrington LB, Doxzen KW, Ma E, Liu JJ, Knott GJ, Edraki A, Doudna JA (2017) A broad-spectrum inhibitor of CRISPR-Cas9. Cell 170(6):1224–1233.e15

Hassan A, Hemeg (2017) Nanomaterials for alternative antibacterial therapy. Int J Nanomed 12:8211–8225

Heilmann S, Krishna S, Kerr B (2015) Why do bacteria regulate public goods by quorum sensing? How the shapes of cost and benefit functions determine the form of optimal regulation. Front Microbiol 6:767. https://doi.org/10.3389/fmicb.2015.00767

Hiramatsu K, Aritaka N, Hanaki H, Kawasaki S, Hosoda Y, Hori S, Fukuchiand Y, Kobayashi I (1997) Dissemination in Japanese hospitals of strains of *Staphylococcus aureus* heterogeneously resistant to vancomycin. Lancet 350:1670–1673

Hiramatsu K, Cui L, Kuroda M, Ito T (2001) The emergence and evolution of methicillin resistant *Staphylococcus aureus*. Trends Microbiol 9:486–493

Hu CB, Malaphan W, Zendo T, Nakayama J, Sonomoto K (2010) Enterocin X, a novel two-peptide bacteriocin from *Enterococcus faecium* KU-B5, has an antibacterial spectrum entirely different from those of its component peptides. Appl Environ Microbiol 76:4542–4545

Hughes JM (1987) Setting priorities: nationwide nosocomial infection prevention and control programs in the USA. Eur J Clin Microbiol 6:348–351

Jiang W, Maniv I, Arain F, Wang Y, Levin BR, Marraffini LA (2013) Dealing with the evolutionary downside of CRISPR immunity: bacteria and beneficial plasmids. PLoS Genet 9(9). https://doi.org/10.1371/journal.pgen.1003844

Jiang Q, Chen J, Yang C, Yin Y, Yao K (2019) Quorum sensing: a prospective therapeutic target for bacterial diseases. Biomed Res Int. https://doi.org/10.1155/2019/2015978

Jogaiah S, Kurjogi M, Abdelrahman M, Nagabhushana H, Tran L-SP (2017) *Ganoderma applanatum*-mediated green synthesis of silver nanoparticles: structural characterization and *in vitro* and *in vivo* biomedical and agrochemical properties. Arab J Chem. https://doi.org/10.1016/j.arabjc.2017.12.002

Jones CL, Sampson TR, Nakaya HI, Pulendran B, Weiss DS (2012) Repression of bacterial lipoprotein production by Francisella novicida facilitates evasion of innate immune recognition. Cell Microbiol 14(10):1531–1543

Kaliwal BB, Sadashiv SO, Kurjogi MM, Sanakal RD (2011) Prevalence and antimicrobial susceptibility of coagulase-negative Staphylococci isolated from Bovine Mastitis. Vet World 4(4):158–161

Kumar N, Palmer GR, Shah V, Walker VK (2015) The effect of silver nanoparticles on seasonal change in arctic tundra bacterial and fungal assemblages. PLoS One 9:e99953

Kumari S, Harjai K, Chhibber S (2011) Bacteriophage versus antimicrobial agents for the treatment of murine burn wound infection caused by *Klebsiella pneumoniae* B5055. J Med Microbiol 60:205–210

Kurjogi MM, Kaliwal BB (2011) Prevalence and antimicrobial susceptibility of bacteria isolated from bovine mastitis. Adv Appl Sci Res 2(6):229–235

Kurjogi MM, Sanakal RD, Kaliwal BB (2010) Antibiotic susceptibility and antioxidant activity of *Staphylococcus aureus* pigment staphyloxanthin on carbon tetrachloride (ccl4) induced stress in Swiss albino mice. Int J Biotechnol Appl 2(2):33–40

Kutateladze M, Adamia R (2008) Phage therapy experience at the Eliava Institute. Med Mal Infect 38:426–430

Lee KX, Shameli K, Miyake M, Kuwano N, Khairudin NBA, Mohamad SEB, Yew YP (2016) Green synthesis of gold nanoparticles using aqueous extract of *Garcinia mangostana* fruit peels. J Nanomater 8489094

Liu W, Ran C, Liu Z et al (2016) Effects of dietary *Lactobacillus plantarum* and AHL lactonase on the control of *Aeromonas hydrophila* infection in tilapia. Microbiol Open 5(4):687–699

Louwen R, Horst-Kreft D, De Boer AG, Van Der Graaf L, De Knegt G, Hamersma M, Van Belkum A (2013) A novel link between Campylobacter jejuni bacteriophage defence, virulence and Guillain-Barré syndrome. Eur J Clin Microbiol Infect Dis 32(2):207–226

Lu TK, Collins JJ (2007) Dispersing biofilms with engineered enzymatic bacteriophage. Proc Natl Acad Sci U S A 104:11197–11202

Lu TK, Collins JJ (2009) Engineered bacteriophage targeting gene networks as adjuvants for antibiotic therapy. Proc Natl Acad Sci U S A 106:4629–4634

Makarova KS, Wolf YI, Alkhnbashi OS, Costa F, Shah SA, Saunders SJ, Koonin EV (2015) An updated evolutionary classification of CRISPR-Cas systems. Nat Rev Microbiol 13(11):722–736

Martin MJ, Clare S, Goulding D et al (2013) The agr locus regulates virulence and colonization genes in clostridium difficile 027. J Bacteriol 195(16):3672–3368

Martin-Visscher LA, van Belkum MJ, Garneau-Tsodikova S, Whittal RM, Zheng J, McMullen LM (2008) Isolation and characterization of carnocyclina, an oval circular bacteriocin produced by *Carnobacterium maltaromaticum* UAL307. Appl Environ Microbiol 74:4756–4763

McAuliffe O, Ryan MP, Ross RP, Hill C, Breeuwer P, Abee T (1998) Lacticin 3147, a broad-spectrum bacteriocin which selectively dissipates the membrane potential. Appl Environ Microbiol 64:439–445

Morello E, Saussereau E, Maura D, Huerre M, Touqui L, Debarbieux L (2011) Pulmonary bacteriophage therapy on Pseudomonas aeruginosa cystic fibrosis strains: first steps towards treatment and prevention. PLoS One 6:e16963

Navaratna MA, Sahl HG, Tagg JR (1998) Two-component anti- Staphylococcus aureus lantibiotic activity produced by Staphylococcus aureus C55. Appl Environ Microbiol 64:4803–4808

Oppegard C, Rogne P, Emanuelsen L, Kristiansen PE, Fimland G, Nissen-Meyer J (2007) The two-peptide class II bacteriocins: structure, production, and mode of action. J Mol Microbiol Biotechnol 13:210–219. https://doi.org/10.1159/000104750

Pandey R, Zahoor A, Sharma S, Khuller GK (2003) Nanoparticle encapsulated antitubercular drugs as a potential oral drug delivery system against murine tuberculosis. Tuberculosis (Edinb) 83:373–378

Pawluk A, Staals RHJ, Taylor C, Watson BNJ, Saha S, Fineran PC, Davidson AR (2016) Inactivation of CRISPR-Cas systems by anti-CRISPR proteins in diverse bacterial species. Nat Microbiol 1(8). https://doi.org/10.1038/nmicrobiol.2016.85

Pawluk A, Davidson AR, Maxwell KL (2018) Anti-CRISPR: discovery, mechanism and function. Nat Rev Microbiol 16(1):12–17

Peacock JE Jr, Marsikand FJ, Wenzel RP (1980) Methicillin-resistant Staphylococcus aureus: introduction and spread within a hospital. Ann Intern Med 93:526–532

Pires DP, Vilas Boas D, Sillankorva S, Azeredo J (2015) Phage therapy: a step forward in the treatment of Pseudomonas aeruginosa infections. J Virol 89:7449–7456

Prasad R (2014) Synthesis of silver nanoparticles in photosynthetic plants. J Nanoparticles, 963961. https://doi.org/10.1155/2014/963961

Prasad R (2016) Advances and applications through fungal nanobiotechnology. Springer International Publishing, Cham. ISBN 978-3-319-42989-2. http://www.springer.com/us/book/9783319429892

Prasad R (2019) Plant nanobionics: approaches in nanoparticles biosynthesis and toxicity. Springer International Publishing, Cham. ISBN 978-3-030-16379-2. https://www.springer.com/gp/book/9783030163785

Prasad R, Jha A, Prasad K (2018) Exploring the realms of nature for nanosynthesis. Springer International Publishing, Cham. ISBN 978-3-319-99570-0. https://www.springer.com/978-3-319-99570-0

Prasad R, Kumar M, Kumar V (2017) Nanotechnology: an agriculture paradigm. Springer Singapore, Singapore. ISBN 978-981-10-4573-8. http://www.springer.com/us/book/9789811045721

Prasad R, Pandey R, Barman I (2016) Engineering tailored nanoparticles with microbes: quo vadis. WIREs Nanomed Nanobiotechnol 8:316–330. https://doi.org/10.1002/wnan.1363

Quester K, Avalos-Borja M, Castro-Longori E (2016) Controllable biosynthesis of small silver nanoparticles using fungal extract. J Biomater Nanobiotechnol 7:118–125

Rauwel P, Rauwel E, Ferdov S, Singh MP (2015) Silver nanoparticles: synthesis, properties, and applications. Adv Mater Sci Eng 624394

Rutherford ST, Bassler BL (2012) Bacterial quorum sending: its role in virulence and possibilities for its control. Cold Spring Harb Perspect Med 2

Saha S, Chattopadhyaya D, Acharya K (2011) Preparation of silver nanoparticles by bio-reduction using *Nigrospora oryzae* culture filtrate and its antimicrobial activity. Dig J Nanomater Biostruct 6(4):1519–1528

Sampson TR, Weiss DS (2014) CRISPR-Cas systems: new players in gene regulation and bacterial physiology. Front Cell Infect Microbiol 4(37). https://doi.org/10.3389/fcimb.2014.00037

Sang Y, Blecha F (2008) Antimicrobial peptides and bacteriocins: alternatives to traditional antibiotics. Anim Health Res Rev 9:227–235. https://doi.org/10.1017/S1466252308001497

Shabbir MAB, Tang Y, Xu Z, Lin M, Cheng G, Dai M, Hao H (2018) The involvement of the Cas9 gene in virulence of *Campylobacter jejuni*. Front Cell Infect Microbiol. https://doi.org/10.3389/fcimb.2018.00285

Shivashankar M, Premkumari B, Chandan N (2013) Biosynthesis, partial characterization and antimicrobial activities of silver nanoparticles from pleurotus species. Int J Int Sci Inn Tech Sec B 2(3):13–23

Smith IM, Vickers AB (1960) Natural history of 338 treated and untreated patients with staphylococcal septicaemia. Lancet:1318–1322

Smith TL, Pearson ML, Wilcox KR, Cruz C, Lancaster MV, Robinson Dunin B, Tenover EC, Zervos ML, Band ID, White E, Larvis WR (1999) Emergence of vancomycin resistance in Staphylococcus aureus. N Engl J Med 340:493–501

Tacconelli E, Carrara E, Savoldi A, Harbarth S, Mendelson M, Monnet DL et al (2018) Discovery, research, and development of new antibiotics: the WHO priority list of antibiotic-

resistant bacteria and tuberculosis. Lancet Infect Dis 18:318–327. https://doi.org/10.1016/S1473-3099(17)30753-3
Torres M, Reina JC, Fuentes-Monteverde JC et al (2018) AHL lactonase expression in three marine emerging pathogenic Vibrio spp. reduces virulence and mortality in brine shrimp (Artemia salina) and Manila clam (Venerupis philippinarum). PLoS One 13(4):e0195176
Tran QH, Nguyen VQ, Le AT (2013) Silver nanoparticles: synthesis, properties, toxicology, applications and perspectives. Adv Nat Sci Nanosci Nanotechnol 4:033001
Tulini FL, Lohans CT, Bordon KC, Zheng J, Arantes EC, Vederas JC (2014) Purification and characterization of antimicrobial peptides from fish isolate Carnobacterium maltaromaticum C2: carnobacteriocin X and carnolysins A1 and A2. Int J Food Microbiol 173:81–88
Vahedi A, Dallal MMS, Douraghi M, Nikkhahi F, Rajabi Z, Yousefi M et al (2018) Isolation and identification of specific bacteriophage against enteropathogenic Escherichia coli (EPEC) and in vitro and in vivo characterization of bacteriophage. FEMS Microbiol Lett 365:fny136. https://doi.org/10.1093/femsle/fny136
Vieira A, Silva YJ, Cunha A, Gomes NC, Ackermann HW, Almeida A (2012) Phage therapy to control multidrug-resistant Pseudomonas aeruginosa skin infections: in vitro and ex vivo experiments. Eur J Clin Microbiol Infect Dis 31:3241–3249
Waters EM, Neill DR, Kaman B, Sahota JS, Clokie MRJ et al (2017) Phage therapy is highly effective against chronic lung infections with Pseudomonas aeruginosa. Thorax 72:666–667
Westwater C, Kasman LM, Schofield DA, Werner PA, Dolan JW et al (2003) Use of genetically engineered phage to deliver antimicrobial agents to bacteria: an alternative therapy for treatment of bacterial infections. Antimicrob Agents Chemother 47:1301–1307
Wildschut JD, Lang RM, Voordouw JK, Voordouw G (2006) Rubredoxin: oxygen oxidoreductase enhances survival of desulfovibrio vulgaris hildenborough under microaerophilic conditions. J Bacteriol 188(17):6253–6260
Xie L, Miller LM, Chatterjee C, Averin O, Kelleher NL, van der Donk WA (2004) Lacticin 481: in vitro reconstitution of lantibiotic synthetase activity. Science 303:679–681
Zhang X, Ou-yang S, Wang J, Liao L, Wu R, Wei J (2018) Construction of antibacterial surface via layer-by-layer method. Curr Pharm Des 24(8):926–935

# Chapter 11
# New and Advanced Technologies in Aquaculture to Support Environmentally Sustainable Development

**Mahipal Singh Sankhla, Rajeev Kumar, and Shefali**

**Abstract** Marine and freshwater organisms have a great purpose in agricultural life: providing nutritional resources and nutritional status within many developing and developed countries. Despite the unquestionable benefits of fish farming, such as providing good-quality food to the population over generations, it is condemned worldwide because of its negative environmental impacts. Aquaculture must push toward expanding to meet the increasing requirements of the present generation without compromising the ability of future generations to meet their own needs and to participate more effectively in the reduction of poverty and malnutrition. The main challenge to aquaculture planners is to attain ecologically safe development, which requires an authority agenda that can easily account for the environmental effects in social and economic terms.

**Keywords** Aquaculture · Fishery · Environmental · Algaculture · Fish farming · Food web · Aquaponics

## 11.1 Introduction

In both the established and the emerging countries of the world, interest in and requests for fish as food are continually increasing. Current advances in development, our increasing population, and rising profits in the emerging world are probably responsible for continuing this scenario. Aquaculture production has significantly increased during the past 20 years, with additional cultivated fish per unit of water and land, and decreasing proportions of fishmeal and fish oil in numerous aquaculture feedstuffs (Jena et al. 2017). Within the food subdivision,

M. S. Sankhla (✉) · R. Kumar
Division of Forensic Science, SBAS, Galgotias University, Greater Noida, India

Shefali
Department of Zoology, DPG Degree College, Gurugram, Haryana, India



J. Singh et al. (eds.), *Microbial Biotechnology: Basic Research and Applications*, Environmental and Microbial Biotechnology,
https://doi.org/10.1007/978-981-15-2817-0_11

aquaculture firms are not unaware of these matters. Aquaculture is among the greatest significant food regions, and its influence on the entire world and all fish manufacturing has scaled up progressively from 20.9% in 1995 to 32.4% in 2005 and 40.3% in 2010; in accumulation, its influence on all world food fish production at that time was 47% in 2010 (FAO 2012). Fresh fish as a food is also a consumable item for which ecological limitations, such as differences in temperature and moisture, must be measured and precisely preserved within recognized parameters (Jedermann et al. 2009; Abad et al. 2009; Tingman et al. 2010). Aquaculture can be created on conservative, low-invention, undeveloped organizations or on extremely developed methods. Within these lies a variety of aquaculture frameworks with numerous components that can be attuned to local monetary circumstances. Fish farming experiences many encounters, in particular, fighting epigastric diseases, brood stock development and taming, and growth of suitable feedstuffs, plus nutritional equipment, grow-out techniques, and hatcheries, as well as water quality organizations. These matters are all currently important opportunities for interventions by biotechnology and other methods. Aquacultural biotechnology is defined as the technical use of organic ideas that improve the efficiency and monetary feasibility of its numerous manufacturing subdivisions (Liao and Chao 1997). Moreover, new aquaculture equipment can increase production, source superior quality fish foods, and provide for the defense of vulnerable aquatic surroundings (Burnell and Allan 2009). In adding to aquaculture technologies and fishing, new knowledge of manufactured fish goods contributes to monetary expansion, for example, through price-counting dispensations such as surimi technology (Martín-Sánchez et al. 2009). Consequently, fishery technology expansion is significant in attaining sustainable fishery reserve administration (Suuronen et al. 2012). Improvements in aquaculture are also challenged with high quality by means of recognized herbivorous/omnivorous principles within widespread or semi-intensive schemes and emerging organizations to meet cumulative manufacturing difficulties. In the same way, such matters as the request for food versus obtainability of marine organisms, efficiency versus ecological concerns, and excellence versus biodiversity, mandate serious attention (Hasan 2001).

## 11.2 Scale and Distribution of Aquaculture

Asian countries have continued to ignore, for an extended period and typically in rural zones, the millions of individuals who are affected by fish farming or aquaculture-related activities: they derive abundant food and revenue from their marine territories. In India, about 2 million persons are directly involved in events connected to fish farming. Typically, hatchery-manufactured seeds are provided by agriculturalists in water-based organizations that assist them in aquatic crop production, whereas prawns and mollusks are gathered from existing territories that

**Table 11.1** World fisheries and aquaculture production and utilization (in million tonnes) (UN 2015)

| Category | 2011 | 2012 | 2013 | 2014 | 2015 | 2016 |
|---|---|---|---|---|---|---|
| *Production* | | | | | | |
| Inland | 10.7 | 11.2 | 11.2 | 11.3 | 11.4 | 11.6 |
| Marine | 81.5 | 78.4 | 79.4 | 79.9 | 81.2 | 79.3 |
| Total capture | 92.2 | 89.5 | 90.6 | 91.2 | 92.7 | 90.9 |
| *Aquaculture* | | | | | | |
| Inland | 38.6 | 42.0 | 44.8 | 46.9 | 48.6 | 51.4 |
| Marine | 23.2 | 24.4 | 25.4 | 26.8 | 27.5 | 28.7 |
| Total aquaculture | 61.8 | 66.4 | 70.2 | 73.7 | 76.1 | 80.0 |

eventually come to be a portion of the aquatic manufacturing. Fisheries contribute about 5.15% of the Farming GDP (Handbook of Fisheries Statistics 2015). Fish farming has provided an increasing influence in the matter of food safety (Table 11.1). Through a median yearly development ratio of 6.2%, aquaculture is the world's major emerging animal food sector (FAO 2012, 2014). There is an important upsurge in aquaculture manufacturing as much of it is provided for human consumption, that is, from 7% in 1974 it has enlarged to 39% in 2004. In 2012, cultivated fish feed contributed 66.6 million tons, which was the highest amount, almost corresponding to 42.2% of the total 158 million tons of fish from both fisheries and aquaculture (FAO 2014). Algae and seaweed are gathered for use as nourishment and in cosmetics; manure determinations are treated to extract the agents that are advanced animal feeds. Last, 22,400 tons of nonfood goods are also cultivated (with a value of 222.4 million dollars), such as gems and seashells for ornamental uses (FAO 2014).

## 11.3 Types of Aquaculture

### *11.3.1 Mariculture*

Mariculture is aquaculture that includes the use of seawater. It can also be bounded next to an oceanic body of water, with portions divided off as ponds or in the ocean distinct from the oceanic water but consisting of marine water. The creatures raised now vary from mollusks to seafood choices such as prawns and other shellfish, and even seaweed. Increasingly, such plants as seaweed are also a portion of mariculture. These animals and sea organisms find usage in manufacturing industries such as in cosmetics and jewelry, wherein collagen from seaweed is used to create facial ointments, and pearls are selected from mollusks and processed into ornamental items (conserve-energy-future 2019).

### 11.3.2 Fish Farming

Fish agriculture is the best mutual variety of aquaculture. It includes the discerning upbringing of fish, in either clean freshwater or seawater, with the objective of manufacturing food. Fish agriculture is extremely controlled as it permits the manufacture of a low-priced source protein. Moreover, fish agriculture is easier than other undeveloped programs because fish care is not rigorous, only requiring food and appropriate water conditions and temperatures. The method also has less land impact as the extent of pools essential to raise some fish varieties, such as tilapia, is abundant, and essential requirements are fewer than those for increasing a similar quantity of protein from beef cattle (conserve-energy-future 2019).

### 11.3.3 Algaculture

Algaculture is a type of aquaculture that includes algae cultivation. Algae are microorganisms that incorporate plant and animal aspects, as they are sometimes stimulated to resemble other microorganisms but also contain chloroplasts, which enable photosynthesis, so that they qualify as green plants. However, for commercial utility, algae have to be developed and reaped in great amounts. Algae are the result of numerous requests by today's shoppers. ExxonMobil has been creating opportunities for algae to emerge as a new basis of energy (conserve-energy-future 2019).

### 11.3.4 Integrated Multitrophic Aquaculture (IMTA)

Integrated multitrophic aquaculture (IMTA) is a progressive organization of aquaculture wherein different trophic levels are varied to meet different nutritional requirements for others in the organization. Particularly, it is a well-organized scheme because it tries to emulate the biological organization that occurs in the normal environment. The IMTA uses these intertrophic allocations of properties to ensure superior supply consumption by using larger underused species as food sources for the smaller ones. The repetition certifies the nutrients are reprocessed: the importance of this method is that it is less extravagant and foodstuffs are extra goods (conserve-energy-future 2019). It includes humanizing of aquatic entities in such a method that their uneaten food, waste material, and nutrients are recalled so they can be better transformed into nourishment and energy for the development of other species. The agriculturalists working in the IMTA system select species that need supplementary feedstuffs, such as mussels, sea urchins, and kelp, using sieves to put nourishing organisms in principal usage as particulate resources, that is, uneaten food and feces, for satisfying the nutrition supplies. The capabilities of these species to certainly reprocess the wastes can assist cultivators in refining the presentation of their aquaculture operations. The species are essentially those which have worth as indicators and eventually benefit the farmers with further financial profits.

## 11.4 Sustainable Aquaculture

Concerning maintainable aquaculture, the obtainable proof as already shown states that throughout the world aquaculture presently enhances the remaining worldwide fish materials, although many kinds of aquaculture cause damage to the remaining fish. The possible influence of aquaculture on fish materials is sharply reduced by the quick growth of species nourished by flesh-eating regimes and by aquaculture that is responsible for coastal environment obliteration, organic contamination, and release of unprocessed sewages. Sustained growth of aquaculture requires well-planned coastline and lake ecological care. Deprived of appreciation by its production impact on normal environments, it is improbable that aquaculture will reach its complete potential or persist in addition to marine fisheries. We consequently propose that aquaculture emphasizes these four main areas: (1) growth of undeveloped low trophic level fish; (2) decrease of fish meal and fish oil contributions in feedstuff; (3) growth of combined farming organizations; and (4) increase of ecologically rigorous aquaculture applications and supply organizations (Naylor et al. 2000).

### *11.4.1 Farming Down the Food Web*

Carp and sea mollusks account for more than three-fourths of the present worldwide aquaculture yield, and catfish, tilapia, and milkfish contribute an additional 5% of the total production. Nourished mostly on herbivorous diets, these varieties offer most of the 19 Mt increase in fish materials from aquaculture. However, marketplace structures and administration strategies in numerous republics favor the quick growth of carnivorous varieties of maximum worth, such as shrimp and salmon. Furthermore, fish food and fish oil are being added to carp and tilapia feedstuffs for weight gain, particularly in Asia where aquatic farming organizations are increasing as a consequence of the increased shortage and costs of land and freshwater supplies. Assuming the enormous capacity of cultivated carp and tilapia in Asia, important upsurges in the fish meal and fish oil additions to feed might place even more burden on pelagic fisheries, with subsequent advanced feed costs and damage to marine environments. New solutions by administrations and worldwide donor organizations are desirable to inspire greater use of undeveloped low trophic level fish with herbivorous foods (Tacon 1997; Baily 1997; El-Sayed 1999; Tacon and De Silva 1997). At the same time, additional systematic study on the feed necessities of herbivores and omnivores is vital to diminish the impetus to enhance feeds by fish food and fish oil. Replacing vegetable oils for fish oils in stream fish diets is theoretically likely: the n-3 fatty acids in fish oil are not vital in such diets (Bell 1998), but the fatty acid scenario and thus taste and marketability may be pretentious (Morris et al. 1995; Steffens 1997). Furthermore, mostly herbivorous fish seem to have extra healthy immune structures when fish oil is included in their diet (Pilarczyk 1995).

## 11.5 Innovations in Aquaculture

The period in which aquaculture has been required to face numerous trials including managing diseases, developing brood stock, water quality, and suitable feed expansion is coming to an end. Aquaculture biotechnology now offers the possibility to increase these experiments to aid in augmenting financial feasibility and efficiency in some production subdivisions.

### *11.5.1 Reproduction*

The hereditary source technologies are still far behindhand in assessing the contribution from aquatic animals compared to the plant and cattle subdivisions. Genetic and biotechnology claims show great promise in improving the potential production to reach environmental sustainability, but only a minor percentage of the marine animals consumed have been improved through the genomic programs (Gjedrem 1997). Biotechnological techniques are practical to improve the degree of imitation and initial growth achievement of cultivated aquatic organisms, and in addition, aid in increasing gamete production and obtainability of fry. In Asia the principles of carp and tilapia are being aided by inheritance examination in some zones that comprise gene sequencing and expansion of precise genetic markers (Cheng et al. 2001).

#### 11.5.1.1 Monosex Culture or Neo-female Technology

This farming exercise is created in fish culture by creating all the male and female populations on the foundation of sex for their effective development and food alteration ratio. Monosex culture is carried out by classifying the sexual dimorphism of the species, and the cultures are fulfilled only when the looked-for market is available. In most areas monosex culture of the carp and salmon female populace and prawn freshwater male inhabitants is accepted as exploiting the production level. In India, freshwater fishes have gained importance, and it is noticed that males reach market size in greater abundance in comparison to females. With progression in these methods, recently a standardized technology known as "neo-female technology" has been established in which sex reversal of males is completed to attain females that consequently will yield all male progeny. The sex of the adolescent

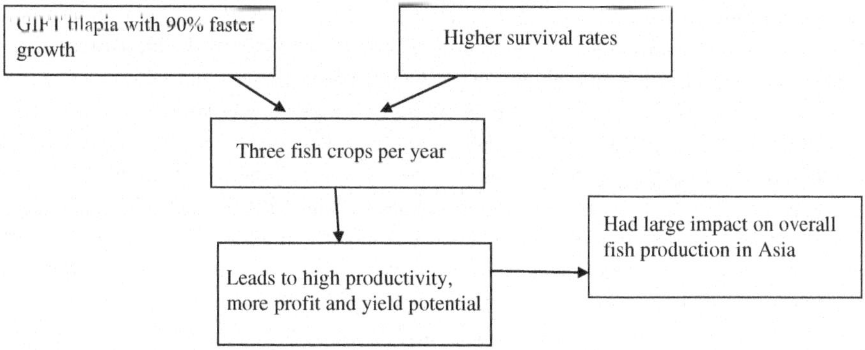


**Fig. 11.1** Programmed purpose of genetic improvement of farmed tilapia (GIFT)

males is altered in this technology by micro-surgically eliminating the androgenic gland or by reducing androgen, which results in female expansion, and when mating with a common male gives rise to all male offspring (Amir 2013). There are some rewards of this method, such as a healthier food change ratio and growth level.

#### 11.5.1.2 GIFT (Genetic Improvement of Farmed Tilapia)

In India, this program has the purpose of the genetic enhancement of cultivated fish species. This scheme was created on hybrids of Nile tilapia, and its strains, targeted for the enlargement of pure-bred lines and enhanced cleansing, are disseminated to the farmers around the area. The better tilapia from this project are still under assessment by fisheries scientists; thus, this project has not reached a completely profitable stage (Fig. 11.1).

### *11.5.2 Disease Management*

In recent times, infectious diseases are the most overwhelming difficulty in fish farming and are mainly responsible for threats to aquaculture sectors. Several new diseases are emerging in aquaculture, which is also a major concern. Further, most of the conventional procedures of controlling diseases in aquaculture are ineffectual, such as the use of chemotherapeutants for new pathogens, mainly viruses; thus, molecular technologies are being developed to be more effective for pathogen identification, information that is anxiously awaited. These molecular biology methods are important for showing the expansion of diseases and then providing procedures for disease control. These methods are also known to offer some treatments for disease, such as DNA vaccines (Bedier et al. 1998). In adding to broadcasting for pathogens, biotechnological techniques are used for determining other fitness limits in aquaculture, such as leukocrits, blood infection, hematocrits, and phagocytic purposes; some methods have been established for these determinations that are useful for the examination of immunoglobulin, lysozymes, cortisol, and plasma (Bachere et al. 1995; Munoz et al. 2000; Nadala and Loh 2000; Shelby et al. 2001). The aquatic organisms are used for some other purposes as well as food production; as they are frequently altered to survive dangerous surroundings, they have a significant part in these investigations as distinctive representations for learning biological and physiological processes to improve our knowledge of disease management and pathogenesis (Wright et al. 2000). At the present time, communicable disease is the main problem in shrimp culture, which is representative of the continuing threats to aquaculture organisms. Further, most of the predictable techniques of regulatory control of these diseases, such as the usage of chemotherapeutants, are known to be ineffective for most pathogens, notably viruses; thus, for these determinations biotechnological and molecular methods provide important means for controlling the diseases. In shrimp brood stock organizations, the programmed production of precise pathogen-free and detailed pathogen-resistant stocks is the chief purpose (OIE 2000, 2001). For the provision of shrimps that are precise pathogen free, those animals are designated which are free of pathogens, and their descendants are raised in controlled hygienic circumstances. The shrimps that are specific pathogen free (SPF) are of great worth for replenishing stocks in the countries or areas wherever disease eruptions have occurred. Detailed pathogen-resistant (SPR) shrimps are formed by breeding from those who have lived through contagions in response to detailed pathogens. These SPR shrimps have great possibilities to be recycled in water bodies in which the precise disease is widespread, but they are unsuitable for usage in nonendemic water bodies because they might transmit subclinical contamination (Bedier et al. 1998). In vitro culture is another method for the discovery and separation of pathogenic viruses, and intracellular bacteria likewise have remained established for numerous crustacean and molluscan organisms (OIE 2000; Groff and La Patra 2000; FAO and NACA 2001). Numerous investigators have established main cell cultures, but maintenance of most of these is unsuccessful, as is also true for mollusks (Cheng et al. 2001).

## 11.5.3 Feed Technologies

### 11.5.3.1 Biofloc Technology

Biofloc technology is based on the technique that emphasizes cumulative water quality by balancing nitrogen and carbon in the aquaculture system (Crab et al. 2012). This technique includes the holding of waste, which is changed into biofloc, which are collections of bacteria, algae, protozoans, and particulate matter such as uneaten food, as a usual food within the aquaculture organization. In this system extra carbon is added, providing an exterior source for carbon in the feed. The biofloc technology delivers a new farming approach as it encourages nitrogen uptake and is also liable for immobilization of ammonium by the bacteria that reduce ammonium concentration more quickly than the nitrification; thus, it is widely used and more maintainable in numerous methods (Avnimelech 2009).

### 11.5.4 Integrated Fish Farming

Integrated fish farming methods involve two or more rural cultivation-related events in which farming is the chief constituent in at least one. Thus, when fish farming is the most important portion of the method, it is called mixed fish agriculture (Ayyappan 2011). Involving fish agriculture with rural features or with animal husbandry is known to be a sustainable farming approach, which is answerable for posing superior productivity in the consumption of possessions and as well shows a significant part in decreasing the hazards of expanding crops. In this classification the excess goods found from the modest structure develop input to the fish culture systems, such as the detritus from agricultural harvests, that is, rice bran, rice polish, and flour, that can be used in diverse administrations as fish feed in aquaculture operations. Thus, combined fish farming aids cumulative food production by enhancing the usage of obtainable natural resources (Jena et al. 2017).

### 11.5.5 Aquaponics Systems

Aquaponics is a current classification that is obligatory for food production by symbiotically merging aquaculture and hydroponics composed in stable recirculatory surroundings. In this system, nutrient-rich water is used as a liquid fertilizer from fish tanks for hydroponic production beds. The nutrients then increase to poisonous levels in the tanks that would harmfully affect the growing of fishes. Aquaponics permits important reduction in the use of water and also delivers an artificial percolation system to fish culture surroundings; in addition, there is a constant flow of nutrients and water to the plants (Azad and Salam 2016).

### 11.5.6 Recirculation Aquaculture System (RAS) Technology

In this technology, aquatic organisms are cultivated with the least water usage, which is renewed successively. This organization is terrestrially based, assists in improving food security, and has a significant effect in reducing ecological effects. In this system sequences of care procedures are completed that are answerable for eliminating organic and other substantial oxygen challenges, for example, suspended solid resources, nutrients, fats, and oils, so that the water can be recycled safely. The larger suspended materials, such as debris and floating matter, for example, wood, paper, and rags, are mostly detached by transition as the wastewater passes through mechanical sieves such as settlement tanks and sand filters; through this course most of the solid materials are removed. After this, the water is purged with biofilters where the liquid is passed over a bacteriological development film comprising bacteria, fungi, protozoa, and algae; the current sifting average is approximately 0.2–2.0 mm thick. These microbes help in conservation of the organic resources and in cleansing the liquid material; for example, ammonia is changed into nitrate, which is excreted by the fish. Oxygen is vital to the fish for food breakdown and growth. The dissolved oxygen levels can be augmented by such approaches as ventilation and oxygenation. In a recirculation aquaculture system, the pH requirements must be prudently monitored, which are measured by adding sodium acetate or sodium hydroxide, etc. This system has some advantages as it needs only a small quantity of water and is ecologically sound (Jena et al. 2017).

## 11.6 Global Aquaculture

Fish remain one of the highly negotiated food products globally, with extra fish being distributed by worth in emerging countries. Current information by high-level specialists, global establishments, production, and public organizations and legislatures all highlight the marvelous possibilities of the oceans and internal waters nowadays, and consequently even more in upcoming times, to contribute meaningfully to food safety and satisfactory nutrition for worldwide inhabitants, predicted to increase to 9.7 billion by 2050. In 2014, fish collected from aquaculture around the world amounted to 73.8 million tonnes, with an projected primary sale value of US $160.2 billion, containing 49.8 million tonnes of finfish (US $99.2 billion), 16.1 million tonnes of shellfish (US $19 billion), 6.9 million tonnes of shellfish (US $36.2 billion), and 7.3 million tonnes of other marine creatures including frogs (US $3.7 billion). Nearly all fish produced from aquaculture are intended for human consumption, although by-manufactured goods might be utilized for nonfood purposes. Measured at the nationwide level, 35 countries produced more cultivated than wild-caught fish in 2014. This grouping of nations has a joint populace of 3.3 billion, or 45% of the world's population. Nations in this collection include five major producers: India, China, Bangladesh, Viet Nam, and Egypt. Worldwide aquaculture production is led by Asia (89%); China by its own financial records is 62%. According to the most current fisheries data of the United States, the U.S. ranks 16th in aquaculture production, not including seaweed. The list of cultivated species introduced to the United States is headed by Atlantic salmon, shrimp, tilapia, and shellfish (mussels, clams, scallops, and oysters). Asian nations and Ecuador supply most of the shrimp to the U.S. market, and Canada, Norway, and Chile provide most of the trade in Atlantic salmon (Fisheries.noaa.gov, 2019).

## 11.7 Ecological Aquaculture

Ecological aquaculture, the farming of vital aquatic proteins essential to human strength, permanency, and public sustainability, is an essential portion of our mutual environmental knowledge and social legacy, a vital part of our past, and a vigorous aspect of our upcoming development as a urban species living in concord with the

Earth's precious, multifaceted aquatic environments. Environmental aquaculture is another aspect of aquaculture expansion that not only brings the methodological aspects of ecosystems design and ecological values to aquaculture but also integrates—at the outset—social environmentalism, preparation for community progress, and concerns for the broader economic, social, and environmental conditions of aquaculture. Ecological aquaculture proposes and calculates both the financial and the societal profit of aquaculture. It uses the various methods of science and practices of environmental and societal ecology to better prepare aquaculture as a way for sustainable society growth and functional shorelines (Costa-Pierce 2002, 2003, 2008). Ecological aquaculture proposes, designs, advances, demonstrates, and evaluates aquatic agriculture ecosystems that protect and improve the form and roles of the environmental and societal environments in which they are situated. Ecological aquaculture fish farms are "aquaculture ecosystems."

Aquaculture hinges on efforts related to numerous food, handling, shipping, and other functions of the world (Table 11.2). In good turn, aquaculture ecosystems can harvest valued product, unpolluted by rubbish waters and fish litter, which can be significant contributions to biologically designed marine and continental ecological agricultural methods. These combined food manufacture methods can be planned at all levels of civilization as communally answerable industries, institutes, private farms, or society-based operations. Biological aquaculture also uses the "aquaculture toolbox" to show vigorous functions in nonfoods, ecological rehabilitation, renovation, and enrichment (Costa-Pierce 2002). Ecological aquaculture takes a worldwide opinion, adding ecological discipline and distribution of technical evidence in an urbane, expertise-based way, endorsing novelty and efficacy in the worldwide market by containing communal and ecological costs, not expressing them (Culver and Castle 2008). Thus, ecological aquaculture tactics are intended

**Table 11.2** Total and per capita apparent fish consumption by region and economic grouping, 2015 (FAO 2018)

| Region/economic grouping | Total food fish consumption (million tonnes live weight equivalent) | Per capita food fish consumption (kg/year) |
|---|---|---|
| World | 148.8 | 20.2 |
| World (excluding China) | 92.9 | 15.5 |
| Africa | 11.7 | 9.9 |
| North America | 7.7 | 21.6 |
| Latin America and the Caribbean | 6.2 | 9.8 |
| Asia | 105.6 | 24.0 |
| Europe | 16.6 | 22.5 |
| Oceania | 1.0 | 25.0 |
| Developed countries | 31.4 | 24.9 |
| Least developed countries | 12.0 | 12.6 |
| Other developing countries | 105.4 | 20.5 |
| Low-income food-deficit countries | 20.8 | 7.7 |

not only for seafood production and financial revenue but also for communal income by emerging societal capital and societal systems that endorse commercial, learning, and communal stewardship performances (Costa-Pierce 2010).

## 11.8 Integrating Aquaculture, Providing Sustainable Seafood, and Restoration

The essential mixing of aquaculture and fisheries development for aquaculture growth is not well integrated into the inclusive planning for maintainable seafood provision, the fisheries, and seaside region management. There are, however, vigorous influences among imprisonment aquaculture and fisheries not only in the biological kingdom but also with fisheries administration and finances, in seafood marketplaces, and in numerous communal events. Although detention fisheries and aquaculture processes are investigated, planned, and achieved as if they were self-governing objects, together they share several fears about water class, genetic variation in hatchery-raised creatures, provender, and the sustainability of the fish meal/oil fisheries. Aquaculture and capture fisheries are indissolubly tangled, a fact supremely perceptible in the seafood market, where, for instance, "white fish" such as tilapia, cod, and haddock are similarly handled. There are also considerable connections among seizure and refined marketplaces, and charge and capacity rivalry among fisheries and aquaculture goods happens regularly in the modern market. Rivalry from imports and from reestablished or seasonally plentiful capture fisheries desirable as a random controlling arrangement are significant reasons why marine fish aquaculture has not been accorded any important courtesy by aquaculture promoters in the United States. Preparation for sustainable seafood supplies for the world needs to include the close connections of aquaculture and fisheries planners together. Some interfacing occurs currently, but conjoined these disciplines can alter strategy choices that could provide more sustainable native food production and creation of jobs. Ensuring viable seafood goods includes development for more than aquaculture because aquaculture is one of the least of five other ways of distributing seafood goods to society: (1) by post-capture fisheries, (2) by cumulative ingestion of underutilized fish, (3) by using bycatch, (4) by increasing administration efficiencies, or (5) by increasing imports. Increasing consequences are a feasible choice for the republics such as the United States and also the European Union, but it is problematic if this level of globalization is maintainable and can endure, particularly as the period of "peak oil" arrives and fuel values continue to increase (UK Energy Research Centre 2009). The UK Energy Investigation Centre (2009) information states that peak oil availability might be reached by 2030 and that people might take up 1228 of the valued 2000 billion containers previously used of the "ultimate recoverable reserve." Worldwide capture fisheries have been decreasing meanwhile since the mid-1980s, and numerous significant food fisheries are overfished (Pauly et al. 2003). However, despite all the undesirable media and buildup that the "collapse" of worldwide capture fisheries has been established, the truths are that capture fisheries are not "dead" ubiquitously, either worldwide or locally, and that the Earth's fisheries administrators are occupied with commerce and administration everywhere, in numerous circumstances, healthier than previously, to reestablish seizure fisheries. Capture fisheries make available the main source of aquatic protein food for life and will do so in the upcoming years. As a consequence, they will endure to afford considerable worth and capacity to the struggle for aquaculture for generations, particularly in the "white fish" seafood marketplaces. Worldwide, the FAO (2009) stated that overall capture fisheries have remained steady during the previous period at about 82 million tons, with 92 million tons from seawater and 10 million tons from internal waters. The entire quantity of abused supplies checked by the FAO has continued stable at about 50% from the mid-1970s to 2007. The proportion of damaged and exhausted stocks has become stable at 25–30% from the mid-1990s to 2007. Of the stocks examined by FAO, 2% were underexploited, 18% were temperately manipulated, 52% were completely manipulated, and 28% were harmed. Capture fisheries are also not "deceased" in the United States. The Countrywide Marine Fisheries Assistance (2009) reported that of the 251 fish reserves or stock improvements it measures, 210 (84%) are not as overfished. Every main fishery in the United States that is overfished is a focus of a severe and frequently contested stock retrieval strategy. Although capture fisheries construction is unlikely to increase but can be continued into the upcoming years at present levels, there are three other choices for organizers for sustainable seafood supplies: (1) increasing the consumption of underused fish, (2) using bycatch as food, and (3) cumulative administration of productivities (Costa-Pierce 2010).

## 11.9 Conclusion

Several biotechnological inventions in aquaculture, along with developments in technology, indicate an optimistic attitude toward aquacultural assets, global expertise, and diversity achievement. Also involved is new and unconventional expertise in aquaculture, providing for fast-growing animals of good physical size that can be used for nutritional resolution, with complete processes that contribute to ecological well-being. Thus, increase will mainly be influenced by the exchange of information among the experts, investigators, and developers from various areas of the state concerning potential harm and advantages, which assists in refining the sector and allowing further emergence of the sustainable use of aquatic animals.

## References

Abad E, Palacio F, Nuin M, González de Zárate A, Juarros A, Gómez JM, Marco S (2009) RFID smart tag for traceability and cold chain monitoring of foods: demonstration in an intercontinental fresh fish logistic chain. J Food Eng 93:394–399

Amir S (2013) Monosex culture of prawns through androgenic gene silencing. Infofish Int 10(4):22–24
Avnimelech Y (2009) Biofloc technology — a practical guide book. World Aquaculture Society, Baton Rouge, p 182
Ayyappan S (2011) Hand book of fisheries and aquaculture. Indian Council of Agricultural Research, New Delhi, pp 439–448
Azad KN, Salam MA (2016) Aquaponics in Bangladesh: current status and future prospects. J Biosci Agric Res 7(2):669–677
Bachere E, Mialhe E, Noel D, Boulo V, Morvan A, Rodriguez J (1995) Knowledge and research prospects in marine mollusc and crustacean immunology. Aquaculture 132(1–2):17–32
Baily C (1997) Aquaculture and basic human needs. World Aquacult 27:28–31
Bedier E, Cochard JC, Le Moullac G, Patrois J, Aquacop (1998) Selective breeding and pathology in penaeid shrimp culture: the genetic approach to pathogen resistance. World Aquacult 29(2):46–51
Bell JG (1998) In: Black KD, Pickering AD (eds) Biology of farmed fish. Sheffield Academic Press, Sheffield, pp 114–145
Burnell G, Allan G (2009) New technologies in aquaculture: improving production efficiency, quality and environmental management. Woodhead Publishing Limited, Oxford
Cheng TC, LaPeyre JF, Buchanan JT, Tiersch TR, Cooper RK (2001) Cryopreservation of heart cells from the eastern oyster. In Vitro Cell Dev Biol Anim 37(4):237–244
conserve-energy-future (2019). https://www.conserve-energy-future.com/aquaculture-types-benefits-importance.php. Accessed 19 Aug 2019
Costa-Pierce B (2002) Ecological aquaculture. Blackwell Science, Oxford
Costa-Pierce BA (2003) Use of ecosystems science in ecological aquaculture. Bull Aquacult Assoc Can 103(2):32–40
Costa-Pierce B (2008) Epilogue. In: Culver K, Castle D (eds) Aquaculture, innovation and social transformation. Springer Science, New York, pp 315–325
Costa-Pierce BA (2010) Sustainable ecological aquaculture systems: the need for a new social contract for aquaculture development. Mar Technol Soc J 44(3):88–112
Crab R, Defoirdt T, Wille M, Bossier P, Verstraete W (2012) Biofloc technology in aquaculture: beneficial effects and future challenges. Aquaculture 356:351–356
Culver K, Castle D (2008) Aquaculture, innovation and social transformation. Springer, New York
El-Sayed AM (1999) Alternative dietary protein sources for farmed Tilapia, Oreochromis spp. Aquaculture 179:149–168
FAO (2012) The State of World Fisheries and Aquaculture 2012. FAO, Rome, 209 pp
FAO (2014) The State of World Fisheries and Aquaculture 2014. FAO, Rome, 223 pp
FAO (2018) The State of World Fisheries and Aquaculture (SOFIA) 2018. FAO, 88 pp
FAO and NACA (2001) Asia diagnostic guide to aquatic animal diseases. FAO Fish. Tech. Pap. No. 402/2, 237 pp
Fisheries.noaa.gov (2019). https://www.fisheries.noaa.gov/national/aquaculture/global-aquaculture. Accessed 19 Aug 2019
Food and Agriculture Organization (2009) The State of World Fisheries and Aquaculture 2008. Fisheries Department, Food and Agriculture Organization of the United Nations, Rome
Gjedrem T (1997) Selective breeding to improve aquaculture production. World Aquacult 28:33–45
Groff JM, La Patra SE (2000) Infectious diseases impacting the commercial culture of salmonids. J Appl Aquacult 10(4):17–90
Handbook of Fisheries Statistics (2015) Department of Animal Husbandry, Dairying and Fisheries, Ministry of Agriculture, Government of India, New Delhi
Hasan MR (2001) Nutrition and feeding for sustainable aquaculture development in the third millennium. In: Aquaculture in the third millennium. Technical proceedings of the conference on aquaculture in the third millennium, pp 193–219
IMTA. https://www.google.com/search?q=IntergratedMultitrophitc+Aquaculture&hl=en&source=lnms&tbm=isch&sa=X&ved=0ahUKEwjumvbY8azkAhXYZCsKHba8D58Q_AUIEigC&biw=1366&bih=657#imgrc=qsS7BLEgB-2mDM. Accessed 19 Aug 2019

Jedermann R, Ruiz Garcia L, Lang W (2009) Spatial temperature profiling by semi-passive RFID loggers for perishable food transportation. Comput Electron Agric 65:145–154
Jena AK, Biswas P, Saha H (2017) Advanced farming systems in aquaculture: strategies to enhance the production. Innov Farm 2(1):84–89
Liao IC, Chao NH (1997) Developments in aquaculture biotechnology in Taiwan. J Mar Biotechnol 5:16–23
Martín-Sánchez AM, Navarro C, Pérez-Álvarez JA, Kuri V (2009) Alternatives for efficient and sustainable production of surimi: a review. Compr Rev Food Sci Food Saf 8:359–374
Morris CA, Haynes KC, Keeton JT, Gatlin DM (1995) Fish oil and dietary pelagics and flavor of channel catfish. J Food Sci 60:1225–1227
Munoz M, Cedeno R, Rodriguez J, van der Knapp WPW, Mialhe E (2000) Measurement of reactive oxygen intermediate production in haemocytes of the penaeid shrimp, *Penaeus vannamei*. Aquaculture 191(1–3):89–107
Nadala ECB Jr, Loh PC (2000) Dot-blot nitrocellulose enzyme immunoassays for the detection of white spot virus and yellow head virus of penaeid shrimp. J Virol Methods 84:175–179
Naylor RL, Goldburg RJ, Primavera JH, Kautsky N, Beveridge MC, Clay J, Troell M (2000) Effect of aquaculture on world fish supplies. Nature 405(6790):1017
OIE (2000) Diagnostic manual for aquatic animal diseases, 3rd edn. Office International des Epizooties, Paris, 237 pp
OIE (2001) International aquatic animal health code, 4th edn. Office International des Epizooties, Paris, 155 pp
Pauly D, Alder J, Bemmet E, Christensen V, Tyedmers P, Watson R (2003) The future of fisheries. Science 203:1359–1361
Pilarczyk A (1995) Changes in specific carp immune reaction caused by addition of fish oil to pellets. Aquaculture 129:425–429
Shelby RA, Shoemaker CA, Evans JJ, Klesius PH (2001) Development of an indirect ELISA to detect humoral response to *Streptococcus iniae* of Nile tilapia, *Oreochromis niloticus*. J Appl Aquacult 11(3):35–44
Steffens W (1997) Effects of variation in essential fatty acids in fish feeds on nutritive value of freshwater fish for humans. Aquaculture 151:97–119
Suuronen P, Chopin F, Glass C, Løkkeborg S, Matsushita Y, Queirolo D, Rihan D (2012) Low impact and fuel efficient fishing—looking beyond the horizon. Fish Res 119–120:135–146
Tacon AJ (1997) Contribution to food fish supplies. FAO Fish Circ 886(Rev. 1):17–21
Tacon AJ, De Silva SS (1997) Feed preparation and feed management strategies within semi-intensive fish farming systems in the tropics. Aquaculture 151:379–404
Tingman W, Jian Z, Xiaoshuan Z (2010) Fish product quality evaluation based on temperature monitoring in cold chain. Afr J Biotechnol 9(37):6146–6151
UK Energy Research Centre (2009) Global Oil Depletion Report (2009). www.ukerc.ac.uk/support/Global%20Oil/%20Depletion
UN (2015) World population prospects – the revision. New York
Wright J Jr, Bonen A, Conlon JM, Pohajdak B (2000) Glucose homeostasis in the teleost fish *Tilapia*: insights from Brockman body xenotransplantation studies. Am Zool 40(2):234–245

# Chapter 12
# Current Trends and Aspects of Microbiological Biogas Production

**Chayanika Putatunda, Abhishek Walia, Rashmi Sharma, and Preeti Solanki**

**Abstract** The whole world today is practically on the verge of severe energy crisis. The industrialization and modernization have improved the overall living conditions of a segment of human population on one hand and also resulted in greater energy requirements as well as a huge burden on the already dwindling land resources. Again, sustaining the ever-increasing human population with the ever-decreasing arable land is becoming a mammoth task. So, there is an urgent need to try to look for possible solutions. Biogas production or biomethanation can be an answer to the twin problem of food and energy since the process leads to generation of biofuel (biogas) as well as effluent slurry, which can act as a very good manure. Another, very significant advantage of the process is that it is very helpful in solid waste management. A wide variety of waste materials like animal excrements, sewage sludge, agricultural residues, industrial wastes, etc. can be used as substrate for biogas production. Moreover, the solid-state biomethanation is also gaining popularity due to negligible water requirements. This chapter presents some of the major developments in the field of biogas production.

**Keywords** Biogas · Methane · Microbiological diversity · Microbial decomposition · Anaerobic digestion · Digestors

C. Putatunda (✉)
Om Sterling Global University, Hisar, Haryana, India

A. Walia
Department of Microbiology, CSK HPKV, Palampur, Himachal Pradesh, India

R. Sharma
Department of Microbiology, DAV University, Jalandhar, Punjab, India

P. Solanki
Multidisciplinary Research Unit (MRU), Pt. BD Sharma PGIMS, Rohtak, Haryana, India



J. Singh et al. (eds.), *Microbial Biotechnology: Basic Research and Applications*, Environmental and Microbial Biotechnology,
https://doi.org/10.1007/978-981-15-2817-0_12

## 12.1 Introduction

Energy and population explosion are two of the major issues being faced by the whole world today. The energy needs are increasing day by day due to the ever-increasing population as well as greater industrialization and modernization (Amigun et al. 2008; Li et al. 2009; Sárvári Horváth et al. 2016). However, we are still heavily dependent on fossil fuels for the energy requirements, which in turn are depleting on an alarming rate (Shafiee and Topal 2009; Nel and Cooper 2009; Abas et al. 2015). Also, these fossil fuels like coal, petroleum products, etc. apart from being nonrenewable are also major source of environmental pollution and are not only impacting the environment but also gravely affecting the human health (Smith et al. 1999; Höök and Tang 2013; Watts et al. 2015; Lelieveld et al. 2015; Perera 2017). Apart from that, the agriculture sector is also under extreme pressure due to the food requirements of the growing human population (FAO 2017). On the other hand, the developmental measures and urbanization are also taking a great toll on the available land resources, thus diminishing the already dwindling arable land (Agus and Irawan 2006; Beesley and Ramsey 2009; Azizan and Hussin 2015). Thus, the farmers are relying on more energy-intensive agricultural practices as well as huge amounts of chemical fertilizers and pesticides which in turn is having a grave impact on the environment (Tilman et al. 2002; Cold and Forbes 2004; Moss 2008; Alamdarlo 2018; Budzinski and Couderchet 2018). Not only this, the greater human activities are also leading to a very large quantity of waste materials which are again posing a serious threat to the environment (Dyson and Chang 2005; Giusti 2009; Fu et al. 2015). So, these problems are very intricately related with each other. Hence, there is an urgent need to address these issues.

A potential solution to these problems is the process of biomethanation. The process involves anaerobic digestion of organic materials, carried out by several groups of microorganisms, leading to production of methane and some other gases in the form of biogas (Dieter and Angelika, 2008). As per Martins das Neves (2009), "The energy content of 1.0 $m^3$ of purified biogas is equal to 1.1 L of gasoline, 1.7 L of bioethanol, or 0.97 $m^3$ of natural gas." Biogas can be used as a fuel for cooking and heating purposes as well as for electricity generation (Xiaohua and Jingfei 2005; Riva et al. 2014; Chang et al. 2015). There is also the possibility of using it as a transportation fuel (Börjesson and Mattiasson 2008; Holm-Nielsen et al. 2009). Moreover, biogas production also yields an effluent slurry which is known to be rich in several plant nutrients (Leela Wati et al. 2008). The effluent can be applied as an organic manure which improves the soil quality as well as plant growth. Sogn et al. (2018) have reported that the biogas plant digestate proved to be a good source of the major plant nutrients like nitrogen, phosphorus, and potassium, and at least for the wheat crop, the digestate alone was sufficient to overcome the requirement of any chemical fertilizer. Diatta et al. (2019) reported that 1:1 mixture of phyto-ash and biogas slurry was found to be effective in improving the quality of soil contaminated with heavy metals. Similarly, Nafees et al. (2018) reported improvement in growth as well as antioxidant properties of *Brassica napus*, grown in chromium-contaminated soil by addition of biogas slurry and *Burkholderia phytofirmans* PsJN. The anaerobic conditions occurring during biomethanation also tend to decrease the load of pathogenic microbes, so the safety concerns associated with the direct soil application of waste materials are also mitigated (Weiland 2010).

Conventionally, cattle dung has been used as the most common substrate for biogas production (Işık and Polat 2018). However, a variety of waste products (either alone or mixed with other wastes) have been reported to be effective for biogas production. Some of the substrates used for biogas production are the excretory materials of various animals like sheep (Sarabia Méndez et al. 2017), goat (Zhang et al. 2013), pigs (Wu et al. 2010), camel (Kheira et al. 2017), and poultry birds (Malik et al. 2008); human excreta (Singh et al. 1993); kitchen wastes (Iqbal et al. 2014; Srinvasa Reddy et al. 2017); agricultural wastes like bagasse (Eshore et al. 2017), wheat straw (Mancini et al. 2018) etc.; industrial wastes like whey (Antonelli et al. 2016); paper industry wastes (Priadi et al. 2014); palm oil industry wastes (Ohimain and Izah 2017); food processing industry wastes (Fang 2010), etc. So, biomethanation can be a boon for both industrial and agricultural sectors, since the industries can use biogas as a possible source of energy and at the same time, the process is helpful in dealing with their waste products. Thus, apart from solving the energy as well as food problem, biogas production is very useful in addressing the environmental issues also. García-González et al. (2019) have concluded that by harvesting the methane in the form of biogas, one can expect mitigation of the greenhouse gas emission. Thus, this can be helpful in reducing the phenomenon of global warming.

Biogas conventionally comprises approximately 60–75% methane, 25–40% carbon dioxide, and traces of other gases like water vapor, hydrogen sulfide, ammonia, etc. (McKendry 2002; Zinoviev et al. 2010). Out of these, methane is the ingredient which acts as the fuel. It is a clean burning fuel and has a high calorific value (Kaltschmitt et al. 2001). Reports have already indicated improvement in health conditions of people working in kitchens supplied with biogas as compared to kitchens where cattle dung cakes, coal, or wood is used as fuel (Dohoo et al. 2012). However, in spite of all these benefits, the process of biomethanation has not gained the desired level of popularity. Lack of awareness among the general public as well as limitation of funds may be the reason for this, but there are some areas of concern with respect to biogas production technology (Surendra et al. 2014). For example, the relatively low efficiency of biogas production especially in areas with too high or too low temperatures is an important drawback of this process. Again, another issue is related to the storage and purification of methane from the rest of the constituents of biogas. Even the availability of substrate and labor requirements has also been thought to impact the popularity of biogas production adversely (Tucho et al. 2016). Apart from these, the conventional biogas production technologies also need input of lots of water (Tucho et al. 2016). This especially becomes challenging in arid and semi-arid regions of the globe.

In order to address these types of concerns, research efforts are being undertaken. Scientists are working on improving the overall process efficiency by modifying the digester designs as well as the process parameters. Similarly, efforts are going on for developing microbial consortia which can not only improve the yield (Zhong et al. 2016; Krzysztof et al. 2016) but also can work efficiently under relatively harsh temperature ranges (Hniman et al. 2011; Kinet et al. 2015). Apart from that, several researchers have shown that by combining different wastes as substrates for biogas production, the process efficiency can be improved (Li et al. 2011a; Munda et al. 2012; Tasnim et al. 2017). Again, several workers have also developed various techniques for improved purification of biogas. Another major step in popularizing the biogas production technology is by the application of solid-state biogas production which minimizes the amount of water requirement for biomethanation (Brown et al. 2012; Brown and Li 2013). These steps are discussed in detail in the following sections.

## 12.2 Conventional Biogas Production

Recent studies have suggested that biogas production using anaerobic digestion (AD) processes delivers compelling edge over the rest of the sources of bioenergy as AD is not only an energy-efficient but also an environmental-friendly technique (Nishio and Nakashimada 2007; van Foreest 2012). Many countries like the USA, China, and India have recently been investing in alternative processes for the production of biogas from cellulosic resources, and they are going to be the future producers of biogas (Lin and Tanaka 2006; Soetaert and Vandamme 2009). The AD technology has been enhanced from the knowledge of the production of compost which is a high-value fertilizer and has today given a boon to biogas economy (European Biogas Association 2011).

### *12.2.1 Principle of Anaerobic Digestion*

AD is based on microbial decomposition of organic matter in the absence of air/oxygen, utilized for metabolism in microbes, leading to their growth and resulting in production of methane. The process can be divided into four characteristic phases depending upon the group of microorganisms involved (Fig. 12.1).

Effective regulation of the AD process is necessary for stable digestion and is maintained by coupling the various biological conversions, i.e., hydrolysis, acidogenesis, acetogenesis, and methanogenesis, that prevents the accumulation of any intermediary compounds. This also helps in the maintenance of physical conditions of digestion like pH which may be affected by the concentration of intermediary VFAs, leading to decrease in pH, which inhibits methanogenic bacteria, and further decreases the pH.

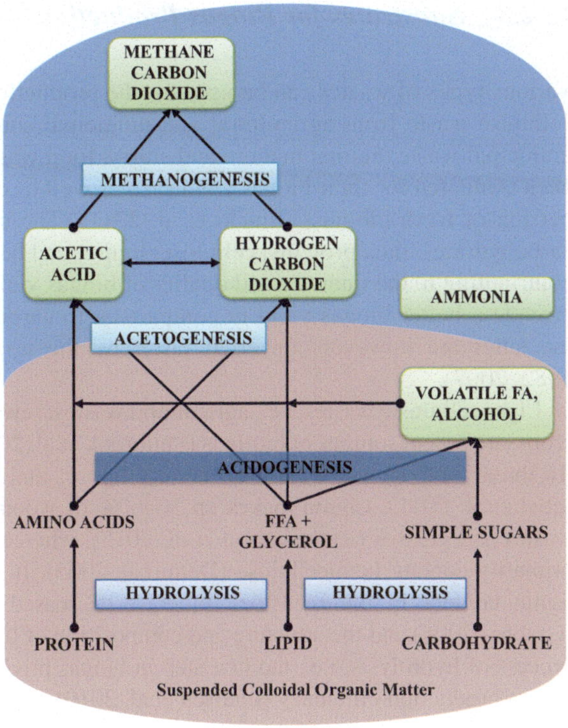


**Fig. 12.1** Principle of anaerobic digestion process: (a) hydrolysis: nonsoluble biomolecules (protein, lipids, carbohydrates) present in the colloidal organic matter of the waste are converted to simpler and soluble organic compounds. (b) Acidogenesis: soluble organic compounds → volatile fatty acids (VFA) + $CO_2$. (c) Acetogenesis: VFA → $CH_3COO^-$ + $H_2$. (d) Methanogenesis: conversion of acetate and $CO_2$ plus $H_2$ to methane gas

### *12.2.2 Types of Biogas Systems*

AD technology is applicable to both wastewater and solid waste treatments with the final product being methane and carbon dioxide. The installation required for AD is both simple and economical in terms of technology, energy, and space and can be divided into two types based on biomass retention (de Mes et al. 2003):

(a) High-rate biogas systems: These systems are characterized by biomass retention and short hydraulic (HRT) but long sludge retention times (SRT) and maybe utilized for many types of wastewater treatments. Examples are anaerobic filter, contact process, upflow anaerobic sludge blanket (UASB) or expanded granular sludge bed (EGSB), and fluidized bed systems. The biomass retention may be possible in two ways:

  (i) Bacterial films are fixed on solid surfaces.
  (ii) Bacterial mass is suspended and biomass is retained by settling process internally or externally.

(b) Low-rate biogas systems: These systems do not retain biomass and are used for solid waste treatment and digestion of slurries. They have long enough HRT to be equal to SRT. Examples are continuously stirred tank reactor (CSTR) system, plug flow system, and batch accumulation system.

### 12.2.3 Substrates for Biogas Production

Various types of wastes can be used for the production of biogas, like the lignocellulose waste from agricultural and municipal sources, sewage sludge, solid municipal waste, animal manure and slurry, food wastes, etc. These wastes have been evaluated for their biogas yield ($m^3$) as well as electricity produced (kW-h) per ton of fresh biomass (Stucki et al. 2011). The molecular constituents, i.e., carbohydrates, fats, proteins, cellulose, lignin, and hemicellulose, are the factors contributing to the quantity and quality of biogas yield. It is also known that fats provide a better biogas yield in comparison to carbohydrates and proteins, but the latter are however converted faster by the microbes (Zubr 1986; Braun 1982, 2007).

Lignocellulosic wastes, i.e., agricultural wastes, sewage, and wastes from energy crops, are potent sources of biofuels (Philbrook et al. 2013) and are made of primarily three components—cellulose, hemicellulose, and lignin (Kumar et al. 2009; Iqbal et al. 2011). Lignin makes up 30–60% of wood and 5–30% of agricultural wastes and grasses (Ratnaweeraa et al. 2015), whereas wastes from energy crops primarily contain hemicellulose (Demirbaş 2005). It has been shown that higher lignin content in the feedstock biomass decreased the degradation efficiency (Grabber 2005) and the structure and composition of lignin positively influence the process of hydrolysis, i.e., the first step in biogas production, hence increasing the biogas production efficiency (Ladisch et al. 2010).

#### 12.2.3.1 Treatment of Slurries and Solid Wastes

The content of total solids (TS) present in the solid wastes and slurries determines the systems to be used for their digestion. Broadly, either wet fermentation systems or dry fermentation systems are employed (de Mes et al. 2003) (Table 12.1).

#### 12.2.3.2 Wastewater Treatments

High-rate AD treatment systems like UASB, contact process, and anaerobic filter are better suited for dilute feedstock like wastewater. In these systems, SRT > HRT and the sludge is either recycled or fixed on support material. Different types of wastewater treatment systems used worldwide are contact process, upflow anaerobic sludge blanket (UASB), Biobulk system by Biothane (Biothane, 2001), fixed film fluidized bed system, hybrid systems, anaerobic filter (AF), anaerobic fixed film reactor (AFFR), and expanded granular sludge bed (EGSB) (Frankin 2001).

**Table 12.1** Fermentation treatment of slurries and solid wastes

| TS 15–25%<br>Low solids AD | TS > 30%<br>High solids AD |
|---|---|
| Wet fermentation | Dry fermentation |
| *CSTR*<br>- Feed introduced in the reactor is continuously stirred for proper mixing of contents, and an equal volume of effluent exits the reactor<br>- RT can be varied<br>- Low operating cost<br>- TS 2–10%<br>- For treatment of agricultural wastes, animal manure, household waste, sewage sludge, feces, kitchen waste, urine, and/or mixtures of these substrates<br>- SRT = HRT<br>- Digestor volume: 6–400 $m^3$/day | *Valogra system*<br>- Waste is screened and crushed to particle size <80 mm<br>- Crushed waste is mixed with some excess process water and heated via steam<br>- Four high-solids mesophilic reactors are combined with incineration of non-digested matter and residues<br>- Mixing in reactor occurs under pressure by reverse circulation with a small volume of the biogas<br>- Methane content in biogas produced after 18–25 days is 55–60% which may be purified to 97% for commercial use |
| | *DRANCO (Dry Anaerobic Composting) system*<br>- Operates with high solid fraction and high temperatures (50–58 °C)<br>- Feed is daily added from top and digested material removed from base after 15–30 days<br>- Part of digested biomass is recycled to be used as inoculum, and the rest is dried to form organic compost<br>- No mixing within the reactor except the plug flow movement of the digested waste |
| *Plug flow digesters*<br>- Basic design made of an underground trough with expandable gas-tight cover, vertical mixing in pipe<br>- Hydrolysis and methanogenesis occur separately through the pipe length<br>- TS 10–12%<br>- SRT = HRT<br>- For treatment of slurries with high TS fractions<br>- Low loading rates | *Kompogas system*<br>- Reactor is a horizontal cylinder; hence, movement of material in reactor is in a horizontal type plug-flow process; an intermittent agitator is also present inside the reactor<br>- Feed is introduced daily from one end and digested biomass removed from the other end after 20 days<br>- Part of digested biomass is recycled to be used as inoculum, and the rest is sent to AD wastewater treatment to produce more biogas |
| | *BIOCEL process*<br>- Batch process with high solids and mesophilic temperatures<br>- Waste is mixed with the inoculum, sealed into the bioreactor without any stirring, and kept till 21 days for biogas production, i.e., till there is no more methane produced<br>- Leachate formed due to AD is heated for recirculation through waste biomass |

### 12.2.4 Types of Biogas Plants

The construction of a biogas plant depends upon its structural strength and fluid dynamics. The best possible solution may be an egg-shaped vessel which is generally used for treatment of sewage on large scale as it is expensive. Digesters in the shape of a cylinder having conical bottoms and covers are much simplified design to build and maybe available as prefabricated units in the market. They are however unfavorable due to surface volume ratio and must have equal height and diameter. Comparison of different types of small-scale biogas plants and their advantages and disadvantages are summarized in Tables 12.2 and 12.3. The basic designs of fixed dome and floating drum biogas plants are given in Figs. 12.2 and 12.3, respectively.

### 12.2.5 Physical Factors Affecting AD

(a) Temperature: AD starts with 0 °C and increases with increasing temperatures reaching a maximum at 35–37 °C, i.e., ideal conditions for mesospheric microorganisms. Maximum methanogenesis occurs at 55 °C or higher, and the choice of temperature will depend upon the biogas yield and the energy demands (Lettinga and Haandel 1993). Some of the important characteristics of different types of anaerobic digestion processes (based on temperature) are presented in Table 12.4.

**Table 12.2** Comparison of different types of small-scale biogas plants (Biogas Digest Volume II: Biogas—Application and Product Development, 2010)

| Factors | Fixed dome | Floating drum | Tubular design | Plastic containers |
|---|---|---|---|---|
| Gas storage | Gas storage internal; large drum sizes up to 20 $m^3$ | Gas storage internal; small drum sizes | Internal and eventually external plastic bags | Gas storage internal; small drum sizes |
| Gas pressure | 60–120 mbar | ~20 mbar | Low, ~ 2 mbar | Low, ~ 2 mbar |
| Skills of contractor | High; masonry, plumbing | High; masonry, plumbing, welding | Medium; plumbing | Low; plumbing |
| Availability of material | Yes | Yes | Yes | Yes |
| Durability | Very high >20 years | High; weak drum | Medium; depends on chosen liner | Medium |
| Agitation | Self-agitated by the biogas pressure | Manual steering | Not possible; plug flow type | Manual steering |
| Sizing | 6–124 $m^3$ digester volume | ~20 $m^3$ | Combination possible | ~6 $m^3$ digester volume |
| Methane emission | High | Medium | Low | Medium |

**Table 12.3** Advantages and disadvantages of different types of small-scale biogas plants (Biogas Digest Volume II: Biogas—Application and Product Development, 2010)

| Biogas reactor type | Advantages | Disadvantages |
|---|---|---|
| (a) Fixed dome biogas plants | • Initial costs are low<br>• Life span is long and useful<br>• Parts involved are neither rusting nor moving<br>• Compact well-insulated underground basic design which saves space<br>• Opportunity for skilled local employment | • Gas-holder masonry requires special sealants<br>• High technical skills required for gas-tight construction<br>• Gas leaks are frequent<br>• Gas utilization is complicated due to fluctuation in gas pressures<br>• Amount of biogas produced may not be immediately visible<br>• Plant operation is not easily understandable<br>• Exact planning of levels is required, and in bedrock areas, the excavation will be difficult<br>• Environmental disadvantage—methane emission from expansion chamber |
| (b) Floating drum biogas plants or gobar gas plant (by Jashu Bhai J Patel, 1956) | • Mode of operation is continuous feed<br>• Can be used for both animal and human feces<br>• *Water-jacket floating-drum plants:*<br>– Easy to maintain<br>– Universally applicable<br>– Do not stuck in scum layer even with high solid content<br>– Long life<br>– Esthetic and more hygienic<br>– Used in fermentation of night soil | • Steel drum is maintenance intensive and expensive<br>• Life span of drum is short, i.e., 5–15 years only<br>• Gas holder has a tendency to be stuck if fibrous substrates are used<br>• A guide is always required for the drum to be removed for repair |
| (c) Low-cost polyethylene tube/balloon biogas digester | • Easy design and low cost<br>• Easy installation of digester bag, replaceable<br>• Can be installed in areas with high water table<br>• Easy monitoring of gas<br>• Ideal for warmer climates | • Shorter life span than traditional biogas plants<br>• Gas pressure is lower than fixed dome and floating drum biogas plants<br>• Production time is highly dependent on the ambient temperature, less insulated |
| (d) Earth pit plants | • Easy design and low cost<br>• Easy installation<br>• Potential for improvements based on self-help approaches | • Short life span<br>• Can be made only in certain impermeable soils<br>• Can be constructed only above groundwater table |

(continued)

**Table 12.3** (continued)

| Biogas reactor type | Advantages | Disadvantages |
|---|---|---|
| (e) Ferro-cement plants | • Smaller volumes (<6 $m^3$)<br>• Can be prefabricated<br>• Can be applied as an earth pit lining or as self-supporting shell<br>• Reliability is proven if made with cemented-on aluminum foil | • Requires substantial amount of cement<br>• Skillful workmanship is required<br>• Uses expensive wire mesh<br>• Adequately not yet time tested<br>• Requires special sealing measures for gas holder |

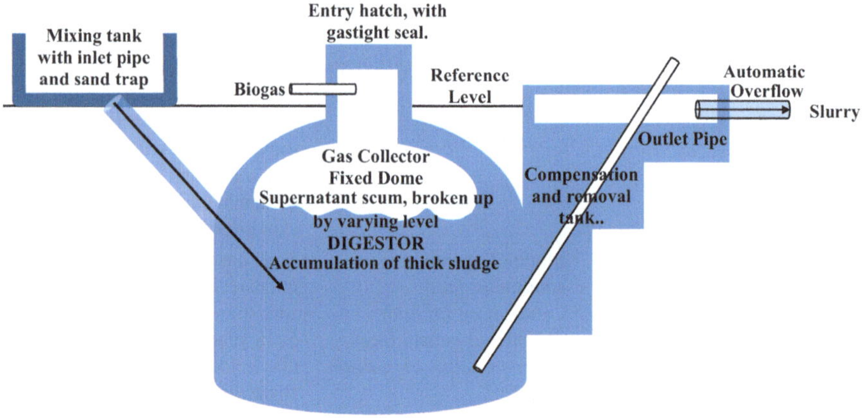


**Fig. 12.2** Basic components of a fixed dome plant (Nicarao design)

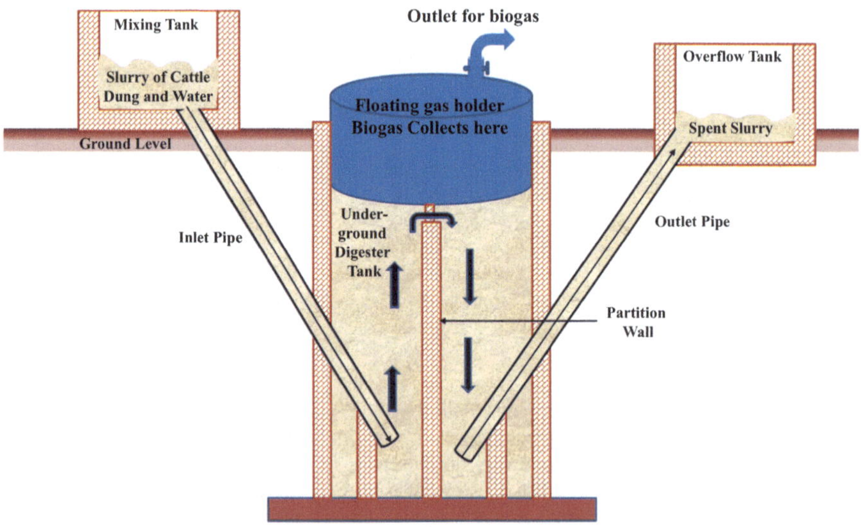


**Fig. 12.3** Floating drum biogas plants or gobar gas plant

Table 12.4 Classification of digestion process based on temperature

| | | | |
|---|---|---|---|
| Psychrophilic | 10–20 °C | Bacterial growth and degradation of the substrate are slower | Requires long retention times and large reactor volumes |
| Mesophilic | 20–40 °C | Moderate bacterial growth and substrate degradation | Lesser retention times and reactor volumes |
| Thermophilic | 50–60 °C | Applicable when wastewater is discharged at high temperature or when pathogen removal is essential | Can be used with high loading rates |

(b) pH: AD start-up processes may occur at a variable pH, but methanogenesis requires neutral pH and is the rate limiting step (Lettinga and Haandel 1993). Hydrogen carbonate ions are required in sufficient amount to maintain the optimal pH for methanogenesis to occur.

(c) Alkalinity and toxicity: Accumulation of intermediaries like VFA and ammonia, or cations like those of sodium, potassium, and calcium, heavy metals, sulfides, and other xenobiotics also adversely affect methanogenesis.

## 12.3 Microbiology of Biogas Production

The process of biogas production from organic matter is a complex and dynamic process and involves intricate interactions between the members of microbial community. The microbial communities of seven different anaerobic sewage sludge digesters were analyzed by using 16S rDNA sequencing (Riviere et al. 2009). They classified the microflora into three categories, viz., the phylotypes common in most of the digesters, the phylotypes found in few digesters, and the phylotypes observed only under certain specific conditions.

The overall anaerobic digestion process has been divided into four major steps, viz., hydrolysis, acidogenesis, acetogenesis, and methanogenesis. During the first step of hydrolysis, the macromolecules like cellulose, starch, hemicellulose, proteins, lipids, etc. present in the organic materials are converted to simple, soluble monomeric compounds like formate, acetate, propionate, butyrate, ethanol, carbon dioxide, and hydrogen (Kelleher et al. 2000). Depending on the type of starting materials, different microbes may become dominant during this phase. According to Rao (1993), cattle dung-fed digesters show higher amylolytic population while the digesters containing poultry waste possess higher proteolytic population. Some of the commonly reported hydrolytic microbes include *Bacteroides succinogenes*, *Butyrivibrio fibrisolvens*, *Clostridium cellobioparum*, *Ruminococcus albus*, *Clostridium* sp., etc. (Khan 1980; Godbole et al. 1981). The small molecules produced by the fermentative and hydrolytic microbes are acted upon by acidogenic microbes which result in the production of volatile fatty acids (VFAs), alcohols, hydrogen, and carbon dioxide along with by-products like ammonia ($NH_3$), hydrogen sulfide ($H_2S$), etc. (Li et al. 2011a, b; De la Torre and Goma 1981; Dinopoulou et al. 1988). Acidogenesis is followed by acetogenesis where organic acids and alcohols are converted into acetate, carbon dioxide, and hydrogen. The obligate hydrogen-producing acetogenic bacteria are one of the most important groups in the biogas digesters. The classical examples of this group of microbes are *Syntrophobacter wolinii* (Boone and Bryant 1980) and *Syntrophomonas wolfei* (McInerney et al. 1981). The most common homoacetogenic bacteria are *Acetobacterium woodii* and *Clostridium aceticum*. These consume hydrogen and carbon dioxide and produce acetate (Balch et al. 1977; Ohwaki and Hungate 1977; Bharathiraja et al. 2018). The last step of anaerobic digestion is carried out by methanogens. There are two broad groups of methanogenic bacteria: aceticlastic bacteria (which convert acetate into methane and carbon dioxide) and hydrogenotrophic methanogens (which convert hydrogen and carbon dioxide into methane). Most of the methanogens belong to the latter group, while the former group is represented by a few like *Methanosarcina barkeri*, *Methanococcus mazei*, and *Methanothrix soehngenii* (Schink 1997). The syntrophic interaction between acetogens and hydrogenotrophic methanogens is very crucial for the performance of an anaerobic digester.

## 12.4 Current Trends in Biogas Production

With the advent of modernization and improvement in the living standards, the energy requirements are skyrocketing. On the other hand, the conventional energy sources are plummeting. So, the world is now trying to look for alternative and renewable energy sources (Nasir et al. 2012). Biomethanation is not a new phenomenon, but nowadays, several improvements or modifications are being done in order to enhance its applicability as a reliable energy source. Some of the approaches are being discussed here.

### *12.4.1 Utilization of Wider Substrate Range*

The substrate feedstocks serve as the nutritional base for the microflora involved in biogas production. The types of nutrients within substrate feedstock determine microbial growth and hence affect degradation process and biogas yield (Cheng 2009; Cooke 2014; De Clercq et al. 2016). Nutrients should be in abundance for an efficient digestion process. Therefore, the substrate should be chosen in a way that meets the nutritional demands of microflora in the digester, produce high biogas and methane yield, and produce a high-quality digestate. Different categories of substrates used for biogas production are depicted in Fig. 12.4, and these categories include various feedstocks such as animal manure, slurries, organic wastes generated from agriculture, dairies, food industries and wastewater sludge, organic inputs from municipal solid wastes, households, as well as energy crops (Cheng 2009).

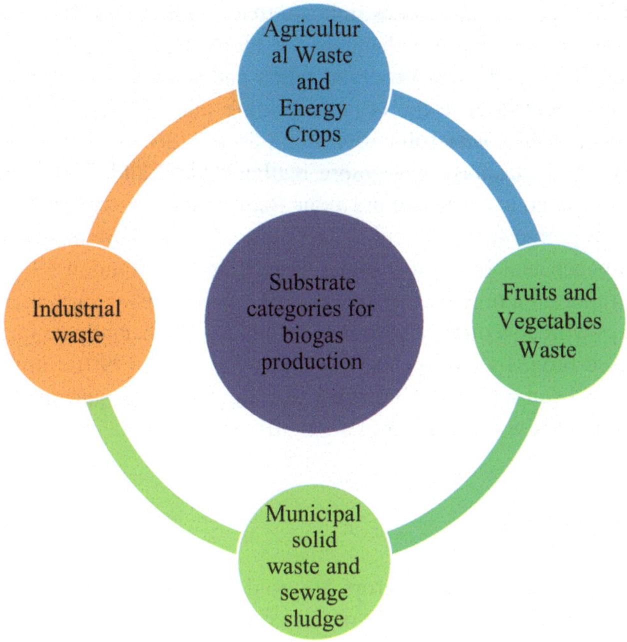


**Fig. 12.4** Substrate categories utilized during anaerobic digestion for biogas production

Cattle dung or cattle manure is the chief organic waste generated within agriculture sector. It has been commonly used as the substrate for biogas production. It is nutrient rich and used by farmers as soil conditioner for increased crop productivity (Cooperband 2002; Yorgey et al. 2014). Cattle manure can also be used as ideal substrate for biogas production, and later the digestate may be used as a soil conditioner. It has been reported that manure digestion increases the availability of nutrients for plants, particularly nitrogen (LeaMaster et al. 1998). Besides beneficial effects, animal manure can pose harm to the environment if not handled properly (Phetyim et al. 2015). High concentration of nitrogen and phosphorus may cause nutrient imbalance; it also contains some fractions of toxic substances such as heavy metals, antibiotics, and some growth regulators. Moreover, the presence of many pathogenic microorganisms in the manure can lead to the outbreak of serious plant and human diseases (LeaMaster et al. 1998). Thus, dumping of the animal manure in an environment-safe way is a matter of prime concern. So, biomethanation offers an attractive avenue of dealing with cattle dung.

#### 12.4.1.1 Agricultural Waste as Substrate Feedstock

Manure could be used as solid phase or slurry; however, more biogas yield has been reported with slurry phase (Sebola 2015). Manure has its own properties and can be mixed with other agricultural substrates during anaerobic digestion for optimum biogas yield. In fact, several studies indicate that combining different wastes with cattle dung can be effective in enhancing the biomethanation process. A 16–65% increase in methane yield was observed when cattle dung was mixed with grass silage and sugar beet litter at the organic loading rate of 2 kg VS/m$^3$ d at 35 °C (Lehtomäki et al. 2007). Anaerobic digestion of a mixture of water hyacinth, cattle manure, algae, and rice husk gave more methane yield and a more nutrient-rich digestate as compared to cattle manure alone (Ghosh and Das 1982). A higher methane yield was reported from anaerobic digestion of cattle dung mixed with lantana slurry, apple-peach leaf litter, and wheat straw (Dar and Tandon 1987). Different studies have reported a high biogas yield with amendments of various organic agricultural wastes with cow dung; a mixture of water hyacinth, cattle dung, and poultry waste increased biogas yield by 100% (Madamwar et al. 1990). Similarly, a high biogas yield has been reported from anaerobic digestion of sugarcane bagasse, cattle dung, and poultry waste (Mallik et al. 1990).

#### 12.4.1.2 Energy Crops

Crop residues and various energy crops are used as feedstocks for digestion during biogas production. Grass is the most preferred energy crop due to easy digestibility, and also its availability is not hindered by seasonal change. A high methane yield (70–80%) has been reported from grass silage digestion, and moreover, methane produced from grass has been found to be suitable for automobile fuel (Gerin et al. 2008; Abu-Dahrich et al. 2011). A significant biogas yield has also been reported from anaerobic digestion of various plant materials such as *Ageratum*, *Calotropis procera*, *Beta vulgaris*, water hyacinth, and even marine macroalgae (Kalia and Kanwar 1990; Traore 1992; Hassan 2003; Singhal and Rain 2003; Hughes et al. 2012).

#### 12.4.1.3 Fruit and Vegetable Waste

Markets and food industries generate tons of fruit and vegetable waste every year. Fruits and vegetables are high moisture-rich (70–90%) substrates, and anaerobic digestion is an ideal way of recycling waste containing 50% or more moisture content. But, the use of fruit and vegetable waste as a sole substrate for biomethanation is an exigent process due to poor substrate composition (Asquer et al. 2013; Bouallagui et al. 2003; Sanjaya et al. 2016). Therefore, different pretreatments are given to increase biogas yield from fruit-vegetable waste (Asquer et al. 2013). The major setback of using fruits during digestion is that they rapidly acidify digestion mixture due to the low pH of waste and accretion of volatile fatty acids, which reduce the activity of methanogens in the digester. Therefore, preliminary treatment is required for efficient digestion of fruit and vegetable wastes (Hall 1995; Appels et al. 2008; Arthur et al. 2011).

#### 12.4.1.4 Municipal Solid Waste and Sewage Waste

Waste generated and collected from residential areas, institutions, and mercantile activities is called municipal solid waste (MSW). MSW is usually collected as mixed stream and then dumped through landfilling sites. Landfilling is not considered an ideal way to get rid of MSW as it is a waste of energy and nutrients. MSW contains a significant biodegradable fraction (65–70%); therefore, it can be subjected to anaerobic digestion for biomethane production. Literature has well documented the production of methane from anaerobic digestion of MSW (Davidson et al. 2008; De Clercq et al. 2016).

Sewage sludge is generated from municipal and industrial wastewater treatment plants, and its production is increasing swiftly with increase in industrialization (Appels et al. 2008). The sludge is a storehouse of nutrients like nitrogen and phosphorus and organic matter; therefore, it is used as manure in crop fields. However, sewage sludge also contains some toxic substances like heavy metals and pathogenic microorganisms; thus, its direct use as manure is environmental sensitive and should be reconsidered (Appels et al. 2008). Sludge treatment through anaerobic digestion is the method of choice for biomethane production, and the digestate produced is nutrient rich and safe and can be used as manure.

Sludge contains significant moisture and other micro- and macronutrients but low in C:N ratio. Therefore, biomethane production is optimized by supplementing with organic fraction of MSW; the process is called co-digestion (Pastor et al. 2013). Organic fractions of MSW are high in C:N ratio, and a significant increase in methane yield is reported with co-digestion of MSW and sewage sludge (Tchobanoglous et al. 1993; Raposo et al. 2012).

#### 12.4.1.5 Industrial Waste

Industrial waste is the waste generated from various manufacturing industrial units such as paper and pulp industries, dairy industries, food industries, and agro-product-based industries. Anaerobic digestion is the method of choice for treatment of industrial waste in an environmental safe way. It has been reported that waste from paper-pulp industries and butcher houses can be utilized for biomethane production (Jia et al. 2013). Waste from paper-pulp industries is rich in organic carbon and can be easily biodegraded during anaerobic digestion; therefore, it is ideally suited as a substrate in biogas plants (Hagelqvist 2013). Anaerobic digestion of waste from paper and pulp industries is also economically beneficial to industries as it includes low transportation cost and operation and maintenance can be easily coordinated with existing organizations (Hagelqvist 2013).

Besides pulp-paper industries, wastes from textile industry, biscuit and chocolate industry, cheese industry, and distilleries have been successfully employed in biogas production (Balasubramanya et al. 1986; Mehla 1986; Ranade et al. 1989; Lebrato et al. 1990). A biogas yield of 2.2 liter per day has been observed when whey, cow dung, and poultry waste (3:1:2) were supplemented at a rate of 6 grams of total

**Table 12.5** Production of biogas from various industrial wastes

| Industrial wastes used for biogas | Reference |
|---|---|
| Dairy effluents | Rani (2001) |
| Spent tea leaves | Goel et al. (2001) |
| Wastewater from food processing unit | Wei et al. (2011) |
| Glycerol as by-product of biodiesel production | Viana et al. (2012) |
| Brewery waste | Tewelde et al. (2012) |
| Palm mill effluent | Thong et al. (2012) |
| Paper mill wastes, brown grease, and corn ethanol | Zhang (2017) |
| Cassava industrial waste | Budiyono et al. (2018) |

solids/day at 40 °C (Desai and Madamwar 1994a). Later, an increase in fuel yield was obtained after adding silica gel (Desai and Madamwar 1994b) and Tween 80 (Desai and Madamwar 1994c). Table 12.5 shows production of biogas from various industrial wastes.

### 12.4.2 *Alterations of Microflora*

Biogas production involves many different groups of microorganisms working together to anaerobically degrade the organic matter and to produce methane and other gaseous components of biogas (Amani et al. 2010). Due to the inherent problems associated with dealing with the anaerobic microorganisms as well as involvement of a variety of microbes growing in close unison in syntrophic manner, it has been somewhat difficult to isolate and identify individual microorganism involved in biomethanation process by the conventional techniques. Still, a lot of work has been done in this direction, and the major groups of microbes involved in anaerobic digestion of organic matter are known.

The molecular biology techniques have been very useful in analyzing the microflora involved in biogas production. Some of the commonly used methods involved 16S rRNA analysis using 454 next-generation sequencing (NGS) technique (Zakrzewski et al. 2012), terminal restriction fragment length polymorphism (Wang et al. 2010), and quantitative real-time polymerase chain reaction (qPCR) (VanGuilder et al. 2008). While the conventional methods target primarily the dominant microflora, the metagenomic and metatranscriptomic approaches allow analysis of even lesser abundant microflora and go a long way in comprehending the microbial community dynamics of biogas production as well as provide insights into the genetic and metabolic capabilities of the microflora (Sárvári Horváth et al. 2016; Bremges et al. 2015). Greater understanding about the microbial community involved in biogas production and their dynamics will help in potentially enhancing the production of biogas.

Some workers have tried using addition of microbial inoculants for enhancing the biogas production. Bagi et al. (2007) reported that the addition of a pure culture of *Caldicellulosiruptor saccharolyticus* (a cellulolytic, $H_2$-producing bacterium) to various substrates like sewage sludge, plant wastes, and animal wastes leads to significant increase in biogas production. Similar results were also obtained when a mixture of these substrates were inoculated with *C. saccharolyticus* culture. However, the use of pure cultures is relatively less common, and most of the workers have focused on utilizing microbial consortia. A thermophilic consortium, obtained from various composting materials like sugarcane dregs, dried straw, and fecal material of chicken, pigs, and cattle, was efficiently degrading ligno-cellulosic substrates like rice straw, corn stalk, cassava residues, etc. (Haruta et al. 2002; Guo et al. 2011). Wei et al. (2010) reported enhanced level of biogas production by utilization of a hemicellulose-degrading microbial consortium comprising *Bacteroides* sp., *Dechlorosoma* sp., and a diverse range of Clostridiales immobilized on zeolite. Dhadse et al. (2012) obtained maximum biogas production (with 76% methane) from a consortium containing four different methanogenic bacteria, while the other two bacterial consortia containing facultative anaerobes result in lower yields of biogas. Gopinath et al. (2014) analyzed the effect of four different microbial consortia obtained from cow dung on biogas production with poultry droppings as a substrate and reported that the consortium number four resulted in maximum methane yield of 79.4%. They reported this consortium to have high concentration of methanogenic archaea. Poszytek et al. (2016) developed a microbial consortium with high cellulolytic activity, comprising 16 strains belonging to the genera *Bacillus*, *Providencia*, and *Ochrobactrum*, which lead to higher efficiency of maize silage degradation as well as higher biogas production under two-phase sequencing reactor. Suksong et al. (2019) compared two thermotolerant microbial consortia, one of which was rich in Lachnospiraceae and the other was rich in Clostridiaceae members for biogas production from palm oil empty fruit bunches. While the former was observed to be more suitable for pre-hydrolysis, the latter was found to be more suitable for direct bioaugmentation during solid-state anaerobic digestion. Tantayotai et al. (2019) have reported 6.5 times enhancement in biogas production from activated wastewater sludge and rice straw residues with the help of an ionic-liquid-tolerant and salt-tolerant consortium primarily comprising Bacteroidetes, Actinobacteria, and Methanosarcinales.

### *12.4.3 Modifications of Biogas Production Process*

The main constituents of the feedstock are lignocellulosic wastes composed of cellulose, hemicellulose, and lignin as discussed before. Kumar and Sharma (2017) have summarized the percentage of these three constituents in various plant sources like sugarcane, wood, newspaper, etc. The yield from a biogas plant is dependent upon:

(a) The type of the biogas plant
(b) The concentration of the biomass fed into the biogas plant
(c) The conditions of the AD process

#### 12.4.3.1 Pretreatment Technology for AD

The present knowledge on pretreatment technology has only recently been investigated and needs to be optimized before use in terms of not only application and efficiency but also in regard to the economic burden. It must be more integrated into the biogas production process rather than being considered as an enhancement or a separate process.

i. *The primary aims of pretreatment of biogas feedstock are the following:*

   a. Avoid the failure of the production process.
   b. Enhance the production efficiency of biogas production—AD is faster.
   c. Lower the methane emission into the surroundings, making the process more environmental friendly.
   d. Increase the yield of biogas from the given feedstock.
   e. Make the feedstock substrates more accessible to the microorganisms for degradation by converting the substrates partially or completely into fermentable sugars.
   f. Overcome the recalcitrance of cellulose-lignocellulosic substrate that is a complex structure to break down by the microorganisms, making its degradation difficult and expensive to the biogas refineries. Recent trends are exploring the potential of genetic engineering approaches to solve recalcitrance problems (Abramson et al. 2013).

ii. *The selection criteria for the pretreatment process are based upon the following factors (Wyman* 1999*):*

   a. Avoid a method that leads to the reduction in particle size of biomass.
   b. The chosen method must aim to preserve the hemicellulose fraction.
   c. There should be minimal formation of the degradation products.
   d. The energy demands of the chosen method must be minimal.
   e. The method must select a cost-effective catalyst for pretreatment and should involve a low-cost pretreatment catalyst and/or inexpensive catalyst recycle and regeneration of high-value lignin co-product.

iii. Types of pretreatment methods for overcoming recalcitrance in biomass can be broadly divided into biochemical and thermochemical methods and more precisely into physical, mechanical, physico-mechanical, and biological methods as summarized in Table 12.6 (ATV-DVWK 2003; Mshandete et al. 2006; Laser et al. 2009; Kumar and Sharma 2017).

#### 12.4.3.2 Multiple Stage AD

Modern-day technology for improving the stability of the biogas production and efficiency of the bioreactor systems has been explored to segregate the processes of AD into hydrolysis–acidogenesis in one chamber and acetogenesis–methanation in

**Table 12.6** Broad classification of types of biomass pretreatment methods

| Biochemical methods | Thermochemical methods |
|---|---|
| *Advantage*:<br>a. High specificity in deconstruction of biomass<br>b. Desired product formation<br>*Disadvantage*:<br>a. Should be coupled with thermochemical pretreatment of low severity | *Advantage*:<br>c. Fast process<br>d. Low residence time<br>e. Uses a wide variety of feedstocks in a continuous manner<br>*Disadvantage*:<br>b. Nonspecific deconstruction of biomass |

*Types of pretreatment methods based on the catalyst used:*

| Physical/mechanical methods | Chemical methods | Physicochemical and thermal methods | Biological and enzymatic methods |
|---|---|---|---|
| a. Mechanical extrusion | a. Dilute acid | a. Steam explosion | a. Fungi (brown, white, soft rot) |
| b. Milling | b. Mild alkali | b. Liquid hot water | b. Bacterial |
| c. Microwave | c. Ozonolysis | c. Ammonia based | c. Archaeal |
| d. Ultrasound | d. Organosolv | d. $CO_2$ explosion | d. Enzymes—cellulase, hemicellulase, cellobiase, pectinase, proteases, etc. (Hosseini Koupaie et al. 2019) |
| e. Pyrolysis | e. Ionic liquids | e. Oxidative pretreatment | |
| f. Pulsed electric field | f. Deep eutectic solvents | f. Wet oxidation | |
| | g. Natural deep eutectic solvents | g. SPORL | |

another chamber of a bioreactor. Such a multiple stage bioreactor system offers the advantage to micro-manipulate the environments and feedstock loading rate for the two chambers of the bioreactor but has the drawback of being more complex and costlier, making it less feasible for commercial use (Vandevivere et al. 2002; US EPA 2006; California EPA 2008; Yu et al. 2013). Recent investigations have reported that multiple stage AD bioreactor shows higher efficiency of COD removal and production of biogas (Colussi et al. 2013), accelerated hydrolysis and faster degradation of biomass (Marín Pérez and Weber 2013), ammonia inhibition (Yabu et al. 2011), higher methane yield (Park et al. 2008), etc. A four-stage AD bioreactor coupled with activated sludge has also been used (Kim et al. 2011) that demonstrated higher AD efficiency than a one-step system. Similarly, higher growth rates of methanogenic bacteria lead to increased biogas production (Blonskaja et al. 2003) and bio-hydrogen production (Nasr et al. 2012).

#### 12.4.3.3 High-Pressure AD

This technique works on the principle of increasing the working pressure within the bioreactor up to 100 bar (1 bar = 100 kPa) resulting in higher methane content (>95%) and less than 5% $CO_2$ in the biogas produced. Hence, the biogas production is integrated with the in situ high-pressure purification as a sole process that produces clean biogas with 99% methane which may be utilized for domestic and commercial applications (Bartlett 2002; Lindeboom et al. 2011; Merkle et al. 2014,

2017). This technology has shown promising results in quality biogas production but requires further research to understand the pressure effects on the growth of the microbiome.

#### 12.4.3.4 Modulating Methanogenesis

Among the four steps of AD, methanogenesis is the most critical and rate-limiting step due to slowest growth rate of the methanogens and their sensitivity to environmental factors like pH, temperature, ionic strength, and various inhibitors (Chen et al. 2008). NGS in combination with genetic engineering approaches to modify the efficiency of the metabolic pathways in the microbes may be considered to decrease the economic burden of the biogas reactors, making their use more commercially applicable. These approaches may also bear potential to improve the quality and energy output of the biofuels (Xu and Koffas 2010). Metabolic redirection has been applied in production of bioethanol in which the production of undesirable metabolic products is limited and the metabolism of the bacterial cell is directed toward the formation of targeted products (Weng et al. 2008).

#### 12.4.3.5 Start-Up Period for the AD

It is critical for every biogas reactor to have a start-up time period in order to produce an efficient, continuous, and stable supply of biogas (Escudié et al. 2011; Kim et al. 2013; Goberna et al. 2015). It is the time required by the microbes to grow and multiply while feeding upon a specific waste, till their population becomes redundant and stable. The time period hence depends upon the microbial population and the type of biomass it feeds on. If the start-up period is neglected, the biomass degradation may be incomplete, leading to accumulated intermediaries like VFA, inhibition of methanogenesis and thus inefficient productivity of biogas, and finally operation failures of the biogas reactor systems (Griffin et al. 1998; Liu et al. 2002; Escudié et al. 2011). Hence, one must keep in mind the calculation of the start-up time period considering the type of organic waste, rate of loading the feedstock, the ratio of inoculum to substrate, temperature conditions of the bioreactor, type of the reactor, etc. Next-generation sequencing (NGS) may be used as an efficient tool to screen the dynamics of complicated microbial community. NGS also aids in monitoring the successful establishment of AD start-up in terms of microbial communities inside the bioreactors, besides elucidating the degradation pathways in the process of biogas production constituted in the microbiome (Appels et al. 2011). Therefore, such tools must be utilized besides the physicochemical monitoring AD start-up.

### 12.4.4 Solid-State Anaerobic Digestion

Anaerobic digestion in a biogas digester is operated in two modes: digestion in slurry phase and solid-state anaerobic digestion. The classification is based upon the available solid content in substrate feedstock (Li et al. 2011b). Slurry phase operates at a total solid content of less than 15%, and solid-state digestion is carried at a total solid content of more than 15% (Brown et al. 2012). The advantages of solid-state anaerobic digestion include requirement of small reactor volume, less moving assembly, low cost-energy requirement, less water wastage, and easy handling (Brown et al. 2012; Guendouz et al. 2008; Cheng et al. 2010). Solid-state fermentation is required with substrates feedstocks rich in lignocellulosic content those contain less moisture (Brown et al. 2012; Singhania et al. 2009). However, due to low moisture availability and complex structure, these substrates are hard to hydrolyze and require pretreatment for easy digestion (Taherzadeh and Karimi 2008; Teghammar et al. 2012).

The crystalline structure of cellulose and the presence of resistant fraction lignin are the important factors determining degradation of lignocellulosic waste (Puri 1984; Chang and Holtzapple 2000; Laureano-Perez et al. 2005). The lignin forms a cross-linked structure in between the carbohydrates, and the presence of this matrix restricts the lignocellulosic substrate digestion. This meshwork is resistant to degradation by enzymes and microbes (Poornejad et al. 2014; Kumar 2014). Therefore, lignin degradation is required for biomethane production from lignocellulosic waste. Ethanol is employed for lignin removal as a pretreatment during biodegradation of lignocellulosic waste (Zhao et al. 2009; Binod et al. 2010). Lignin is itself a valuable by-product; therefore, the use of organic solvent as a pretreatment method helps to separate lignin in unaltered form, increase the biogas yield, and ultimately improve economy of the process (Zhao et al. 2009; Obama et al. 2012).

Biogas production was examined over a range of total solid concentration, and most efficient digestion was reported at a total solid content of 13.5% (Singh et al. 1984). Similarly, Pathak et al. (1985) observed biogas yield per gram of solid consumed with solid content of manure slurry at 7.7, 10.2, and 14.8% respectively. Various studies have also highlighted a high biogas yield at total solid contents ranging between 5 and 20% (Itodo and Awulu 1999; Itodo et al. 2001; Malik et al. 2008; Leela Wati et al. 2008).

### 12.4.5 Improvement in Biogas Purification

The anaerobic digestion of organic materials leads to production of biogas which is basically a mixture of gases, methane being the most dominant as well as the desired component. Apart from methane, other gases include carbon dioxide ($CO_2$), hydrogen sulfide ($H_2S$), ammonia ($NH_3$), and water vapors. In order to enhance the efficiency of energy yield from biogas, it is imperative that the unwanted components of biogas be removed and methane be enriched because, ultimately, it is methane which is going to act as the energy source (Lohani et al. 2010). Moreover, the corrosiveness of hydrogen sulfide will interfere with the storage and transfer of biogas through metallic components. Similarly, water may also interfere with the efficient utilization of biogas and additionally can cause rusting of metallic components. The presence of $CO_2$ lowers the energy content of the biogas, and its compression leads to higher energy inputs (Bari 1996; Appels et al. 2008). Biogas upgrading is especially needed for utilization in vehicles and fuel cells (Kapadi et al. 2005).

Some of the methods which have been reported to be used for $CO_2$ removal include water scrubbing, pressure swing adsorption (PSA) with activated carbon or molecular sieves, physical absorption, chemical absorption, adsorption on a solid surface, membrane separation, cryogenic separation, etc. (Ryckebosch et al. 2011; Morero et al. 2017).

Water scrubbing is regarded as one of the relatively simple, economic, and practical methods for $CO_2$ removal from biogas especially in the rural areas. This method has the capability of removing $CO_2$ and $H_2S$ as well (Wellinger and Lindeberg 1999). Organic solvents like methanol and dimethyl ethers of polyethylene glycol (DMPEG) have also been be employed. These also have the capacity of removing $CO_2$, $H_2S$, and $H_2O$ (Tock et al. 2010). Methane concentrations as high as 96–98.5% have been reported in these techniques (Bauer et al. 2013a; Sun et al. 2015).

Chemical absorption generally uses either aqueous solution of amines or aqueous solution of alkaline salts (Al-Baghli et al. 2001; Razi et al. 2013). Some of the amines which have been used include monoethanolamine, diglycolamine, diethanolamine, triethanolamine, methyldiethanolamine, and piperazine. Another category of chemicals includes caustic solvents (sodium hydroxide, potassium hydroxide, and calcium hydroxide). In this process, $CO_2$ content of biogas is reduced from about 40% to 0.5–1.0%. However, the processes are still to be standardized for large-scale operation in biogas purification, and there are many technical problems like high energy requirements as well as solvent recovery issues (Abdeen et al. 2016). The PSA process makes use of vertical columns packed with absorbents under adsorption, depressurization, desorption, and pressurization sequences (Yeh et al. 2001). The most commonly used adsorbents are zeolite, activated carbon, activated charcoal, silica gel, and synthetic resins. This can be used to separate $CO_2$, $N_2$, $O_2$, and $H_2S$ (Ryckebosch et al. 2011). Linking of several columns has been reported to reduce energy need for operation (Bauer et al. 2013b). Membrane separation process makes use of a thin membrane, which is more permeable to some biogas components than the others. These may be operated at higher pressure ranges of greater than 20–40 bar or at lower pressure range of 8–10 bar and are capable of removing $CO_2$, $H_2S$, $H_2O$, and $O_2$ (Bauer et al. 2013a). Some workers have reported that efforts to obtain higher purity lead to methane losses (Persson et al. 2007; Sun et al. 2015); that is why multi-phase membrane separation systems have been suggested by Scholz et al. (2013). The cryogenic separation is a relatively newer technology and relies on the difference in the boiling points of various gaseous components of biogas. In this process, biogas is cooled with chillers nearly to −45 °C at elevated pressure. The process is especially useful for obtaining liquid biomethane with high purity and relatively with less than 1% losses (Hosseini and Wahid 2014). The product may be directly used for vehicles or injected to grid as gas. However, the energy consumption of the process is relatively high (Johnston 2014; Sun et al. 2015).

For removal of $H_2S$, some of the suggested methods include iron oxide adsorption, liquid phase oxidation process, lime scrubbing, air injection, iron chloride addition, dry oxidation process, and liquid phase oxidation (Shah et al. 2016). A variety of chemicals like NaOH, $FeCl_2$, $Fe^{3+}/MgO$, $Fe(OH)_3$, $Fe^{3+}/CuSO_4$ and $Fe^{3+}/$EDTA (ethylenediaminetetraacetate), activated carbon, and zeolite have been used in different techniques (Cosoli et al. 2008; Ryckebosch et al. 2011; Sun et al. 2015). However, a major obstacle in case of $H_2S$ is its toxic and corrosive nature. So, more research efforts are needed to tackle this problem. One possible approach is to develop a system that can combine multiple technologies for removal of $H_2S$, $CO_2$, and other contaminants. Another important development in the direction of making the overall process more efficient is the employment of lithotrophic sulfur oxidizing microbial agents for desulfurization and biofiltration of $H_2S$. Some of the possible microbial candidates include *Thiobacillus*, *Paracoccus*, *Acidithiobacillus*, and *Halothiobacillus* (Montebello 2013; Mora et al. 2014).

For removal of water vapors, both physical drying and chemical drying have been used. The physical drying methods by condensation are demisters, cyclone separators, moisture traps, and water traps (Persson et al. 2007; Bauer et al. 2013b). The chemical drying can be carried out with the help of glycol, silica gel, magnesium oxide, activated carbon, and alumina (Bailón Allegue and Hinge 2012; Awe et al. 2017).

## 12.5 Conclusions and Future Perspectives

The anaerobic digestion processes for the production of biogas are poorly understood due to their complexity and are often linked to the risk of failure in terms of large-scale investments. But the growing need for the development of biofuels has shifted the R&D toward the exploration of transportation fossil fuels' replacements like biomethane. The present-day research on biofuel technology faces the challenges in terms of technical understanding, economical burden, and ecological impacts. The microbial strains and their inoculum/substrate ratio, the types of catalyst chosen, the kind of AD bioreactor, the substrate composition, the start-up period, and the pretreatment process all have to be optimized in terms of efficiency, process stability, and cost-effectiveness. So, it can be concluded that although biomethanation offers a wide range of advantages, a lot of research efforts are required in this direction to enhance the applicability and appeal of this process. It is high time that we put in maximum efforts in developing and modernizing the biogas production for the benefit of not only the mankind but the entire globe.

## References


Abas N, Kalair A, Khan N (2015) Review of fossil fuels and future energy technologies. Futures 69:31–49

Abdeen FRH, Mel M, Jami MS, Ihsan SI, Ismail AF (2016) A review of chemical absorption of carbon dioxide for biogas upgrading. Chin J Chem Eng 24:693–702

Abramson M, Shoseyov O, Hirsch S, Shani Z (2013) Genetic modifications of plant cell walls to increase biomass and bioethanol production. In: Lee JW (ed) Advanced biofuels and bioproducts. Springer, New York, pp 315–338

Abu-Dahrich J, Orozco A, Ahmad M, Rooney D (2011) The potential of biogas production from grass. In: Proceedings of the Jordan International Energy Conference, Amman

Agus F, Irawan D (2006) Agricultural land conversion as a threat to food security and environmental quality. Jurnal Litbang Pertanian 25:90–98

Alamdarlo HN (2018) The economic impact of agricultural pollutions in Iran, spatial distance function approach. Sci Total Environ 616–617:1656–1663

Al-Baghli NA, Pruess SA, Yesavage VF, Selim MS (2001) A rate-based model for the design of gas absorbers for the removal of CO2 and H2S using aqueous solutions of MEA and DEA. Fluid Phase Equilib 185:31–43

Amani T, Nosrati M, Sreekrishnan TR (2010) Anaerobic digestion from the viewpoint of microbiological, chemical, and operational aspects- a review. Environ Rev 18:255–278

Amigun B, Sigamoney R, Von Blottnitz H (2008) Commercialisation of biofuel industry in Africa: a review. Renew Sustain Energy Rev 12:690–711

Antonelli J, Lindino CA, Azevedo JCR, de Souza SNM, Cremonez PA, Ross E (2016) Biogas production by the anaerobic digestion of whey. Revista de Ciências Agrárias 39(3):463–467

Appels L, Baeyens J, Degrève J, Dewil R (2008) Principles and potential of the anaerobic digestion of waste-activated sludge. Prog Ener Combus Sci 34:755–781

Appels L, Lauwers J, Degrève J, Helsen L, Lievens B, Willems K et al (2011) Anaerobic digestion in global bio-energy production: potential and research challenges. Renew Sustain Energy Rev 15:4295–4301

Arthur R, Baidoo MF, Brew-Hammond A, Bensah EC (2011) Biogas generation from sewage in four public universities in Ghana: a solution to potential health risk. Biomass Bioenergy 35(7):3086–3093

Asquer C, Pistis A, Scano EA (2013) Characterization of fruit and vegetable wastes as a single substrate for the anaerobic digestion. Environ Eng Manage J 12:89–92

ATV-DVWK (2003) Thermische, chemische und biochemische Desintegrationsverfahren 3. Arbeitsbericht der Arbeitsgruppe AK-1.6 "Klärschlammdesintegration". Corresp Wastewater 50:796–804. Germany

Awe OW, Zhao Y, Nzihou A, Minh DP, Lyczko N (2017) A review of biogas utilisation, purification and upgrading technologies. Waste Biomass Valor. https://doi.org/10.1007/s12649-016-9826-4

Azizan MU, Hussin K (2015) Understanding the pressure on agriculture land as a safeguard for food security in Malaysia. Int J Built Environ Sustain 2:278–283

Bagi Z, Ács N, Bálint B, Horváth L, Dobó K, Perei RK, Rákhely G, Kovács KL (2007) Biotechnological intensification of biogas production. Appl Microbiol Biotechnol 76:473–482

Bailón Allegue L, Hinge J (2012) Biogas and bio-syngas upgrading. Danish Technological Institute, Aarhus, pp 1–97

Balasubramanya RH, Khandeparkar VG, Sundaram V (1986) Production of biogas and biomanure from the textile processing residue, willow dust, by dry anaerobic fermentation. Agric Wastes 16:295–302

Balch WE, Schoberth S, Tanner RS, Wolfe RS (1977) *Acetobacterium*, a new genus of H– oxidizing, CO2 reducing anaerobic bacteria. Int J Syst Bact 27:355–361

Bari S (1996) Effect of carbon dioxide on the performance of biogas/diesel dual – fuel engine. In: World Renewable Energy Congress

Bartlett DH (2002) Pressure effects on in vivo microbial processes. Biochim Biophys Acta 1595(1–2):367–381
Bauer F, Hulteberg C, Persson T, Tamm D (2013a) Biogas upgrading-review of commercial technologies. SGC Rapport 270:83
Bauer F, Persson T, Hulteberg C, Tamm D (2013b) Biogas upgrading-technology overview, comparison and perspectives for the future. Biofuels Bioprod Biorefining 7:499–511
Beesley KB, Ramsey D (2009) Agricultural land preservation. Int Enc Hum Geogr 25:65–69
Bharathiraja B, Sudharsanaa T, Jayamuthunagai J, Praveenkumar R, Chozhavendhand S, Iyyappana J (2018) Biogas production – a review on composition, fuel properties, feed stock and principles of anaerobic digestion. Renew Sustain Energy Rev 90:570–582
Binod P, Sindhu R, Singhania RR et al (2010) Bioethanol production from rice straw: an overview. Bioresour Technol 101:4767–4774
Blonskaja V, Menert A, Vilu R (2003) Use of two-stage anaerobic treatment for distillery waste. Adv Environ Res 7(3):671–678
Boone DR, Bryant MP (1980) Propionate-degrading bacterium, *Syntrophobacter wolinii* sp. nov. gen. nov., from methanogenic ecosystems. Appl Environ Microbiol 40:626–632
Börjesson P, Mattiasson B (2008) Biogas as a resource-efficient vehicle fuel. Trends Biotechnol 26:7–13
Bouallagui H, Cheikh RB, Marouani L, Hamdi M (2003) Mesophilic biogas production from fruit and vegetable waste in a tubular digester. Bioresour Technol 86:85–89
Braun R (1982) Biogas-methane treatment of organic waste. Springer, Wien
Braun R (2007) Anaerobic digestion: a multi-faceted process for energy, environmental management and rural development. In: Ranalli P (ed) Improvement of crop plants for industrial end uses. Springer, Dordrecht, pp 335–415
Bremges A, Maus I, Belmann P, Eikmeyer F, Winkler A, Albersmeier A, Pühler A, Schlüter A, Sczyrba A (2015) Deeply sequenced metagenome and metatranscriptome of a biogas-producing microbial community from an agricultural production-scale biogas plant. Giga Sci 4:33
Brown D, Li Y (2013) Solid state anaerobic co-digestion of yard waste and food waste for biogas production. Bioresour Technol 127:275–280
Brown D, Shi J, Li Y (2012) Comparison of solid-state to liquid anaerobic digestion of lignocellulosic feedstocks for biogas production. Bioresour Technol 124:379–386
Budiyono PAD, Ardhannari L, Matin HHA, Sumardiono S (2018) Study of biogas production from cassava industrial waste by anaerobic process. MATEC Web Conf 156:30–52
Budzinski H, Couderchet M (2018) Environmental and human health issues related to pesticides: from usage and environmental fate to impact. Environ Sci Pollut Res 25:14277–14279
California Environmental Protection Agency (2008) Current anaerobic digestion technologies used for treatment of municipal organic solid waste. Report. California Integrated Waste Management Board, California
Chang VS, Holtzapple MT (2000) Fundamental factors affecting biomass enzymatic reactivity. In: Proceedings of the 21st Symposium on Biotechnology for Fuels and Chemicals. Springer
Chang C-W, Lee T-H, Lin W-T, Chen C-H (2015) Electricity generation using biogas from swine manure for farm power requirement. Int J Green Energy 12:339–346
Chen Y, Cheng JJ, Creamer KS (2008) Inhibition of anaerobic digestion process: a review. Bioresour Technol 99:4044–4064
Cheng J (2009) Biomass to renewable energy processes. CRC Press, United State of America, p 151
Cheng YS, Zheng Y, Yu CW, Dooley TM, Jenkins BM, Vandergheynst JS (2010) Evaluation of high solids alkaline pretreatment of rice straw. Appl Biochem Biotechnol 162(6):1768–1784
Cold A, Forbes VE (2004) Consequences of a short pulse of pesticide exposure for survival and reproduction of *Gammarus pulex*. Aquat Toxicol 67:287–299
Colussi I, Cortesi A, Piccolo CD, Galloa V, Fernandeza ASR, Vitanza R (2013) Improvement of methane yield from maize silage by a two-stage anaerobic process. Chem Eng Trans 32:151–156
Cooke E (2014) Gas production from waste and crops. City of Johannesburg gas for mobility summit, Johannesburg, South Africa

Cooperband L (2002) Building soil organic matter with organic amendments. Center for Integrated Agricultural Systems
Cosoli P, Ferrone M, Pricl S, Fermeglia M (2008) Hydrogen sulphide removal from biogas by zeolite adsorption. Part I. GCMC molecular simulations. Chem Eng J 145:86–92
Dar GH, Tandon SM (1987) Biogas production from pretreated wheat straw, lantana residue, apple and peach leaf litter with cattle dung. Bio Wastes 21:75–83
Davidson A, Lövstedt C, la Cour Jansen J, Gruvberger C, Aspegren H (2008) Co-digestion of grease trap sludge and sewage sludge. Waste Manage 28:986–992
De Clercq D, Wen Z, Fan F, Caicedo L (2016) Biomethane production potential from restaurant food waste in megacities and project level-bottlenecks: a case study in Beijing. Renew Sustain Energy Rev 59:1676–1685
De la Torre I, Goma G (1981) Characterization of anaerobic microbial culture with high acidogenic activity. Biotechnol Bioeng 23:185–199
de Mes TZD, Stams AJM, Reith JH, Zeeman G (2003) Methane production by anaerobic digestion of wastewater and solid wastes. In: Bio-methane & bio-hydrogen: status and perspectives of biological methane and hydrogen production. Dutch Biological Hydrogen Foundation, Petten, pp 58–102
Demirbaş A (2005) Bioethanol from cellulosic materials: a renewable motor fuel from biomass. Energy Sources 27:327–337
Desai M, Madamwar D (1994a) Effect of temperature and retention time on biomethanation of cheese whey –poultry waste –cattle dung. Environ Pollut 83(3):311–315
Desai M, Madamwar D (1994b) Anaerobic digestion of a mixture of cheese whey, poultry waste, and cattle dung: a study of the use of adsorbents to improve digester performance. Environ Pollut 86:337–340
Desai M, Madamwar D (1994c) Surfactants in anaerobic digestion of cheese whey, poultry waste and cattle dung for improved biomethanation. Trans ASAE 37(3):959–962
Dhadse S, Kankal NC, Kumari B (2012) Study of diverse methanogenic and non-methanogenic bacteria used for the enhancement of biogas production. Int J Life Sci Biotechnol Pharma Res 1:176–191
Diatta J, Grzebisz W, Bzowski Z, Spychalski W, Biber M (2019) Ameliorative effect of phyto-ash and biogas digestate improvers on soil contaminated with heavy metals. Arch Environ Protect 45:73–83
Dieter D, Angelika S (2008) Biogas from waste and renewable resources: an introduction. Wiley-VCH, Weinheim
Dinopoulou G, Rudd T, Lester JN (1988) Anaerobic acidogenesis of a complex wastewater: 1. The influence of operational parameters on reactor performance. Biotechnol Bioeng 31:958–968
Dohoo C, Guernsey JR, Critchley K, Van Leeuwen J (2012) Pilot study on the impact of biogas as a fuel source on respiratory health of women on rural Kenyan smallholder dairy farms. J Environ Public Health 636298
Dyson B, Chang N-B (2005) Forecasting municipal solid waste generation in a fast-growing urban region with system dynamics modeling. Waste Manage 25:669–679
Escudié R, Cresson R, Delgenès J-P, Bernet N (2011) Control of start-up and operation of anaerobic biofilm reactors: an overview of 15 years of research. Water Res 45:1–10
Eshore S, Mondal C, Das A (2017) Production of biogas from treated sugarcane bagasse. Int J Sci Eng Technol 6:224–227
European Biogas Association (2011) Biogas: simply the best. Report. European Biogas Association, Brussels
Fang C (2010) Biogas production from food-processing industrial wastes by anaerobic digestion. Technical University of Denmark (DTU), Kgs. Lyngby
FAO (2017) The future of food and agriculture – trends and challenges. Rome
Frankin RJ (2001) Full scale experiences with anaerobic treatment of industrial wastewater. In: Anaerobic digestion for sustainable development; Papers of the Farewell seminar of Prof. Dr. Ir. Gatze Lettinga. Wageningen, The Netherlands

Fu H, Li Z, Wang R (2015) Estimating municipal solid waste generation by different activities and various resident groups in five provinces of China. Waste Manage 41:3–11

García-González MC, Hernández D, Molinuevo-Salces B, Riaño B (2019) Positive impact of biogas chain on GHG reduction. In: Treichel H, Fongaro G (eds) Improving biogas production, biofuel and biorefinery technologies, vol 9. Springer, Cham

Gerin PA, Vliegen F, Jossart JM (2008) Energy and $CO_2$ balance of maize and grass as energy crops for anaerobic digestion. Bioresour Technol 99:2620–2262

Ghosh TK, Das D (1982) Maximization of energy recovery in the biomethanation process. Process Biochem 17:39–42

Giusti L (2009) A review of waste management practices and their impact on human health. Waste Manage 29:2227–2239

Goberna M, Gadermaier M, Franke-Whittle IH, García C, Wett B, Insam H (2015) Start-up strategies in manure-fed biogas reactors: process parameters and methanogenic communities. Biomass Bioenergy 75:46–56

Godbole SH, Gore JA, Ranade DR (1981) Associated action of various groups of microorganisms in the production of biogas. Biovigyan 7:107–113

Goel B, Pant DC, Kishan VVN (2001) Two phase anaerobic digestion of spent tea leaves for biogas and manure generation. Bioresour Technol 80(2):153–156

Gopinath LR, Christy PM, Mahesh K, Bhuvaneswari R, Divya D (2014) Identification and evaluation of effective bacterial consortia for efficient biogas production. IOSR J Environ Sci Toxicol Food Technol 8:80–86

Grabber JH (2005) How do lignin composition, structure, and cross-linking affect degradability? A review of cell wall model studies. Crop Sci 45:820–831

Griffin ME, McMahon KD, Mackie RI, Raskin L (1998) Methanogenic population dynamics during start-up of anaerobic digesters treating municipal solid waste and biosolids. Biotechnol Bioeng 57:342–355

Guendouz J, Buffière P, Cacho J, Carrère M, Delgenes J (2008) High-solids anaerobic digestion: comparison of three pilot scales. Water Sci Technol 58:1757–1763

Guo P, Mochidzuki K, Cheng W, Zhou M, Gao H, Zheng D, Wang X, Cui Z (2011) Effects of different pretreatment strategies on corn stalk acidogenic fermentation using a microbial consortium. Bioresour Technol 102(16):7526–7531

Hagelqvist P (2013) Batchwise mesophilic anaerobic co-digestion of secondary sludge from pulp and paper industry and municipal sewage sludge. Waste Manage 33(4):820–824

Hall JE (1995) Sewage sludge production, treatment and disposal in the European Union. Water Environ J 9:335–343

Haruta S, Cui Z, Huang Z, Li M, Ishii M, Igarashi Y (2002) Construction of a stable microbial community with high cellulose-degradation ability. Appl Microbiol Biotechnol 59(4–5):529–534

Hassan EA (2003) Biogas production from forage and sugar beets: process control and optimization, ecology and economy. Doctoral Thesis submitted to University of Kassel, Kassel, Germany

Hniman A, O-Thong S, Prasertsan P (2011) Developing a thermophilic hydrogen-producing microbial consortia from geothermal spring for efficient utilization of xylose and glucose mixed substrates and oil palm trunk hydrolysate. Int J Hydrogen Energy 36:8785–8793

Holm-Nielsen JB, Al Seadi T, Oleskowicz-Popiel P (2009) The future of anaerobic digestion and biogas utilization. Bioresour Technol 100:5478–5484

Höök M, Tang X (2013) Depletion of fossil fuels and anthropogenic climate change—a review. Energy Policy 52:797–809

Hosseini Koupaie E, Dahadha S, BazyarLakeh AA, Azizi A, Elbeshbishy E (2019) Enzymatic pretreatment of lignocellulosic biomass for enhanced biomethane production-a review. J Environ Manage 233:774–784

Hosseini SE, Wahid MA (2014) Development of biogas combustion in combined heat and power generation. Renew Sustain Energy Rev 40:868–875

Hughes AD, Kelly MS, Black KD, Stanley MS (2012) Biogas from macroalgae: is it time to revisit the idea? Biotechnol Biofuels 5:86

Iqbal HMN, Ahmed I, Zia MA, Irfan M (2011) Purification and characterization of the kinetic parameters of cellulase produced from wheat straw by *Trichoderma viride* under SSF and its detergent compatibility. Adv Biosci Biotechnol 2:149–156
Iqbal SA, Rahaman S, Rahman M, Yousuf A (2014) Anaerobic digestion of kitchen waste to produce biogas. Procedia Eng 90:657–662
Işık EHB, Polat F (2018) Effects of pretreatments on the production of biogas from cow manure. Int Adv Res Eng J 2:48–52
Itodo IN, Awulu JO (1999) Effect of total solid concentrations of poultry, cattle and piggery waste slurries on biogas yield. Trans ASAE 42:1853–1855
Itodo IN, Awulu JO, Philip T (2001) A comparative analysis of biogas yield from poultry, cattle and piggery wastes. In: Livestock Environment VI. Proceedings of the 6th International Symposium ASAE, pp 402–405
Jia J, Zhao S, Kong X, Li Y, Zhao G, He W, Jing R (2013) Aegilops tauschii draft genome sequence reveals a gene repertoire for wheat adaptation. Nature 496:91–95
Johnston MW (2014) Breaking down renewable natural gas injection barriers. Biocycle 55:60
Kalia AK, Kanwar SS (1990) Anaerobic fermentation of *Ageratum* for biogas production. Biol Wastes 32(2):155–158
Kaltschmitt M, Thran D, Smith KR (2001) Renewable energy from biomass. In: Meyersn RA (ed) Encyclopedia of physical science and technology. Academic Press, Cambridge, pp 203–228
Kapadi SS, Vijay VK, Rajesh SK, Prasad R (2005) Biogas scrubbing, compression and storage: perspective and prospectus in Indian context. Renew Energy 30:1195–1205
Kelleher BP, Leahy JJ, Henihan AM, O'Dwyer TF, Sutton D, Leahy MJ (2000) Advances in poultry litter disposal technology— a review. Bioresour Technol 83:27–36
Khan AW (1980) Degradation of cellulose to methane by a coculture of *Acetivibrio cellulolyticus* and *Methanosarcina barkeri*. FEMS Microbiol Lett 9:233–235
Kheira BN, Dadamoussa B, Bendraoua A, Mel M, Labed B (2017) Effects of co-digestion of Camel Dung and municipal solid wastes on quality of biogas, methane and biofertilizer production. J Adv Res Fluid Mech Therm Sci 40:7–17
Kim J, Novak JT, Higgins MJ (2011) Multi-staged anaerobic sludge digestion processes. J Environ Eng 137:0000372
Kim J, Lee S, Lee C (2013) Comparative study of changes in reaction profile and microbial community structure in two anaerobic repeated-batch reactors started up with different seed sludges. Bioresour Technol 129:495–505
Kinet R, Destain J, Hiligsmann S, Thonart P, Delhalle L, Taminiau B, Daube G, Delvigne F (2015) Thermophilic and cellulolytic consortium isolated from composting plants improves anaerobic digestion of cellulosic biomass: toward a microbial resource management approach. Bioresour Technol 189:138–144
Krzysztof P, Martyna C, Aleksandra S, Lukasz D (2016) Microbial Consortium with High Cellulolytic Activity (MCHCA) for enhanced biogas production. Front Microbiol 7:324
Kumar S (2014) Hydrothermal processing of biomass for biofuels. Biofuel Res J 2:43
Kumar AK, Sharma S (2017) Recent updates on different methods of pretreatment of lignocellulosic feedstocks: a review. Bioresour Bioprocess 4:7
Kumar P, Barrett DM, Delwiche MJ, Stroeve P (2009) Methods for pretreatment of lignocellulosic biomass for efficient hydrolysis and biofuel production. Ind Eng Chem Res 48:3713–3729
Ladisch R, Mosier NS, Youngmi KIM, Ximenes E, Hogsett D (2010) Converting cellulose to biofuels. Chem Eng Prog 106:56–63
Laser M, Larson E, Dale B, Wang M, Greene N, Lynd LR (2009) Comparative analysis of efficiency, environmental impact, and process economics for mature biomass refining scenarios. Biofpr 3:247–270
Laureano-Perez L, Teymouri F, Alizadeh H, Dale BE (2005) Understanding factors that limit enzymatic hydrolysis of biomass: characterization of pretreated corn stover. Appl Biochem Biotechnol 124:1081–1099
LeaMaster B, Hollyer JR, Sullivan JL (1998) Composted animal manures: precautions and processing

Lebrato J, Perez-Rodriguez JL, Maqueda C, Morrillo E (1990) Cheese factory wastewater treatment by anaerobic semicontinuous digestion. Resour Conserv Recycl 3:193

Leela Wati, Malik RK, Putatunda C (2008) Poultry manure enriched biogas effluent slurry – a potential organic manure. Geobios 35:133–136

Lehtomäki A, Huttunena S, Rintala JA (2007) Laboratory investigations on co-digestion of energy crops and crop residues with cow manure for methane production: effect of crop to manure ratio. Resour Conserv Recycl 51(3):591–609

Lelieveld J, Evans JS, Fnais M, Giannadaki D, Pozze A (2015) The contribution of outdoor air pollution sources to premature mortality on a global scale. Nature 525:367–371

Lettinga G, Haandel AC (1993) Anaerobic digestion for energy production and environmental protection. In: Johansson TB et al (eds) Renewable energy; sources for fuels and electricity. Island Press, California, pp 817–839

Li G, Niu S, Ma L, Zhang X (2009) Assessment of environmental and economic costs of rural household energy consumption in loess hilly region, Gansu province, China. Renew Energy 34:1438–1444

Li C, Champagne P, Anderson BC (2011a) Evaluating and modeling biogas production from municipal fat, oil, and grease and synthetic kitchen waste in anaerobic co-digestions. Bioresour Technol 102:9471–9480

Li Y, Park SY, Zhu J (2011b) Solid-state anaerobic digestion for methane production from organic waste. Renew Sustain Energy Rev 15:821–826

Lin Y, Tanaka S (2006) Ethanol fermentation from biomass resources: current state and prospects. Appl Microbiol Biotechnol 69:627–642

Lindeboom REF, Fermoso FG, Weijma J, Zagt K, van Lier JB (2011) Autogenerative high pressure digestion: anaerobic digestion and biogas upgrading in a single step reactor system. Water Sci Technol 64(3):647–653

Liu WT, Chan OC, Fang HHP (2002) Microbial community changes during start-up of acidogenic anaerobic reactors. Water Res 36:3203–3210

Lohani SP, Pandey S, Baral B (2010) Biogas purification, compression and storage: a promising energy recovery and delivery technology for Nepal. Linnaeus ECO-TECH '10. https://doi.org/10.15626/Eco-Tech.2010.060

Madamwar D, Patel V, Patel A (1990) Effect of agricultural and other wastes on anaerobic digestion of water hyacinth– cattle dung. J Ferment Bioeng 70:343–345

Malik RK, Wati L, Putatunda C (2008) Effect of adding poultry waste to cattle dung on biogas production in Hisar, Haryana. Environ Ecol 26:1728–1731

Mallik MK, Singh UK, Ahmad N (1990) Batch digester studies on biogas production from *Cannabis sativa*, water hyacinth and crop wastes mixed with dung and poultry litter. Bio Wastes 31:315–319

Mancini G, Papirio S, Lens PNL, Esposito G (2018) Increased biogas production from wheat straw by chemical pretreatments. Renew Energy 119:608–614

Marín Pérez C, Weber A (2013) Two stage anaerobic digestion system: hydrolysis of different substrate. Landtechnik 68:252–255

Martins das Neves LC, Converti A, Vessoni Penna TC (2009) Biogas production: new trends for alternative energy sources in rural and urban zones. Chem Eng Technol 32:1147–1153

McInerney MJ, Bryant MP, Hespell RB, Costerton WJ (1981) *Syntrophomonas wolfei* gen. nov. sp. nov., an anaerobic, syntrophic, fatty acid–oxidizing bacterium. Appl Environ Microbiol 41:1029–1039

McKendry P (2002) Energy production from biomass (part 2): conversion technologies. Bioresour Technol 83:47–54

Mehla RD (1986) Methane production from distillery wastewater. M.Sc. thesis submitted to Haryana Agric. Univ.

Merkle W, Zielonka S, Oechsner H, Lemmer A (2014) High-pressure anaerobic digestion up to 180 bar: the effects on biogas production and upgrading. In: Proceedings of the Progress in Biogas III Conference; 2014 Sep 10–11; Stuttgart, Deutschland

Merkle W, Baer K, Haag NL, Zielonka S, Ortloff F, Graf F et al (2017) High-pressure anaerobic digestion up to 100 bar: influence of initial pressure on production kinetics and specific methane yields. Environ Technol 38:337–344

Montebello AM (2013) Aerobic biotrickling filtration for Andrea Monzón Montebello. J Hazard Mater 280:200–208

Mora M, Fernández M, Gómez JM, Cantero D, Lafuente J, Gamisans X, Gabriel D (2014) Kinetic and stoichiometric characterization of anoxic sulfide oxidation by SO-NR mixed cultures from anoxic biotrickling filters. Appl Microbiol Biotechnol 99:77–87

Morero B, Groppelli ES, Campanella EA (2017) Evaluation of biogas upgrading technologies using a response surface methodology for process simulation. J Clean Prod 141:978–988

Moss B (2008) Water pollution by agriculture. Philos Trans R Soc Lond B Biol Sci 363:659–666

Mshandete A, Björnsson L, Kivaisi AK, Rubindamayugi MST, Matthiasson B (2006) Effect of particle size on biogas yield from sisal fibre waste. Renew Energy 31:2385–2392

Munda US, Pholane L, Kar DD, Meikap BC (2012) Production of bioenergy from composite waste materials made of corn waste, spent tea waste, and kitchen waste co-mixed with cow dung. Int J Green Energy 9:361–375

Nafees M, Ali S, Naveed M, Rizwan M (2018) Efficiency of biogas slurry and *Burkholderia phytofirmans* PsJN to improve growth, physiology, and antioxidant activity of *Brassica napus* L. in chromium-contaminated soil. Environ Sci Pollut Res 25:6387–6397

Nasir IM, Ghazi TIM, Omar R (2012) Production of biogas from solid organic wastes through anaerobic digestion: a review. Appl Microbiol Biotechnol 95:321–329

Nasr N, Elbeshbishy E, Hafez H, Nakhla G, El Naggar MH (2012) Comparative assessment of single-stage and two-stage anaerobic digestion for the treatment of thin stillage. Bioresour Technol 111:122–126

Nel WP, Cooper CJ (2009) Implications of fossil fuel constraints on economic growth and global warming. Energy Policy 37:166–180

Nishio N, Nakashimada Y (2007) Recent development of anaerobic digestion processes for energy recovery from wastes. J Biosci Bioeng 103:105–112

Obama P, Ricochon G, Muniglia L, Brosse N (2012) Combination of enzymatic hydrolysis and ethanol organosolv pretreatments: effect on lignin structures, delignification yields and cellulose-to-glucose conversion. Bioresour Technol 112:156–163

Ohimain E, Izah SC (2017) A review of biogas production from palm oil mill effluents using different on figurations of bioreactors. Renew Sustain Energy Rev 70:242–253

Ohwaki K, Hungate RE (1977) Hydrogen utilization of Clostridia in sewage sludge. Appl Environ Microbiol 33:1270–1274

Park Y, Hong F, Cheon J, Hidaka T, Tsuno H (2008) Comparison of thermophilic anaerobic digestion characteristics between single-phase and two-phase systems for kitchen garbage treatment. J Biosci Bioeng 105:48–54

Pastor LL, Ruiz AP, Ruiz B (2013) Co-digestion of used oils and urban landfill leachates with sewage sludge and the effect on the biogas production. Appl Energy 107:438–445

Pathak BS, Jain AK, Dev DS (1985) Biogasification of cattle dung and cattle dung – rice straw at different solid concentrations. Agric Wastes 13:251–259

Perera F (2017) Pollution from fossil-fuel combustion is the leading environmental threat to global pediatric health and equity: solutions exist. Int J Environ Res Public Health 15:16

Persson M, Jonsson O, Wellinger A (2007) Biogas upgrading to vehicle fuel standards and grid. IEA Bioenergy:1–32

Phetyim N, Wanthong T, Kannika P, Supngam A (2015) Biogas production from vegetable waste by using dog and cattle manure. Energy Procedia 79:436–444

Philbrook A, Alissandratos A, Easton CJ (2013) Biochemical processes for generating fuels and commodity chemicals from lignocellulosic biomass, environmental biotechnology. In: Marian P (ed) New approaches and prospective applications. InTech, Rijeka, pp 39–64

Poornejad N, Karimi K, Behzad T (2014) Ionic liquid pretreatment of rice straw to enhance saccharification and bioethanol production. J Biomass Biofuel 2:8–15

Poszytek K, Ciezkowska M, Sklodowska A, Drewniak L (2016) Microbial Consortium with High Cellulolytic Activity (MCHCA) for enhanced biogas production. Front Microbiol 7:324
Priadi C, Wulandari D, Rahmatika I, Moersidik SS (2014) Biogas production in the anaerobic digestion of paper sludge. APCBEE Procedia 9:65–69
Puri VP (1984) Effect of crystallinity and degree of polymerization of cellulose on enzymatic saccharification. Biotechnol Bioeng 26:1219–1222
Ranade DR, Yeole TY, Meher KK, Gadre RV, Godbole SH (1989) Biogas from solid waste originated during biscuit and chocolate production: a preliminary study. Bio Wastes 28:157–161
Rani K (2001) Biomethanation of dairy effluent. M.Sc. thesis submitted to CCS Haryana Agric. Univ.
Rao PP, Shivraj D, Seenayya C (1993) Succession of microbial population in cow dung and poultry litter waste digesters during methanogenesis. Ind J Microbiol 33:185–189
Raposo FMA, la Rubia D, Fernández-Cegrí V, Borja R (2012) Anaerobic digestion of solid organic substrates in batch mode: an overview relating to methane yields and experimental procedures. Renew Sustain Energy Rev 16(1):861–877
Ratnaweeraa DR, Saha D, Pingali SV, Labbé N, Naskar AK, Dadmun M (2015) The impact of lignin source on its self-assembly in solution. RSC Adv 5:67258–67266
Razi N, Svendsen HF, Bolland O (2013) Validation of mass transfer correlations for CO2 absorption with MEA using pilot data. Int J Greenhouse Gas Control 19:478–491
Riva C, Schievano A, D'Imporzano G, Adani F (2014) Production costs and operative margins in electric energy generation from biogas. Full-scale case studies in Italy. Waste Manage 34:1429–1435
Riviere D, Desvignes V, Pelletier E, Chaussonnerie S, Guermazi S et al (2009) Towards the definition of a core of microorganisms involved in anaerobic digestion of sludge. ISME J 3:700–714
Ryckebosch E, Margriet D, Han V (2011) Techniques for transformation of biogas to biomethane. Biomass Bioenergy 35:1633–1645
Sanjaya AP, Cahyanto MN, Millati R (2016) Mesophilic batch anaerobic digestion from fruit fragments. Renew Energy 98:135–141
Sarabia Méndez MA, Laines Canepa JR, Sosa Olivier JA, Escalante Espinosa E (2017) Production of biogas by anaerobic codigestion of excreta of lamb and rumen added with sludge from a sewage plant. Revista Internacional de Contaminación Ambiental 33:109–116
Sárvári Horváth I, Tabatabaei M, Karimi K, Kumar R (2016) Recent updates on biogas production - a review. Biofuel Res J 3:394–402
Schink B (1997) Energetics of syntrophic cooperation in methanogenic degradation. Microbiol Rev 61:262–280
Scholz M, Melin T, Wessling M (2013) Transforming biogas into biomethane using membrane technology. Renew Sustain Energy Rev 17:199–212
Sebola R (2015) Studies on the improvement of biogas production from anaerobic digestion of animal waste. Department of Chemical Engineering, University of Johannesburg, Master of technology, Dissertation, p 23
Shafiee S, Topal E (2009) When will fossil fuel reserves be diminished? Energy Policy 37:181–189
Shah DR, Nagarsheth HJ, Acharya P (2016) Purification of biogas using chemical scrubbing and application of purified biogas as fuel for automotive engines. Res J Recent Sci 5:1–7
Singh R, Malik RK, Jain MK, Tauro P (1984) Biogas production at different solid concentration in daily fed cattle waste digesters. Agric Wastes 11:253–257
Singh L, Maurya MS, Ram MS, Alam SI (1993) Biogas production from night soil — effects of loading and temperature. Bioresour Technol 45:59–61
Singhal V, Rain JPN (2003) Biogas production from water hyacinth, channel grass used for phytoremediation of industrial effluents. Bioresour Technol 86:221–225
Singhania RR, Patel AK, Soccol CR, Pandey A (2009) Recent advances in solid-state fermentation. Biochem Eng J 44:13–18
Smith KR, Corvalan CF, Kjellstrom T (1999) How much global ill health is attributable to environmental factors? Epidemiology 10:573–584

Soetaert W, Vandamme EJ (2009) Biofuels in perspective. In: Soetaert W, Vandamme EJ (eds) Biofuels. Wiley, New Jersey, pp 1–8
Sogn TA, Dragicevic I, Linjordet R, Krogstad T, Eijsink VGH, Eich-Greatorex S (2018) Recycling of biogas digestates in plant production: NPK fertilizer value and risk of leaching. Int J Recycl Org Waste Agric 7:49
Srinvasa Reddy N, Satyanarayana SV, Sudha G (2017) Bio gas generation from biodegradable kitchen waste. Int J Environ Agric Biotechnol 2:689–694
Stucki M, Jungbluth N, Leuenberger M (2011) Life cycle assessment of biogas production from different substrates. Final report. Bern: Federal Department of Environment, Transport, Energy and Communications, Federal Office of Energy; 2011 Dec
Suksong W, Kongjan P, Prasertsan P, O-Thong S (2019) Thermotolerant cellulolytic Clostridiaceae and Lachnospiraceae rich consortium enhanced biogas production from oil palm empty fruit bunches by solid-state anaerobic digestion. Bioresour Technol 291:121851
Sun Q, Li H, Yan J, Liu L, Yu Z, Yu X (2015) Selection of appropriate biogas upgrading technology-a review of biogas cleaning, upgrading and utilisation. Renew Sustain Energy Rev 51:521–532
Surendra K, Takara D, Hashimoto AG, Khanal SK (2014) Biogas as a sustainable energy source for developing countries: opportunities and challenges. Renew Sustain Energy Rev 31:846–859
Taherzadeh MJ, Karimi K (2008) Pretreatment of lignocellulosic wastes to improve ethanol and biogas production: a review. Int J Mol Sci 9:1621–1651
Tantayotai P, Rattanaporn K, Tepaamorndech S, Cheenkachorn K, Sriariyanun M (2019) Analysis of an ionic liquid and salt tolerant microbial consortium which is useful for enhancement of enzymatic hydrolysis and biogas production. Waste Biomass Valoriz 10:1481–1491
Tasnim F, Iqbal SA, Chowdhury AR (2017) Biogas production from anaerobic co-digestion of cow manure with kitchen waste and Water Hyacinth. Renew Energy 109:434–439
Tchobanoglous G, Theisen H, Vigil S (1993) Integrated solid waste management: engineering principles and management issues. McGraw-Hill, Inc.
Teghammar A, Karimi K, Sárvári Horváth I, Taherzadeh MJ (2012) Enhanced biogas production from rice straw, triticale straw and softwood spruce by NMMO pretreatment. Biomass Bioenergy 36:116–120
Tewelde S, Eyalarasan K, Radhamani R, Karthikeyan K (2012) Biogas production from co-digestion of brewery waste and cattle dung. Int J Latest Trends Agr Food Sci 2:90–93
Thong SO, Boe K, Angelindaki I (2012) Thermophilic anaerobic co-digestion of oil palm empty fruit bunches with palm mill effluent for efficient biogas production. Appl Energy 93:648–654
Tilman D, Cassman KG, Matson PA, Naylor R, Polasky S (2002) Agricultural sustainability and intensive production practices. Nature 418:671–677
Tock L, Gassner M, Maréchal F (2010) Thermochemical production of liquid fuels from biomass: thermo-economic modeling, process design and process integration analysis. Biomass Bioenergy 34:1838–1854
Traore AS (1992) Biogas production from *Calotropis procera*: a latex plant found in west Africa. Bioresour Technol 41:105–109
Tucho GT, Moll HC, Schoot Uiterkamp AJM, Nonhebel S (2016) Problems with biogas implementation in developing countries from the perspective of labor requirements. Energies 9:750
US Environmental Protection Agency (EPA) (2006) Biosolids technology fact sheet: multi-stage anaerobic digestion. Report. Office of Water, EPA, Washington, DC
van Foreest F (2012) Perspectives for biogas in Europe. Oxford Institute for Energy Studies, Oxford
Vandevivere P, De Baere L, Verstraete W (2002) Types of anaerobic digesters for solid wastes. In: Mata-Alvarez J (ed) Biomethanization of the organic fraction of municipal solid wastes. IWA Publishing, Barcelona, pp 111–140
VanGuilder HD, Vrana KE, Freeman WM (2008) Twenty-five years of quantitative PCR for gene expression analysis. Biotechniques 44:619–626
Viana MB, Freitas AV, Leitão RC, Santaella ST (2012) Biodegradability and methane production potential of glycerol generated by biodiesel industry. Water Sci Technol 66:2217–2222

Wang H, Vuorela M, Keränen A-L, Lehtinen TM, Lensu A, Lehtomäki A, Rintala J (2010) Development of microbial populations in the anaerobic hydrolysis of grass silage for methane production. FEMS Microbiol Ecol 72:496–506
Watts N, Adger WN, Agnolucci P, Blackstock J, Byass P, Cai W, Chaytor S, Colbourn T, Collins M, Cooper A et al (2015) Health and climate change: policy responses to protect public health. Lancet 386:1861–1914
Wei S, Tauber M, Somitsch W, Meincke R, Müller H, Berg G, Guebitz GM (2010) Enhancement of biogas production by addition of hemicellulolytic bacteria immobilised on activated zeolite. Water Res 44:1970–1980
Wei C, Zhang T, Feng C, Wu H, Deng Z, Wu C, Lu B (2011) Treatment of food processing wastewater in a full-scale jet biogas internal loop anaerobic fluidized bed reactor. Biodegradation 22:347–357
Weiland P (2010) Biogas production: current state and perspectives. Appl Microbiol Biotechnol 85:849–860
Wellinger A, Lindeberg A (1999) Biogas upgrading and utilization. Task 24: energy from biological conversion of organic wastes, pp 1–19
Weng JK, Li X, Bonawitz ND, Chapple C (2008) Emerging strategies of lignin engineering and degradation for cellulosic biofuel production. Curr Opin Biotechnol 19:166–172
Wu X, Yao W, Zhu J, Miller C (2010) Biogas and CH(4) productivity by co-digesting swine manure with three crop residues as an external carbon source. Bioresour Technol 101:4042–4047
Wyman CE (1999) Biomass ethanol: technical progress, opportunities, and commercial challenges. Ann Rev Energy Environ 24:189–226
Xiaohua W, Jingfei L (2005) Influence of using household biogas digesters on household energy consumption in rural areas—a case study in Lianshui County in China. Renew Sustain Energy Rev 9:229–236
Xu P, Koffas MAG (2010) Metabolic engineering of *Escherichia coli* for biofuel production. Biofuels 1:493–504
Yabu H, Sakai C, Fujiwara T, Nishio N, Nakashimada Y (2011) Thermophilic two-stage dry anaerobic digestion of model garbage with ammonia stripping. J Biosci Bioeng 111:312–319
Yeh JT, Pennline HW, Resnik KP (2001) Study of CO2 absorption and desorption in a packed column. Energy Fuel 15:274–278
Yorgey G, Frear C, Kruger C, Zimmerman T (2014) The rationale for recovery of phosphorus and nitrogen from dairy manure
Yu L, Ma J, Frear C, Zaher U, Chen S (2013) Two-stage anaerobic digestion systems wherein one of the stages comprises a two-phase system. United States Patent, US 20130309740
Zakrzewski M, Goesmann A, Jaenicke S, Jünemann S, Eikmeyer F, Szczepanowski R, Al-Soud WA, Sørensen S, Pühler A, Schlüter A (2012) Profiling of the metabolically active community from a production-scale biogas plant by means of high-throughput metatranscriptome sequencing. J Biotechnol 158:248–258
Zhang P (2017) Biogas recovery from anaerobic digestion of selected industrial wastes. Adv Biofuels Bioenergy. https://doi.org/10.5772/intechopen.72292
Zhang T, Liu L, Song Z et al (2013) Biogas production by co-digestion of goat manure with three crop residues. PLoS One 8:e66845
Zhao X, Cheng K, Liu D (2009) Organosolv pretreatment of lignocellulosic biomass for enzymatic hydrolysis. Appl Microbiol Biotechnol 82:815–827
Zhong C, Wang C, Wang F, Jia H, Weia P, Zhaob Y (2016) Enhanced biogas production from wheat straw with the application of synergistic microbial consortium pretreatment. RSC Adv 6:60187–60195
Zinoviev S, Müller-Langer F, Das P, Bertero N, Fornasiero P, Kaltschmitt M, Centi G, Miertus S (2010) Next-generation biofuels: survey of emerging technologies and sustainability issues. Chem Sus Chem 3:1106–1133
Zubr J (1986) Methanogenic fermentation of fresh and ensiled plant materials. Biomass 111:159–171

# Chapter 13
# Utilization of Biosensors for Environment Monitoring

**Shalini Singh and Robinka Khajuria**

**Abstract** Rampant release of substances like heavy metals, polychlorinated biphenyls (PCBs), and phenolic and nitrogen compounds in the environment that pollute the ecosystem has made environment security a major global concern. Though various initiatives and legislative actions have been formulated and accepted to overcome the threats of environmental pollution, due to release of such substances in the environment, the control and regulation are still limited and remain a challenge for regulatory bodies, policy makers, and researchers. Therefore, there is a need for developing sensitive and rapid techniques that can detect and screen the pollutants for effective remediation processes. Moreover, the application requirements of most of the traditional methods of detection of contaminants are limited by the obstacles faced in their regular applications. The need for disposable systems for environmental monitoring has thus encouraged the development of new technologies. In this context, biosensors, a category of chemical sensors that utilize biological methods and interventions for detecting the analytes, appear as a suitable alternative tool. They are cost-effective, portable, rapid, and in situ and provide real-time results. The availability of commercial biosensors and technologies to detect pollutants on-site, in place, has greatly enhanced the applicability of biosensors in environment monitoring. This chapter discusses different biosensors used for detection of various pollutants in the environment.

**Keywords** Biosensors · Electrochemical sensors · Enzymatic sensors · Microbial sensor · Tyrosinase

S. Singh (✉)
School of Bioengineering and Biosciences, Lovely Professional University, Phagwara, Punjab, India

R. Khajuria
Department of Biotechnology, Harlal Institute of Management and Technology, Greater Noida, India



J. Singh et al. (eds.), *Microbial Biotechnology: Basic Research and Applications*, Environmental and Microbial Biotechnology,
https://doi.org/10.1007/978-981-15-2817-0_13

## 13.1 Introduction

Environmental security is the most important requirement of human health that has greatly enhanced the value of revealing environmental pollutants through suitable techniques and integral component of environmental monitoring. Stricter regulations by regulatory agencies and a greater consciousness among the general public for environmental issues and environmental safety have further created much interest in both regulated community and the general public in maintaining environmental quality (Rogers 1995). This is particularly true for harmful pollutants including chemicals, toxic substances, and even microbial pathogens, which can cause health hazards for humans as well as decrease the quality of the environment they exist in (Silva et al. 2011). The frequent and accurate monitoring of a broader variety of analytes in the ecosystem (D'Souza 2005) has thus become one of the prime areas of concern for environmentalists all across the globe for overall safety and security of all life forms.

A number of methods, including a range of portable analytical techniques, are available with analysis for monitoring environmental quality (D'Souza 2005), but the search for improved analytical methods still continues. Traditional approaches for environmental monitoring include various chromatographic methods, coupled with electrophoretic and spectrometric methods (Lang et al. 2016; Hassani et al. 2016) (Table 13.1). Still, regular application of most of these traditional analytical methods in environmental monitoring faces multiple obstacles. Such conventional methods usually require expensive reagents and equipment and are time-consuming (Lang et al. 2016; Hassani et al. 2016). The presence of different concentrations of pollutants in case of co-reactive contaminants also poses limitations to the use of conventional methods. In situ measurements, especially in accidental releases of pollutants into the environment, require fast, miniaturized, and portable equipment (Arduini et al. 2013; Guo et al. 2017; Zhang et al. 2015) for analysis, which the traditional methods unfortunately fail to offer.

These analyses thus call for fast, substantial, operational, and economical detection methods in environment applications. So, the requirement for more rapid, efficient one-use systems or techniques for environment monitoring, especially in reference to monitor the environment continuously, has encouraged the development of newer effective techniques and better suited methods. Biosensors, a subgroup of chemical sensors that uses a biological system for analyte detection, thus, are found to be a promising substitute or a supplement to the conventional analytical techniques used for the same purpose (Rodriguez-Mozaz et al. 2004a, b; Rogers 2006; Rogers and Gerlach 1996) for the analyst's armory, especially for continuous monitoring of the environmental pollutants.

In simple terms, a biosensor is a device that incorporates biological features/components as the sensing element, connected to a transducer (D'Souza 2005). Biosensors are considered better than the conventional methods of similar/related analysis owing to their ease in transport, being much smaller and compact than many conventional equipment, use of a small amount of samples for the analysis, etc.

**Table 13.1** Examples of biosensors commercially available for use in environmental monitoring (Adapted from Bahadir and Sezgintürk 2015)

| Instrument | Transducing and recognition element | Company |
|---|---|---|
| Biacore | Optical BI | Biacore AB, Uppsala, Sweden. |
| SPR-CELLIA | Optical whole cells or macromolecules | Nippon Laser and Electronics Lab, Japan |
| Ibis | Optical BI | Windsor Scientific Limited, Berkshire, UK |
| Spreeta | Optical BI | Texas Instruments, Dallas, TX, USA |
| BIOS-I | Optical BI | Artificial Sensing Instruments, Zurich, Switzerland |
| Kinomics Plasmoon | Optical BI | BioTul AG (Munich, Germany) |
| IASys Plus | Optical antibody | Affinity Sensors (UK) |
| Remedios | Optical whole cell | Remedios (Aberdeen, Scotland) |
| Cellsense | Electrochemical *E. coli* | Euroclon Ltd. (Yorkshire, UK) |
| ToxSen | Electrochemical BI | Abtech Scientific, Inc. (Richmond, VA USA) |
| NEC nitrate biosensor | Amperometric enzyme | Nitrate Elimination Co. Inc. (Lake Linden, MI, USA) |
| eTag assay system | Optical eTag reporters | Aclara Bioscience (Mountain View, CA, USA) |
| AgriPo II | Electrochemical | ApexBio (Taiwan) |
| Micredox | Amperometric, whole cell based | Lincoln Ventures Ltd., New Zealand |

## 13.2 Biosensor: Structure and Working

A biosensor is an analytical device that contains a biological or biologically derived sensor that in turn is connected to a transducer system. Analyte to be detected specifically binds to the biological element creating a biochemical signal that is translated into an electronic signal by the transducer (Karim and Fakhruddin 2011; D'Souza 2001). The concept of the biosensor was first explained by M. Cremer in 1906. He demonstrated the concentration of an acid in a liquid is directly proportional to the electric potential arising between two parts of the fluid that are separated by a glass membrane. However, it was in 1956 that Leland C. Clark, Jr., designed the first biosensor for the detection of oxygen which was eventually named Clark electrode. And it is because of this invention that he is known as the "father of biosensors" (Cremer 1906). In 1962, Clark developed an amperometric enzyme electrode for detecting glucose which was followed by development of the first potentiometric biosensor for urea detection in 1969 by Guilbault and Montalvo. Eventually in 1975, Yellow Springs Instrument designed the first commercial biosensor (Bhalla et al. 2016). Since then, remarkable work has been done in the field of biosensors.

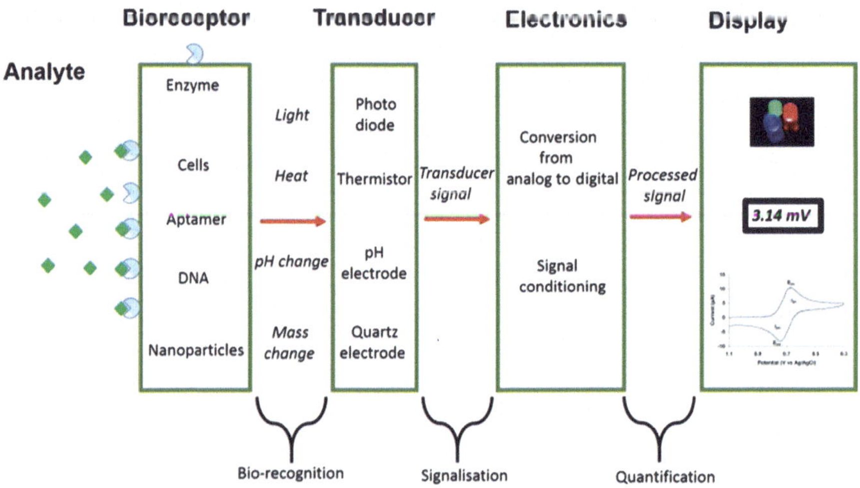


**Fig. 13.1** Conceptual diagram of a typical biosensor (Adapted from Bhalla et al. 2016)

Typically, a biosensor is comprised of the following components (Fig. 13.1):

*Analyte*: It is the molecule of interest that is to be detected. For example, for a glucose biosensor, glucose is an "analyte."

*Bioreceptor*: It is a biological or biologically derived molecule that specifically detects the analyte. It can be a cell, enzyme, antibody, or DNA molecule. The interaction of a bioreceptor with an analyte known as bio-recognition generates a signal. This signal can be in the form of a change in the mass, charge, pH, light, or heat.

*Transducer*: A transducer is responsible for converting the biochemical signals into either an optical or electrical signal. It converts the bio-recognition event into a measurable element that is usually proportionate to the amount of analyte–bioreceptor interactions.

*Electronics*: Process the signal produced by the transducer and convert into a form that can be displayed. It comprises complex electronic circuitry that conditions the signal and converts it from analogue to digital form. The processed signals are then quantified by the display unit of the biosensor.

*Display*: It generally consists of a computer screen or a printer that displays signal in the form of numbers, graphs, tables, or images, depending on the requirements of the end user (Kumar and D'Souza 2012; Bhalla et al. 2016).

## 13.3 Biosensors: Classification

There are two ways of classifying biosensors: either on the basis of the transducer type or on the basis of the bioreceptor type .The following section discusses both classifications briefly.

1. On the basis of transducer, biosensors are classified into four general categories:

   a. Electrochemical transducers: They are portable, highly sensitive, economical, and compatible with micro-fabrication technologies. Up to 80% of the biosensors use electrochemical transducers. There are further various types:

      - *Potentiometric transducers*: In these biosensors, the analytical signal comprises the fall in potential between reference and working electrodes or between two reference electrodes separated from each other by a selectively permeable membrane. The most commonly used transducer is an ion-selective electrode.
      - *Voltammetric transducers*: They measure the oxidation or reduction potential in reference to electroactive species. This current is induced by the production of a preset potential drop between the electrodes. The current generated is found to be proportional to the concentration of electroactive species. The current generated can be proportional to the rate of formation/disappearance of electroactive species in the biocatalytic layer.
      - Impedimetric transducers: They calculate the effective resistance of an electrochemical cell and the differences of the effective resistance.
      - *Conductometric transducers*: They measure the ability of the current to flow through the specific material and hence are used to determine electrical conductivity of the solution during a reaction. Such sensors are not in common use for developing biosensors, especially in case of an enzyme-based biosensor.

   b. Transducers based on field-effect transistors: They employ the use of ion-sensitive silicon field-effect transistors. The bio-sensitive layer is placed above the surface of an ion-sensitive membrane surface that forms a part of the gate of the field-effect transistor. These transducers have enhanced resolving power, thereby imparting higher sensitivity to the biosensor.
   c. Optical transducers: They can use phenomena like fluorescence, absorption, luminescence, reflection, surface plasmon resonance, and light scattering spectroscopy. For example, an immune sensor for detection of casein in milk is based on localized surface plasmon resonance on gold nanoparticles.
   d. Piezoelectric devices: Piezoelectric transducer consists of crystals that undergo elastic deformation when an electric potential is applied. An alternating potential at a specific frequency generates a standing wave in the crystal. The analyte is adsorbed on the surface of the crystal, covered with a bioreceptor, and alters the resonance frequency, indicating binding. Piezoelectric immunosensors are among the most sensitive sensors as they are can detect antigens in the picogram range.
   e. Thermometric transducers: They determine the quantity of heat with the help of a thermistor and use it to determine the analyte concentration. They are not used commonly (Kumar and D'Souza 2012; Korotkaya 2014; Hassani et al. 2016; Malhotra et al. 2017; Alhadrami 2017) for application under discussion (as biosensors).

2. On the basis of bioreceptor, biosensors are categorized as;

   a. *Microbial/whole cell biosensors*: They comprise the application of immobilized cells as the bioreceptor. The whole cell detects the analyte, and the biochemical signal produced is converted by the transducer into a readable form (Fig. 13.2). As compared to a conventional biosensor, whole cell-based biosensors can detect a wider range of analytes and are more sensitive to change in the electrochemical state of a sample or in the environment. Another advantage of microbial biosensors is that they can be genetically modified and thus can be used over a wider range of conditions such as different temperature and pH. Due to their high sensitivity, selectivity, and their capability for in situ detection, they have been used extensively for environmental monitoring, food analysis, pharmacology, and drug screening (Gui et al. 2017). One of the examples of microbial sensors is toxin determination systems based on the inhibition of luciferase, a microbial enzyme that generates luminescence during the oxidation of some substrates. These biosensors are also employed in toxicological studies to determine the lethal concentration of toxicants and in the optimization of doses of antibiotics and the amounts of antimicrobial and antifungal additives for paints and finishing materials. Microbial biosensors are also used to estimate the condition of natural microorganism to monitor the performance of biological wastewater treatment systems.
   b. *Enzymatic sensors*: In these sensors, the bioreceptor can be an enzyme that has been purified. It can also be homogenized tissues or microbial cultures. The simplest enzymatic biosensors are those where the product/substrate of the enzymatic reaction is electrochemically active. The analyte reversibly undergoes oxidation or reduction on the electrode in the presence of a suitable potential. Based on the roles of this category of sensors, enzyme-based sensors are further divided into two categories:

      - Substrate-based biosensors: These are designed to determine specific substrates involved in enzymatic reactions. For example, enzyme-based sensor using glucose oxidase activity for glucose determination or urea determination by urease sensor.

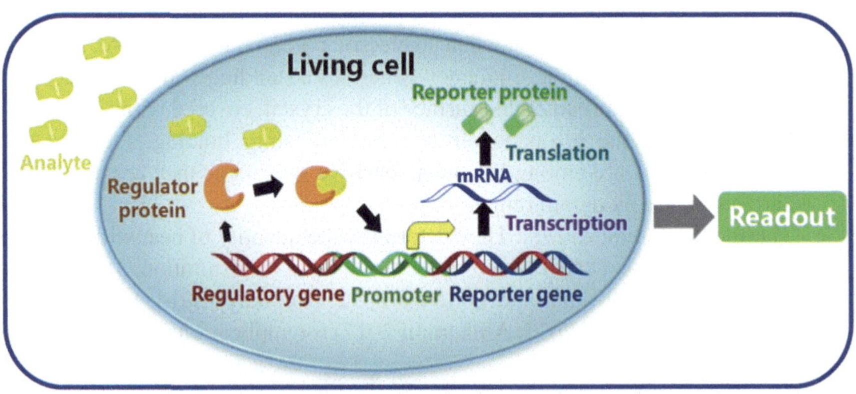


**Fig. 13.2** Schematic representation of a whole cell-based biosensor (Gui et al. 2017)

- Inhibitor sensors: These are designed to determine the presence of substances that reduce the enzymatic activity and organophosphorus pesticides that inhibit breakdown of acetylcholine by the enzyme; acetylcholine esterase is an example of inhibitor biosensor (Kitova et al. 2004; Zaitsev 2013; Korotkaya 2014). Due to high surface area, biocompatibility, and reduced in vivo toxicity, gold nanostructures provide an efficient platform for immobilization of enzymes (Lang et al. 2016; Zhao et al. 2013).
  - *Immunosensors*: Immunosensors, also called affinity ligand-based sensors, determine the communications between immobilized antibodies or immobilized antigens that are present on the surface of transducing element. The exclusive specificity of the immobilized antibodies against the specific immobilized antigen actually dictates the working of these biosensors. The antibodies offer high specificity and ability to regenerate and are found to be highly sensitive in detection. This enhances the degree of reliability and efficacy of such sensors. The binding of the specific antibody onto the solid matrix eases repeated application of the antibody-based sensors (Kandimalla et al. 2004). The detection of specific antibodies in blood samples indicates some microbial infection or presence of certain toxic substances. The antigens can be measured both in biological samples and other types of samples like environment-related ones. Based on detection of a unique antibody, this particular class of sensors can be used for measuring a wide variety of compounds with great selectivity as well as specificity (Korotkaya 2014; Mauriz et al. 2007; OWLS 2006; Soh et al. 2003).
  - *DNA sensors*: They consist of DNA-based probes using single-stranded DNA molecules. The DNA strand is directly bound to transducer's surface and determines the particular binding with DNA. They generally employ electricity, heat, and light as signals as a part of transducing element. However, electrochemical transducer is used most commonly. Electrochemical DNA biosensors, which undergo formation of the interface for recognizing the DNA molecule, followed by occurrence of DNA hybridization and translation or transformation of hybridized system into a readable electric signal (Teles and Fonseca 2008; Vercoutere and Akeson 2002), are the most commonly used ones in this category of biosensors.

## 13.4 Application of Biosensors in Environment Monitoring

### *13.4.1 Heavy Metals*

Heavy metal-induced environment pollution is a serious threat, where anthropogenic activities leading to release of metals like copper, mercury, cadmium, zinc, chromium, lead, zinc, arsenic, etc., into the environment disrupt the metal balance in the ecosystem. Human activities like combustion greatly add these heavy metals into the environment, far more in concentration than naturally present in the ecosystem. The uptake of these metals by life forms thus exceeds the limits of required intake for organisms. Due to the recalcitrant nature of these metals, they tend to accrue in the environment through food chain and become one of the causes of toxicity in living organisms. Early detection of metal contaminants guided by strict regulations on heavy metal pollution in the environment is thus required, and biosensors can significantly contribute toward this direction. The enzyme-based biosensors, especially those utilizing oxidative enzymes as recognition elements, have been prominently applied for detection of heavy metals (Amine et al. 2006). Inhibition-based enzyme sensors involving enzymes like alkaline phosphatase, urease, horseradish peroxidase, etc., with immobilized enzyme systems are commonly applied for detection of cadmium, copper, chromium, lead, etc. (Rodriguez-Mozaz et al. 2006; Rodriguez-Mozaz et al. 2004a, b; Amine et al. 2006). Nomngongo et al. (2011) and Silwana et al. (2014) reported the use of amperometric sensors using horseradish peroxidase for detecting cadmium, lead, mercury, and copper. The detection was based on measuring cathodic currents from water samples containing trace metals and hydrogen peroxide. Inhibition of enzyme activity in the presence of metal, subsequently leading to decrease in cathodic current, helped to determine the metal concentration in the water sample. It was also confirmed that the enzyme inhibition process was both reversible and noncompetitive. These sensors were important as they could screen even very low concentration of heavy metals in water. The use of carbon nanotubes and horseradish peroxidase has also been successfully used for detecting copper, lead, and cadmium in aqueous solutions, where the decrease in release of signal was found to be directly proportional to the concentration of the inhibitor in the tested solutions (Moyo and Okonkwo 2014; Moyo et al. 2014). Tyrosinase, immobilized over a glass-carbon electrode, detected Cr(III) in aqueous solutions as well (Domínguez Renedo et al. 2004), where enzyme inhibition was again found to decrease electrical signals in inverse ratio of contaminant present in the water sample. These biosensors were successfully applied for chromium recognition in a variety of samples.

### *13.4.2 Pesticides*

Chemical pesticides are one of the most dreadful environmental pollutants, as they show acute toxicity and are recalcitrant to degradation, persisting in the environment for long durations. Strict regulations by regulatory bodies for the release of pesticides are in place and are to comply with, and hence, fast and competent recognition of such environmental pollutants is very important (Rebollar-Perez et al. 2015) for further application of faster and improved control of their release into the environment. The enzymatic sensors are the most used for finding out the presence of pesticides in soil and water. Enzyme inhibition-based methods involving acetylcholinesterase and cholin oxidase and enzyme-based sensors using laccases, horseradish peroxidase, and peroxidases as recognition elements have all been reported to be used for detection of different types of pesticides (Van Dyk and Pletschke 2011; Mercurio et al. 2014; Hsu and Whang 2009; Oliveira et al. 2012; Songa et al. 2009; Qian et al. 2009; Hanke et al. 2008; Pundir and Chauhan 2012; Miao et al. 2010). Glyphosate has been detected using immobilized peroxidase and horseradish peroxidase (Oliveira et al. 2012; Songa et al. 2009), while carbamates like formetanate hydrochloride, commonly used as an insecticide for fruit crops and reported to have serious health and environment hazards, have been detected using laccase-based biosensors (Ribeiro et al. 2014). Chlorophenols have been found to be associated with high toxicity and biorefractive ability and tend to accumulate in the ecosystem. Immobilized horseradish peroxidase in an amperometric biosensors was successfully used for detection of 4-chlorophenol. The decrease in 4-chlorophenol mediator on the surface of the electrode was used to indicate concentration of the given chlorophenol. The electrode response was proportional to the concentration of 4-chlorophenol (Qiu et al. 2013).

### *13.4.3 Biological Oxygen Demand (BOD)*

Biochemical oxygen demand (BOD) is a commonly used parameter for determination of environmental pollution. The conventional 5-day BOD analysis, where the BOD of effluent waters is measured over a period of 5 days at specific temperature (20 °C), is commonly used in analysis of water samples. Still, because of the time-consuming, complex, variable nature of this monitoring system, conventional BOD analysis is now quickly being replaced by BOD sensors using biological elements, called biosensors, and they have now become an integral part of wastewater analysis (Bahadır and Sezginturk 2015; Chee 2013). Since the time the first microbial BOD sensor was introduced in 1983 by Nisshin Denki (Electric) Co. Ltd., different types of BOD biosensors (like Ra-BOD, AppliTek, Belgium; Biox-1010, Endress+Hauser, Switzerland, etc.) have been commercialized, including both biofilm and bioreactor types. Some of the pollutants successfully detected by such biosensors include nitrate, dioxins and dioxin-like compounds, etc. Even microbial pollutant indicators (*E. coli*) have been successfully detected by BOD biosensors (Liu and Mattiasson 2002; Jouanneau et al. 2014; Rodriguez-Mozaz et al. 2005a, b).

### *13.4.4 Nitrogen Compounds*

Nitrogen compounds are used extensively as fertilizers in agricultural fields. Their excessive use had led to increased levels of nitrates in groundwater and surface water that can have a negative impact on aquatic and human life. Keeping this in mind, the regulations for controlling the nitrate levels in domestic and industrial sewage treatment plants have been implemented (Silva et al. 2011). Amperometric biosensors containing immobilized cytochrome c nitrite reductase (ccNiR) from *Desulfovibrio desulfuricans* were developed to determine the nitrate concentrations between the concentrations of nitrite 0.015 and 2.35 μM and a detection limit of 4 nM (Chen et al. 2007) was achieved. Conductometric electrodes with nitrate reductase from *Aspergillus niger* could detect nitrates in the range of 0.02 and 0.25 mM with detection limits of 0.005 mM nitrate (Khadro et al. 2008). Albanesea et al. (2010) developed an amperometric screen printed biosensor containing nitrate reductase by *Escherichia coli* to detect nitrate ions in water. Gokhale et al. (2015) reported a disposable nitrate biosensor consisting of a bioreceptor made up of nitrate reductase that has been immobilized on a conductive polymer matrix to generate a quantifiable amperometric response. A variety of biosensors for nitrite detection using viologen-modified sulfonated polyaminopropylsiloxane (PAPS-SO3H-V), myoglobin (Mb), copper-containing nitrite reductase (Cu-NiR, from *Rhodopseudomonas sphaeroides* sp. *denitrificans*), cytochrome c (Cyt c), and viologen-modified chitosan (CHIT-V) have been designed (Yilong et al. 2015).

### 13.4.5 Phenolic Compounds

They comprise of a large group of pollutants present in the effluents of a variety of industries associated with the manufacturing of paper, plastics, pesticides, herbicides, dyes, and drugs (Kumar and D'Souza 2012). Chloro- and nitrophenols are also produced during the degradation of organophosphorus pesticides and chlorinated phenoxyacids. These compounds have been reported to exhibit severe toxicity including genotoxicity and mutagenicity in animals and decreased rates of life processes like photosynthesis, respiration, and enzyme-catalyzed reactions. Hence, phenols are listed as hazardous materials by the European Commission and the US Environmental Protection Agency (Nigam and Shukla 2015). A number of phenol-detecting biosensors have been developed for monitoring the presence of phenols in the environment. The biosensors either use microorganisms or enzymes in free or immobilized forms.

### 13.4.6 Microbial Biosensors

The use of live microorganisms in biosensors for determining the amount of analytes is established upon the detection of specific type of enzyme systems in microbial cells. Whole cell biosensors have the benefit of low cost and improved stability compared to enzyme-based sensors. In addition, Microbes can be easily manipulated and have better stability under harsh conditions. Enzyme-based sensors require difficult enzyme purification methods, while cells can be produced in large amounts through a simple cell culturing. Under aerobic conditions, the first step of phenol consumption involves the phenol being oxygenated by hydroxylases to yield catechol which is succeeded by breakdown of the ring found near –OH groups of the catechol or breakdown of the ring existing between the two –OH groups of catechol. The enzyme catechol 1,2-dioxygenase is considered to be an important enzyme in the degradation of aromatic pollutants such as phenols (Park et al. 2013).

Majority of the microorganism-based sensors are based on the amperometric detection. A number of bacteria such as *Rhodococcus*, *Trichosporon*, *Moraxella*, and *Pseudomonas* have been reported to detect phenol and chlorophenol (Mulchandani et al. 2005). Riedel et al. (1995) developed a sensor based on amperometric detection for finding out the presence of phenols and chlorinated phenols using the fungus *Trichosporon beigelii*. *Pseudomonas putida* DSM 50026 has been used in an amperometric detection of phenol too (Timur et al. 2003). Other amperometric biosensors using *Rhodococcus* sp., *Pseudomonas putida* GFS-8, and *Pseudomonas* sp. 83-IV with a detection limit of 4 μM, 1 μM, and 10 μM, respectively, have also been developed for monitoring phenol concentrations in the environment (Skladal et al. 2002). Another amperometric biosensor comprising *Arthrobacter* sp., bound to a Nafion polymer deposited on a carbon electrode as transducing element, has also been reported. This sensor measured the oxidation current of 4-nitrocatechol and 1,2,4-benzenetriol, which are the intermediates formed, produced by the oxidation of p-nitrophenol by bacterial species under study (Lei et al. 2004). Lyophilized *Lactobacillus* immobilized on a Teflon-membrane oxygen electrode have been developed for determining the presence of catechols in dairy products and wastewater (Sagiroglu et al. 2011). The sensing element determines the concentration of catechol, which in turn indicates dissolved oxygen value changes. Shin (2010) developed a sensor based on bioluminescent detection of salicylate using genetically modified *E. coli* sp. transformed using a pNRSAL plasmid that contained nahR and reporter (luciferase) genes. The modifications lead to significant changes in the microbial activity in response to salicylic acid, with highly enhanced sensitivity. Sakti et al. (2016) explained the use of the bacterium *Pseudomonas putida* in a potentiometric phenol biosensor with a glass electrode transducer. Kolahchi et al. (2018) developed a conductometric bacterial sensor using *Pseudomonas* sp. that was obtained from contaminated soil of a local oil refinery. The detection limit for phenol in the given study was 2 μM.

Enzymatic biosensors: Enzymes are frequently used biological systems in biological sensors (Rodriguez-Mozaz et al. 2006). The pure enzymes are generally used in the assembly of these sensors owing to high specificity related with such biological molecules. Enzyme biosensors offer various advantages such as ability to genetically change the catalytic properties and amplification of the biosensor response by modulating the enzyme activity (Rogers and Mascini 2009). A number of enzymes such as tyrosinase, laccase, and horseradish peroxidase (HRP) are reported for the development of phenol detecting biosensors. These enzymes have been choice of research as they have good ability to oxidize a variety of phenolic as well as non-phenolic aromatic hydrocarbons. These enzymes are produced extracellularly by many bacteria, fungi, plants, and even insects. Each of these enzymes has their own specific reaction mechanisms for detecting different phenolics. In case of laccase and tyrosinase, the enzymes are first oxidized followed by reduction of phenolic compounds. Biosensors using tyrosinase can be used for phenol containing compounds with phenol ring free of substituent group at ortho position, while the laccase-based biosensors are usually used for phenolic compounds with para position free of substituent group or the meta-position free of substituent group. On the other hand, HRP gets reduced by phenolics followed by oxidation using hydrogen peroxide (Karim and Fakhruddin 2011).

Various types of enzyme sensors have been described by various workers for finding the presence of phenols in the environment. An electrochemical biosensor containing tyrosinase entrapped in a matrix of agarose and guar gum has also been reported. It was developed for the detection of catechol. In another study, tyrosinase was entrapped on oxidized porous silicon, and a conductometric biosensor was developed for quantitative estimation of catechol (Kumar and D'Souza 2012). A bienzymatic biosensor consisting of tyrosinase and laccase was developed using titania sol–gel matrix by Kochana et al. (2008). The biosensor has a significantly high sensitivity for various phenol containing compounds like chlorophenol and methylcatechols. Similarly, an amperometric biosensor containing horseradish peroxidase-modified electrodes was designed for the detection of phenolic compounds (Korkut et al. 2008). An amperometric biosensor using HRP bound to Au film modified electrode has been reported for the detection of 4-chlorophenol. The determination is based on the reduction of a specific mediator on the electrode, and the amperometric response is found to be proportional to the concentration (Qiu et al. 2013) of the analyte.

## 13.5 Current Status and Future of Biosensors

Biosensors are attracting a lot of attention as agents to monitor environment quality, but a significant number of limitations associated with biosensors still limit their applications. A large number of biosensors have been evaluated in closely controlled conditions where distilled water or buffered solutions have been used. Such testing fails to give real estimates of the working efficiency of biosensors, real time. The huge variety of compounds to be tested and complexity of samples also bring about limitations to the efficiency of biosensors. Cost is another factor that has been affecting biosensor technology. Some of the major cost-intensive areas include the development cost for single analyte systems, developing sufficient marker to decrease cost of bringing a laboratory scale prototype to real-time market scale-up, cost involving conducting field trials, and expenses incurred due to evaluation and validation by regulatory bodies for approval of application-oriented applications. A lot of time is spent in bringing a laboratory prototype to the market, through trials and approvals from regulatory agencies, which needs to be reduced for large-scale applicability of the technology for environment monitoring. There is a need to increase the range of detectable analytes with potable device structure, as the concerned area still needs to expand. Limited shelf and operational life times for pre-manufactured bio-recognition components and complexity in devising potentially portable biosensor systems are other areas of concern in the use of biosensors. Improved and enhanced integration of biosensor technology with other allied ones including biochemistry, polymer chemistry, electronics, micro-fluidics, and separation technology will further be needed for more comprehensive and robust systems to develop, as this fusion of biological sciences with other disciplines will help to realize the full potential of biosensors in the coming times (Arduini et al. 2013; Guo et al. 2017; Zhang et al. 2015).

At the same time, changing times have brought newer hopes and promises with appearance of more biosensors that can be applied to real samples in the past few years. With availability of newer and improved genetic modification of enzymes and microorganisms, improvement in recognition element immobilization, and sensor interfaces, the applicability of biosensors has increased manifold. Developments in multi-pollutant screening have enhanced the significance of biosensors in environmental monitoring. Micro-Electro-Mechanical Systems that integrate mechanical elements, sensors, actuators, and electronics on a common silicon substrate through micro-fabrication technology, yielding improved biochip and sensor array system, are one of the emerging and promising areas in biosensor technology.

The development of aptasensors in recent times further adds to the efficacy of biosensors. These aptamers are easy to modify, are stable to heat, and exhibit stability during in vitro synthesis, apart from many other desirable characteristics (Justino et al. 2015).

As highlighted above, nanotechnology, as with other fields, can play a crucial role in biosensor technology as well (Maduraiveeran and Jin 2017). Thus, development of fast and smart bio-sensing devices, for improved availability of commercial biosensors, is very much required for adding to the success of biosensor-based technology for environment monitoring. The commercially available biosensors have been successfully used for carrying out toxicity assay, evaluating quality of liquids, and determining concentration of specific analytes, including those in waste streams, detecting volatile hydrocarbons, non-volatile hydrocarbons, heavy metals, and whole cells, detecting microbial pathogens, etc.

These new age biosensors shall be good enough for in situ measurements and for overcoming current technical and future challenges in the given field. Also, binding and stabilization of biomolecules on such nano-devices also needs extensive exploration and applications, as these areas are still in their nascent stages (Gau Jr et al. 2001; Mello and Kubota 2002) of development and usage.

## 13.6 Conclusion

It is clearly evident that biosensors offer the following advantages: they perform evaluations faster than the conventional methods, are simple in operation, are cheaper as compared to conventional techniques, which tend to be costlier, and even offer high level of sensitivity to the detection. The past few decades have seen a tremendous rise in development and applications of biosensors owing to increased interests among the stakeholders. Further improvements in biosensor technology have been attributed to the innovative efforts by researchers to integrate inputs from nanotechnology, bioinformatics, and bioelectronics into biosensors in a compatible system, which has further brought down the overall cost of usage of biosensors in detection of environmental pollutants. In spite of the huge development of biosensor instrumentation, the applicability of biosensors in monitoring and detection of environment related pollutants, the real-time applications are still higher in medical and food-based applications field is still restricted, as compared to medical and food-related applications. Limitations with detection and specificity in reference to different analytes will hold the key to further applications of biosensing in environment monitoring. The interference caused by complex nature of environmental matrix in analyte detection further needs to be improvised upon. Additional research is needed to manufacture reliable devices that are not only good at detecting pollutants but are also a commercial success. Thus, the looming gaps in screening and monitoring technology matrix need to be bridged up for overcoming environmental problems through well-defined and highly competitive monitoring systems.

## References

Albanesea D, Matteoa MD, Alessioa C (2010) Screen printed biosensors for detection of nitrates in drinking water. Computer Aided Chemical Engineering 28:283–288

Alhadrami HA (2017) Biosensors: classifications, medical applications, and future prospective. Biotechnol Appl Biochem 65(3):497–508

Amine A, Mohammadi H, Bourais I, Palleschi G (2006) Enzyme inhibition-based biosensors for food safety and environmental monitoring. Biosens Bioelectron 21:1405–1423

Arduini F, Guidone S, Amine A, Palleschi G, Moscone D (2013) Acetylcholinesterase biosensor based on self-assembled monolayer-modified gold-screen printed electrodes for organophosphorus insecticide detection. Sens Actuators B Chem 179:201–208

Bahadir EB, Sezgintürk MK (2015) Applications of commercial biosensors in clinical, food, environmental, and biothreat/biowarfare analyses. Anal Biochem 478:107–120

Bahadır EB, Sezginturk MK (2015) Applications of commercial biosensors in clinical, food, environmental, and biothreat/biowarfare analyses. Anal Biochem 478:107–120

Bhalla N, Jolly P, Formisano N, Estrela P (2016) Introduction to biosensors. Essays Biochem 60(1):1–8. https://doi.org/10.1042/EBC20150001

Chee G (2013) Development and characterization of microbial biosensors for evaluating low biochemical oxygen demand in rivers. Talanta 117:366–370

Chen H, Mousty C, Cosnier S, Silveira C, Moura JJG, Almeida MG (2007) Highly sensitive nitrite biosensor based on the electrical wiring of nitrite reductase by [ZnCr-AQS] LDH. Electrochem Commun 9:2240–2245

Cremer M (1906) Über die Ursache der elektromotorischen Eigenschaften der Gewebe, zugleich ein Beitrag zur Lehre von den polyphasischen Elektrolytketten. Z Biol 47:562–608

D'Souza SF (2001) Biosens Bioelectron 16:337

D'Souza SF (2005) Environmental biosensors. IANCAS Bull 4:54–59

Domínguez Renedo O, Alonso Lomillo MA, Arcos Martinez MJ (2004) Optimisation procedure for the inhibitive determination of chromium(III) using an amperometric tyrosinase biosensor. Anal Chim Acta 521:215–221

Gau J Jr, Lan JEH, Dunn B, Ho CM, Woo JCS (2001) MEMS based amperometric detector for *E. coli* bacteria using self-assembled monolayer. Biosens Bioelectron 16:745

Gokhale AA, Lu J, Weerasiri RU, Yu J, Lee I (2015) Amperometric detection and quantification of nitrate ions using a highly sensitive nanostructured membrane electrocodeposited biosensor array. Electroanalysis 27:1127–1137

Gui Q, Lawson T, Shan S, Yan L, Liu Y (2017) The application of whole cell-based biosensors for use in environmental analysis and in medical diagnostics. Sensors 17(7):1623. https://doi.org/10.3390/s17071623

Guo L, Li Z, Chen H, Wu Y, Chen L, Song Z, Lin T (2017) Colorimetric biosensor for the assay of paraoxon in environmental water samples based on the iodine-starch color reaction. Anal Chim Acta 967:59–63

Hanke I, Singer H, Hollender J (2008) Ultratrace-level determination of glyphosate, aminomethylphosphonic acid and glufosinate in natural waters by solid-phase extraction followed by liquid chromatography–tandem mass spectrometry: performance tuning of derivatization, enrichment and detection. Anal Bioanal Chem 391:2265–2276

Hassani S, Momtaz S, Vakhshiteh F, Maghsoudi AS, Ganjali MR, Norouzi P, Abdollahi M (2016) Biosensors and their applications in detection of organophosphorus pesticides in the environment. Arch Toxicol 91(1):109–130. https://doi.org/10.1007/s00204-016-1875-8-2016

Hsu C-C, Whang C-W (2009) Microscale solid phase extraction of glyphosate and aminomethylphosphonic acid in water and guava fruit extract using alumina-coated iron oxide nanoparticles followed by capillary electrophoresis and electrochemiluminescence detection. J Chromatogr 1216:8575–8580

Jouanneau S, Recoules L, Durand MJ, Boukabache A, Picot V, Primault Y, Lakel A, Sengelin M, Barillon B, Thouand G (2014) Methods for assessing biochemical oxygen demand (BOD): a review. Water Res 49:62–82

Justino CIL, Freitas AC, Pereira R, Duarte AC, Rocha-Santos TAP (2015) Recent developments in recognition elements for chemical sensors and biosensors. Trends Anal Chem 68:2–17

Kandimalla V, Neeta N, Karanth N (2004) Regeneration of ethyl parathion antibodies for repeated use in immunosensor: a study on dissociation of antigens from antibodies. Biosens Bioelectron 20(4):903–906

Karim F, Fakhruddin ANM (2011) Recent advances in the development of biosensor for phenol: a review. Rev Environ Sci Biotechnol 11:261–274. https://doi.org/10.1007/s11157-012-9268-9

Khadro B, Namour P, Bessueille F, Leonard D, Jaffrezic-Renault N (2008) Enzymatic conductometric biosensor based on PVC membrane containing methyl viologen/Nafion®/nitrate reductase for determination of nitrate in natural water samples. Sens Mater 20:267–279

Kitova AE, Alferov VA, Ponamoreva ON (2004) Microbial biocatalysts and outlooks for the development of enzyme technologies in processing sectors of agribusiness. Pishchepromizdat, Moscow, pp 255–262

Kochana J, Nowak P, Jarosz-Wilkołazka A, Bieron M (2008) Tyrosinase/laccase bienzyme biosensor for amperometric determination of phenolic compounds. Microchem J 89:171–174

Kolahchi N, Braiek M, Ebrahimipour G, Ranaei-Siadat SO, Lagarde F, Jaffrezic-Renault N (2018) Direct detection of phenol using a new bacterial strain-based conductometric biosensor. J Environ Chem Eng 6:478–484. https://doi.org/10.1016/j.jece.2017.12.023

Korkut S, Keskinler B, Erhan E (2008) An amperometric biosensor based on multiwalled carbon nanotube-poly(pyrrole)-horseradish peroxidase nanobiocomposite film for determination of phenol derivatives. Talanta 76(5):1147–1152

Korotkaya V (2014) Biosensors: design, classification, and applications in the food industry. Foods Raw Mater 2(2):161–171

Kumar J, D'Souza SF (2012) Biosensors for environmental and clinical monitoring. BARC Newsl 324:34–38

Lang Q, Han L, Hou C, Wang F, Liu AA (2016) Sensitive acetylcholinesterase biosensor based on gold nanorods modified electrode for detection of organophosphate pesticide. Talanta 156:34–41

Lei Y, Mulchandani P, Chen W, Wang J, Mulchandani A (2004) Arthrobacter sp. JS443-based whole cell amperometric biosensor for p-nitrophenol. Electroanalysis 6:24

Liu J, Mattiasson B (2002) Microbial BOD sensors for wastewater analysis. Water Res 36:3786–3802

Maduraiveeran G, Jin W (2017) Nanomaterials based electrochemical sensor and biosensor platforms for environmental applications. Trends Environ Anal Chem 13:10–23

Malhotra S, Verma A, Tyagi N, Kumar V (2017) Biosensors: principle, types and applications. Int J Adv Res Innovative Ideas Edu 3(2):3639–3644

Mauriz E, Calle A, Manclús JJ, Montoya A, Hildebrandt A, Barceló D, Lechuga LM (2007) Optical immunosensor for fast and sensitive detection of DDT and related compounds in river water samples. Biosens Bioelectron 22:1410–1418

Mello LD, Kubota LT (2002) Review of the use of biosensors as analytical tools in the food and drink industries. Food Chem 77:237

Mercurio P, Flores F, Mueller JF, Carter S, Negri AP (2014) Glyphosate persistence in seawater. Mar Pollut Bull 85:385–390

Miao Y, He N, Zhu J-J (2010) History and new developments of assays for cholinesterase activity and inhibition. Chem Rev 110:5216–5234

Moyo M, Okonkwo JO (2014) Horseradish peroxidase biosensor based on maize tassel–MWCNTs composite for cadmium detection. Sens Actuators B Chem 193:515–521

Moyo M, Okonkwo JO, Agyei NM (2014) An amperometric biosensor based on horseradish peroxidase immobilized onto maize tassel-multi-walled carbon nanotubes modified glassy carbon electrode for determination of heavy metal ions in aqueous solution. Enzym Microb Technol 56:28–34

Mulchandani P, Hangarter C, Lei Y, Chen W, Mulchandani A (2005) Amperometric microbial biosensor for p-nitrophenol using Moraxella Sp.-modified carbon paste electrode. Biosens Bioelectron 21(3):523–527

Nigam VK, Shukla P (2015) Enzyme based biosensors for detection of environmental pollutants-a review. J Microbiol Biotechnol 25(11):1773–1781

Nomngongo PN, Ngila JC, Nyamori VO, Songa EA, Iwuoha EI (2011) Determination of selected heavy metals using amperometric horseradish peroxidase (HRP) inhibition biosensor. Anal Lett 44:2031–2046

Oliveira GC, Moccelini SK, Castilho M, Terezo AJ, Possavatz J, Magalhaes MRL, Dores EFGC (2012) Biosensor based on atemoya peroxidase immobilised on modified nanoclay for glyphosate biomonitoring. Talanta 98:130–136

OWLS (2006) OWLS application notes no-002. Label-free immunosensor for z trifluralin detection with OWLS. Available online http://www.owls-sensors.com/pdf/Application-Note-2-Labelfree-immunosensor-for-herbicide-trifluralin-detection.pdf

Park M, Tsai SL, Wilfred C (2013) Microbial biosensors: engineered microorganisms as the sensing machinery. Sensors 13(5):5777–5795

Pundir CS, Chauhan N (2012) Acetylcholinesterase inhibition based biosensors for pesticide determination: a review. Anal Biochem 429:19–31

Qian K, Tang T, Shi T, Wang F, Li J, Cao Y (2009) Residue determination of glyphosate in environmental water samples with high-performance liquid chromatography and UV detection after derivatization with 4-chloro-3,5-dinitrobenzotrifluoride. Anal Chim Acta 635:222–226

Qiu C, Chen T, Wang X, Li Y, Ma H (2013) Application of horseradish peroxidase modified nanostructured Au thin films for the amperometric detection of 4-chlorophenol. Colloids Surf B Biointerfaces 103:129–135

Rebollar-Perez G, Campos-Teran J, Ornelas-Soto N, Mendez-Albores A, Torres E (2015) Biosensors based on oxidative enzymes for detection of environmental pollutants. Biocatalysis 1:118–129

Ribeiro FWP, Barroso MF, Morais S, Viswanathan S, de Lima-Neto P, Correia AN, Oliveira MBPP, Delerue-Matos C (2014) Simple laccase-based biosensor for formetanate hydrochloride quantification in fruits. Bioelectrochemistry 95:7–14

Riedel K, Beyersdorf-Radeck B, Neumann B, Schaller F (1995) Microbial sensors for determination of aromatics and their chloro derivatives. Part III: determination of chlorinated phenols using a biosensor containing Trichosporon beigelii (cutaneum). Appl Microbiol Biotechnol 43(1):7–9

Rodriguez-Mozaz S, Marco M-P, Alda MJL, Barceló D (2004a) Biosensors for environmental applications: future development trends. Pure Appl Chem 76:723–752

Rodriguez-Mozaz S, Marco M-P, Lopez de Alda MJ, Barcelo D (2004b) Biosensors for environmental applications: future development trends. Pure Appl Chem 76:723–752

Rodriguez-Mozaz S, Lopez de Alda MJ, Marco M, Barcel D (2005a) Biosensors for environmental monitoring: a global perspective. Talanta 65:291–297

Rodriguez-Mozaz S, Alda MJLD, Marco M-P, Barcelo D (2005b) Biosensors for environmental monitoring: A global perspective. Talanta 65:291–297

Rodriguez-Mozaz S, Lopez de Alda M, Barcelo D (2006) Biosensors as useful tools for environmental analysis and monitoring. Anal Bioanal Chem 386:1025–1041

Rogers KR (1995) Biosensors for environmental applications. Biosens Bioelectron 10:533–541

Rogers KR (2006) Recent advances in biosensor techniques for environmental monitoring. Anal Chim Acta 568:222–231

Rogers KR, Gerlach CL (1996) Environmental biosensors: a status report. Environ Sci Technol 30:486–491

Rogers KR, Mascini M (2009) Biosensors for analytical monitoring-general introduction and review. US Environmental Protection Agency. http://www.epa.gov/heasd/edrb/biochem/intro.html. Accessed 1 May 2011

Sagiroglu A, Paluzar H, Ozcan HM, Okten S, Sen B (2011) A novel biosensor based on lactobacillus acidophilus for determination of phenolic compounds in milk products and wastewater. Prep Biochem Biotechnol 41:321–336. https://doi.org/10.1080/10826068.2010.540607

Sakti SP, Sugiarti S, Juswono P (2016) Development of simple bacterial biosensor for phenol detection in water at medium concentration using glass microelectrode. J Pure Appl Chem Res 5:1–8

Shin HJ (2010) Development of highly-sensitive microbial biosensors by mutation of the nahR regulatory gene. J Biotechnol 150:246–250. https://doi.org/10.1016/j.jbio-tec.2010.09.936

Silva LMC, Melo AF, Salgado A (2011) Biosensors for environmental applications. In: Environmental biosensors. InTech, Croatia

Silwana B, Van Der Horst C, Iwuoha E, Somerset V (2014) Amperometric determination of cadmium, lead, and mercury metal ions using a novel polymer immobilised horseradish peroxidase biosensor system. J Environ Sci Heal A 49:1501–1511

Skladal P, Morozova NO, Reshetilov AN (2002) Amperometric biosensors for detection of phenol using chemically modified electrodes containing immobilized bacteria. Biosens Bioelectron 17:867–873

Soh N, Tokuda T, Watanabe T, Mishima K, Imato T, Masadome T, Asano Y, Okutani S, Niwa O, Brown SA (2003) Surface plasmon resonance immunosensor for detecting a dioxin precursor using a gold binding polypeptide. Talanta 60:733–745

Songa EA, Arotiba OA, Owino JHO, Jahed N, Baker PGL, Iwuoha EI (2009) Electrochemical detection of glyphosate herbicide using horseradish peroxidase immobilized on sulfonated polymer matrix. Bioelectrochemistry 75:117–123

Teles F, Fonseca L (2008) Trends in DNA biosensors. Talanta 77(2):606–623

Timur S, Pazarlioglu N, Pilloton R, Telefoncu A (2003) Detection of phenolic compounds by thick film sensors based on *Pseudomonas putida*. Talanta 61:87–93

Van Dyk JS, Pletschke B (2011) Review on the use of enzymes for the detection of organochlorine, organophosphate and carbamate pesticides in the environment. Chemosphere 82:291–307

Vercoutere W, Akeson M (2002) Biosensors for DNA sequence detection. Curr Opin Chem Biol 6(6):816–822

Yilong Z, Dean Z, Daoliang L (2015) Electrochemical and other methods for detection and determination of dissolved nitrite: a review. Int J Electrochem Sci 10:1144–1168

Zaitsev MG (2013) Extended abstract of Cand. Sci. (Chem.) dissertation. Moscow State University of Fine Chemical Technologies, Moscow

Zhang W, Asiri AM, Liu D, Du D, Lin Y (2015) Nanomaterial-based biosensors for environmental and biological monitoring of organophosphorus pesticides and nerve agents. Trends Anal Chem 54:1–10

Zhao Y, Zhang W, Lin Y, Du D (2013) The vital function of Fe3O4@Au nanocomposites for hydrolase biosensor design and its application in detection of methyl parathion. Nanoscale 5:1121–1126

# Chapter 14
# Biological Biosensors for Monitoring and Diagnosis

**Simranjeet Singh, Vijay Kumar, Daljeet Singh Dhanjal, Shivika Datta, Ram Prasad, and Joginder Singh**

**Abstract** Quantification and detection of various contaminants in the ecosystem have become critically important in the past few decades due to their exhaustive use in soil and aquatic ecosystems. The contamination by both organic and inorganic contaminants in the ecosystem has drawn attention due to their persistence, biological accumulation, and toxicity. Organic contaminants reach the air, water, food, soil, and other systems through drift mechanism and have detrimental effect on various life systems after entering the food chain, thus interfering the normal biological process of the ecosystem. Inorganic contaminants have less solubility, primarily get adsorbed, and accumulate on lower sediments. The sources of both organic and inorganic contaminants include anthropogenic activities which dispose industrial and sewage effluent directly into water bodies. Most of the contaminants are very much toxic and have tumorigenic, carcinogenic, and mutagenic effect on various life-forms. Biosensors have various prospective and existing applications in the detection of these compounds in the environment by transducing a signal. It also has immense applications in the detection of different contaminants in the food industry, environmental monitoring, disease diagnosis, etc. where reliable and precise

S. Singh
Regional Advanced Water Testing Laboratory, S.A.S. Nagar, India

Punjab Biotechnology Incubator (PBTI), S.A.S. Nagar, India

Department of Biotechnology, Lovely Professional University, Phagwara, India

V. Kumar
Regional Ayurveda Research Institute for Drug Development, Gwalior, India

D. S. Dhanjal · J. Singh (✉)
Department of Biotechnology, Lovely Professional University, Phagwara, India
e-mail: joginder.15005@lpu.co.in

S. Datta
Department of Zoology, Doaba College Jalandhar, Jalandhar, India

R. Prasad
Department of Botany, Mahatma Gandhi Central University, Motihari, Bihar, India

 
J. Singh et al. (eds.), *Microbial Biotechnology: Basic Research and Applications*, Environmental and Microbial Biotechnology,
https://doi.org/10.1007/978-981-15-2817-0_14

analyses are required. This chapter points out a comprehensive glimpse on different biosensors and their characteristics, operating principles, and their designs, based on transduction types and biological components. Efforts have been made to summarize various applications of biosensors in food industry, environmental monitoring, drug delivery systems, and clinical diagnostics etc.

**Keywords** Biosensors, environment monitoring · Disease diagnosis · Drug delivery

## 14.1 Introduction

Biosensors are devices comprising of a biological and physicochemical component to detect an analyte by producing a signal which can be measured (Mishra et al. 2019; Rovira and Domingo 2019; Mehrotra 2016). The first biosensor was invented by American biochemist "L.L Clark" in the year 1950 and the term "biosensor" was first introduced by "Cammann" in 1977 (Bhalla et al. 2016; Krishnaperumal and Lakshmanan 2013; Tchounwou et al. 2012). According to IUPAC nomenclature, biosensors are integrated receptor–transducer devices, which are able to provide selective quantitative or semiquantitative analytical information using a biological recognition element (Tangahu et al. 2011; Thévenot et al. 2001). A typical biosensor usually comprises of a biosensing element and a transducer. It has various biological applications and is used for the detection of several components such as pollutants, microbial load, metabolites, control parameters, and various other substances (Neethirajan et al. 2018). It also has immense applications in food industry, clinical diagnostics, and various other areas where reliable and precise analyses are required (Rasheed et al. 2019; Mishra et al. 2018). During the last few decades, numerous biosensing elements and devices have been developed (Grieshaber et al. 2008). Biosensors have numerous applications in various fields such as bio-monitoring of pollutants, disease diagnostics etc. The most common used biosensor is blood glucose biosensor which is used to check blood glucose levels (Nilsen et al. 2019; Yoo and Lee 2010).

Detection of various contaminants such as chemical and hazardous pollutants, drug detection, and detection of toxins in food, water, and soil ecosystems are some applications where biosensors are regularly used (Kimmel et al. 2007). Recent advancement in recombinant DNA technology led to the development of DNA-based or aptamer-based biosensors which act as a diagnostic tool in clinical assessment (Zhu et al. 2015). Incorporation of nanoparticles in biosensors helps in improving its parameters such as reliability, validity, lower detection limit, residence time, stability, sensitivity, etc. (Malekzad et al. 2017) (Fig. 14.1).

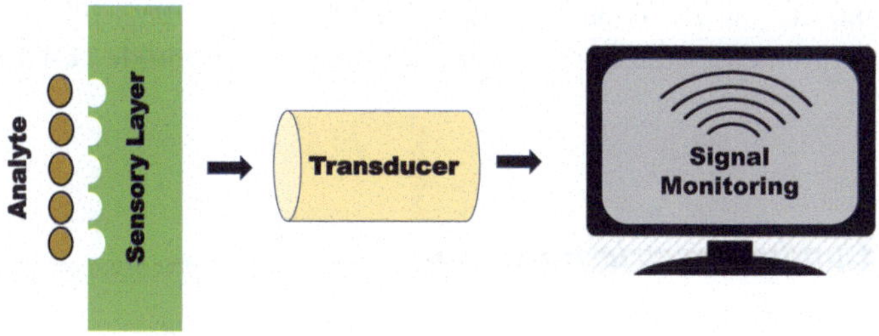


**Fig. 14.1** General components of a biosensor

## 14.2 Types of Biosensors

The biological sensing material can be DNA probes, antibodies and enzymes, and cell receptors which interact with the analyte. The transducer may be optical, physicochemical, or piezoelectric material that translates the biological signals to optical and electrical signals (Nguyen et al. 2019). Biosensors are divided into different groups depending on signal transductions (Rocchitta et al. 2016). The electrochemical sensors were first introduced in 1962 by Leland C. Clark during the NASS symposium (Bhalla et al. 2016). In this sensor, molecules interact with the reactants and transduce an electrical signal which is directly proportional to the concentration of the analyte (Grieshaber et al. 2008). Based on this principle, it employs impedimetric, amperometric, and potentiometric sensors converting sensing information into a measurable signal (Malhotra et al. 2017).

### *14.2.1 Voltammetric Biosensors*

Voltammetric biosensors measure the current produced during reduction or oxidation of electro-active reactant or product (Hung Tzang et al. 2001). In voltammetric biosensors, the electroactive species reduction or oxidation current is measured by producing the potential drop (Goyal et al. 2010). Here, the reference electrode and constant potential is applied to the working electrode which measures the current which is either directly proportional to the formation of a biocatalytic layer or to the volume concentration of electroactive species (Sangamithirai et al. 2018).

### *14.2.2 Potentiometric Biosensors*

They usually measure the potential of the biosensor electrode with respect to a reference electrode (Bundschuh et al. 2018; Pisoschi 2016). The potential drop carries the analytical signal between the two reference electrodes or between the reference electrode and the analytical electrode separated by the membrane. The transducer used in potentiometric transducers is the ion-selective electrode (ISE) (Cuartero et al. 2019).

### *14.2.3 Conductometric Biosensors*

These biosensors involve the change in conductance arising due to the biochemical reaction (Jaffrezic-Renault and Dzyadevych 2008). These biosensors measure the electrical conductivity of the solution during a biochemical reaction. The sensing element is an enzyme and less discounted for the detection of affine interactions (Velychko et al. 2016).

### *14.2.4 Optical Biosensors*

It measures light absorbed or emitted during a biochemical reaction (Damborský et al. 2016). In these biosensors, the optical fibers detect the analytes on the basis of light scattering fluorescence or absorption (Long et al. 2013). These biosensors measure both the affinity and catalytic reactions. The sensing element causes a change in absorbance or fluorescence which changes the refractive index between two media having different densities (Pospíšilová et al. 2015). They are superior to nonelectrical biosensors as they allow multiple analyte detection using various monitoring wavelengths (Dey and Goswami 2011). The adaptability of using optic probes is because of their ability to transmit signals that account on changes in polarity, time, wavelength, wave propagation, distribution of the spectrum, or intensity of the light (Peltomaa et al. 2018). They are solely based on the principle of light scattering absorption, internal reflection, fluorescence, surface plasmon resonance, or luminescence spectroscopy (Fan et al. 2008).

### *14.2.5 Calorimetric Biosensors*

Also known as thermal biosensors, they work on the change in enthalpy during a reaction (Y. Zhang and Tadigadapa 2004). These are developed by integrating biosensor components into a physical transducer. It is used for the detection of a pathogen in water and food by measuring the change in optical density or color of the test sample upon a chemical reaction (Park et al. 2007).

### 14.2.6 Enzymatic Sensors

These sensors include biological material having certain antibiological activities (Hwang et al. 2018). The simplest form of enzymatic biosensors is capable of reversible reduction or oxidation on the electrode upon application of electrochemically active potential (Bernards et al. 2008). Enzymatic sensors were further categorized into inhibitor and substrate sensors. Inhibitor sensors often tend to determine the reducing activity of the enzyme or substances, while substrate biosensors tend to determine selected substrates and its enzymatic reactions (Campanella et al. 2000).

### 14.2.7 Impedimetric Biosensors

It measures the variation impedance of an electrochemical cell with AC frequency (Guan et al. 2004). These involve ion-sensitive silicon field-based sensors which enhance the resolving power of the transducer by raising its sensitivity. Mostly, biologically modified impedimetric biosensors are used to determine small proteins and peptides on the basis of a net charge on it (Kim et al. 2019).

### 14.2.8 Piezoelectric Biosensors

It involves measurement of mass change during biomolecular interaction. They are also considered as mass-based biosensors and are based on the principle of sound vibrations and also called as acoustic biosensors (Tombelli 2012). They produce an electrical signal when mechanical force is applied (Pohanka 2017). Sensing molecules are directly attached to a piezoelectric surface which is piezoelectric in nature. Hence, mechanical vibrations arise from the interaction between the sensing molecules and analyte and translate them to electrical signals (Pohanka 2018).

### 14.2.9 Immunosensors

They are widely used to detect the immunochemical reaction which occurs between antigens and antibodies (Piro and Reisberg 2017). Hence, they are employed to detect the presence of antibodies and used as a diagnostic indication for toxic substances. They determine the antigens in both media (biological liquids and natural environment) (Balahura et al. 2019). They detect any compound having high selectivity and specificity against specific antibodies (Wen et al. 2017).

### 14.2.10 DNA Sensors

The major component of DNA sensors is nucleic acids, mostly DNA. These sensing materials are the fragments commonly called DNA primers or DNA probes which reflect specificity of the whole DNA structure. These probes or primers are synthesized by amplification of DNA by PCR (polymerase chain reaction) (Campàs i Homs 2002). They are modified to increase the stability or to facilitate introduction of probes into biosensors. This type of biosensors helps in revealing non-macromolecular and protein compounds which interact with specific DNA fragments (Diculescu et al. 2005). On the basis of the type of biorecognition unit used, they are classified as nucleic acid-based, enzymatic, whole cell-based, antibody-based, or aptamer-based biosensors (Rasheed and Sandhyarani 2017) (Table 14.1).

### 14.2.11 Biosensors Based on Supramolecular Structures of a Cell

These biosensors inhabit intermediate between DNA enzyme and sensors, since they have a hierarchical structure. Sometimes these entities are composed of lipid membranes, poly-enzyme complexes, and cell organelles (chloroplasts and mitochondria) (Duan et al. 2013). These biosensors are less stable as they are derived from their natural component and cannot maintain operating parameters during procedures that are time consuming (Wajs et al. 2016).

## 14.3 Biosensor Design and Operation

It is an analytical device that employs certain biological compounds to recognize some molecules by providing their presence and concentration as a signal for processing and recording (Vigneshvar et al. 2016). Usually, biosensors contain three components: the first one is the recognition element which is a membrane having various biological structures. The second one is the transducer and the last one is the electronic system which amplifies the signal and records the signal for data presentation (Kozitsina et al. 2018).

The basic part of any sensor is the recognition element which responds to one or many analytes among various substances (Luka et al. 2015). The commonly used recognition elements include biological structures such as living cells, nucleic acids, receptors, antibodies, and enzymes (Rahaie and Kazemi 2010). Transducers convert the variations into optical or electric signal when a reaction occurs between the analytes and the selective biological layer. This signal is further measured using an electronic or light-sensitive device (Rahaie and Kazemi 2010). The biological component is immobilized by membrane or physical entrapment, covalent binding, or

**Table 14.1** Different types of biosensors, their transducers, specifications, and applications

| Biosensor | Transducer type | Specifications | Applications | References |
|---|---|---|---|---|
| Amperometric | Electrochemical | Usually redox reaction which brings change in current between reference and working electrode | Quantification and detection of alcohols, cholesterol, urea, amino acids, and glucose | Goriushkina et al. (2009) |
| Baroxymeter | Others | Measures respiration of bacterial cells by pressure measurements | Detection of toxic components in wastewater | Tzoris et al. (2002) |
| Bioluminescent | Optical | Light emission principle in which viable bacteria responds to any physical chemical or biological change | Quantification and detection of heavy metal, food toxicants, and environmental monitoring | Alloush et al. (2006) |
| Colorimetric | Optical | Monitors the change in optical density of a sample during a reaction | Water- and foodborne pathogens | Radhakrishnan et al. (2014) |
| Conductometric | Electrochemical | Measures change in the electrical conductivity of the medium upon entry of any analyte that changes the concentration of ionic species by measuring electrical conductivity when analyte reacts with medium | Detection of protein markers, chemicals, heavy metals | Jaffrezic-Renault and Dzyadevych (2008) |
| DNA sensors | DNA primers or DNA probes | Synthesized by amplification of DNA by PCR (polymerase chain reaction) | Measures non-macromolecular and protein compounds | Rasheed and Sandhyarani (2017) |

(continued)

**Table 14.1** (continued)

| Biosensor | Transducer type | Specifications | Applications | References |
|---|---|---|---|---|
| Fluorescence | Optical | Emit fluorescence during immobilization of fluorescent-tagged biomolecules during reaction with analyte | Water or iron availability in plants or cell populations, BOD measurement, water in microbial habitat | Lei et al. (2006) |
| FET-based biosensor | Others | Measures change in conductance of field effect biosensor | IV blood pH recording, clinical investigation | Xu et al. (2005) |
| Impedance | Electrochemical | Measures the changes in conductivity or resistivity due to biorecognition event in medium | Polychlorinated biphenyls, milk toxins, PCBs, endocrine-disrupting hormones | Jaffrezic-Renault and Dzyadevych (2008) |
| Immunosensors | Based on ELISA technique | Amplifies and detect an antigen–antibody reaction | Clinical chemistry, antigen–antibody reaction | Balahura et al. (2019) |
| Piezoelectric | Others | Monitors changes in resonating frequency due to absorption or desorption of analyte results in current generation in piezoelectric material | Cellular studies, nucleic acid sensing | Skládal (2016) |
| Potentiometric | Electrochemical | Measures charge or potential accumulation between a reference and an ion-selective electrode | Determination of carbon dioxide, urea, pesticides, neurotransmitters, sugars, etc. | Pisoschi (2016) |

(continued)

**Table 14.1** (continued)

| Biosensor | Transducer type | Specifications | Applications | References |
|---|---|---|---|---|
| Pyroelectric | | Measurement of change in current induced by an analyte upon a temperature difference | Diagnostics | Spain and Venkatanarayanan (2014) |
| Supramolecular | | Entities composed of lipid membranes, poly-enzyme complexes, cell organelles, immobilization of organic thin film | Inhabit intermediate between DNA enzyme and sensor | Wajs et al. (2016) |

noncovalent interactions. Analytes bind with biological material resulting in the generation of an electronic response. Sometimes these reactions may be exogenous or release oxygen, hydrogen, or electrons ions (Nguyen et al. 2019). The transducer amplifies the changes in the product linked into the signal. Signal processing involves minimizing the reference signal arising from a similar transducer without any biological component and also smoothens the unwanted signal noise (Semenova et al. 2019).

## 14.4 Biosensors for Monitoring and Diagnostic Purposes

Biosensors exhibit numerous promising applications in various fields such as environmental monitoring, molecular diagnostics, pathogen detection, food industries, etc. (Mehrotra 2016). Biosensors monitor the presence of various contaminants in order to ensure the quality of drinking water, food, and soil. Biosensors detect contaminants at a low concentration which is a matter of priority for environmental protection and disease prevention as well (Rodriguez-Mozaz et al. 2006) (Fig. 14.2).

### *14.4.1 Biosensors for Monitoring Water Quality*

Presently, water quality monitoring has become a primary environmental concern due to anthropogenic activities that are deteriorating the quality of water (Bi et al. 2018). To improve it, different preventive measures need to be taken like enhancing

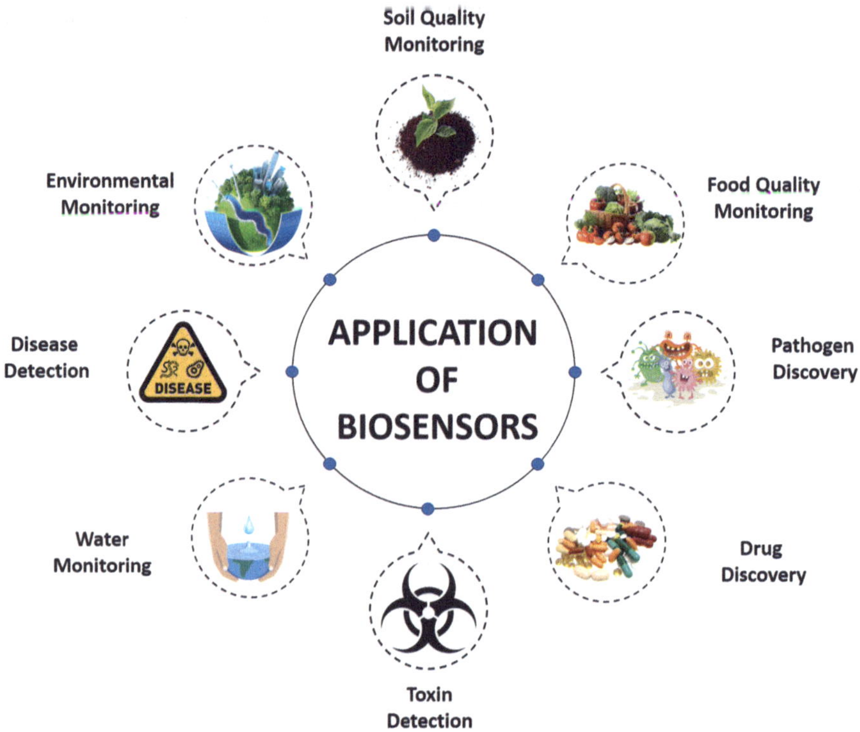


**Fig. 14.2** Different applications of biosensors

the methods for sustaining natural resources, controlling the release of toxic compounds into the environment, and treating the localized sources of contamination (Ezeonu et al. 2012), as different contaminants like insecticides, pesticides, and surface-active molecules (SAM) enter the water sources and affect the life of marine organisms. These interventions disturb the water equilibrium and reduce the natural ability of self-purification. As a result, it induces harmful effects on humans upon consuming the contaminated water (Agrawal et al. 2010). Hence, there is need for the monitoring of toxic compounds in water. Regrettably, conventional methods like chromatography, spectrometry, etc. have their limitations as they work on chemical and physical principles. Moreover, these approaches are high-priced and labor-intensive, have narrow range of sensitivity, and do not robustly provide an effective result of the ecological situation (Starodub et al. 2005). Therefore, sensors involving a biological component/organism have gained attention and can serve as reliable and effective analytical techniques for assessing environmental contamination (Kovalchuk and Kovalchuk 2008).

Animals, culture of different tissues, microbes, protozoa, and water plants can function as biological indicators for checking toxicity (Parmar et al. 2016). At the same instance, it may also aid in determining group-specific substances that are highly sensitive to specific groups present on the contaminant (Parmar et al. 2016). From different living organisms, *Daphnia magna St.* is an effective candidate for determining the total toxicity level of ecological contaminants as it has chemiluminescence (ChL) property. In an experiment, 5 organisms of *Daphnia* in 10 mL were found to be optimum for producing ChL response in the presence of $H_2O_2$. Using this idea, a *Daphnia*-based ChL biosensor has been created, which is aided with a computer controlling device for automatic detection of ChL signal. This biosensor especially detects the presence of SAM and is able to give measurement in the range of 20 $mgL^{-1}$ or less (Starodub 2009). Another biosensor has been developed using bioluminescent bacteria. For this, bacteria were isolated from the Azov Sea and Black Sea according to their tolerance to SAM and other toxic entities to assess the toxicity of water samples. To develop this computer-controlled biosensor, bioluminescent bacteria like *Photobacterium phosphoreum* K3 and *Vibrio fischeri* F1/Sh1 are used. This biosensor works on bioluminescence inhibition as majority of SAM and toxic compounds inhibit bioluminescence activity. This inhibition response determines the presence of toxic compounds (Kuznetsov et al. 2002). Various other biosensors that possess the ability to determine specific toxic substances like chlororganics, cyanides, and phosphororganics have also been developed working on the principle of the electrolyte–insulator–semiconductor (EIS) system (Starodub et al. 2012). One such system is surface plasmon resonance (SPR)-based biosensor, which involves ethoxylated nonylphenol (NphEO) and immunoglobulin gamma (IgG) antibodies for detection. This approach measures the response of an SPR sensor against the NphEO concentration in the test sample (Bakhmachuk et al. 2017). Another powerful system using same component is ISFET biosensor, which has five times higher sensitivity than SPR-based biosensor and is the preferred biosensor for water assessment (Grieshaber et al. 2008).

### 14.4.2 Biosensor for Infectious Disease Detection

The spread of various infectious diseases like avian influenza, Hendra, Nipah, and SARS has become a global threat, which demands considerable effort to regulate their proliferation (Karesh et al. 2012). As there are various challenges associated with these infectious diseases, there is need for the development of diagnostic tools for eradicating/minimizing the chances of virus outbreak beforehand (Reyes et al. 2013). Biosensors have emerged as an attractive tool for providing robust information on these diseases. Usually, biosensors are characterized on the basis of their biological component and nature of the process like biocatalytic agent (such as enzyme), immunological agent (such as antibody), and nucleic acid material (such as DNA) (Mehrotra 2016). Majority of the biosensors developed for detecting pathogens involved in causing infectious disease are based on the principle of electrochemical reaction. This type of biosensor holds the majority as they are cost effective, independent of solution turbidity, low power requirement, high sensitivity, and simple instrumentation (Srinivasan and Tung 2015).

Different electrochemical methods like amperometric, impedance, and potentiometric are used to examine the changes which take place during disease detection (Hammond et al. 2016). The amperometric biosensor generally involves biosensor marker, antibody–antigen, and DNA hybridization reactions with an electrochemical transducer which amplifies the signal to a significant level for detection (Belluzo et al. 2008). One of the most common amperometric sensors is a glucometer (Yoo and Lee 2010). Previously, Gong and his colleagues have developed an amperometric-based immunosensor for detecting Newcastle disease (Gong et al. 2003). Another amperometric-based immunosensor has also been developed to diagnose forest-spring encephalitis with great precision (Brainina et al. 2003). Biosensors based on label-free amperometric immunosensor have been developed for Japanese B encephalitis vaccine (Yuan et al. 2005). Another group of researchers developed an optical biosensor to check the presence of Newcastle disease virus with sensitivity to 10 ng/ml (Lee and Thompson 1996). SPR-associated immunosensors have also been developed to detect the coronavirus responsible for severe acute respiratory syndrome (SARS) (Huang et al. 2009). Another type of biosensor, i.e., piezoelectric biosensors, has also been developed by the research group to assess the food and mouth disease virus (Gajendragad et al. 2001). Moreover, piezoelectric-associated DNA biosensor has also been developed by a group of researchers to detect hepatitis B virus infection with limited concentration range of about 0.02–0.14 μg/ml (Yao and Fu 2014).

### 14.4.3 Biosensor for Pathogen Detection

Foodborne illnesses have become a major health issue worldwide and raised the health safety concern in both food industries and regulatory bodies (Fung et al. 2018). Microbes have also been found to perform a function that is beneficial to food and its production, whereas some of them have been found to be associated with food spoilage (Rawat 2015). Few of food pathogen microbes are *Aeromonas* spp., *Bacillus anthracis, Bacillus subtilis, Brucella* spp., *Campylobacter* spp., *Clostridium* spp., *Escherichia coli, Listeria monocytogenes, Salmonella* spp., *Shigella* spp., *Staphylococcus aureus,* and *Yersinia enterocolitica* (Mody and Griffin 2015). These microbes have been found to produce cell metabolites and toxins which can cause severe diseases. These issues have prompted the researchers to develop a cheap, portable, and robust detector for pathogens (Abbasian et al. 2018). They developed different biosensors to detect these pathogenic microbes in food sources. Therefore, robust detection approaches for assessing foodborne pathogens are classified into antigen–antibody-based, bacteriophage-based, and nucleic acid-based biosensors (Mehrotra 2016). The rapid detection of foodborne pathogens has increased the popularity of these biosensors in the food sector, as these biosensors are effective in examining the food quality and microbial contamination in food products (Law et al. 2014).

The different types of immunosensors have been developed which measure the signal induced on the interaction of specific antibodies with a targeted antigen. An immunosensor has been found effective in detecting two species of *Salmonella* (i.e., *S. gallinarum* and *S. pullorum*) in chicken meat and eggs (Cinti et al. 2017). Another immunosensor based on a screen-printed interdigitated microelectrode has been developed to detect *E. coli* O157:H7 as well as *S. typhimurium* in chicken and milk sample within the range of $10^3$–$10^6$ CFU/mL (Xu et al. 2016). An enzyme-based biosensor is another category, in which an enzyme works as a bioreceptor for detection. One of the most common enzyme-based biosensors is a glucometer which measures the level of glucose with the help of the immobilized glucose oxidase enzyme (Yoo and Lee 2010). Hesari and his colleagues developed an enzyme-based biosensor to robustly detect the *E. coli* contamination in drinking water (Hesari et al. 2016). Another category of biosensor, i.e., optical biosensor, has also been developed in which the toxins or pathogens are labeled with fluorescent compound, which on interaction with surface biosensor gets triggered by laser wave and induces a signal for detection (Bosch et al. 2007). Based on this principle, Adak and his group have developed an optical biosensor for *S. aureus* detection with detection limit in the range of $10^2$–$10^3$ CFU/mL (Adak et al. 2013). Fluorescence resonance energy transfer (FRET)-based biosensors are also a type of optical biosensor. This type of biosensor has been found to be effective in detecting *S. aureus* in spilled milk and buffer (He et al. 2014). Moreover, Zhang with his colleagues developed a surface plasmon resonance (SPR) biosensor for the rapid detection of *E. coli O157:H7, L. monocytogenes,* and *S. enteritidis* in food products (Zhang et al. 2014). A colorimetric biosensor, another type of optical biosensor method, has allowed us to robustly detect pathogens like *Listeria* spp. and *S. aureus* in food products as well as the environment (Oluwaseun et al. 2018).

On the other hand, different electrochemical biosensors have also been developed to detect the microbes in contaminated food. Recently, amperometric biosensors with high sensitivity have been developed for identifying different foodborne pathogens like *C. jejuni, E. coli* O157:H7, *L. monocytogenes,* and *Salmonella* species (Arora et al. 2018), although a field effect transistor-based potentiometric biosensor has been developed to detect the presence of *E. coli* with detection limit as low as 10 cells/mL (Moran et al. 2016). Another electrochemical impedimetric biosensor based on aptasensors has been developed by Sheikhzadeh and his colleagues to detect the presence of *Salmonella typhi* in apple juice having a detectable range of $10^2$–$10^8$ CFU/mL (Sheikhzadeh et al. 2016), whereas Zhang and his team developed aptasensors based on surface-enhanced Raman spectroscopy to detect *Staphylococcus aureus* and *Salmonella typhimurium* in pork with a working range of $10^2$–$10^7$ CFU/mL (Zhang et al. 2015). Moreover, magnetoelastic sensors have also been developed to detect the foodborne pathogens like *Staphylococcus aureus* and *Salmonella typhimurium* in food items like spinach and tomato (Byeon et al. 2015; Li et al. 2010).

## 14.5 Conclusion

Biosensors have various applications in different fields such as disease diagnosis, environment monitoring, food control, drug discovery, biomedical research, forensics, etc. These devices need the interaction of various disciplines and are dependent on very special features like interaction of biomolecular analytes with recognition elements, device fabrication and design, on-chip electronics, sampling techniques, microfluidics, etc. Incorporation of nanoparticles in biosensors provides an opportunity to build a new generation of sensing technologies. Nanoparticles improve the magnetic, optical, electrochemical, and mechanical properties of the biosensors. No studies have been reported on understanding the mechanism of interaction between biomolecules and nanomaterials on nanofilms or surface of electrodes for fabrication of new-generation biosensors. However, nanoparticle-based biosensors show attractive prospects which will be applied in process control, food analysis, environmental monitoring, and clinical diagnosis in the near future.

## References

Abbasian F, Ghafar-Zadeh E, Magierowski S (2018) Microbiological sensing technologies: a review. Bioengineering 5(1):1–33. https://doi.org/10.3390/bioengineering5010020

Adak AK, Boley JW, Lyvers DP, Chiu GT, Low PS, Reifenberger R, Wei A (2013) Label-free detection of Staphylococcus aureus captured on immutable ligand arrays. ACS Appl Mater Interfaces 5(13):6404–6411. https://doi.org/10.1021/am4016236

Agrawal A, Pandey RS, Sharma B (2010) Water pollution with special reference to pesticide contamination in India. J Water Resour Prot 02(05):432–448. https://doi.org/10.4236/jwarp.2010.25050

Alloush HM, Lewis RJ, Salisbury VC (2006) Bacterial bioluminescent biosensors: applications in food and environmental monitoring. Anal Lett 39(8):1517–1526. https://doi.org/10.1080/00032710600713172

Arora S, Ahmed N, Siddiqui S (2018) Detecting food borne pathogens using electrochemical biosensors: an overview. Int J Chem Stud 6(1):1031–1039

Bakhmachuk A, Gorbatiuk O, Rachkov A, Dons'koi B, Khristosenko R, Ushenin I et al (2017) Surface Plasmon resonance investigations of bioselective element based on the recombinant protein a for immunoglobulin detection. Nanoscale Res Lett 12(1). https://doi.org/10.1186/s11671-017-1903-5

Balahura LR, Stefan-Van Staden RI, Van Staden JF, Aboul-Enein HY (2019) Advances in immunosensors for clinical applications. J Immunoass Immunochem 40(1):40–51. https://doi.org/10.1080/15321819.2018.1543704

Belluzo MS, Ribone MÉ, Lagier CM (2008) Assembling amperometric biosensors for clinical diagnostics. Sensors 8(3):1366–1399

Bernards DA, MacAya DJ, Nikolou M, Defranco JA, Takamatsu S, Malliaras GG (2008) Enzymatic sensing with organic electrochemical transistors. J Mater Chem 18(1):116–120. https://doi.org/10.1039/b713122d

Bhalla N, Jolly P, Formisano N, Estrela P (2016) Introduction to biosensors. Essays Biochem 60(1):1–8. https://doi.org/10.1042/EBC20150001

Bi W, Weng B, Yuan Z, Ye M, Zhang C, Zhao Y et al (2018) Evolution characteristics of surface water quality due to climate change and LUCC under scenario simulations: a case study in

the Luanhe River Basin. Int J Environ Res Public Health 15(8):1724. https://doi.org/10.3390/ijerph15081724
Bosch ME, Sánchez AJR, Rojas FS, Ojeda CB (2007) Recent development in optical fiber biosensors. Sensors 7(6):797–859. https://doi.org/10.3390/s7060797
Brainina K, Kozitsina A, Beikin J (2003) Electrochemical immunosensor for forest-spring encephalitis based on protein a labeled with colloidal gold. Anal Bioanal Chem 376(4):481–485. https://doi.org/10.1007/s00216-003-1912-3
Bundschuh M, Filser J, Lüderwald S, McKee MS, Metreveli G, Schaumann GE et al (2018) Nanoparticles in the environment: where do we come from, where do we go to? Environ Sci Eur 30(1). https://doi.org/10.1186/s12302-018-0132-6
Byeon HM, Vodyanoy VJ, Oh JH, Kwon JH, Park MK (2015) Lytic phage-based magnetoelastic biosensors for on-site detection of methicillin-resistant Staphylococcus aureus on spinach leaves. J Electrochem Soc 162(8):B230–B235. https://doi.org/10.1149/2.0681508jes
Campanella L, Persi L, Tomassetti M (2000) New tool for superoxide and nitric oxide radicals determination using suitable enzymatic sensors. Sensors Actuators B Chem 68(1):351–359. https://doi.org/10.1016/S0925-4005(00)00415-9
Campàs i Homs M (2002) DNA sensors. Anal Lett 35(12):1875–1894. https://doi.org/10.1081/AL-120014280
Cinti S, Volpe G, Piermarini S, Delibato E, Palleschi G (2017) Electrochemical biosensors for rapid detection of foodborne salmonella: a critical overview. Sensors 17(8). https://doi.org/10.3390/s17081910
Cuartero M, Parrilla M, Crespo GA (2019) Wearable potentiometric sensors for medical applications. Sensors (Switzerland) 19(2):1–24. https://doi.org/10.3390/s19020363
Damborský P, Švitel J, Katrlík J (2016) Optical biosensors. Essays Biochem 60(1):91–100. https://doi.org/10.1042/EBC20150010
Dey D, Goswami T (2011) Optical biosensors: a revolution towards quantum nanoscale electronics device fabrication. J Biomed Biotechnol 2011. https://doi.org/10.1155/2011/348218
Diculescu VC, Paquim AMC, Brett AMO (2005) Electrochemical DNA sensors for detection of DNA damage. Sensors 5(6–10):377–393. https://doi.org/10.3390/s5060377
Duan X, Rajan NK, Routenberg DA, Huskens J, Reed MA (2013) Regenerative electronic biosensors using supramolecular approaches. ACS Nano 7(5):4014–4021. https://doi.org/10.1021/nn306034f
Ezeonu CS, Tagbo R, Anike EN, Oje OA, Onwurah INE (2012) Biotechnological tools for environmental sustainability: prospects and challenges for environments in Nigeria—a standard review. Biotechnol Res Int 2012:1–26. https://doi.org/10.1155/2012/450802
Fan X, White IM, Shopova SI, Zhu H, Suter JD, Sun Y (2008) Sensitive optical biosensors for unlabeled targets: a review. Anal Chim Acta 620(1–2):8–26. https://doi.org/10.1016/j.aca.2008.05.022
Fung F, Wang H-S, Menon S (2018) Food safety in the 21st century. Biom J 41(2):88–95. https://doi.org/10.1016/j.bj.2018.03.003
Gajendragad MR, Kamath KNY, Anil PY, Prabhudas K, Natarajan C (2001) Development and standardization of a piezo electric immunobiosensor for foot and mouth disease virus typing. Vet Microbiol 78(4):319–330. https://doi.org/10.1016/S0378-1135(00)00307-2
Gong JL, Gong FC, Zeng GM, Shen GL, Yu RQ (2003) An amperometric immunosensor for the Newcastle disease antibody assay. Anal Lett 36(2):287–302. https://doi.org/10.1081/AL-120017691
Goriushkina TB, Soldatkin AP, Dzyadevych SV (2009) Application of amperometric biosensors for analysis of ethanol, glucose, and lactate in wine. J Agric Food Chem 57(15):6528–6535. https://doi.org/10.1021/jf9009087
Goyal RN, Gupta VK, Chatterjee S (2010) Voltammetric biosensors for the determination of paracetamol at carbon nanotube modified pyrolytic graphite electrode. Sensors Actuators B Chem 149(1):252–258. https://doi.org/10.1016/j.snb.2010.05.019
Grieshaber D, MacKenzie R, Vörös J, Reimhult E (2008) Electrochemical biosensors - sensor principles and architectures. Sensors 8:1400–1458. https://doi.org/10.3390/s80314000

Guan JG, Miao YQ, Zhang QJ (2004) Impedimetric biosensors. J Biosci Bioeng 97(4):219–226. https://doi.org/10.1263/jbb.97.219

Hammond JL, Formisano N, Estrela P, Carrara S, Tkac J (2016) Electrochemical biosensors and nanobiosensors. Essays Biochem 60(1):69–80. https://doi.org/10.1042/EBC20150008

He X, Li Y, He D, Wang K, Shangguan J, Shi H (2014) Aptamer-fluorescent silica nanoparticles bioconjugates based dual-color flow cytometry for specific detection of Staphylococcus aureus. J Biomed Nanotechnol 10(7):1359–1368. https://doi.org/10.1166/jbn.2014.1828

Hesari N, Alum A, Elzein M, Abbaszadegan M (2016) A biosensor platform for rapid detection of E. coli in drinking water. Enzym Microb Technol 83:22–28. https://doi.org/10.1016/j.enzmictec.2015.11.007

Huang JC, Chang Y-F, Chen K-H, Su L-C, Lee C-W, Chen C-C et al (2009) Detection of severe acute respiratory syndrome (SARS) coronavirus nucleocapsid protein in human serum using a localized surface plasmon coupled fluorescence fiber-optic biosensor. Biosens Bioelectron 25(2):320–325. https://doi.org/10.1016/j.bios.2009.07.012

Hung Tzang C, Yuan R, Yang M (2001) Voltammetric biosensors for the determination of formate and glucose-6-phosphate based on the measurement of dehydrogenase-generated NADH and NADPH. Biosens Bioelectron 16(3):211–219. https://doi.org/10.1016/S0956-5663(00)00143-3

Hwang DW, Lee S, Seo M, Chung TD (2018) Recent advances in electrochemical non-enzymatic glucose sensors – a review. Anal Chim Acta 1033:1–34. https://doi.org/10.1016/j.aca.2018.05.051

Jaffrezic-Renault N, Dzyadevych SV (2008) Conductometric microbiosensors for environmental monitoring. Sensors 8(4):2569–2588. https://doi.org/10.3390/s8042569

Karesh WB, Dobson A, Lloyd-Smith JO, Lubroth J, Dixon MA, Bennett M et al (2012) Ecology of zoonoses: natural and unnatural histories. Lancet 380(9857):1936–1945. https://doi.org/10.1016/S0140-6736(12)61678-X

Kim M, Iezzi R, Shim BS, Martin DC (2019) Impedimetric biosensors for detecting vascular endothelial growth factor (VEGF) based on poly(3,4-ethylene dioxythiophene) (PEDOT)/gold nanoparticle (Au NP) composites. Front Chem 7:1–11. https://doi.org/10.3389/fchem.2019.00234

Kimmel DW, LeBlanc G, Meschievitz ME, Cliffel DE (2007) Chemical sensors and biosensors. Anal Chem 84(2):685–707. https://doi.org/10.1021/ac202878q

Kovalchuk I, Kovalchuk O (2008) Transgenic plants as sensors of environmental pollution genotoxicity. Sensors 8(3):1539–1558. https://doi.org/10.3390/s8031539

Kozitsina AN, Svalova TS, Malysheva NN, Okhokhonin AV, Vidrevich MB, Brainina KZ (2018) Sensors based on bio and biomimetic receptors in medical diagnostic, environment, and food analysis. Biosensors 8(2):1–34. https://doi.org/10.3390/bios8020035

Krishnaperumal I, Lakshmanan R (2013) An approximate analytical method for the evaluation of the concentrations and current for hybrid enzyme biosensor. ISRN Phys Chem 2013:1–12. https://doi.org/10.1155/2013/202781

Kuznetsov AM, Rodicheva EK, Medvedeva SE, Gitelson JI (2002) Bioluminescent bioassays based on luminous bacteria marker system. Biolumin Chemilumin 4:323–326. https://doi.org/10.1142/9789812776624_0072

Law JWF, Mutalib NSA, Chan KG, Lee LH (2014) Rapid metho ds for the detection of foodborne bacterial pathogens: principles, applications, advantages and limitations. Front Microbiol 5:1–19. https://doi.org/10.3389/fmicb.2014.00770

Lee WE, Thompson HG (1996) Detection of newcastle disease virus using an evanescent wave immune-based biosensor. Can J Chem 74(5):707–712. https://doi.org/10.1139/v96-076

Lei Y, Chen W, Mulchandani A (2006) Microbial biosensors. Anal Chim Acta 568(1–2):200–210. https://doi.org/10.1016/j.aca.2005.11.065

Li S, Li Y, Chen H, Horikawa S, Shen W, Simonian A, Chin BA (2010) Direct detection of Salmonella typhimurium on fresh produce using phage-based magnetoelastic biosensors. Biosens Bioelectron 26(4):1313–1319. https://doi.org/10.1016/j.bios.2010.07.029

Long F, Zhu A, Shi H (2013) Recent advances in optical biosensors for environmental monitoring and early warning. Sensors (Switzerland) 13(10):13928–13948. https://doi.org/10.3390/s131013928

Luka G, Ahmadi A, Najjaran H, Alocilja E, Derosa M, Wolthers K et al (2015) Microfluidics integrated biosensors: a leading technology towards lab-on-A-chip and sensing applications. Sensors (Switzerland) 15(12):30011–30031. https://doi.org/10.3390/s151229783

Malekzad H, Sahandi Zangabad P, Mirshekari H, Karimi M, Hamblin MR (2017) Noble metal nanoparticles in biosensors: recent studies and applications. Nanotechnol Rev 6(3):301–329. https://doi.org/10.1515/ntrev-2016-0014

Malhotra S, Verma A, Tyagi N, Kumar V (2017) Biosensors: principle, types and applications. Int J Adv Res Innov Ideas Edu 3(2):3639–3644

Mehrotra P (2016) Biosensors and their applications - a review. J Oral Biol Craniofac Res 6(2):153–159. https://doi.org/10.1016/j.jobcr.2015.12.002

Mishra GK, Barfidokht A, Tehrani F, Mishra RK (2018) Food safety analysis using electrochemical biosensors. Foods 7(9). https://doi.org/10.3390/foods7090141

Mishra S, Bharagava RN, More N, Yadav A, Zainith S, Mani S, Chowdhary P (2019) Heavy metal contamination: an alarming threat to environment and human health. Environ Biotechnol Sustain Future:103–125. https://doi.org/10.1007/978-981-10-7284-0_5

Mody RK, Griffin PM (2015) 103 - Foodborne disease. In: Bennett JE, Dolin R, Blaser MJ (eds) Mandell, douglas, and Bennett's principles and practice of infectious diseases, 8th edn. Content Repository Only!, Philadelphia, pp 1283–1296.e3. https://doi.org/10.1016/B978-1-4557-4801-3.00103-X

Moran KLM, Fitzgerald J, McPartlin DA, Loftus JH, O'Kennedy R (2016) Biosensor-based technologies for the detection of pathogens and toxins. In: Scognamiglio V, Rea G, Arduini F, Palleschi G (eds) Biosensors for sustainable food - new opportunities and technical challenges, vol 74. Elsevier, pp 93–120. https://doi.org/10.1016/bs.coac.2016.04.002

Neethirajan S, Ragavan V, Weng X, Chand R (2018) Biosensors for sustainable food engineering: challenges and perspectives. Biosensors 8(1). https://doi.org/10.3390/bios8010023

Nguyen HH, Lee SH, Lee UJ, Fermin CD, Kim M (2019) Immobilized enzymes in biosensor applications. Materials 12(1):1–34. https://doi.org/10.3390/ma12010121

Nilsen E, Smalling KL, Ahrens L, Gros M, Miglioranza KSB, Picó Y, Schoenfuss HL (2019) Critical review: grand challenges in assessing the adverse effects of contaminants of emerging concern on aquatic food webs. Environ Toxicol Chem 38(1):46–60. https://doi.org/10.1002/etc.4290

Oluwaseun AC, Phazang P, Sarin NB (2018) Biosensors: a fast-growing technology for pathogen detection in agriculture and food sector. In: Rinken T, Kivirand K (eds) Biosensing technologies for the detection of pathogens - a prospective way for rapid analysis, 1st edn. INTECH, pp 37–52. https://doi.org/10.5772/57353

Park SC, Cho EJ, Moon SY, Yoon SI, Kim YJ, Kim DH, Suh JS (2007) A calorimetric biosensor and its application for detecting a cancer cell with optical imaging. IFMBE Proc 14(1):637–640

Parmar TK, Rawtani D, Agrawal YK (2016) Bioindicators: the natural indicator of environmental pollution. Front Life Sci 9(2):110–118. https://doi.org/10.1080/21553769.2016.1162753

Peltomaa R, Glahn-Martínez B, Benito-Peña E, Moreno-Bondi MC (2018) Optical biosensors for label-free detection of small molecules. Sensors (Basel, Switzerland) 18(12). https://doi.org/10.3390/s18124126

Piro B, Reisberg S (2017) Recent advances in electrochemical immunosensors. Sensors (Switzerland) 17(4). https://doi.org/10.3390/s17040794

Pisoschi AM (2016) Potentiometric biosensors: concept and analytical applications-an editorial. Biochem Anal Biochem 5(3):19–20. https://doi.org/10.4172/2161-1009.1000e164

Pohanka M (2017) The piezoelectric biosensors: principles and applications, a review. Int J Electrochem Sci 12(1):496–506. https://doi.org/10.20964/2017.01.44

Pohanka M (2018) Overview of piezoelectric biosensors, immunosensors and DNA sensors and their applications. Materials 11(3). https://doi.org/10.3390/ma11030448

Pospíšilová M, Kuncová G, Trögl J (2015) Fiber-optic chemical sensors and fiber-optic biosensors. Sensors (Switzerland) 15(10):25208–25259. https://doi.org/10.3390/s151025208

Radhakrishnan R, Suni II, Bever CS, Hammock BD (2014) Impedance biosensors: applications to sustainability and remaining technical challenges. ACS Sustain Chem Eng 2(7):1649–1655. https://doi.org/10.1021/sc500106y

Rahaie M, Kazemi SS (2010) Lectin-based biosensors: as powerful tools in bioanalytical applications. Biotechnology 9(4):428–443. https://doi.org/10.3923/biotech.2010.428.443

Rasheed PA, Sandhyarani N (2017) Electrochemical DNA sensors based on the use of gold nanoparticles: a review on recent developments. Microchim Acta 184(4):981–1000. https://doi.org/10.1007/s00604-017-2143-1

Rasheed T, Bilal M, Nabeel F, Adeel M, Iqbal HMN (2019) Environmentally-related contaminants of high concern: potential sources and analytical modalities for detection, quantification, and treatment. Environ Int 122:52–66. https://doi.org/10.1016/j.envint.2018.11.038

Rawat S (2015) Food spoilage: microorganisms and their prevention. Pelagia Res Libr Asian J Plant Sci Res 5(4):47–56. www.pelagiaresearchlibrary.com

Reyes R, Ahn R, Thurber K, Burke TF (2013) Urbanization and infectious diseases: general principles, historical perspectives, and contemporary challenges. In: Fong IW (ed) Challenges in infectious diseases. Springer, New York, pp 123–146. https://doi.org/10.1007/978-1-4614-4496-1_4

Rocchitta G, Spanu A, Babudieri S, Latte G, Madeddu G, Galleri G et al (2016) Enzyme biosensors for biomedical applications: strategies for safeguarding analytical performances in biological fluids. Sensors (Switzerland) 16(6). https://doi.org/10.3390/s16060780

Rodriguez-Mozaz S, Lopez De Alda MJ, Barceló D (2006) Biosensors as useful tools for environmental analysis and monitoring. Anal Bioanal Chem 386(4):1025–1041. https://doi.org/10.1007/s00216-006-0574-3

Rovira J, Domingo JL (2019) Human health risks due to exposure to inorganic and organic chemicals from textiles: a review. Environ Res 168:62–69. https://doi.org/10.1016/j.envres.2018.09.027

Sangamithirai D, Munusamy S, Narayanan V, Stephen A (2018) A voltammetric biosensor based on poly(o-methoxyaniline)-gold nanocomposite modified electrode for the simultaneous determination of dopamine and folic acid. Mater Sci Eng C 91:512–523. https://doi.org/10.1016/j.msec.2018.05.070

Semenova D, Silina YE, Koch M, Micheli L, Zubov A, Gernaey KV (2019) Sensors for biosensors: a novel tandem monitoring in a droplet towards efficient screening of robust design and optimal operating conditions. Analyst 144(8):2511–2522. https://doi.org/10.1039/c8an02261e

Sheikhzadeh E, Chamsaz M, Turner APF, Jager EWH, Beni V (2016) Label-free impedimetric biosensor for salmonella Typhimurium detection based on poly [pyrrole-co-3-carboxyl-pyrrole] copolymer supported aptamer. Biosens Bioelectron 80:194–200. https://doi.org/10.1016/j.bios.2016.01.057

Skládal P (2016) Piezoelectric biosensors. TrAC Trends Anal Chem 79:127–133. https://doi.org/10.1016/j.trac.2015.12.009

Spain E, Venkatanarayanan A (2014) Review of physical principles of sensing and types of sensing materials. Comprehensive materials processing, vol 13. Elsevier. https://doi.org/10.1016/B978-0-08-096532-1.01302-9

Srinivasan B, Tung S (2015) Development and applications of portable biosensors. J Lab Autom 20(4):365–389. https://doi.org/10.1177/2211068215581349

Starodub N (2009) Biosensors in the system of express control of chemicals, regularly used as terrorist means, to prevent non-desirable consequences. In: Bahadir AM, Duca G (eds) The role of ecological chemistry in pollution research and sustainable development, 1st edn. Springer, pp 275–284. https://doi.org/10.1007/978-90-481-2903-4_27

Starodub NF, Katzev AM, Starodub VM, Levkovetz IA, Goncharuk VV, Klimenko NA et al (2005) Biosensors for water quality monitoring. In: Alexander Omelchenko AAPWJS (ed) Modern tools and methods of water treatment for improving living standards. Springer, Netherlands, pp 51–70

Starodub NF, Ogorodniichuk YA, Sitnik YA, Slishik NF (2012) Biosensors for the control of some toxins, viral and microbial infections to prevent actions of bioterrorists. In: Nikolelis D (ed) Portable chemical sensors: weapons against bioterrorism, 1st edn. Springer, pp 95–117. https://doi.org/10.1007/978-94-007-2872-1

Tangahu BV, Sheikh Abdullah SR, Basri H, Idris M, Anuar N, Mukhlisin M (2011) A review on heavy metals (As, Pb, and Hg) uptake by plants through phytoremediation. Int J Chem Eng 2011. https://doi.org/10.1155/2011/939161

Tchounwou PB, Yedjou CG, Patlolla AK, Sutton DJ (2012) Molecular, clinical and environmental toxicicology volume 3: environmental toxicology. Mol Clin Environ Toxicol 101:133–164. https://doi.org/10.1007/978-3-7643-8340-4

Thévenot DR, Toth K, Durst RA, Wilson GS (2001) Electrochemical biosensors: recommended definitions and classification1International Union of Pure and Applied Chemistry: Physical Chemistry Division, Commission I.7 (Biophysical Chemistry); Analytical Chemistry Division, Commission V.5 (Electroanalytical). Biosens Bioelectron 16(1–2):121–131. https://doi.org/10.1016/s0956-5663(01)00115-4

Tombelli S (2012) Piezoelectric biosensors for medical applications. Biosens Med Appl. Elsevier Masson SAS. https://doi.org/10.1016/B978-1-84569-935-2.50002-4

Tzoris A, Fearnside D, Lovell M, Hall EAH (2002) Direct toxicity assessment of wastewater: Baroxymeter, a portable rapid toxicity device and the industry perspective. Environ Toxicol 17(3):284–290. https://doi.org/10.1002/tox.10059

Velychko TP, Soldatkin OO, Melnyk VG, Marchenko SV, Kirdeciler SK, Akata B et al (2016) A novel conductometric urea biosensor with improved analytical characteristic based on recombinant urease adsorbed on nanoparticle of silicalite. Nanoscale Res Lett 11(1):1–6. https://doi.org/10.1186/s11671-016-1310-3

Vigneshvar S, Sudhakumari CC, Senthilkumaran B, Prakash H (2016) Recent advances in biosensor technology for potential applications - an overview. Front Bioeng Biotechnol 4:1–9. https://doi.org/10.3389/fbioe.2016.00011

Wajs E, Fernández N, Fragoso A (2016) Supramolecular biosensors based on electropolymerised pyrrole-cyclodextrin modified surfaces for antibody detection. Analyst 141(11):3274–3279. https://doi.org/10.1039/c6an00532b

Wen W, Yan X, Zhu C, Du D, Lin Y (2017) Recent advances on electrochemical Immunosensors. Anal Chem 89:138–156. https://doi.org/10.1021/acs.analchem.6b04281

Xu J, Luo X, Chen H (2005) Analytical aspects of FET-based biosensors. Front Biosci 10:420–430. https://doi.org/10.2741/1538

Xu M, Wang R, Li Y (2016) Rapid detection of Escherichia coli O157:H7 and Salmonella typhimurium in foods using an electrochemical immunosensor based on screen-printed interdigitated microelectrode and immunomagnetic separation. Talanta 148:200–208. https://doi.org/10.1016/j.talanta.2015.10.082

Yao CY, Fu WL (2014) Biosensors for hepatitis B virus detection. World J Gastroenterol 20(35):12485–12492. https://doi.org/10.3748/wjg.v20.i35.12485

Yoo EH, Lee SY (2010) Glucose biosensors: an overview of use in clinical practice. Sensors 10(5):4558–4576. https://doi.org/10.3390/s100504558

Yuan R, Zhang L, Li Q, Chai Y, Cao S (2005) A label-free amperometric immunosenor based on multi-layer assembly of polymerized o-phenylenediamine and gold nanoparticles for determination of Japanese B encephalitis vaccine. Anal Chim Acta 531(1):1–5. https://doi.org/10.1016/j.aca.2004.10.072

Zhang H, Ma X, Liu Y, Duan N, Wu S, Wang Z, Xu B (2015) Gold nanoparticles enhanced SERS aptasensor for the simultaneous detection of Salmonella typhimurium and Staphylococcus aureus. Biosens Bioelectron 74:872–877. https://doi.org/10.1016/j.bios.2015.07.033

Zhang X, Kitaoka H, Tsuji S, Tamai M, Kobayashi H, Honjoh KI, Miyamoto T (2014) Development of a simultaneous detection method for foodborne pathogens using surface plasmon resonance biosensors. Food Sci Technol Res 20(2):317–325. https://doi.org/10.3136/fstr.20.317

Zhang Y, Tadigadapa S (2004) Calorimetric biosensors with integrated microfluidic channels. Biosens Bioelectron 19(12):1733–1743. https://doi.org/10.1016/j.bios.2004.01.009

Zhu Q, Liu G, Kai M, Miller AOA (2015) DNA aptamers in the diagnosis and treatment of human diseases. Molecules 20(12):20979–90997. https://doi.org/10.3390/molecules201219739

# Chapter 15
# Aflatoxin: Occurrence, Regulation, and Detection in Food and Feed

**Abdulhadi Yakubu and Ashish Vyas**

**Abstract** The carcinogenic nature of aflatoxins produced mainly by *Aspergillus flavus* and *Aspergillus parasiticus* is of great threat to humans and animals as well as an economic concern especially in tropical and subtropical countries where environmental conditions favor fungal growth. These conditions raises the possibility of contamination of aflatoxin in many agricultural foodstuffs like peanut, cereals, maize grains, and animal feed. Due to their low concentration in food and feeds, detection of aflatoxins in food and animal feeds must require highly sensitive, rapid, specific, portable, and inexpensive technique or device to meet the international maximum residue level (MRL). Several analytical techniques such as chromatography, mass spectrometry, infrared spectroscopy, capillary electrophoresis, immunoassays, and biosensors were used for the detection of aflatoxins. In this chapter, some aspects of aflatoxin occurrence in food and feeds, its current regulations by national and international regulatory bodies, and trends of some rapid and highly sensitive devices for easy detection of aflatoxins in food and feeds were discussed.

**Keywords** Aflatoxins · *Aspergillus* · Food · Feed · Occurrence · Biosensors · Regulations · Detection

A. Yakubu
Department of Microbiology, School of Bioscience and Bioengineering, Lovely Professional University, Phagwara, Punjab, India

Department of Science Laboratory Technology, College of Science and Technology, Jigawa State Polytechnic, Dutse, Nigeria

A. Vyas (✉)
Department of Microbiology, School of Bioscience and Bioengineering, Lovely Professional University, Phagwara, Punjab, India
e-mail: ashish.vyas@lpu.co.in



J. Singh et al. (eds.), *Microbial Biotechnology: Basic Research and Applications*, Environmental and Microbial Biotechnology,
https://doi.org/10.1007/978-981-15-2817-0_15

## 15.1 Introduction

Mycotoxins are fungal secondary metabolites which may exhibit their effect as teratogens, carcinogens, mutagens, and estrogens. The presence of these toxins in foods may pose a serious health hazard to consumers and may lead to economic loss in food and feed industries. These include aflatoxins, patulin, ochratoxins, fumonisin, citrinin, trichothecenes, and zearalenone. In all these mycotoxins, aflatoxins are the most toxic and highly carcinogenic and therefore have a serious impact on human health (Nfossi et al. 2008; Paniel et al. 2010). Aflatoxins are produced by fungal species of Aspergilli, especially *Aspergillus flavus* and *Aspergillus parasiticus* (Nfossi et al. 2008; Paniel et al. 2010; Leong et al. 2010). Their contamination in food and feed has received public attention since few decades ago. Their presence in agricultural produce has great consequences on the economy of many affected areas mainly in the developing countries where there are poor pre- and post-harvest techniques (van Egmond 1983; Applebaum et al. 1982; Kumar et al. 2017). Up to now, 18 aflatoxins (AFs) have been identified, but only AFB1, AFB2, AFG1, and AFG2 are the most common of which among them, AFB1 is the most toxic (Hansmann et al. 2009). When a cow ingested aflatoxin $B_1$ ($AB_1$) from contaminated feed, enzymatic hydroxylation will transform it to aflatoxin $M_1$ ($AFM_1$), now classified as group 1 carcinogenic agent by the International Agency for Research on Cancer (IARC 2002; Krishnamachari et al. 1975).

Aflatoxins were first discovered in the 1960s in England when an outbreak of Turkey "X" disease killed around 100,000 turkeys and other farm animals. Heavy ingested peanut containing *Aspergillus flavus* was found to be the feed component that caused the disease (van Egmond 1983; Hansmann et al. 2009). In India, Rajasthan and Gujarat states also recorded a case of hepatitis that resulted in the death of about 106 people due to the intake of food containing aflatoxin (Bhat and Vasanthi 2003). Preliminary analysis confirmed the presence of *Aspergillus flavus* in maize which is the major food staple of these states (Kumar et al. 2017; Krishnamachari et al. 1975). Fungal growth and production of aflatoxins are generally found in tropical regions where there are high environmental conditions temperature, moisture, relative humidity, unseasonal rain during harvest as well as flood. Fungal proliferation in food is mainly due to bad harvesting practices, lack of good storage facilities, and poor conditions in transportation and marketing (Mohamadi and Alizadeh 2010; Matabaro et al. 2017).

## 15.2 Global Aflatoxin Occurrence in Food and Feed

As defined by CODEX Alimentarius (2011), any substances that are accidentally found in human food or feed of food-producing animals due to production, manufacturing, processing, preparation, treatment, packaging, packing, transport, or holding of such food or feed or as a result of environmental conditions are called contaminants. Contaminations in food and feeds with aflatoxins have a great negative impact economically and have received a lot of attentions since the previous decades. Global detection of these toxins in food commodities mainly in developing countries where pre- and post-harvest equipment are not enough to curtail the growth of these fungi is of great economic concern. Aflatoxins B1, B2, G1, and G2 occur naturally in foods and contaminate a large number of foods such as rice, wheat, corn, and peanuts (Schatzmayr and Streit 2013; Han et al. 2013) (Fig. 15.1). Table 15.1 shows recent investigations with different types of method performed globally for the detection of aflatoxins in food and feeds. The most toxic among them is AFB1 which is classified as group 1 liver carcinogen by the International Agency for Research on Cancer (IARC). Direct contact and indirect contact of human to AFs occur by consuming AF-contaminated foods and products from animals initially exposed to AF-contaminated feeds, respectively. Most developed countries set up a permissible level of AFs as low as possible because of their carcinogenic nature. A maximum permissible level of 2 μg/kg for AFB1 as well as 4 μg/kg for total AFB1, AFB2, AFG1, and AFG2 was approved by the European Union in a variety of products.

The urgent need for control measures against toxicogenic fungi especially aflatoxins was suggested in a research conducted in Eastern Ethiopia when a high concentration of total aflatoxin level was detected in 93 out of 120 samples of groundnut analyzed using ELISA test. As per variation of total aflatoxin level, between 15 mg/kg and 11,900 mg/kg is an indication of its high occurrence in Ethiopian groundnuts which is by far beyond the limits set by the European Union (EU) (4–15 mg/kg), Food and Agriculture Organization of the United Nations (FAO) (15 μg/kg), and World Health Organization (FAO/WHO) standard (15 μg/kg) (Chala et al. 2013). From a survey of 200 feeds and 200 milk samples in China, 40% of the feed samples have AFB1 in the range of 0.05–3.53 mg/kg, while 36% were positive for AFB2 in the range of 0.03–0.84 mg/kg. Although the amount of aflatoxin B1 was slightly

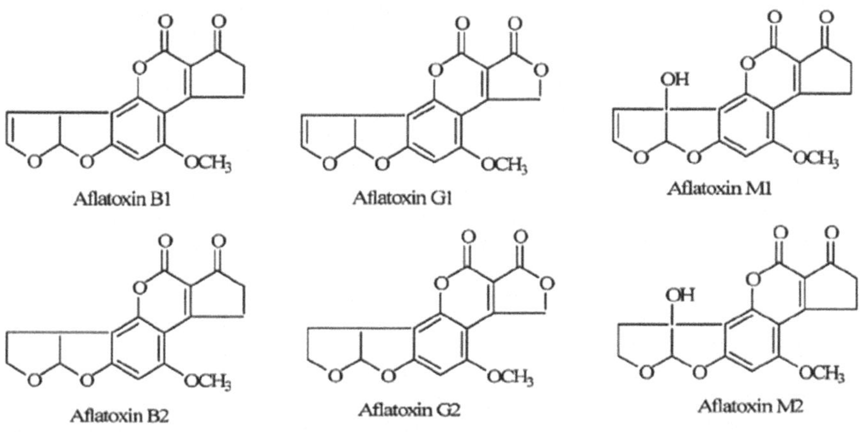


**Fig. 15.1** Different types of aflatoxins (source: Zhang et al. 2014)

**Table 15.1** Global occurrences of aflatoxins in food and feed in most countries

| Country | Sample type | Type of aflatoxin | Number of samples (*N*) | Method of detection | Year | Range (ng/kg) | Mean | Reference |
|---|---|---|---|---|---|---|---|---|
| Portugal | Baby foods | M1 | 6 | HPLC | 2007 | 1.4 | 4.6 | Alvito et al. (2010) |
| | | B1 | 6 | | | 0.52 | 1.7 | |
| Brazil | Skimmed yoghurt | M1 | 23 | Fluorescence | 2010 | 0.00–2.00 | | Iha et al. (2011) |
| Malaysia | Rice based | B1 | 13 | ELISA | N/S | 0.68–3.79 | 1.75 | Reddy et al. (2011) |
| | Wheat based | B1 | 14 | | | 0.55–5.07 | 1.60 | |
| | Peanut | B1 | 13 | | | 1.47–15.33 | 4.25 | |
| India | UHT-treated milk | M1 | 45 | ELISA | 2010 | 5.00–15.00 | | Siddappa et al. (2016) |
| | Pasteurized milk | M1 | 7 | | | 3.00–4.00 | | |
| Serbia | Cow milk | M1 | 150 | ELISA | 2013 | 2.00–11.00 | 0.01–1.2 | Kos et al. (2014) |
| Ethiopia | Stored groundnut | AFs | 120 | ELISA | N/S | 15–1980<br>15–5560 | 905<br>1890 | Chala et al. (2013) |
| China | Dairy cow feed | B1 | 200 | ELISA | 2010 | 3.53 | 0.31 | Han et al. (2013) |
| | | B2 | 200 | | | 0.84 | 0.14 | |
| | | M1 | 200 | | | 59.60 | 15.30 | |
| | Semi-skimmed yoghurt | M1 | 21 | | | 0.00–12.00 | | |
| Zimbabwe | Peanut | B1 | | HPLC | 2011 | | | Mupunga et al. (2014) |
| | | B2 | | | | | | |
| | | G1 | | | | | | |
| | | G2 | | | | | | |
| China | Rice | B1 | 370 | HPLC | 2009–2011 | 0.03–20.00 | 0.65 | Lai et al. (2015) |

higher than aflatoxin B2 in the feeds, it was still below the acceptable limits of set aside by European Union 5 μg/kg as well as 10 μg/kg for China, respectively. The total amount of aflatoxins was also below the US acceptable limit of 20 μg/kg (Han et al. 2013). In Malawi, Matumba et al. (2014) collected samples of locally processed and imported maize as well as groundnut-based food products. The extent of aflatoxin contamination was analyzed with the help of immunoaffinity-reversed-phase liquid chromatography. All imported baby cereal foods and locally processed de-hulled maize have low contents of AFs below acceptable limit, while that of locally processed maize-based foods was above the EU maximum acceptable limit of 0.1 mg/kg; monitoring of AFs in locally processed foods will likely reduce AF amount and also reduce the risk of eating AF-contaminated food and feeds.

In Zimbabwe, fungal contaminations and aflatoxin were detected using high-performance liquid chromatography-fluorescence and standard mycology culture methods, respectively. Four out of six peanuts examined for fungal contamination were infected with *Aspergillus flavus* or *Aspergillus parasiticus* ranging from 3 to 20% of the seeds studied, while 27% of the peanut butter samples were also infected with either *Aspergillus flavus* or *Aspergillus parasiticus*. The result also indicated that, 91% of peanut butter and 17% of peanut samples are contaminated with aflatoxins with mean values of 75.66 ng/kg, respectively. It was found to exceed the EU acceptable level and hence advice sensitization for manufacturers so as to reduce contamination level (Mupunga et al. 2014). A study of 45 samples of ultra-high treatment (UHT) milk from the Indian states of Karnataka and Tamil Nadu were analyzed for the detection of aflatoxin M1 by reversed-phase HPLC using fluorescent detector. All the UHT milk samples tested were positive for AFM1, and 38% of these contained AF levels higher than the acceptable limit of 0.5 μg/kg prescribed by the Codex and Indian regulatory commission (FSSAI 2011; Siddappa et al. 2016).

In Lebanon, Elkak et al. (2012) conducted a research to detect AFM1 in 111 randomly selected cheese samples from local small dairy farms and dairy industries as well as imported cheese. From the cheese samples analyzed, AFM1 was detected in 67.56% of which AFM1 levels in 17.33% of the samples exceeded the European Commission (EC) acceptable limit of 250 ng/kg. Frequent supervision of locally processed cheese in Lebanon may drastically reduce the health risk associated with AFM1. Vagef and Mahmoudi (2013) provide an update on the level of AFM1 in 144 fresh and pasteurized milk samples from western region of Iran using ELISA technique. They concluded that the amount of AFM1 in both fresh and pasteurized milk was higher than the tolerable level of 0.5 ppb where the contamination level was significantly higher in winter than in summer. In a similar research, cow milk samples were found to contain AFM1 in an amount ranging from 0.01 to 1.2 mg/kg, out of which, 86.0% of the milk samples contained high quantity of AFM1 higher than the tolerable limit of 0.05 mg/kg set by European Union (EU) regulations. Other types of milk samples indicated a percentage of AFM1 as 80.0%, 60.0%, and 60.0% in goat, donkey, and breast milk, respectively (Kos et al. 2014).

In Brazil, a research was conducted to check the incidence and occurrence of aflatoxin M1 in cheese, yoghurt, and dairy drinks. A total of 123 samples were collected and analyzed using different methods in which all the samples tested have

AFM1 higher than the acceptable level. Although there is lack of regulatory limit of aflatoxins in Brazil, this survey offered some useful information on the occurrence of AFM1 in Brazilian dairy products with potential risk to consumers as well as an insight into the need for establishment of Brazilian Maximum Residue Level (MRL) of AFs in food and feeds (Iha et al. 2011). Also in Croatia, Bilandzir et al. (2014) found a high concentration of AFM1 above tolerable limit when different animals were studied for the presence of AFM1 in their milk from July to September 2013. The result indicated that high level of this toxin in cow milk shows the use of contaminated feedstuff in some farms within the studied period.

## 15.3 Regulations

International and national regulatory bodies set maximum tolerable limits of aflatoxins as their carcinogenic and hazardous nature was detected in food and feed of humans and other animals, respectively. The joint FAO/WHO expert committee stated that the presence of aflatoxins in food should be limited to tolerable limit defined as the amount of a substance that cannot be removed from food or feed without discarding that particular food or compromising the exact availability of main food supplies. The first legal act on aflatoxin was established by United States Food and Drug Administration (USFDA) in 1965 when a maximum residual level (MRL) of 30 pg/kg for total aflatoxin (B1 + B2 + G1 + G2) was proposed. Since then, the maximum residual level has been regularly revised. The current regulations for aflatoxin approved by the Joint FAO/WHO Committee (CODEX), FDA, and some other countries are given in Table 15.2.

Many countries have set their maximum acceptable limit of aflatoxins in food and feeds. Industrialized nations set lower limit and regularly monitor and update their acceptable limit for import and export of food commodities likely to contain

**Table 15.2** Standard ranges of aflatoxins set by international organizations

| Regulatory body | Type of AF | Food/feed items | Standard range (ppb) | Year | Source |
|---|---|---|---|---|---|
| CODEX (WHO/FAO) | Aflatoxin total and AFM1 | Milk, almonds, Brazil nuts, hazelnuts, peanuts, etc. | 0.5–15 | 2015 | Stan (2015) |
| EU | B1 (sum of B1, B2, G1, and G2) and M1 | Peanuts and other oilseeds, almonds, dried fruits, corn, infant formula milk, etc. | 0.5–15 | 2006 | European (2006) |
| FDA | B1, total, and M1 | Milk, peanut, peanut products, foods, Brazil and pistachio nuts | 0.5–300 | 2011 | USFDA (2011) |
| FSSAI | Total and M1 | Milk, cereals and cereal products, pulses, nuts, etc. | From 0.5 in milk to 30 in spices | 2011 | FSSAI (2011) |

**Table 15.3** Maximum regulation limits for aflatoxins in food and feed in some countries

| Country | Aflatoxin | MRL (ppb) | Types of food | Reference |
|---|---|---|---|---|
| India | B1 | 30 | All | Anukul et al. (2013) |
| Switzerland | B1<br>B1 | n/a | All<br>Maize and cereals | Creppy (2002) |
| Malaysia | Total | 35 | All | Anukul et al. (2013) |
| Taiwan | Total | 15 | Peanut, corn | Anukul et al. (2013) |
| South Korea | B1 | 0.1 | Baby food | Anukul et al. (2013) |
| Sweden | M1 | n/a | Liquid milk products | Creppy (2002) |
| Czech Republic | M1<br>M1 | n/a | Children milk<br>Adult milk | Creppy (2002) |
| China | B1<br>M1 | 20<br>0.5 | Maize, peanuts<br>Infant food | Anukul et al. (2013) |
| Philippines | Total | 20 | Human foods | Anukul et al. (2013) |
| Austria | B1 | n/a | All<br>Cereals and nuts | Creppy (2002) |
| Japan | Total<br>B1<br>B1 | 10<br>10<br>5 | All<br>Rice<br>Other grains | Anukul et al. (2013) |

aflatoxin than developing and underdeveloped countries. In Table 15.3, for example, countries like China and the Philippines as reported by Anukul et al. (2013) set a tolerance level of 20 ppb for total aflatoxins in all human foods, while 30 and 35 ppb are the tolerance levels of aflatoxin B for all human foods in India and Malaysia, respectively. However, this difference of acceptable limit among countries brings about difficulties in the trades of commodities from one country to another. Even though in 2011, the Serbian Government has changed and harmonized their tolerable limit for AFM1 in milk with the European Union (EU) regulation, the 2013 occurrence of AFM1 in Serbian milk led to change in regulation where MRL was changed from 0.05 to 0.5 mg/kg. Kos et al. (2014) suggest permanent and harmonious regulations of AFM1 in milk and that of other aflatoxins in animal feeds with that of the EU taking into account that milk is one of the major food staples for Serbians.

## 15.4 Aflatoxin Detection in Food and Feed

### *15.4.1 Sampling and Sample Preparation*

The main problem with aflatoxin analysis is obtaining a representative sample from the said commodity. A large amount of commodity may be required to increase the chances of toxin detection since a very small quantity may differ widely within any batch of food and feed. Detailed procedures for sampling and preparation of aflatoxin analysis can be found in FAO/WHO standard regulation 20/2015. Based on USFDA (2011) criterion, three steps which include sample size reduction, sample particle reduction, and sample homogenization for uniformity are vital in preparing any sample for aflatoxin detection.

### 15.4.2 *Extraction of Aflatoxin from Food and Feed*

An efficient extraction step is required in order to detect and quantify aflatoxins in any food samples. In polar protic solvents like methanol, acetone, chloroform, and acetonitrile, aflatoxins are normally and easily soluble. Aflatoxin extraction involves the use of organic solvents like methanol, acetonitrile, or acetone mixed in different proportion with a small quantity of water (Bertuzzi et al. 2011).

Determination of aflatoxins based on immunoassay technique requires extraction using mixture of ethanol and water. This is due to the fact that methanol has lesser negative effect on antibodies compared to acetone (Stroka et al. 1999; Lee et al. 2004). The clean-up technique which uses immune-affinity column (IAC) chromatography is usually followed after extraction (Ma et al. 2013). This immune-affinity column chromatography can bind with antigen and antibody with high reversibility and specificity which can separate as well as purify target analytes from matrices (Shelver et al. 1998). The crude sample is mainly applied to the column having a specific antibody to aflatoxins immobilized on a solid support like silica during clean-up. Aflatoxin binds onto the column and is retained with the sample moving beneath the column. The other washing step is usually required to remove impurities and unbound proteins when conducted with appropriate buffers and ionic strengths.

### 15.4.3 *Trends in the Methods Used for Aflatoxin Detection*

Over some decades, a lot of analytical methods and devices have been developed for the detection and separation of aflatoxins in food and feeds. Monitoring its presence in various commodities is important for the protection of consumer as well as producing raw materials prior to cost intensive processing or transport. Most of the devices or methods described below have advantages and disadvantages over one another in terms of their mode of operation, utilization, purchase, reliability, duration, and acceptability explained in many literatures.

#### 15.4.3.1 Chromatographic Methods

A separation technique that involves the physical interaction between a mobile phase and a stationary phase is known as chromatographic technique. The separated components are to be distributed between mobile phase which is usually fluid passing along stationary phase (Braithwaite and Smith 2000). In practical point of view, the analyzed sample usually dissolved in mobile phase and applied as a spot on the stationary phase. Sorbents are the partitions between solid and liquid stationary phase when samples are analyzed in mobile phase. The most common chromatographic methods for evaluations of aflatoxins are described below.

#### Thin-Layer Chromatography (TLC)

De Iongh first used thin-layer chromatography (TLC) in 1964, and TLC was regarded as the best method for aflatoxin detection in 1990 by the Association of Official Analytical Chemists (AOAC). This separation technique usually depends on silica, aluminum, or cellulose as stationary matrix, while the mobile phase consists of a mixture of methanol, acetonitrile, and water immobilized on plastic or glass. Aflatoxin movement within these phases is based mainly on changes of aflatoxin solubility in the two phases. This technique has an application for the measurement of aflatoxins in agricultural produce and can also detect as small as 1–20 ppb of aflatoxin. It can however need a well-trained personnel and tedious sample pretreatment, and sometimes, it is not accurate since there is probability that errors may occur at multiple points along the process. This technique has application in measurement of already known aflatoxins at high concentrations, especially if new equipment is not readily available (Wacoo et al. 2014).

#### High-Performance Thin-Layer Chromatography (HPTLC)

This technique is like thin-layer chromatography, but unlike TLC, all the separation processes such as plate development, application of sample, as well as interpretation of data were carried out automatically in a precise and efficient way. This method is time-consuming and laborious and requires well-trained personnel and expensive instrumentations (Badea et al. 2004). Other limitations of this method are requirement of complex gradient mobile phase, large amount of organic solvent, and regular maintenance of equipment (Wacoo et al. 2014). However, some of these limitations were overcome by the use of gas chromatography.

#### Gas Chromatography (GC)

Gas chromatography mainly separates aflatoxins by the movement of carrier gas acting as mobile phase through the column acting as stationary phase that has a liquid coated onto inert solid particles (Cunha and Fernandes 2010). Electron capture detector (ECD) or flame ionization detectors (FID) are usually used to detect aflatoxins while gases are separated from other samples as they move along the column. Separation of aflatoxins in gas chromatography requires molecule derivatization to a detectable volatile form since most of the toxins are not volatile.

High-Performance Liquid Chromatography (HPLC)

Just like similar chromatographic methods, this method also separates aflatoxins as mobile phase moves along stationary one. This movement mainly consists of adsorbents depending on the chemical and physical structure of aflatoxins (Gamliel et al. 2017). The sample normally in liquid form moves along the column by carrier solvents where the aflatoxins separate from the main components during extraction. The procedure applied in this separation technique is usually differentiated from other techniques based on their column types and carrier liquid since only few detectors like ultraviolet and fluorescent light are coupled to HPLC.

#### 15.4.3.2 Spectroscopic Methods

Fluorescence Spectrophotometry

This method can quantitatively measure from 5 to 5000 ppb of aflatoxin in 5 minutes and remains to be the most pivotal technique in the analysis and characterization of molecules that emit energy at a certain wavelength in peanuts and grains for aflatoxin analysis. As per Gamliel et al. (2017), derivatization may be required for best analysis of aflatoxin using fluorometry for improved aflatoxin fluorescence. As per approval by European High commission, aflatoxin detection limit using this method is moderately high than the limit of 4 μg/kg.

Frontier Infrared Spectroscopy

Fourier transform infrared spectroscopy (FTIR) is the preferred method of infrared spectroscopy. In infrared spectroscopy, IR radiation is passed through a sample. Some of the infrared radiation is absorbed by the sample, and some of it is passed through (transmitted). The resulting spectrum represents the molecular absorption and transmission, creating a molecular fingerprint of the sample. Like a fingerprint, no two unique molecular structures produce the same infrared spectrum. This makes infrared spectroscopy useful for several types of analysis (Mirghani et al. 2001) reported the use of frontier infrared spectroscopy for the analysis of aflatoxins in peanuts and peanut cake which use total internal reflectance. Detection of aflatoxins was also reported by Pearson et al. (2001) using transmittance and reflectance in single corn kernels.

Infrared Spectroscopy

Infrared (IR)-based methods require little sample preparation with extensive calibration. Near-infrared (NIR) and mid-infrared (MIR) are rapid and nondestructive analytical techniques and, hence, usually used for food quality (McMullin et al.

2015). Due to their limited sensitivity, commonly used IR-based method cannot detect mycotoxin directly in a food and plant tissues. However, it can be used as a detection tool following appropriate separation procedures like HPLC. The major limitation of this method is the limited sensitivity of detection.

#### 15.4.3.3 Immunochemical Methods

According to Wacoo et al. (2014), this method of detection depends mainly on specific binding between antigens and antibodies. Various immunochemical methods have been developed because of the high affinity and specificity of antibodies on antigens. However, this binding specificity is not limited to antibodies and antigens but also on ligands and receptors which also have such affinity and high specificity (Sargent and Sadik 1999).

##### Radioimmunoassay (RIA)

Rauch et al. (1987) invented RIA with its application in the qualitative determination of insulin in human blood with further extension for of aflatoxin in contaminated food items. Langone and Vunakis (1976) confirmed through their studies the use of solid phase radioimmunoassay technique in determining aflatoxin B1 in peanut, and a limit of detection of 1 μg/kg was achieved. In addition, Rauch et al. (1987) reported the use of radioimmunoassays for the qualitative and quantitative determination of aflatoxin B1 and aflatoxin M1 levels. This immunochemical method depends on competitive binding among radioactive-labelled and nonradioactive antigens. Berson and Yalow (1968) reported that for a set number of antibody or antigen binding sites on the same antibody, radioactive-labelled antigen takes part with unlabelled nonradioactive antigen. A measured amount of labelled antigen and an unknown quantity of unlabelled antigen with that of standards react competitively with a known and small amount of the antibody. These labelled amounts of antigen are in inverse proportion to the amount of unlabelled antigen in the test. The advantage of this method is its capacity to run many examinations simultaneously with high reactivity and precision. One negative aspect of RIA is it requires antigen in pure form and used label isotopes associated with possible endangerment as well as problem associated with storage and disposing radioactive waste (Wacoo et al. 2014).

##### Enzyme-Linked Immunosorbent Assay (ELISA)

Avrameas (1969) invented the ELISA using enzyme-antigen conjugates and enzyme-antibody conjugates. ELISA method relies on the preciseness of antibodies for antigens, and the reactiveness of the assay is enhanced by tagging either the antibodies or the antigens with an enzyme that can be simply evaluated by use of specific substrates. Hence, an antibody which is immobilized onto a stable support may take an unlabelled antigen in the analyte, which is subsequently distinguished by a labelled antibody. As per Devi et al. (1999), mean immobilized antibody onto a solid support may capture an unlabelled antigen in the analyte, which later identified by a labelled antibody. ELISA method is presently used in the identification of aflatoxin in agricultural products (Anjaiah et al. 1989; Thirumala-Devi et al. 2002; Ondieki et al. 2014). Some commercially available ELISA kits that use enzymes alkaline phosphatase and horseradish peroxidase as labels in analysis of aflatoxins based on a competitive immunoassay format are extensively used (Ostadrahimi et al. 2014) and (Huybrechts 2011). The ELISA technique has advantages; that is, it is feasible to achieve the test on a 96-well assay platform resulting in the analysis of a large number of concurrent samples; the ELISA kits are cheap and easy to use, and most importantly, there is no possibility of health hazards associated with enzyme label.

#### 15.4.3.4 Immunosensors

An international scientific endeavor, i.e., International Union of Pure and Applied Chemistry (IUPAC), defined a biosensor as any device that can provide précised quantitative and semi-quantitative interpretive data using biological understanding element in explicit connection with transducers (Shruthi et al. 2014). This device is based on interaction between biological components and transducers. Biological components such as enzymes, antibodies, and tissue slices are used to recognize and interact with a specific analyte, while transducers convert this interaction into a signal that can be amplified with respect to the concentration of the analyte (Shruthi et al. 2014). Amperometric, optical, potentiometric, magnetic, and colorimetric devices are normally used as transducers. Magliulo et al. (2005) reported a rapid and highly sensitive chemiluminescent enzyme immunoassay for the determination of AFM1 in a milk sample. Similar work using these transducers was described by Parker et al. (2009), Zangheri et al. (2014), Vdovenko et al. (2014), Mavrikou et al. (2017), and Stepurska et al. (2015). It is found to be highly sensitive, accurate, cost-effective, sensitive, and throughput in the screening of AFM1 as compared with other immunoassay. Cuccioloni et al. (2008) designed an assay for analytical test of aflatoxins B1 and G1 which is an alternate screening technique for mycotoxins. This determination approach to monitor toxins was based on surface plasmon resonance using neutrophil porcine elastase as bait. Its applications include moderate speed, recycling of the capturing surface and cost effective. Stepurska et al. (2015) has designed a potentiometric biosensor based on a pH-sensitive field-effect transistor and an enzyme acetylcholinesterase for the detection of aflatoxin B1 in real samples. It was proved to be very stable and highly sensitive when tested in the determination of AFB1 in walnut, sesame, and peanut. The application of protein for creating a highly sensitive site against AFB1 produced through bioimprinting as a means of detecting AFB1 by capacitive biosensors was reported by Gutierrez et al. (2016). This biosensor has the ability to generate specific interactions with aflatoxin B1 demonstrated in a linear relation between log concentration and signal registered of the target aflatoxin in a concentration ranges between $3.2 \times 10^6$ to $3.2 \times 10^9$ M when using ovalbumin as framework for bioimprinting. Other biosensors developed for aflatoxin detections include an aptamer for detection of AFB1 (Castillo et al. 2015) and an electrical immunosensor for detection of ultra-trace quantity of AFM1 in food products (Paniel et al. 2010). This immunosensor has a low detection limit of 0.01 ppb which is under the recommended level of 0.05 ppb and has good reproducibility.

## 15.5 Conclusion and Future Challenges

The occurrence of aflatoxins in food and feeds is of global concern both in terms of health implication and economic consequences especially in developing countries where pre- and post-harvesting practices are poor. These lead to the establishment of various regulatory bodies in different countries for the sole purpose of regulating and controlling risk associated with consumption of aflatoxins in food or feeds. Different analytical methods that are usually applied for identification of toxins in food and feeds include thin-layer chromatography (TLC), high-performance liquid chromatography (HPLC), and gas chromatography (GC), which are largely time-consuming and expensive and require trained personnel and also a series of sample preparation. Based on these mentioned problems, it is important to develop better methods based on sensitivity and cost. This leads to the development of more improved analytical techniques that are rapid, more sensitive, specific, and on-site immunoassay. RIA and ELISA suffer a setback as they require pure state of antigen, used label isotopes associated with possible health hazard, and multiple washing steps, which sometimes prove laborious and time-consuming. The development of immunosensors like biosensors has brought about an opportunity for more rapid, highly sensitive, inexpensive and rapid on-site technique for easy detection of aflatoxins in food and animal feeds. This technique is based on interaction between biological components such as enzymes, antibodies, microorganisms, and tissue slices which recognize and react with a transducer like amperometric, optical, potentiometric, and magnetic devices. The biological component reacts and recognizes a specific analyte, while transducers convert this interaction into signal. However, these biosensors have some setbacks as some transducers are expensive and lose activity after some time except when stored under a better condition, require purification and isolation cost, and have slow response to time and longer recovery time.

Future research on recent recognition elements such as bacteriophages and aptamers should be focused where more robust, rapid, highly sensitive, cost-effective, and miniaturized biosensors for on-site detection of aflatoxins can be develop. The development of biosensors based on interactions between nanomaterials and biomolecules on the surface of nanofilms may also attract attention in future researches for aflatoxin detection in food and animal feeds.

## References

Alvito PC, Sizoo EA, Almeida CMM, Van Egmond HP (2010) Occurrence of aflatoxins and ochratoxin A in baby foods in Portugal. Food Anal Methods 3:22–30. https://doi.org/10.1007/s12161-008-9064-x

Anjaiah V, Mehan VK, Jayanthi S. Reddy DVR, Mcdonald D (1989) Enzyme-linked immunosorbent assay (ELISA) for aflatoxin B1 estimation in groundnuts. In ICRISAT Conference Paper no: CP 432, Patancheru, India, pp 183–189. https://doi.org/10.1057/9781137313256

Anukul N, Vangnai K, Mahakarnchandkul W (2013) Significance of regulation limits in mycotoxin contamination in Asia and risk management programs at the national level. J Food Drug Anal 21(3):227–241. https://doi.org/10.1016/jjfda.2013.07.009

Applebaum RS, Brackett RE, Wiseman DW, Marth EH (1982) Aflatoxin: toxicity to dairy cattle and occurrence in milk and milk products - a review. J Food Prot 45(8):752–777. https://doi.org/10.4315/0362-028X-45.8.752

Avrameas S (1969) Coupling of enzymes to proteins with glutaraldehyde. Use of the conjugates for the detection of antigens and antibodies. Immunochemistry 6(1):43–52. https://doi.org/10.1016/0019-2791(69)90177-3

Badea M, Micheli L, Cristina M, Candigliota T, Marconi E, Mottram T, Palleschi G (2004) Aflatoxin M1 determination in raw milk using a flow-injection immunoassay system. Analysis 520:141–148. https://doi.org/10.1016/j.aca.2004.05.068

Berson SA, Yalow RS (1968) General principles of radioimmunoassay. Clin Chim Acta 369(2):125–143. https://doi.org/10.1016/j.cca.2006.05.002

Bertuzzi T, Rastelli S, Mulazzi A, Donadini G, Pietri A (2011) Mycotoxin occurrence in beer produced in several European countries. Food Control 22(12):2059–2064. https://doi.org/10.1016/j.foodcont.2011.06.002

Bhat RV, Vasanthi S (2003) Mycotoxin food safety risk in developing countries. Food Saf Food Secur Food Trade. https://doi.org/10.1016/j.fm/2004.01.012

Bilandzir N, Bozic D, Dokic M, Sedak M, Kolanovic BS, Varenina I, Cvetnic Z (2014) Assessment of aflatoxin M1 contamination in the milk of four dairy species in Croatia. Food Control 43:18–21. https://doi.org/10.1016/j.foodcont.2014.02.044

Braithwaite A, Smith FJ (2000) Chromatographic methods, vol 11, 5th edn. Springer, Cham. https://doi.org/10.1007/978-94-011-0599-6

Castillo G, Spinella K, Poturnayová A, Šnejdárková M, Mosiello L, Hianik T (2015) Detection of aflatoxin B1 by aptamer-based biosensor using PAMAM dendrimers as immobilization platform. Food Control 52:9–18. https://doi.org/10.1016/j.foodcont.2014.12.008

Chala A, Mohammed A, Ayalew A, Skinnes H (2013) Natural occurrence of aflatoxins in groundnut (Arachis hypogaea L.) from eastern Ethiopia. Food Control 30(2):602–605. https://doi.org/10.1016/j.foodcont.2012.08.023

Creppy EE (2002) Update of survey, regulation and toxic effects of mycotoxins in Europe. Toxicol Lett 127:19–28

Cuccioloni M, Mozzicafreddo M, Barocci S, Ciuti F, Pecorelli I, Eleuteri AM, Angeletti M (2008) Biosensor-based screening method for the detection of aflatoxins B1–G1. Anal Chem 80(23):9250–9256

Cunha SC, Fernandes JO (2010) Development and validation of a method based on a QuEChERS procedure and heart-cutting GC-MS for determination of five mycotoxins in cereal products. J Sep Sci 33(4–5):600–609. https://doi.org/10.1002/jssc.200900695

Devi KT, Mayo MA, Reddy KLN, Delfosse P, Reddy G, Reddy SV, Reddy DVR (1999) Production and characterization of monoclonal antibodies for aflatoxin B1. Lett Appl Microbiol 29(5):284–288. https://doi.org/10.1046/j.1472-765X.1999.00685.x

Elkak A, El O, Habib J, Abbas M (2012) Occurrence of aflatoxin M1 in cheese processed and marketed in Lebanon. Food Control 25(1):140–143. https://doi.org/10.1016/j.foodcont.2011.10.033

European CR (2006) European Commission Regulation (EC): setting maximum levels for certain contaminants in foodstuffs. Eur Union Law 364:5–24. https://doi.org/10.2203/dose-response.06-012.Hanekamp

FSSAI (2011) Food safety and standards (contaminants, toxins and residues) regulations. New Delhi. Retrieved from https://fssai.gov.in/hi/dam/jcr...5b17.../Compendium_Contaminants_Regulations.pdf. Accessed 12 Mar 2018

Gamliel A, Dehne HW, Karlovsky P, Fletcher J (2017) Detection of mycotoxins in food: applications of rapid and reliable tools in a biosecurity context. Springer, Cham. https://doi.org/10.1007/978-3-319-46897-6

Gutierrez AVR, Hedstrum M, Mattiasson B (2016) Bioimprinting as a tool for the detection of aflatoxin B1 using a capacitive biosensor. Biotechnol Rep 11:12–17. https://doi.org/10.1016/j.btre.2016.05.006

Han RW, Zheng N, Wang JQ, Zhen YP (2013) Survey of aflatoxin in dairy cow feed and raw milk in China. Food Control 34:35–39. https://doi.org/10.1016/j.foodcont.2013.04.008

Hansmann T, Sanson B, Stojan J, Weik M, Marty JL, Fournier D (2009) Kinetic insight into the mechanism of cholinesterasterase inhibition by aflatoxin B1 to develop biosensors. Biosens Bioelectron 24(7):2119–2124. https://doi.org/10.1016/j.bios.2008.11.006

Huybrechts B (2011) Evaluation of immunoassay kits for aflatoxin determination in corn & rice. Tervuren, Belgium

IARC (2002) International Agency for Research on Cancer: monographs on the evaluation of carcinogenic risks to humans. Int Agency Res Cancer 96:1–390. https://doi.org/10.1002/food.19940380335

Iha MH, Barbosa CB, Okada IA, Trucksess MW (2011) Occurrence of aflatoxin M1 in dairy products in Brazil. Food Control 22:1971–1974. https://doi.org/10.1016/j.foodcont.2011.05.013

Kos J, Levi J, Đuragi O, Koki B, Miladinovi I (2014) Occurrence and estimation of aflatoxin M1 exposure in milk in Serbia. Food Control 38:41–46. https://doi.org/10.1016/j.foodcont.2013.09.060

Krishnamachari KA, Bhat RV, Nagarajan V, Tilak TBG (1975) Hepatitis due to aflatoxicosis: an outbreak in Western India. Lancet 5:1061–1063. https://doi.org/10.1016/S0140-6736(75)91829-2

Kumar P, Mahato DK, Kamle M, Mohanta TK, Kang SG (2017) Aflatoxins: a global concern for food safety, human health and their management. Front Microbiol 7(1):1–10. https://doi.org/10.3389/fmicb.2016.02170

Lai X, Zhang H, Liu R, Liu C (2015) Potential for aflatoxin B1and B2 production by *Aspergillus flavus* strains isolated from rice samples. Saudi J Biol Sci 22(2):176–180. https://doi.org/10.1016/j.sjbs.2014.09.013

Langone JJ, Vunakis H (1976) Aflatoxin B1: specific antibodies and their use in radioimmunoassay. J Natl Cancer Inst 56(3):591–595

Lee NA, Wang S, Allan RD, Kennedy IR (2004) A rapid aflatoxin B 1 ELISA: development and validation with reduced matrix effects for peanuts, corn, pistachio and soybeans. J Agric Food Chem 52:2746–2755

Leong Y, Ismail N, Latif AA, Ahmad R (2010) Aflatoxin occurrence in nuts and commercial nutty products in Malaysia. Food Control 21(3):334–338. https://doi.org/10.1016/j.foodcont.2009.06.002

Ma F, Chen R, Li P, Zhang Q, Zhang W, Hu X (2013) Preparation of an immunoaffinity column with amino-silica gel microparticles and its application in sample cleanup for aflatoxin detection in agri-products. Molecules 18(2):2222–2235. https://doi.org/10.3390/molecules18022222

Magliulo M, Mirasoli M, Simoni P, Lelli R, Portanti O, Roda A (2005) Development and validation of an ultrasensitive chemiluminescent enzyme immunoassay for Aflatoxin M1 in milk. J Agric Food Chem 53(Cl):3300–3305

Matabaro E, Ishimwe N, Uwimbabazi E, Lee BH (2017) Current immunoassay methods for the rapid detection of aflatoxin in milk and dairy products. Compr Rev Food Sci Food Saf 11:1–13. https://doi.org/10.1111/1541-4337.12287

Matumba L, Monjerezi M, Biswick T, Mwatseteza J, Makumba W, Kamangira D, Mtukuso A (2014) A survey of the incidence and level of aflatoxin contamination in a range of locally and imported processed foods on Malawian retail market. Food Control 39:87–91. https://doi.org/10.1016/j.foodcont.2013.09.068

Mavrikou S, Flampouri E, Iconomou D, Kintzios S (2017) Development of a cellular biosensor for the detection of aflatoxin B1, based on the interaction of membrane engineered Vero cells with anti AFB1 antibodies on the surface of gold nanoparticle screen printed electrodes. Food Control 73:64–70. https://doi.org/10.1016/j.foodcont.2016.06.002

McMullin D, Mizaikoff B, Krska R (2015) Advancements in IR spectroscopic approaches for the determination of fungal derived contaminations in food crops. Anal Bioanal Chem 407(3):653–660. https://doi.org/10.1007/s00216-014-8145-5

Mirghani MES, Che Ma YB, Jinap S, Baharin BS, Bakar J (2001) A new method for determining aflatoxins in groundnut and groundnut cake using Fourier transform infrared spectroscopy with attenuated total reflectance. JAOCS 78(10):985–992. https://doi.org/10.1007/s11746-001-0376-y

Mohamadi H, Alizadeh M (2010) A study of the occurrence of aflatoxin M1 in dairy products marketed in Urmia. Iran J Agr Sci Tech 12:579–593

Mupunga I, Lebelo SL, Mngqawa P, Rheeder JP, Katerere DR (2014) Natural occurrence of aflatoxins in peanuts and peanut butter from Bulawayo, Zimbabwe. J Food Prot 77(10):1814–1818. https://doi.org/10.4315/0362-028X.JFP-14-129

Nfossi LA, Alderara MAC, Aggiani CLB, Iovannoli CRG, Rletti ENA, Iraudi GIG (2008) Development and application of solvent-free extraction for the detection of aflatoxin M1 in dairy products by enzyme immunoassay. J Agric Food Chem 56:1852–1857

Ondieki D, Lutta ST, Okoth MO, Kering P (2014) Rapid assessment of aflatoxin contamination of groundnut by thin layer chromatography and competitive enzyme linked immunosorbent assay from selected divisions of Busia County in Kenya. Afr J Food Sci Technol 5(1):12–20. https://doi.org/10.14303/ajfst.2014.008

Ostadrahimi A, Ashrafnejad F, Kazemi A, Sargheini N, Mahdavi R, Farshchian M, Mahluji S (2014) Aflatoxin in raw and salt-roasted nuts (pistachios, peanuts and walnuts) sold in markets of Tabriz, Iran. Jundishapur J Microbiol 7(1):1–6. https://doi.org/10.5812/jjm.8674

Paniel N, Radoi A, Marty J (2010) Development of an electrochemical biosensor for the detection of aflatoxin M1 in milk. Sensors 10:9439–9448. https://doi.org/10.3390/s101009439

Parker CO, Lanyon YH, Manning M, Arrigan DWM, Tothill IE (2009) Electrochemical immunochip sensor for aflatoxin M1 detection. Anal Chem 81(13):5291–5298

Pearson TC, Wicklow DT, Maghirang EB, Xie F, Dowell FE (2001) Detecting aflatoxin in single corn kernels by transmittance and reflectance spectroscopy. Trans ASAE 44(5):1247–1254. https://doi.org/10.13031/2013.6418

Rauch P, Fukal L, Prosek J, Brezina P, Kas J (1987) Radioimmunoassay of aflatoxin M1. J Radioanal Nucl Chem 117(3):163–169. https://doi.org/10.1007/BF02165370

Reddy KRN, Farhana NI, Salleh B (2011) Occurrence of Aspergillus spp. and aflatoxin B1 in Malaysian foods used for human consumption. J Food Sci 76(4):T99–T104. https://doi.org/10.1111/j.1750-3841.2011.02133.x

Sargent A, Sadik OA (1999) Monitoring antibody-antigen reactions at conducting polymer-based immunosensors using impedance spectroscopy. Electrochim Acta 44(26):4667–4675. https://doi.org/10.1016/S0013-4686(99)00265-0

Schatzmayr G, Streit E (2013) Global occurrence of mycotoxins in the food and feed chain: facts and figures. World Mycotoxin J 6:213–222. https://doi.org/10.3920/WMJ2013.1572

Shelver WL, Larsen GL, Huwe JK (1998) Use of an immunoaffinity column for tetrachlorodibenzo-p-dioxin serum sample cleanup. J Chromatogr B Biomed Appl 705(2):261–268. https://doi.org/10.1016/S0378-4347(97)00540-9

Shruthi GS, Amitha CV, Blessy BM (2014) Biosensors a modern day achievement. J Instrum Technol 2(1):26–39. https://doi.org/10.12691/jit-2-1-5

Siddappa V, Nanjegowda DK, Viswanath P (2016) Occurrence of aflatoxin M1 in some samples of UHT, raw & pasteurized milk from Indian states of Karnataka and Tamil Nadu. Food Chem Toxicol 50(11):4158–4162. https://doi.org/10.1016/j.fct.2012.08.034

Stan C (2015) General standard for contaminants and toxins in food and feed (Codex Stan 193-1995). New York City. Retrieved http://www.fao.org/input/download/standards/17/CXS_193e_2015.pdf. Accessed 12 Mar 2018

Stepurska KV, Soldatkin OO, Arkhypova VM, Soldatkin AP, Lagarde F, Jaffrezic-Renault N, Dzyadevych SV (2015) Development of novel enzyme potentiometric biosensor based on pH-sensitive field-effect transistors for aflatoxin B1 analysis in real samples. Talanta 144:1079–1084. https://doi.org/10.1016/j.talanta.2015.07.068

Stroka J, Petz M, Joerissen U, Anklam E (1999) Investigation of various extractants for the analysis of aflatoxin B1 in different food and feed matrices. Food Addit Contam 16(8):331–338. https://doi.org/10.1080/026520399283902

Thirumala-Devi K, Mayo MA, Hall AJ, Craufurd PQ, Wheeler TR, Waliyar F, Reddy D (2002) Development and application of an indirect competitive enzyme-linked immunoassay for aflatoxin M1 in milk and milk-based confectionery. J Agric Food Chem 50(4):933–937. https://doi.org/10.1021/jf011139b

USFDA (2011) Mycotoxin regulatory guidance: a guide for grain elevators, feed manufacturers, grain processors and exporters. Washington DC. Retrieved https://www.ngfa.org/wp-content/uploads/NGFAComplianceGuide-FDARegulatoryGuidanceforMycotoxins8-2011.pdf. Accessed 12 Mar 2018

Vagef R, Mahmoudi R (2013) Occurrence of aflatoxin M1 in raw and pasteurized milk produced in west region of Iran (during summer and winter). Int Food Res J 20(3):1421–1425

van Egmond HP (1983) Mycotoxins in dairy products. Food Chem 11(4):289–307. https://doi.org/10.1016/0308-8146(83)90076-6

Vdovenko MM, Lu C, Yu F, Yu I (2014) Development of ultrasensitive direct chemiluminescent enzyme immunoassay for determination of aflatoxin M1 in milk. Food Chem 158:310–314. https://doi.org/10.1016/j.foodchem.2014.02.128

Wacoo AP, Wendiro D, Vuzi PC, Hawumba JF (2014) Methods for detection of aflatoxins in agricultural food crops. J Appl Chem 2014:1–15. https://doi.org/10.1155/2014/706291

Zangheri M, Di Nardo F, Anfossi L, Giovannoli C, Baggiani C, Roda A, Mirasoli M (2014) Multiplex chemiluminescent biosensor for type B-fumonisins and aflatoxin B1 quantitative detection in maize flour. R Soc Chem 140:1–9. https://doi.org/10.1039/b000000x

Zhang X, Kuča K, Dohnal V, Dohnalová L, Wu Q, Wu C (2014) Military potential of biological toxins. J Appl Biomed 12(2):63–77. https://doi.org/10.1016/j.jab.2014.02.005

# Chapter 16
# Recent Approaches Used in Environmental Monitoring Methods

**Anjuvan Singh and Joginder Singh**

**Abstract** Environmental pollution poses a great threat to society. Sometimes, it becomes difficult to determine the level of pollutants present in the atmosphere in spite of numerous methods available for determining pollutants in the atmosphere. Environmental monitoring helps to determine the presence of pollutants and monitor the quality of the environment. The information collected can help implement control measures for reducing the level of pollutant concentration to an accessible permissible limit. Sometimes, environmental changes are so slow that their determination requires careful monitoring over long periods. However, different analytical and chemical methods available to determine a pollutant do not determine the real effect of the pollutants on the organism that makes up the environment. Bioindicator and biomarker have the advantage, as they measure the changes in the biological systems that respond to exposure of pollutants that lead to biological effects. This chapter describes the use of plants, organisms, and microorganisms as indicator organisms which can reveal the presence of pollutants in the atmosphere by acquiring certain changes in itself, which may be ecological, behavioral, and physiological, along with the effect of pollutants on the organisms in the environment that can be determined using a wide range of biomarkers.

**Keywords** Bioindicator · Biomarker · Indicator organisms · Pollutants

## 16.1 Introduction

Environment means the surroundings or the conditions in which people or animals live and interact with each other. The changes in the environment are a result of interaction between the living and nonliving components present on the earth's

A. Singh (✉) · J. Singh
Department of Biotechnology, School of Bioengineering and Biosciences, Lovely Professional University, Phagwara, Punjab, India
e-mail: anjuvan.15950@lpu.co.in



J. Singh et al. (eds.), *Microbial Biotechnology: Basic Research and Applications*, Environmental and Microbial Biotechnology,
https://doi.org/10.1007/978-981-15-2817-0_16

surface that generates a different kind of contaminants over natural levels into the environment called pollutants. However, anthropogenic activities such as deforestation, agriculture, and industrialization cause imbalance in the environment and are the root causes of various environmental problems (Fig. 16.1). Pollution is a severe problem in the environment. Any substance (contamination) that causes an adverse or undesirable change in an environment is termed as a pollutant, and the presence of this substance and its effects causes pollution.

There are two types of pollution on the basis of origin:

1. Point source pollution—The pollution released from a stationary source. For example, oil spillage from refinery and ships, industrial unit, and chemical leakage.
2. Non-point source pollution—Here, the origin of pollution cannot be traced back to its source. For example, heaps of trash, pesticides, and fertilizers.

Pollution can be categorized into three major categories as:

1. Air pollution—It is caused by particulates and chemical in the atmosphere, and its concentrations are higher than the acceptable quantity. These concentrations are harmful to living beings. Examples are ground-level ozone, nitrogen dioxide, and carbon monoxide.
2. Water pollution—It is caused when ions and other chemicals are present in higher concentrations than normal, and this affects living beings in the water ecosystem.

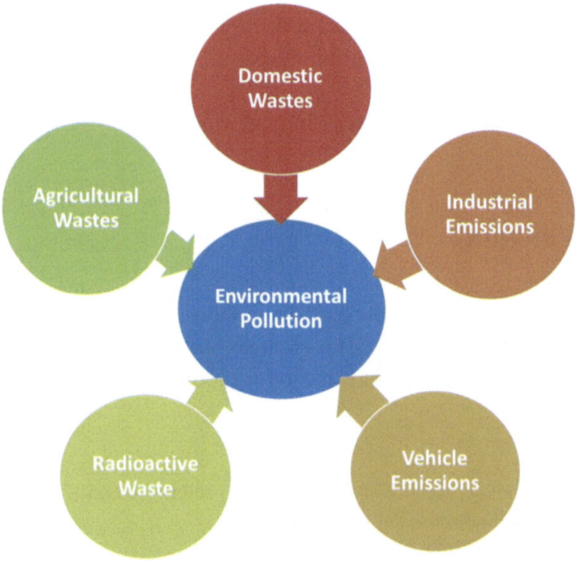


**Fig. 16.1** Different types of anthropogenically generated waste causing environmental pollution

3. Soil pollution—It is caused by the undesirable substance that is added to the soil such as pesticides, herbicides, fertilizers, etc. These substances adversely affect soil and decrease its quality.

Environmental pollution refers to the injection of harmful substances in the atmosphere, and it takes place only when the environment cannot neutralize the toxic by-products generated as an outcome of human activities (anthropogenic pollutants). However, such pollutants will last over many years during which nature will attempt to decompose the pollutants. Environmental pollution poses a great threat to society. Sometimes, it becomes difficult to determine the level of pollutants present in the atmosphere in spite of numerous methods available for determining pollutants in the atmosphere. Environmental monitoring can be used to determine the presence of pollutants as a measurement and then by comparing the obtained data with the regulated standards. The information obtained can be helpful for implementing control measures for reducing the level of pollutant concentration to an accessible permissible limit. The rise of pollutants in the atmosphere ranges from a slow increase in global temperature over the year to rapid accumulation of heavy metals and xenobiotics (Kumar and Saini 2017). In certain cases, environmental changes are too slow, and their investigations require careful monitoring over long periods. However, different analytical and chemical methods available to determine a pollutant do not determine the real effect of the pollutants on the organism that makes up the environment. In this specific situation, a huge consideration is given to the improvement of biosensor techniques for control that empower a basic gauge of co ntamination thickness, impressively increase operational effectiveness of the examination, and decrease its expense. Biochemical oxygen demand (BOD) and chemical oxygen demand (COD) are certain parameters in order to determine the quality of the water, which gives a measure of how much water is polluted. BOD (biochemical oxygen demand) is the amount of oxygen required by the aerobic microorganisms to digest or degrade the organic material at certain temperatures in a specific period of time, whereas COD (chemical oxygen demand) is the amount of oxygen consumed by the microorganism in order to digest the organic matter. The level of contamination in water can be determined by their BOD and COD measurements. An elevated level of BOD breaks downstream water quality by fast disintegration of a biodegradable natural issue and the exhaustion of oxygen, while COD generally speaks to the all-out natural issue.

Bioindicator and biomarker have the advantage as they measure the changes in the biological systems that respond to the exposure of metals that lead to biological effects. A bioindicator can be any organism which can state about qualitative status of the environment, whereas biomarker explains the importance of understanding relationships between exposure to environmental chemicals and subsequent development of chronic human diseases that lead to the route of establishment of human disease (Chammem et al. 2015). This chapter describes the use of different organisms as indicator organisms which can reveal the presence of pollutants in the atmosphere by acquiring certain changes in itself, which may be ecological, behavioral, and physiological, along with the effect of pollutants on the organisms in the environment that can be determined using a wide range of biomarkers.

## 16.2 Different Measurement Techniques by Broad Categories

Environmental monitoring involves the collection of one or more measured values of pollutants that are employed to evaluate the status of an environment (Artiola and Brusseau 2019). In other words, it is the collection of data from which information is gathered. Environmental monitoring requires the integrated use of different disciplines of subjects like physics, chemistry, mathematics, biology, statistics, and computer science (Artiola et al. 2004). Therefore, all science-based disciplines are involved in generating information with much accuracy. The conclusion regarding the presence of a pollutant in the environment cannot be made just by observation. Figure 16.2 presents flow chart for the interpretation of the data in a sequential manner from incident occurring at a site, sample collection and its analysis, to final interpretation of the data and the conclusion derived from it.

## 16.3 Different Methods of Environmental Monitoring

Environmental monitoring is the collection of one or more quantities that are used to measure the quality of an environment.

Areas of monitoring include:

- Water quality monitoring
- Soil monitoring
- Air quality monitoring

These monitoring techniques involve processes, including contamination, erosion, salinity, chemical, biological, radiological, and microbiological monitoring.

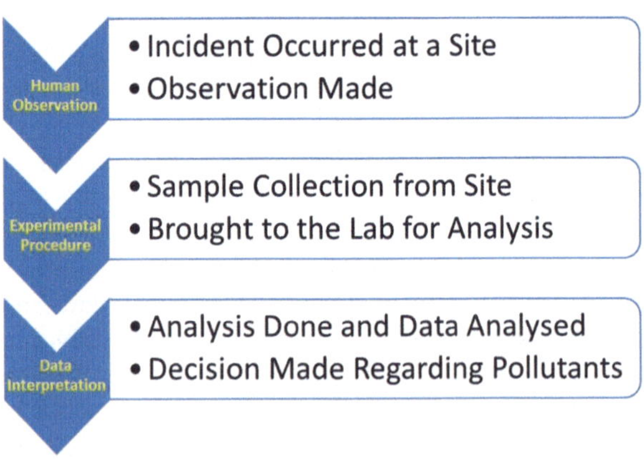


**Fig. 16.2** Interpretation of data for environmental monitoring

**Table 16.1** The indicators used for soil and water quality measurement

| *Indicators used for soil quality measurement* | | |
|---|---|---|
| Physical | Chemical | Biological |
| Electrical conductivity | Aggregate stability | Earthworm |
| Soil nitrate | Available water capacity | Particulate organic matter |
| Soil electrical conductivity | Bulk density | Respiration |
| Reactive carbon | Infiltration | Soil enzyme |
| Phosphorus | Soil crusts | Total organic carbon |
| *Indicators used for water quality measurement* | | |
| Temperature | | |
| Dissolved oxygen | | |
| Nutrient | | |
| Metals | | |
| Hydrocarbon | | |
| Industrial chemicals | | |
| Fecal coliform | | |

Table 16.1 shows different types of indicators that can be used for measuring water and soil quality.

## 16.4 Drones or Unmanned Aerial Vehicles as Environmental Monitors

Drones can reach places that are sometimes inaccessible by other means. They can take photos and videos from the air and provide us with information about the places without us visiting them manually. Drones can also be equipped with various sensors like thermometers, wind gauges, and pressure and humidity sensors. These sensors will allow us to gather climate data of the place.

### *16.4.1 Gravity Sedimentation Methods*

#### 16.4.1.1 Sedimentation from Still Air

In this method, a box is used to study polluted, heavy air particles. It has two hinged slides and a coated receptacle at rock bottom for inserting a microscopic slide or petri dish throughout air sampling the hinged slides square measure raised horizontally, and the wind is allowed to blow through the box. The hinged slides when closed, still air, present inside sink with characteristic and constant 'terminal velocity'. Here, sampling is discontinuous and a little volume of air being sampled at a time.

#### 16.4.1.2 Sedimentation from Wind

The method of examining suspended materials present in air released by industrial operations falling freely on the surface due to gravity is measured. Similar aspects of studies on microscopic slides were carried by earlier scientists like Pasteur, Pouchet and even are derived back to van Leeuwenhoek however it fails to assess air quantitatively.

### *16.4.2 Filtration*

In this technique, the polluted dust, smoke, volatile toxins present in the air are removed by suction pressure. The air is allowed to pass through a fibrous or porous medium that sieves the toxic particles. For this purpose, a suitable filter with a sleek surface like molecular membranes having unit area appropriate for the microscopic examination of the entrapped particles is generally used.

### *16.4.3 Precipitation*

#### 16.4.3.1 Electrostatic Precipitation

It is helpful for little particles. Air is drawn in the filters and soot & ash from the polluted air get charged. These charged particles are attracted to the opposite charge and separated on the charged plates.

#### 16.4.3.2 Thermal Precipitation

The polluted aerosols having the size below 10μ in the air can be analysed in precision by this method. It works on the thermal repulsion phenomenon.

## 16.5 Biological Oxygen Demand (BOD) and Chemical Oxygen Demand (COD)

The Royal Commission on Sewage Disposal established BOD in 1912 as an important parameter in water quality. Biological oxygen demand (BOD) is a measure of the amount of consumption of oxygen in a sample as a result of its metabolizable organic content by the microorganisms, whereas COD measures all the oxidizable material in the waste (Sawyer et al. 2003). The values obtained from BOD and COD will enable us to know about the presence of organic matter in solid and liquid samples, also total hydrocarbon (TH) in air samples. BOD biosensors have some limitations that impede their use. BOD sensors are simple and cheap devices that can be used for controlling aqueous ecosystems along with the old BOD determination methods. The COD is more efficient than BOD as it takes less time to interpret the results. COD test only takes 2–3 h, while BOD test takes 5 days. COD levels also can be used for estimation of BOD levels. Therefore, COD tests are used if the strength of wastes is too strong or toxic for BOD test. The estimated water can be used for various utilization purposes, if the BOD levels are safe, i.e., they are in limits of standard or permissible levels and can be utilized safely for irrigation, drinking, or for other domestic uses.

Total hydrocarbon (TH) is used to describe the quantity of the measured hydrocarbon impurities present in the sample. It was very important in determining total organic carbon content (TOC) which was recognized in 1931. There are several ways by which organic compounds are converted to carbon dioxide that may include (Mopper and Qian 2006):

(i) High-temperature combustion
(ii) High-temperature catalytic oxidation (HTCO method)
(iii) Chemical oxidizing agent
(iv) Wet chemical oxidation (WCO method)
(v) Ultraviolet irradiation (UV)

The ratio of BOD and COD value provides useful information about the types of organic material present in wastewater, but because of the rapidity of obtaining the results, TOC is often used to define highly contaminated wastes. Figure 16.3 shows the analysis of BOD from the given sample. Sample water is taken in a sealed bottle and is incubated in the dark at 20 °C, so that algae or bacteria present in wastewater may not acquire photosynthetic activity.

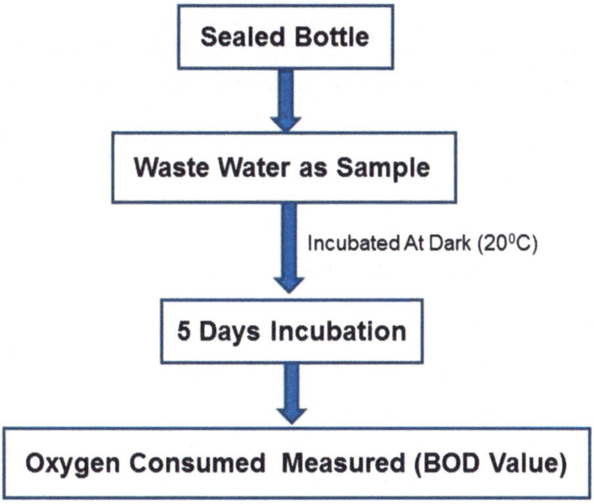


**Fig. 16.3** Biological oxygen demand (BOD) analysis

Further, the amount of oxygen consumed by the aerobic bacteria as a result of its metabolizable organic content present in water is the measurement of its BOD value (APHA 1992). However, a conventional BOD test requires 5 days of the incubation period, and the result still depends on the skill of the operator. There are certain disadvantages of monitoring environmental pollution using BOD and COD as in the case of wastewater containing a high amount of nitrogenous material, in which more than 5 days is required for complete oxidation. It is time-consuming and expensive for a large number of samples. Apart from it, there are certain compounds which require the presence of certain microorganisms for the oxidation and are toxic to many microorganisms. Due to these concerns, BOD estimation by using BOD sensor offers much advantage than the existing method.

### *16.5.1 Biochemical Oxygen Demand (BOD) Sensor*

A rapid test with biosensor uses yeast *Trichosporon cutaneum* organism which is sandwiched between an oxygen-permeable Teflon membrane and a porous membrane. The membrane is then directly fixed on the surface of the platinum cathode of an oxygen probe which measures the value of oxygen being consumed. A continuous flow system using the new microbial sensor exists for automatic estimation of 5 days BOD. To evaluate the amount of oxygen being consumed, sample solution containing glutamic acid was placed into the system; organic compounds that pass through the porous membrane were consumed by the immobilized microorganisms. After some time due to consumption of dissolved oxygen present in the sample solution by immobilized microorganisms, there is a decrease in the amount of dissolved oxygen around the membranes, as a result of which current of electrode decreased with time. The steady-state current depends on the BOD of the sample solution. BOD of various other effluents can be estimated with this sensor. Nowadays, BOD sensors are available in the market, but they have several drawbacks such as limited run lengths due to the need for reactivation and their high maintenance costs (Praet et al. 1995). Owing to these limitations nowadays, biosensor and biomarker offer much advantage than using BOD sensor.

### *16.5.2 Oxygen Electrode-Based Sensor*

Oxygen electrode-based biosensor consists of a layer of microbial film that structures the organic recognition component sandwiched between a component of permeable (regularly, cellulose) layer and the gas porous layer of the oxygen terminal. Some portion of oxygen present in the layer of immobilized microorganisms is expended in the oxidation of natural mixes contained in the example. Remaining oxygen enters through the gas porous Teflon layer and is diminished at the cathode of the oxygen terminal. The quality of current in the framework is straightforwardly relative to the size of oxygen decreased at the anode. After a harmony is set up between the dispersion of oxygen to the layer of immobilized microorganisms and the metabolic respiration process of immobilized microorganisms, the balance current is recorded. When a sample wastewater is taken in the measuring cuvette, the organic components of the analyzed sample are consumed by immobilized microorganisms, which lead to increase in the rate of respiration and eventually consumption rate of oxygen. In this situation, a little measure of oxygen infiltrates through the Teflon layer to be diminished. This situation will remain until another harmony is set up. For washing the biosensor the buffer solution is fed in the cuvette, it restores the endogenous respiratory rate of the microbes and the initial equilibrium of the oxygen flows in the system is re-established.

An alternative method to measure BOD is the development of biosensors. These devices are used for the detection of a substance and include the combination of a biological component and a physicochemical detector component. Enzymes are used in most of the cases. The biological sensing elements are majorly used in the construction of biosensors. On account of time-consuming and costly enzyme purification methods, their application in biosensor construction is limited. Microorganisms provide an ideal alternative in this case. Many microorganisms useful for BOD detection are easy to maintain in pure cultures, grow, and harvest at low cost. A number of pure cultures such as *Bacillus cereus*, *Trichosporon cutaneum*, and *Klebsiella oxytoca* species have been used by many workers for the construction of BOD biosensor. Many workers have immobilized activated sludge or a mixture of two or three bacterial species and on the membranes for the construction of BOD biosensor. One of the most widely used membranes was polyvinyl alcohol and porous hydrophilic membranes. Biosensors can be used to indirectly to measure BOD quickly, i.e., usually less than 30 min.

## 16.6 Environmental Monitoring Through Bioindicator and Biomarker

The use of living organisms called biomonitors is preferable for pollutant quantification. These organisms tend to accumulate pollutants usually from water and food; an increase level of such pollutants in such living organisms like insects or microorganisms will reflect the presence of pollutants in the environment (Fig. 16.4). The use of biomonitors to evaluate the pollution impact is called biomonitoring.

### *16.6.1 Bioindicator*

Bioindicator organisms are those that can be used to identify and quantify the effects of pollutants on the environment. An "ideal" bioindicator is expected to have several characteristics.

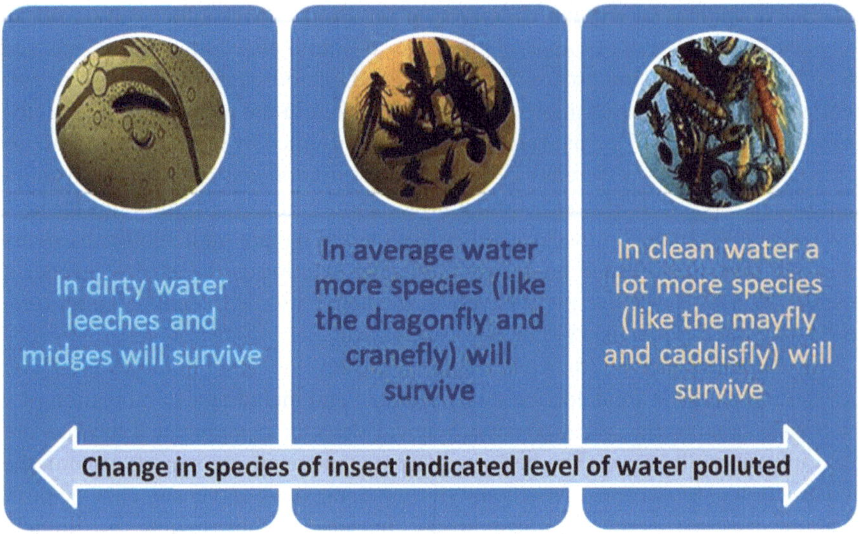


**Fig. 16.4** Monitoring water quality by looking at the species of insect in water

The most important of them are as follows:

(i) They are distributed widely.
(ii) Easy to collect.
(iii) Sensitive to pollutant and have a high concentration factor for the specific pollutant.

They are useful in three situations:

(i) Where environmental pollutants cannot be measured
(ii) Where the pollutant factor is difficult to measure, e.g., pesticides and their residues or complex toxic effluent
(iii) Where the environmental pollutant can be measured but difficult in the interpretation of data.

There are three types of changes observed:

- Ecological (including population density, key species, and species diversity)
- Behavioral (feeding activities, bacterial mobility)
- Physiological (accumulation of heavy metals, carbon dioxide production, BOD)

Some examples of bioindicators are explained below:

- Lichens can be sensitive to air pollutants (Fig. 16.5) and have been used widely as a bioindicator of air quality by the change of their color (Conti 2008).
- Honeybees have been used as a bioindicator for the contamination of the atmosphere by heavy metals (Cd, Pb).

**Fig. 16.5** Lichens can be used as an air pollution indicator

- Honey was assayed for measuring pollutants, and the differences were found between the reference cities and cities (Conti and Botre 2001).
- The Mediterranean mussel *Mytilus galloprovincialis* was used to monitor heavy metal contamination (Stellio and Cédric 2006).
- Marine pollution has been monitored by an increase in malformations in embryos of dab (*Limanda limanda*) which has been linked with high input of DDT and PCBs.
- The behavioral and lethal test has been applied to the concentration of 2,4,6-trinitrotoluene (TNT) in contaminated soils using earthworm and crabs are indicators of lead pollution in the aquatic environment.

### *16.6.2 Biomarkers*

Molecular biomarkers provide early warning signals for the presence of pollutants before effects at higher levels of the organization took place. Nowadays, new technologies to enhance environmental monitoring are continually improving, often becoming cheaper to use. Adoption of new technologies or novel monitoring approaches and indicators have included the use of dispersal and fish body condition, genetic measures of fish population viability, use of underwater video to sample fish composition in places with crocodile safety risks, and analyses of a time series of satellite images to detect ecological changes over time. Biomarkers were primarily had applications in the medical and animal sciences, and nowadays, it has been used in invertebrate and particularly as bivalve biomarkers to assess marine pollution (Nicholson 2003). The response made by the living forms against such pollutants would be noticed at the molecular or cellular (biochemical), tissue/organ and whole-body levels. However, a successful biomarker should be sensitive enough to detect the pollutants at an early stage of the process of toxicity, should be specific to a particular contaminant, and also should respond in a concentration-dependent manner to change in ambient levels of the contaminant.

The advantages of using biomarker for environmental monitoring include:

- Determination can be made over a period of time and not the single sample often used in chemical analysis.
- It indicates the risk of exposure to particular chemicals.
- It determines the effect in various habitats; the routes of exposure can be established.
- It can provide information about the toxicity of a single compound or mixture.

#### 16.6.2.1 Classification of Biomarkers

Biomarkers can be classified into three groups depending on their analysis method:

- Biochemical indicators: They are based on the ability of the pollutants to generate a response at the gene level.
- Immunochemistry: Antibodies against PCBs have been developed and used in enzyme-linked immunosorbent assay (ELISA).
- Genetic indicators: These include the introduction of genetically engineered indicator organisms.

##### Biochemical Indicators

They are one of the advanced techniques used for monitoring pollutants as per the response generated at the gene level. The difficulty of performing analysis to survey general populations is that there is a change in enzyme activity in unexposed individuals, and even in the same individual, the enzyme activity will change with time.

The most common example, reflecting the use of biochemical indicator was observed for the exposure to organophosphorus and carbamate pesticides in humans, which leads to increase in the level of cholinesterase enzyme activity in human blood plasma and erythrocytes (Ladhar-Chaabouni et al. 2009). Another example is an exposure of lead, resulting in the decrease in the enzyme δ-aminolevulinic acid dehydratase in human blood (La-Llave-León et al. 2017).

### Immunochemistry

Environmental monitoring can also be done by using the interaction between antigen and antibody in the presence of specific pollutants like xenobiotic compounds in environmental samples, but there are some disadvantages as well. One of the major drawbacks is the nature of selectivity. Besides, there are certain enzymes that are extremely unstable molecules and can be inhibited by a large number of compounds such as sodium chloride. Antibodies against polychlorinated biphenyls (PCBs) and polychlorinated dibenzofurans (PCDFs) have been developed and used in enzyme-linked immunosorbent assay (ELISA) system to determine the presence of toxicity in the environmental sample.

### Genetic Indicator

With the advancement in the field of technology, recombinant DNA technology is playing a major role with an advanced molecular technique that can be used to monitor or diagnose pollutants present in the atmosphere. Figure 16.6 shows one of the most suitable methods of screening for gene expression by fusing the promoter sequence to those for a product that can be easily detected. The detection for the presence of pollutant would be in terms of production of light.

- One example involves using *Pseudomonas fluorescens* (genetically engineered for bioluminescence) to monitor pollutants (Edward and Selvam 2011).
- Whenever any pollutant is present in the environment, the promoter region will sense and would be switched on, which would turn on the reporter gene responsible for giving the signal in the form of green fluorescent protein (GFP).
- If the signal in the form of pollutants is present, GFP will be produced.

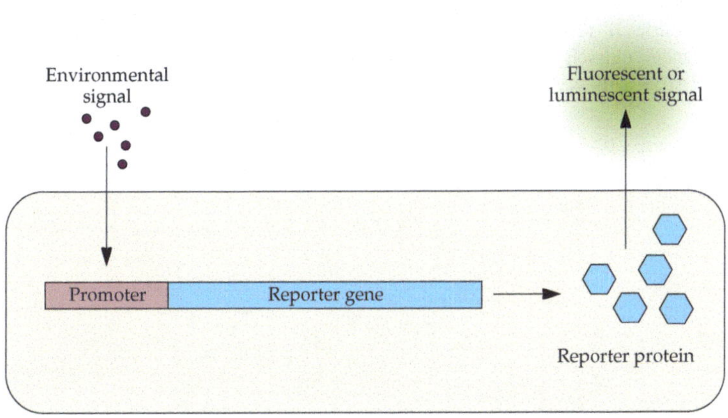


**Fig. 16.6** Genetically engineered bacteria to monitor environmental pollutants

## References


APHA (1992) Standard methods for the examination of water and wastewater, 18th edn. American Public Health Association, Washington, DC

Artiola JF, Brusseau ML (2019) The role of environmental monitoring in pollution science. Environmental and Pollution Science, Academic Press, Ch 10, pp 149–162

Artiola JF, Pepper IL, Brasseau ML (2004) Monitoring and characterization of the environment. Elsevier, Environmental Monitoring and Characterization, Ch-20, 01-09

Chammem H, Hafaid I, Ardhaoui M, Mora L, Meilhac O (2015) Functionalised magnetic beads on gold surface for C-reactive protein detection. Int J Nanotechnol 12:552–561

Conti ME (2008) Lichens as bioindicators of air pollution. In: Biological monitoring theory and applications, vol 30. WIT Press, pp 111–142

Conti ME, Botre F (2001) Honeybees and their products as potential bioindicators of heavy metals contamination. Environ Monit Assess 69:267–282

Edward R, Selvam GS (2011) Construction of green fluorescent protein based bacterial biosensor for heavy metal remediation. Int J Environ Sci Technol 8:793–798

Kumar A, Saini P (2017) Testing and monitoring of biodegradable contaminants in bioremediation technique. In: Handbook of research on inventive bioremediation techniques. Springer, Ch 20, pp 470–491

Ladhar-Chaabouni R, Smaoui-Damak W, Hamza-Chaffai A (2009) In vivo variation of some biomarkers with time and cadmium concentration in the cockle Cerastoderma glaucum. Mar Biol Res 5:487–495

La-Llave-León O, Méndez-Hernández EM, Castellanos-Juárez FX, Esquivel-Rodríguez E, Vázquez-Alaniz F, Sandoval-Carrillo A, García-Vargas G, Duarte-Sustaita J, Candelas-Rangel JL, Salas-Pacheco JM (2017) Association between blood lead levels and delta-aminolevulinic acid dehydratase in pregnant women. Int J Environ Res Pub Hlth 14(4):432

Mopper K, Qian J (2006) Water analysis: organic carbon determinations. In: Encyclopedia of analytical chemistry in 2006. Wiley, pp 1–9

Nicholson S (2003) The mytilid mussel Perna viridis (Mytilidae: Bivalvia) as a pollution monitor in Hong Kong. In: Shin PKS (ed) Turning the tides, Marine Biological Association of Hong Kong. University Press, Hong Kong, pp 201–228

Praet E, Reuter V, Gaillard T, Vasel JL (1995) Bioreactors and biomembranes for biochemical oxygen demand estimation. Biosens Bioelectron 14:371–378

Sawyer CN, McCarty PL, Parkin GF (2003) Chemistry for environmental engineering and science, 5th edn. McGraw-Hill, New York. ISBN 978-0-07-248066-5

Stellio C, Cédric B (2006) Modelling trace metal (Hg and Pb) bioaccumulation in the Mediterranean mussel, Mytilus galloprovincialis, applied to environmental monitoring. J Sea Res 56(2):168–181

# Index




J. Singh et al. (eds.), *Microbial Biotechnology: Basic Research and Applications*, Environmental and Microbial Biotechnology,
https://doi.org/10.1007/978-981-15-2817-0

The manufacturer's authorised representative in the EU is Springer Nature Customer Service Centre GmbH, Europaplatz 3, 69115 Heidelberg, Germany. If you have any concerns regarding our products, please contact ProductSafety@springernature.com

Printed and bound by CPI Group (UK) Ltd, Croydon, CR0 4YY

30/09/2026

02225557-0011